Materials Science for Engineers

Materials Science for Engineers

Edited by **Andrew Green**

NY RESEARCH PRESS
New York

Published by NY Research Press,
23 West, 55th Street, Suite 816,
New York, NY 10019, USA
www.nyresearchpress.com

Materials Science for Engineers
Edited by Andrew Green

International Standard Book Number: 978-1-63238-489-8 (Hardback)

Printed in China.

Contents

Preface

This book was inspired by the evolution of our times; to answer the curiosity of inquisitive minds. Many developments have occurred across the globe in the recent past which has transformed the progress in the field.

Materials Science is a multidisciplinary field which involves exploration and design of new materials, especially with respect to solids. It plays a significant role in various fields such as nanotechnology, biomaterials, metallurgy, etc. This discipline has gained significance over the years due to its applicability in a large number of industries such as aviation, manufacturing, etc. This book contains some path-breaking studies in the area of materials science. The various advancements in this field have been glanced at. Those with an interest in materials science would find this book useful. It will help new researchers by foregrounding their knowledge in this area and also provide innovative insights for future researches and progress.

This book was developed from a mere concept to drafts to chapters and finally compiled together as a complete text to benefit the readers across all nations. To ensure the quality of the content we instilled two significant steps in our procedure. The first was to appoint an editorial team that would verify the data and statistics provided in the book and also select the most appropriate and valuable contributions from the plentiful contributions we received from authors worldwide. The next step was to appoint an expert of the topic as the Editor-in-Chief, who would head the project and finally make the necessary amendments and modifications to make the text reader-friendly. I was then commissioned to examine all the material to present the topics in the most comprehensible and productive format.

I would like to take this opportunity to thank all the contributing authors who were supportive enough to contribute their time and knowledge to this project. I also wish to convey my regards to my family who have been extremely supportive during the entire project.

Editor

1

Effect of Mortar Coating's Bond Quality on the Structural Integrity of Prestressed Concrete Cylinder Pipe with Broken Wires

Shaoqing Ge[1] & Sunil Sinha[1]

[1] Department of Civil and Environmental Engineering, Virginia Tech, USA

Correspondence: Shaoqing Ge, Department of Civil and Environmental Engineering, Virginia Tech, USA. E-mail: geshaoqing@gmail.com

Abstract

In North America, 90 out of 100 large water utilities utilize Prestressed Concrete Cylinder Pipe (PCCP) in their water transmission mains or force mains since 1942. This is mainly due to its low cost and relatively stable performance compared with other large-diameter pipe materials working under high-pressure. Though PCCP has relatively low failure rate, its failures usually occur without warning and can cause catastrophic consequences, threatening public safety. Most PCCP failures are caused by the breakage of prestressing wires that hold the pipe in compression. To be proactive and avoid PCCP failures, it is important to know the pipe condition and then take actions before it fails. Hence, numerical models have been developed to evaluate the pipe condition. However, current numerical models do not consider the bond between the presressing wires and mortar coating. Instead all broken wires are removed in the model. This is over-conservative and many good pipes had been misjudged to be "bad" pipes, causing unnecessary and very costly pipe excavation and replacement. To effectively and efficiently manage PCCP pipeline, it is very important to have a realistic modeling of the broken prestressing wires by considering the bond quality of the mortar coating.

This paper analyzes the effect of mortar coating's bond quality on PCCP's structural integrity. A theoretical approach is proposed to determine the length of broken wires losing prestress by considering the bond quality of the mortar coating. Based on the proposed method, a numerical model is developed and the pipe's structural integrity is analyzed. Several cases with different bond qualities of mortar coatings are analyzed in this paper. The results are then compared and summarized. The simulations results indicate that PCCPs with good mortar bond are not compromised much, while the assumptions of current models are too conservative. The research presented in this paper provides a better understanding of the importance of considering bond quality in modeling PCCP. The findings of this research can help utilities better management their PCCP pipelines with different bond qualities.

Keywords: Prestressed Concrete Cylinder Pipe (PCCP), PCCP failure, condition evaluation, numerical model, bond quality

1. Introduction

Prestressed Concrete Cylinder Pipe (PCCP) is a composite pipe material mainly composed of concrete, steel cylinder, mortar coating and prestressing wires (see Figure 1). The high-strength prestressing wire is drawn to a certain level of prestress to keep concrete core in great compression which can offset the tensile stress under normal working loadings. The mortar coating protects the prestressing wires from corrosion and determines the long-term durability of the pipe (Price et al., 1998; Wardany, 2008). The steel cylinder seal water from leaking. There are two type of PCCP (see Figure 2 and Figure 3): Lined Cylinder Pipe (LCP), and Embedded Cylinder Pipe (ECP). In LCP, the prestressing wires are wrapped around the steel cylinder which is outside of concrete core. In ECP, the steel cylinder is embedded within concrete core, and the prestressing wires are wrapped around the concrete core (AWWA C301-07, 2007; ANSI/AWWA C304-07, 2007). PCCP has been widely used in water industry and more than 22,000 miles of PCCP has been installed in North America (Semanuik & Mergelas, 2006; Romer et al., 2008).

PCCP has lower break rate compared with other pipe materials, however, its failure usually occurs suddenly without warning, leaving almost no time for utilities to respond. Due to the large size and high pressure inside the pipe, the failure can be catastrophic: flushing highways, houses, and other important neighboring infrastructures, causing millions dollars of loss. Moreover, the pipeline inspection and repairs following pipe failure is very costly as well (Bell, 2001; Essamin et al., 2005; Galleher and Stift, 1998). Miami-Dade had one 72-inch PCCP force main burst in 2010 and millions gallons of untreated wastewater flew into Biscayne Canal. They had to spend over 1 million dollars to fix the pipe, and received a penalty of approximately 1 million dollars (UIM, 2011; USEPA, 2013). In December 2008, a 66-inch PCCP failed on River Road, Maryland and trapped cars in flooding water. Helicopters had to be sent to rescue the drivers. Over 1.7 million dollars were spent to repair the pipe and over millions gallons of water was lost (WSSC, 2012).

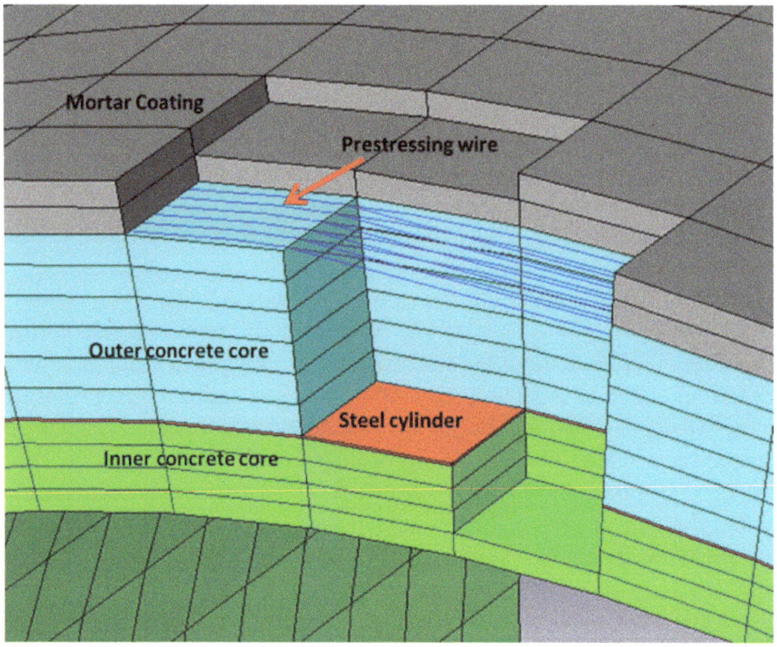

Figure 1. Pipe components in PCCP FE Model

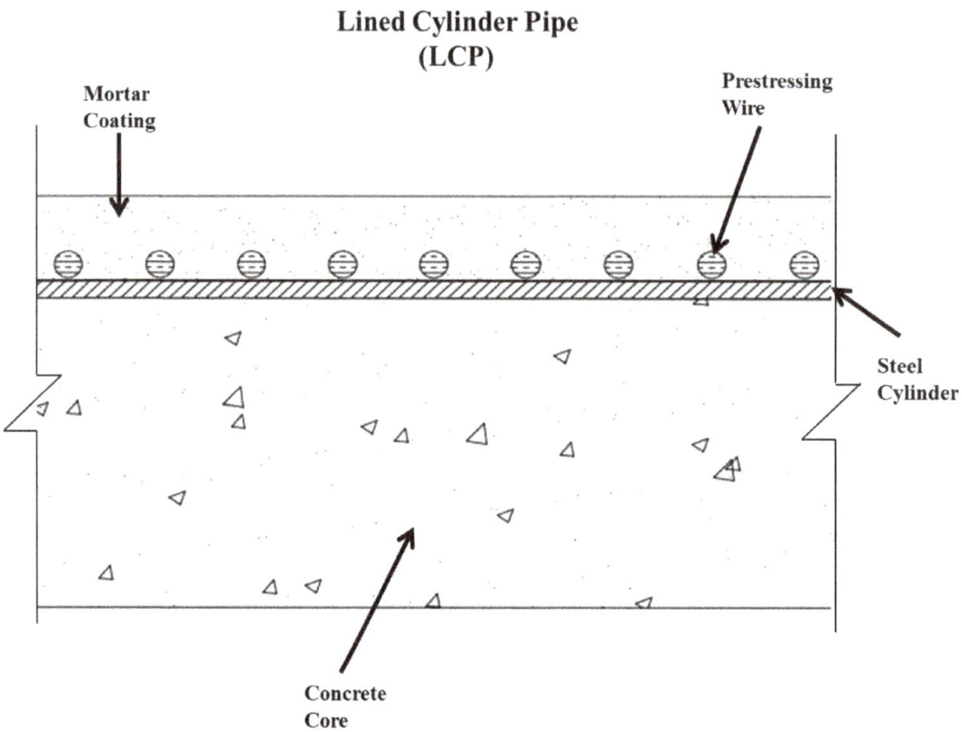

Figure 2. Lined Cylinder Pipe (LCP)

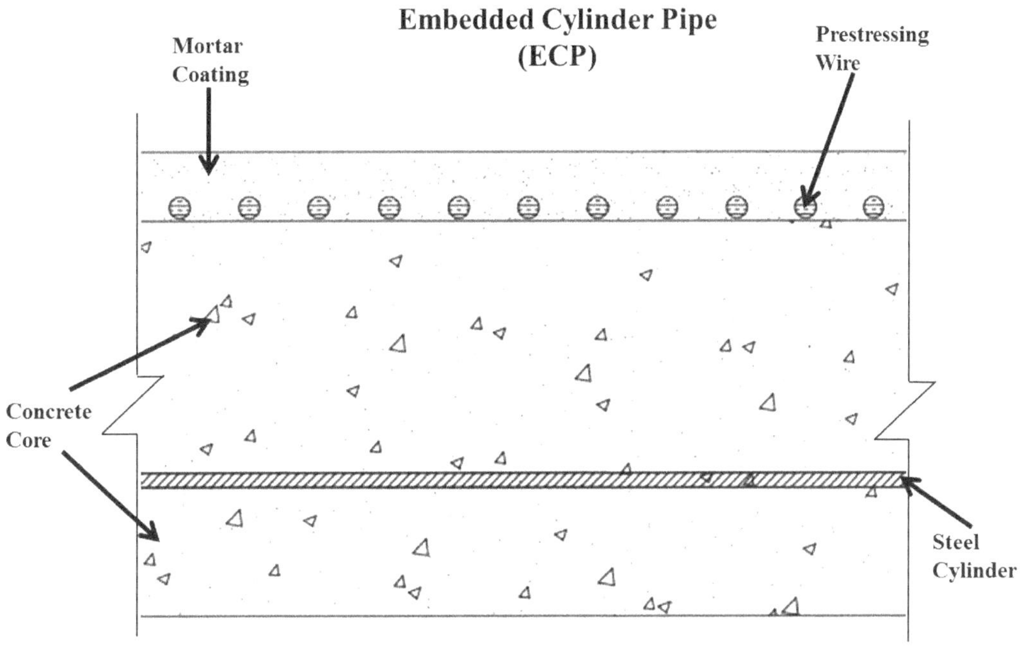

Figure 3. Embeded Cylinder Pipe (ECP)

PCCP failures are complicated and can be caused by factors that are grouped into three categories (AwwaRF, 2007; Walsh & Hodge, 1998; Ojdrovic et al., 2001; Parks et al., 2001; Galleher & Stift, 1998; Villalobos, 1998; Lewis, 2002): manufacture and design (poor bond, low prestress, poor material, etc.), construction and operation (poor bedding, construction damage, etc.), and environmental (corrosive soil, chloride, etc.). PCCP failure is usually initiated with "birth defects" such as cracks in coating caused in manufacture and construction, or overloading in operation. These cracks provide passage for chemicals in soil to reach the prestressing wires and steel cylinder and then corrosion starts. The hydrogen atom produced in corrosion process accumulates in prestressing wires and then it causes hydrogen embrittlement which makes the wire brittle and fail before it reach the ultimate tensile strength. The breakage of prestressing wires weakens the pipe structure and a critical number of breakages can cause concrete core to collapses and cylinder to burst (Price et al., 1998; Woodcock, 2008; Romer et al., 2008; Kola, 2010; Ge & Sinha, 2012).

In practice, it was found that the prestress in the wire resumes beyond a certain length from the broken point of the wire (Gomez et al., 2004). Therefore, the broken wire still has prestress in certain portion between the broken point and the point where full prestress is resumed. Though most PCCP failures are caused by the breakage of prestressing wires, however, the bond quality between the prestressing wire and mortar coating is a critical factor, as it has been found in practice that PCCPs with good bond are not much compromised even though there are many broken wires. In contrast, for PCCPs with very poor bond, even fewer than normal number of broken wires could weaken pipe's structural integrity.

To mitigate catastrophic failures of PCCP, it is important to understand the remaining structure integrity of the pipe, so that proactive actions can be taken before it fails. To this end, considerable research has been conducted on developing numerical models to evaluate the structural integrity of the pipe. Zarghamee et al. (1993, 2001, 2002, 2003), Erbay et al. (2007), and Weare et al. (2007) established 3-dimetional models of PCCP using finite element method. These models take the nonlinear properties of the pipe materials into account. Loadings such as internal pressure, weight of soil cover, pipe and water are considered. However, all these models assume that broken wires have completely lost prestress and are removed from the simulation.

In some linear models developed by Diab and Bonierbale (2001), Gomez et al. (2004), and Alavinasab et al. (2010), the broken wires are either completely removed from the model or assumed to have a certain length removed based on rule of thumb. No robust analytical analysis or experimental verification has been conducted to account for the effect of bond quality of the mortar coating on PCCP's structural integrity.

Xiong et al. (2010) simulates the contact between the wire and concrete core for ECP using a new wire contact model to simulate the whole wire wrapping process. However, this model does not consider the bond between the prestressing wire and mortar coating. In other recent models (Lee, 2011; Rauniyar, 2013), the bond quality of mortar coating is not considered.

Current condition assessment technologies such as sonic/ultrasonic can provide qualitative information on prestress loss in prestressing wires. Further destructive tests are required to quantitatively measure the level of prestress in wires (Marshall & Fisk, 2015).

In this paper, a theoretical approach is proposed to determine the length of prestressing wire losing prestressing once it breaks. Based on this approach, a numerical model is developed simulating different cases in which the mortar coating has different bond qualities.

2. Methodology Proposed to Considerthe Bond Quality of MortarCoating

To evaluate the pipe condition of PCCP with broken wires using numerical model, the key is to determine the range of broken wires losing prestress (or the length of broken wire needed to resume prestress in the wire) once the wire breaks, which depends on the bond quality of mortar coating. As is mentioned in introduction section, current numerical models for PCCP do not consider the bond quality of the mortar coating and all broken wires are assumed to have completely lost the prestress. In fact, because of the bond between the prestressing wire and mortar coating, after the wire breaks, it resumes prestress beyond a certain length (Gomez et al., 2004). So far, no lab tests have been conducted to determine the length needed to resume prestress in the broken wire due to the complexity, trickiness and high cost of the tests.

In this paper, an analytical methodology is proposed to calculate the length that a broken wire needs to resume prestress. It is assumed that the average bond strength of the mortar coating is σ_{bond}, the prestress in the prestressing wire is σ_{wire}, and the diameter of the prestressing wire is D_{wire}. Then, the length needed to resume the prestress is expressed below:

The analysis of this length is illustrated in Figure 4 and Figure 5. Based on the equilibrium of the broken wire, it can be expressed as:

$$\begin{cases} \sum F_x = 0 \\ \sum F_y = 0 \\ \sum M = 0 \end{cases} \tag{1}$$

From $\sum F_x = 0$,

$$-F_1 + \int_0^\theta dF_{bond} \cdot cos\frac{d\varphi}{2} \cdot + \int_0^\theta dN_r \cdot sin\frac{d\varphi}{2} = 0 \tag{2}$$

Where,

$dF_{bond} = \sigma_{bond} \cdot \pi \cdot D_{wire} \cdot ds$

$dN_r = q_r \cdot ds$ (radial stress increment)

$ds = R \cdot \varphi$, R is the diameter of the turns of prestressing wires

$F_1 = \sigma_{wire} \cdot \frac{\pi}{4} \cdot D_{wire}^2$

From $\sum F_y = 0$, it is expressed as

$$\int_0^\theta dF_{bond} \cdot sin\frac{d\varphi}{2} \cdot - \int_0^\theta dN_r \cdot cos\frac{d\varphi}{2} = 0 \tag{3}$$

The sum of all the moments about point O is zero, from $\sum M = 0$,

$$F_1 \cdot R - \int_0^\theta dF_{bond} \cdot R = 0 \tag{4}$$

Since the purpose is to calculate the range of the wire lost prestress, equation (4) can be used to get the results.

From equation (4), it can be derived as

$$\sigma_{wire} \cdot \frac{\pi}{4} \cdot D_{wire}^2 \cdot R - \int_0^\theta \sigma_{bond} \cdot \pi \cdot D_{wire} \cdot R \cdot d\varphi \cdot R = 0 \tag{5}$$

$$\sigma_{bond} \cdot R \cdot \theta = \sigma_{wire} \cdot \frac{1}{4} \cdot D_{wire} \tag{6}$$

Thus,

$$\theta = \frac{\sigma_{wire}}{\sigma_{bond}} \cdot \frac{D_{wire}}{4 \cdot R} \tag{7}$$

$$L = \theta \cdot R \tag{8}$$

The σ_{wire} is generally around 1.09e10 Pa (158,250 psi) and the average bond strength can be determined by direct tension pull-out lab test (Tastani & Pantazopoulou, 2010).

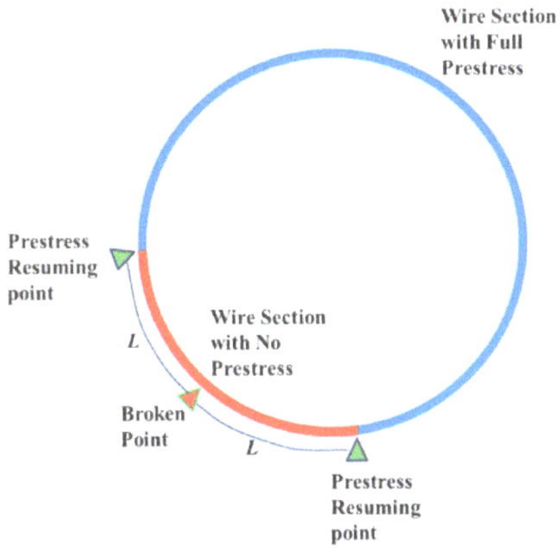

Figure 4. Prestress resuming in the broken wire of PCCP

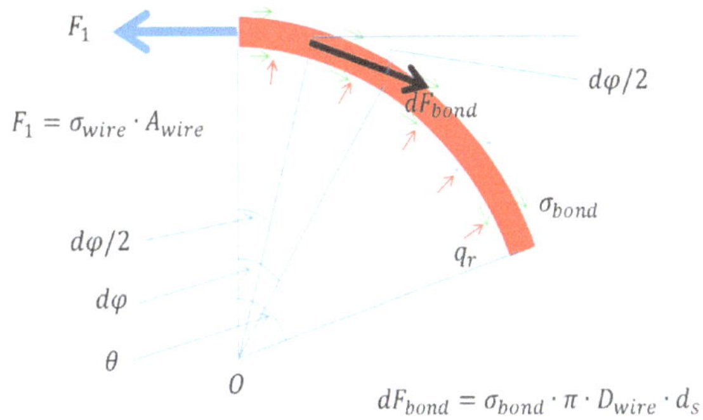

Figure 5. External forces acting on the broken wire

3. Numerical Model for PCCP Considering Different Bond Quality

Based on methodology proposed in section 2, a numerical model is developed using finite element method (FEM) to evaluate the effect of various bond quality of the mortar coating on the structural integrity of PCCP. In this numerical model, pipe components such as mortar coating, concrete core, prestressing wires, and steel cylinder are considered. These components are considered to stay connected without relative displacement along the interface of materials under loading conditions used in this paper. Loadings including weight of soil, pipe, and internal/surge pressure are considered. The FE model is shown in Figure 6. For boundary conditions, the bottom surface of the model is fully fixed; the top surface of the model is free; the surfaces of other four sides of the model are fixed in their normal directions.

Figure 6. FE model of PCCP

In this research, different wire bond strengths (σ_{bond}) are compared to evaluate the effect of different bond qualities on PCCP's structural integrity. In the numerical model, the broken wires without full prestress are cut from the model, as is shown in Figure 7. Six cases studies are analyzed as below:

1) Case 1: Zero broken wires
2) Case 2: 40 broken wires, only prestress loss at broken point
3) Case 3: 40 broken wires, with 1/12 of the broken wire lost prestress ($\theta = 30°$)
4) Case 4: 40 broken wires, with 1/6 of the broken wire lost prestress ($\theta = 60°$)
5) Case 5: 40 broken wires, with 1/3 of the broken wire lost prestress ($\theta = 120°$)
6) Case 6: 40 broken wires, the whole broken wire lost prestress, as is the assumption used in current numerical models.

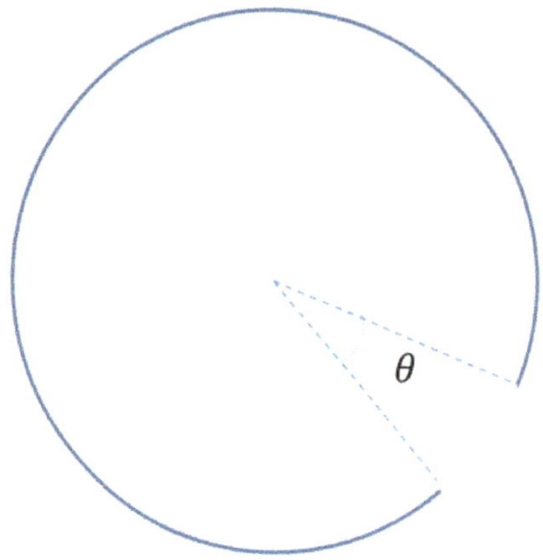

Figure 7. Broken wire model

In these six cases, the pipe (ECP) is subject to the loadings such as: soil, pipe and inside water weight, internal pressure 1.38e6 Pa (200 psi), and surge pressure 5.52e5 Pa (80 psi). The pipe dimensions are listed below:

- Pipe diameter: 1.68 m (66 inch)
- Cylinder thickness: 1.40 mm (0.05 inch)
- Concrete core thickness: 0.13 m (5.00 inch)
- Coating thickness: 0.02 m (0.81 inch)
- Wire spacing: 0.01 m (0.43 inch)
- Buried depth: 2.44 m (8 ft)

The material properties used are listed in table below (Table 1):

Table 1. Mechanical property of materials

Material	Young's Modulus (Pa)	Poisson's Ratio	Density (kg/m^3)
Concrete	2.45e+10	0.17	2322.61
Soil	6.90e+06	0.25	1840.00
Prestressing wire	1.93e+11	0.30	7832.80
Cylinder	2.07e+11	0.30	7832.80

Other properties of the materials include: Yield strength of the cylinder is 2.28e+8 Pa (33,000 psi) and its ultimate strength is 3.87e+8 Pa (56,100 psi). The wire ultimate strength is 1.45e+9 Pa (211,000 psi); and its wrapping stress is 1.09e+9 Pa (158,250 psi). Compressive strength of concrete is 48 MPa (6,962 psi) and its tensile strength is 4.02 MPa (583 psi). The compressive strength of mortar coating is 37.90 MPa (5,500 psi), and its tensile strength is 3.57 MPa (518 psi). In this research only elastic analysis is conducted due to unavailability of plastic properties of coating and concrete core.

In the analysis of the six cases, the maximum principal stress at nodes in four strings (see Figure 8) are analyzed and compared to see the impacts caused by broken wires embedded in mortar coating with different bond qualities. The relative deformations in horizontal direction is analyzed and summarized to see how the deformation of the pipe is affected in each case (see Figure 9). The vertical deformation change is not provided because the change in horizontal direction can represent the deformation of the pipe.

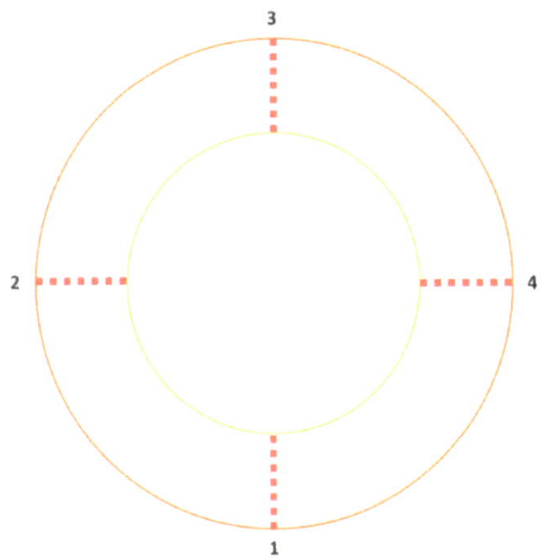

Figure 8. Locations of pipe for stress evaluation

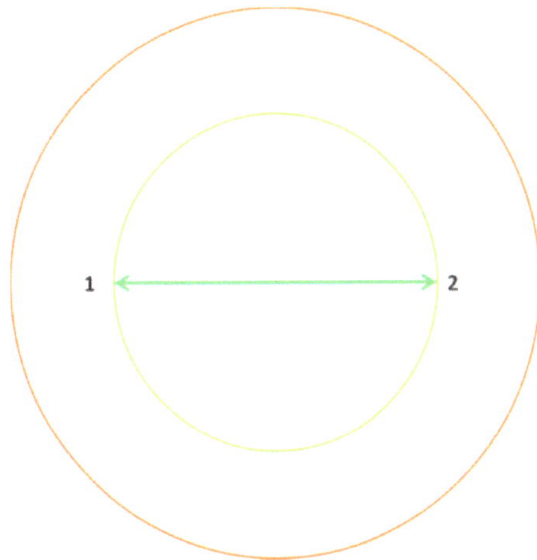

Figure 9. Relative deformations

3.1 Case 1: Zero Broken Wires

In this case, there are no broken wires in the model. This case is analyzed to compare with the results of other cases in which the PCCP has broken wires and its mortar coating has different bond strengths. The maximum principal stress in mortar coating and concrete core is shown in Figure 10. The stress distribution in the middle section of the pipe is shown in Figure 11.

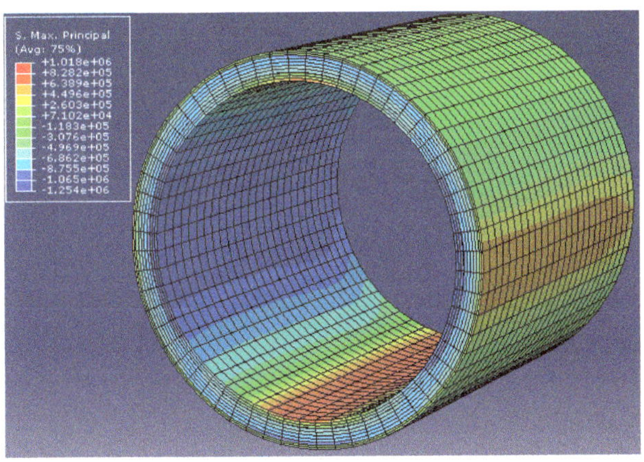

Figure 10. Maximum principal stress in PCCP (Case 1)

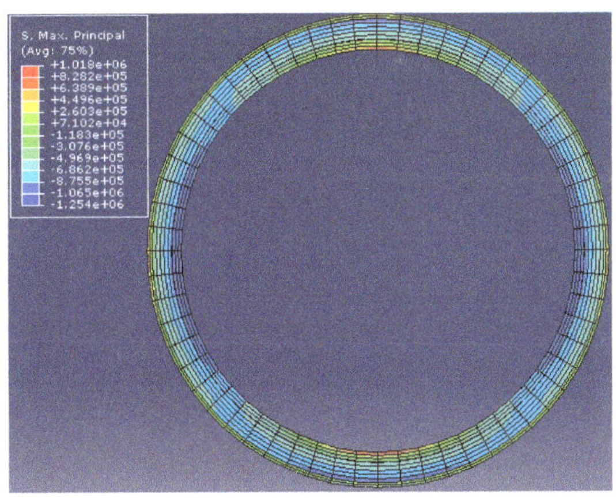

Figure 11. Maximum Principal stress in the middle section of PCCP (Case 1)

The maximum principal stress at nodes in four strings (Figure 8) of middle section of the pipe is shown Figure 12. According to Figure 10 – 12, for PCCP without any broken wire, under the specified loading conditions, the maximum principal stress is 1.00 MPa and it is located at invert of the pipe. Since the maximum principal stress is lower than the tensile strength of concrete (4.02 MPa), there is no cracks in the pipe. The deformation of the whole model in horizontal direction is shown in Figure 13 and the horizontal deformation in middle section is shown in Figure 14. According to Figure 13 and Figure 14, the maximum horizontal deformation is 0.77 mm. In Figure 14, because of symmetry, the stress at nodes in String 2 and String 4 are exactly the same. String 1 and String 3 have slight difference in the nodes close to the interior side of the pipe, and String 1 (invert) has bigger tensile stress compared to String 3 (crown).

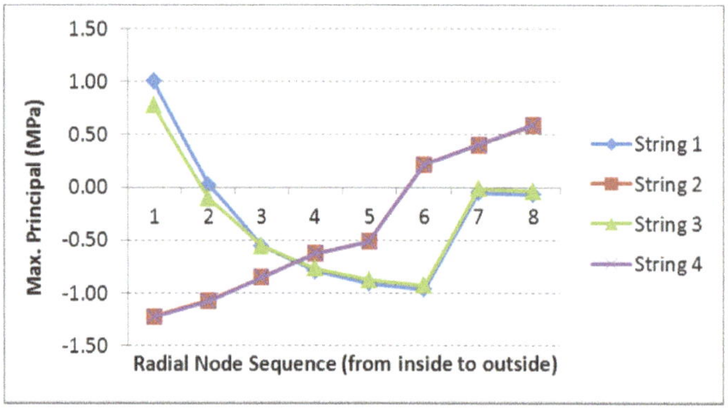

Figure 12. Maximum Principal stress at nodes in four strings in the middle section of PCCP (Case 1)

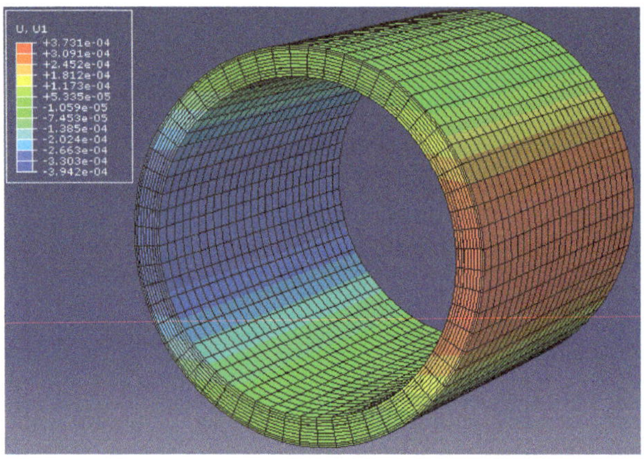

Figure 13. Horizontal deformation of PCCP (Case 1)

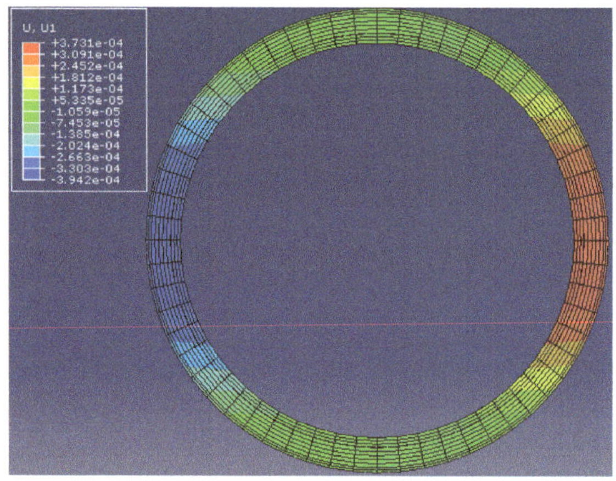

Figure 14. Horizontal deformation in the middle section of PCCP (Case 1)

3.2 Case 2: 40 Broken Wires, Only Prestress Loss at Broken Point

In this case, there are 40 broken wires in the middle of the model. However, because of the strong bond between the prestressing wire and mortar coating, there is prestress loss only at the point where the wire breaks. The maximum principal stress in mortar coating and concrete core is shown in Figure 15. The stress distribution in the middle section of the pipe is shown in Figure 16.

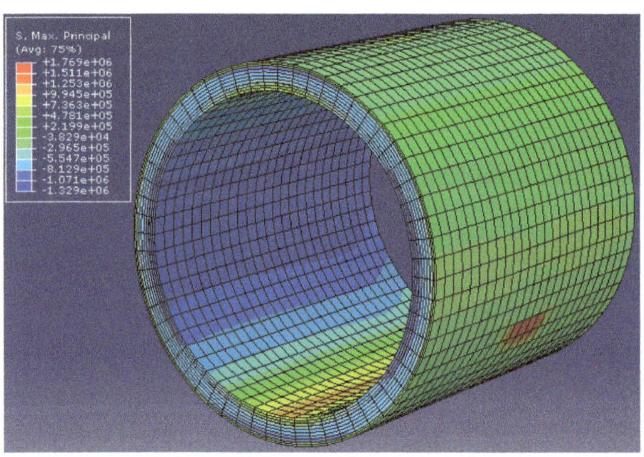

Figure 15. Maximum principal stress in PCCP (Case 2)

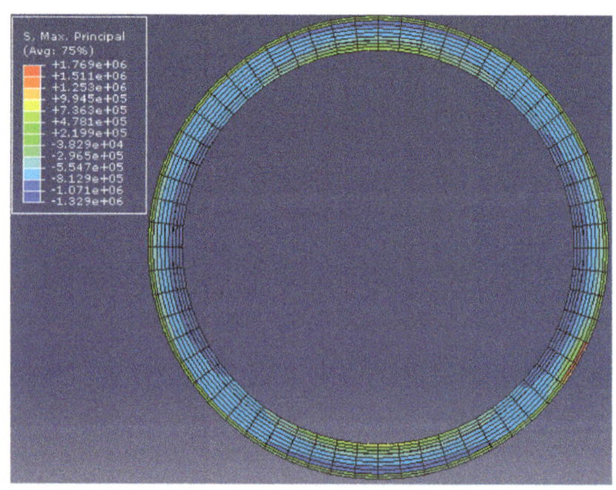

Figure 16. Maximum Principal stress in the middle section of PCCP (Case 2)

The maximum principal stress of the four strings in the middle section of the pipe is shown in Figure 17. According to Figure 15-17, the maximum principal stress is 1.00 MPa and it is located in mortar coating near the broken point of the prestressing wire. As the maximum principal stress is lower than the tensile strength of mortar coating (3.57 MPa), there is no crack in the coating. The deformation of the whole model and in middle section is shown in Figure 18 and Figure 19. The maximum horizontal deformation is 0.77 mm, almost the same compare with Case 1. In Figure 19, the stress at nodes in String 2 and String 4 are almost the same. In general, the stress and deformation distribution is very close to that in Case 1. This is because the pipe is not much compromised since there is only prestress loss at the broken point of the prestressing wires.

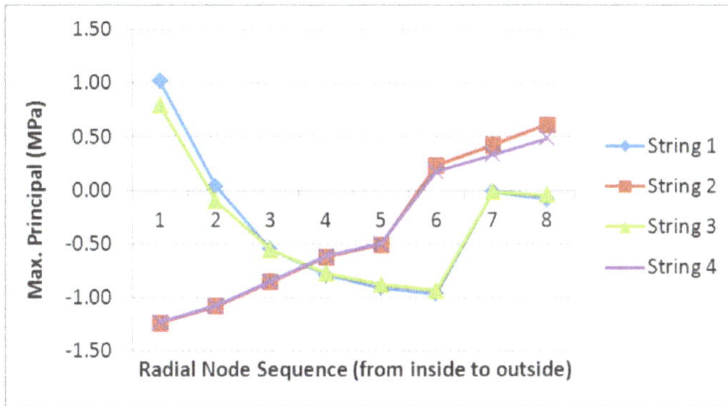

Figure 17. Maximum Principal stress at nodes in four strings in the middle section of PCCP (Case 1)

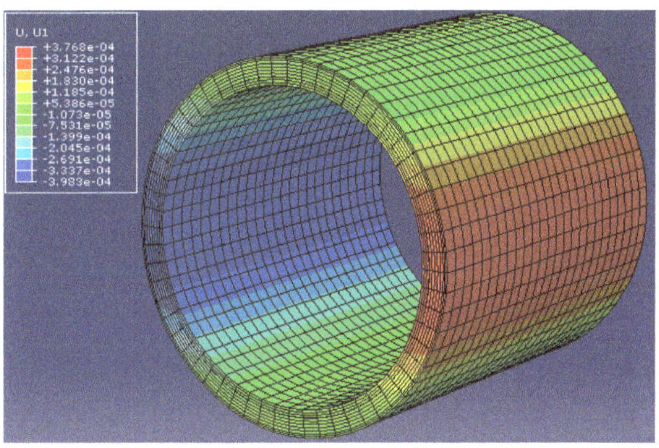

Figure 18. Horizontal deformation of PCCP (Case 2)

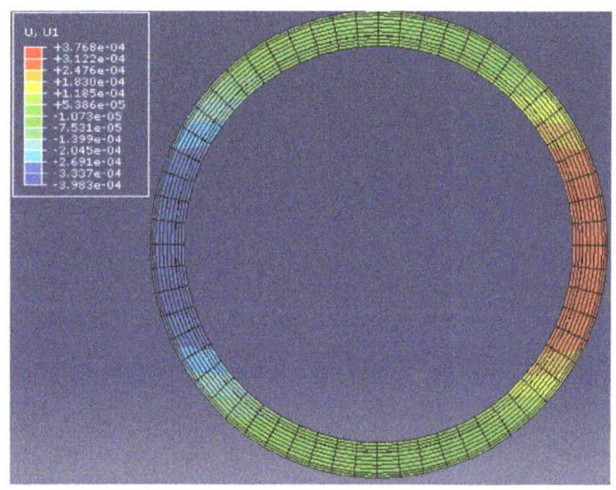

Figure 19. Horizontal deformation in the middle section of PCCP (Case 2)

3.3 Case 3: 40 Broken Wires, with 1/12 of the Broken Wire Lost Prestress

In this case, the pipe has 40 broken wires and because of lower mortar bond strength, the range of prestress wire losing prestress increased to 1/12 of the whole wire turn. The distribution of maximum principal stress is shown in Figure 20 and Figure 21. The maximum principal stress of the four strings in the middle section of the pipe is shown in Figure 22. According to Figure 20-22, the maximum principal stress is 19.72 MPa and it is located in mortar coating near the broken point of the prestressing wire. This stress is much higher than the tensile strength of mortar coating (3.57 MPa), so the coating will crack. As is shown in Figure 21, the concrete core in the broken wire zone near the prestressing wire has very high tensile stress (higher than the tensile strength of concrete), therefore, there could also cracks in the concrete core. Figure 22 shows that nodes in string 4 generally have higher stress than nodes in string 2, this is because it is closer to broken wire zone and received more transferred load.

The deformation of the whole model and in middle section is shown in Figure 23 and Figure 24. The maximum horizontal deformation is 0.88 mm, a little higher than that in Case 1 and Case 2. From Figure 23 and Figure 24, it can be seen that there is much higher deformation in the broken wire zone compared with other locations of the pipe.

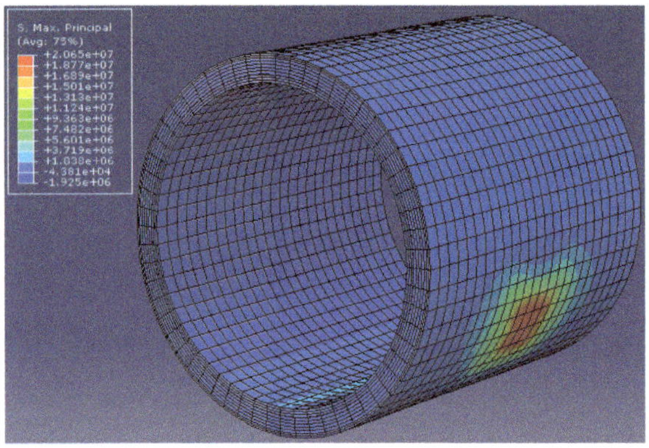

Figure 20. Maximum principal stress in PCCP (Case 3)

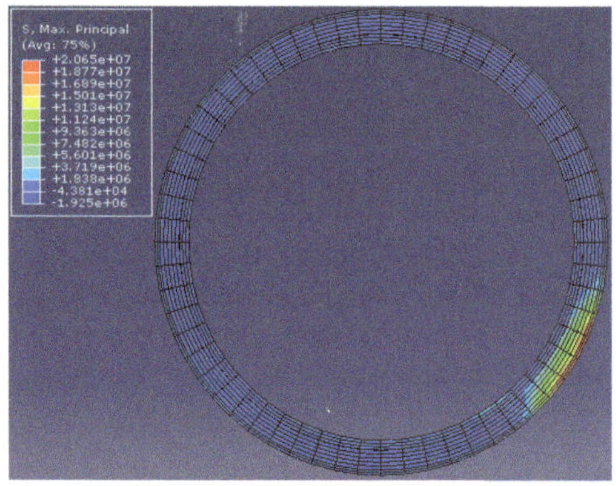

Figure 21. Maximum Principal stress in the middle section of PCCP (Case 3)

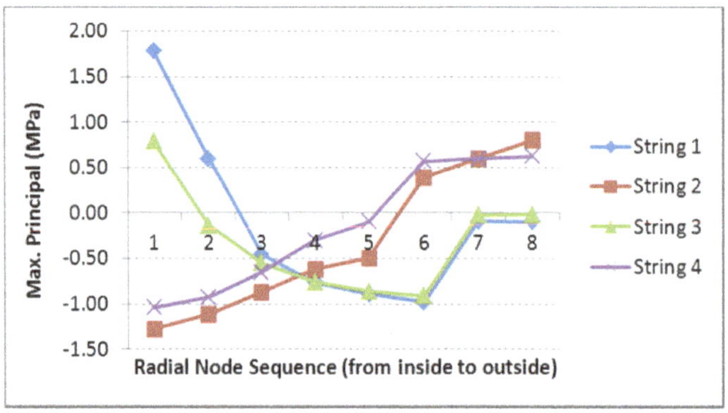

Figure 22. Maximum Principal stress at nodes in four strings in the middle section of PCCP (Case 3)

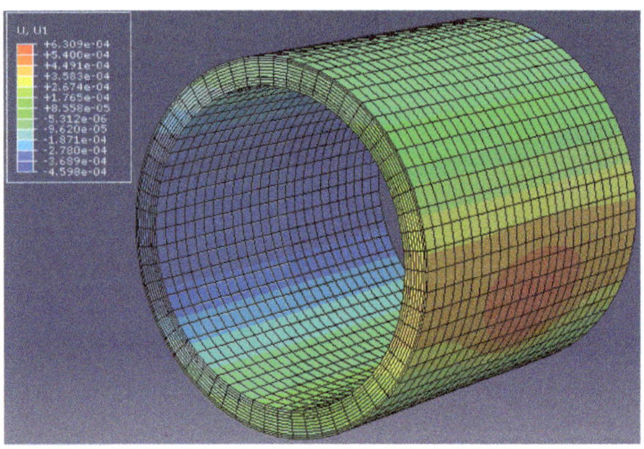

Figure 23. Horizontal deformation of PCCP (Case 3)

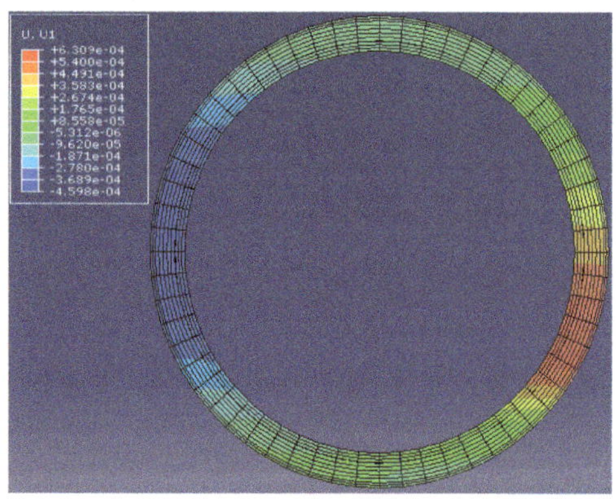

Figure 24. Horizontal deformation in the middle section of PCCP (Case 3)

3.4 Case 4: 40 Broken Wires, with 1/6 of the Broken Wire Lost Prestress

In this case, the pipe has 40 broken wires and the range of prestress wire losing prestress increased to 1/6 of the whole wire turn. The distribution of maximum principal stress is shown in Figure 25 and Figure 26. The maximum principal stress of the four strings in the middle section of the pipe is shown in Figure 27. According to Figure 25-27, the maximum principal stress is 27.33 MPa and it is located in mortar coating near the broken point of the prestressing wire. This stress is much higher than the tensile strength of mortar coating (3.57 MPa), so the coating will crack. As is shown in Figure 26, the concrete core in the broken wire zone near the prestressing wire also has very high tensile stress (higher than the tensile strength of concrete), and there could also be cracks in the concrete core. Figure 27 shows that nodes in string 4 has very high tensile stress due to the load transferred from the broken wires.

The deformation of the whole model and in middle section is shown in Figure 28 and Figure 29. The maximum horizontal deformation is 1.10 mm, much higher than that in Case 1, Case 2 and Case 3. Figure 28 and Figure 29 show that there is much higher deformation in the broken wire zone compared with other locations of the pipe.

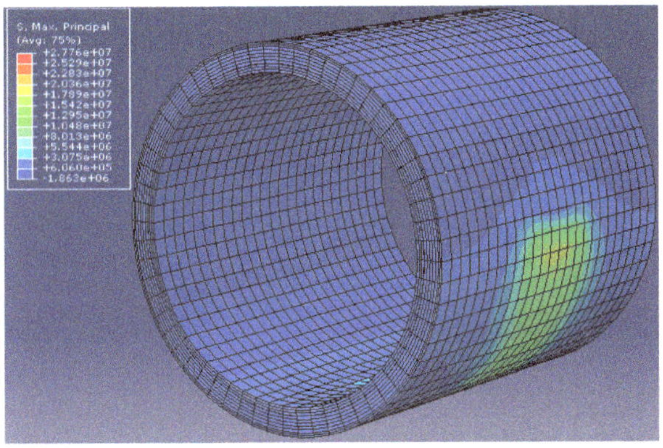

Figure 25. Maximum principal stress in PCCP (Case 4)

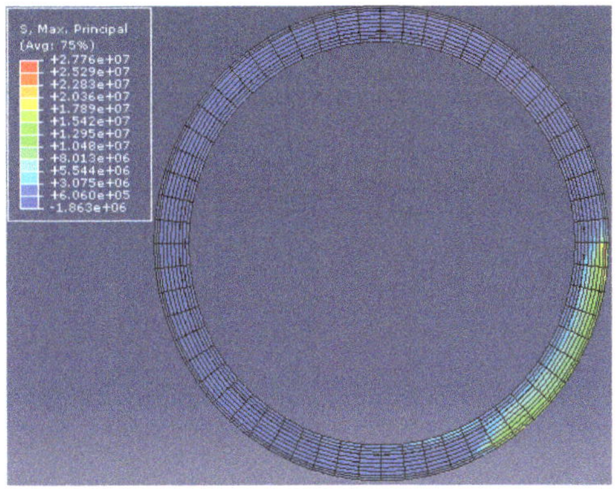

Figure 26. Maximum Principal stress in the middle section of PCCP (Case 4)

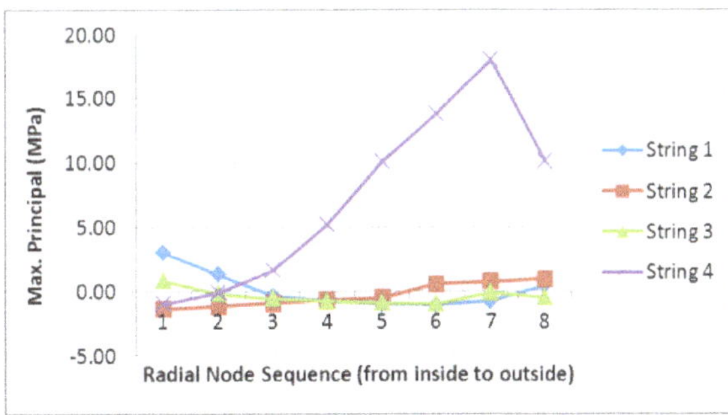

Figure 27. Maximum Principal stress at nodes in four strings in the middle section of PCCP (Case 4)

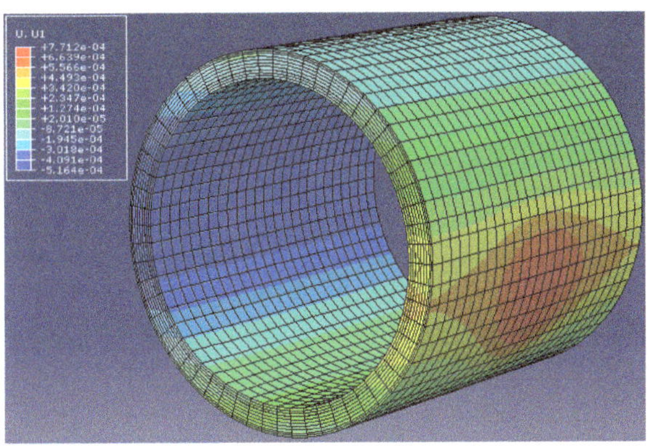

Figure 28. Horizontal deformation of PCCP (Case 4)

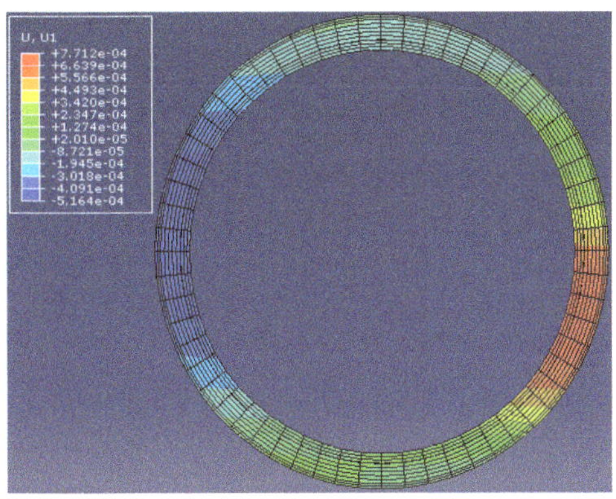

Figure 29. Horizontal deformation in the middle section of PCCP (Case 4)

3.5 Case 5: 40 Broken Wires, with 1/3 of the Broken Wire Lost Prestress

In this case, the pipe has 40 broken wires and the range of prestress wire losing prestress increased to 1/3 of the whole wire turn. Figure 30 and Figure 31 shows the distribution of maximum principal stress. Figure 32 shows the maximum principal stress of the four strings in the middle section of the pipe. From Figure 30-32, it can be seen that the maximum principal stress is 25.31 MPa and it is located in mortar coating near the broken point of the prestressing wire. This stress is much higher than the tensile strength of mortar coating (3.57 MPa), so the coating will crack. From Figure 31, it can be seen that the concrete core in the broken wire zone near the prestressing wire also has very high tensile stress (higher than the tensile strength of concrete), and there could also be cracks in the concrete core. Figure 32 shows that nodes in String 1 and 4 have very high tensile stress because they are very close to the broken point of the prestressing wire.

The deformation of the whole model and in middle section is shown in Figure 33 and Figure 34. The maximum horizontal deformation is 1.26 mm and there is much higher deformation in the broken wire zone compared with other locations of the pipe.

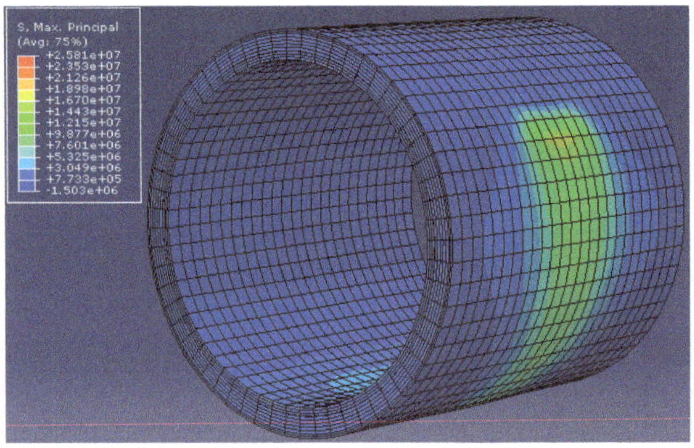

Figure 30. Maximum principal stress in PCCP (Case 5)

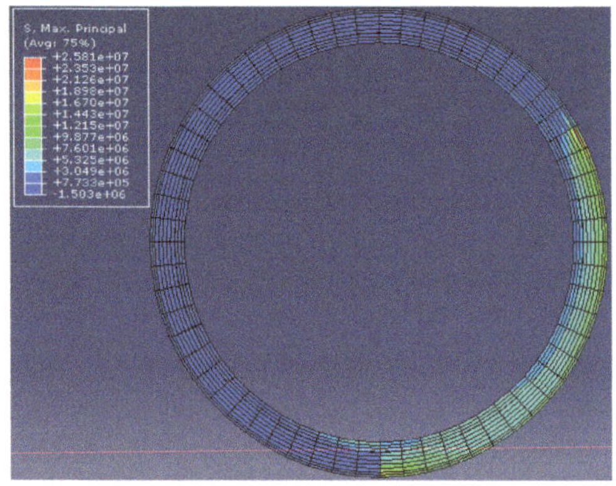

Figure 31. Maximum Principal stress in the middle section of PCCP (Case 5)

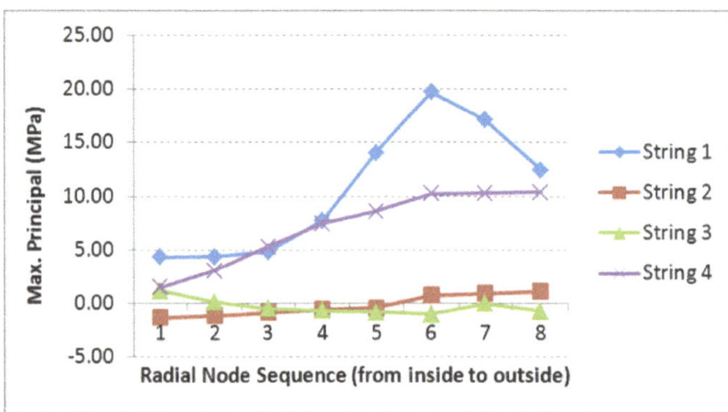

Figure 32. Maximum Principal stress at nodes in four strings in the middle section of PCCP (Case 5)

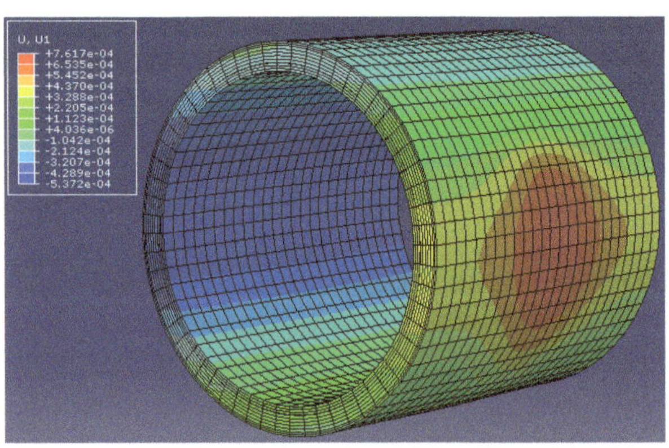

Figure 33. Horizontal deformation of PCCP (Case 5)

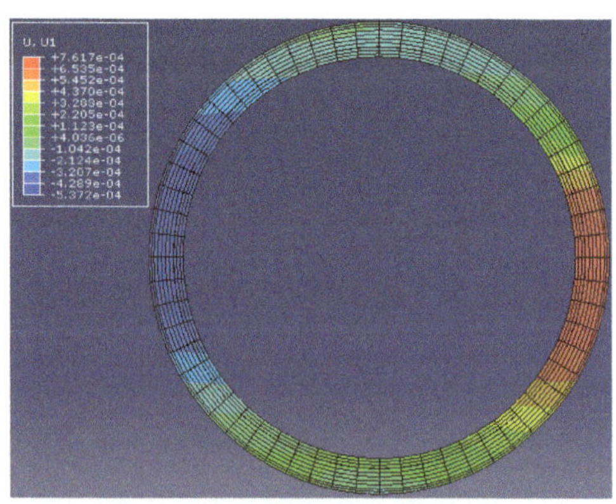

Figure 34. Horizontal deformation in the middle section of PCCP (Case 5)

3.6 Case 6: 40 Broken Wires, with the Whole Broken Wire Lost Prestress

In this case, all the 40 broken wires lost prestress and they are removed out of the model. This is exactly what the current models assume and do. Figure 35 and Figure 36 shows the distribution of maximum principal stress. Figure 37 shows the maximum principal stress of the four strings in the middle section of the pipe. From Figure 35-37, it can be seen that the maximum principal stress is 9.39 MPa and it is located at the invert of the pipe in the broken wire zone. This stress is much higher than the tensile strength of concrete (4.02 MPa), so the coating will crack here. Figure 36 shows that much of the concrete core in the broken wire zone has very high tensile stress (higher than the tensile strength of concrete). Figure 37 shows that nodes in String 2 and 4 have the same stress because of symmetry. Compared with other cases, this case shows cracks at invert, crown inside the pipe and springlines outside the pipe. Though the maximum principal stress is lower compared with other cases, the case shows the most severe condition of the pipe.

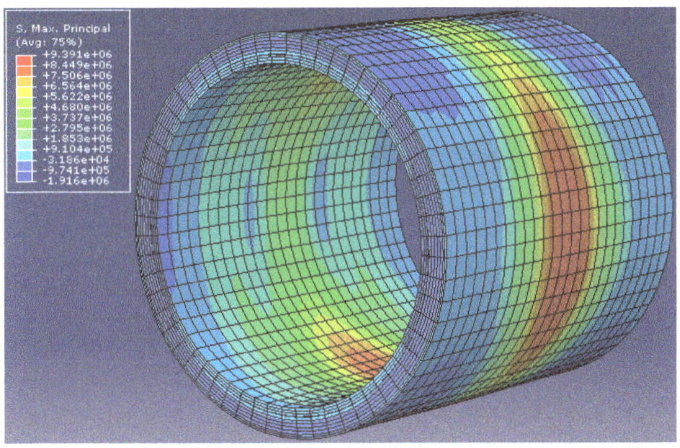

Figure 35. Maximum principal stress in PCCP (Case 6)

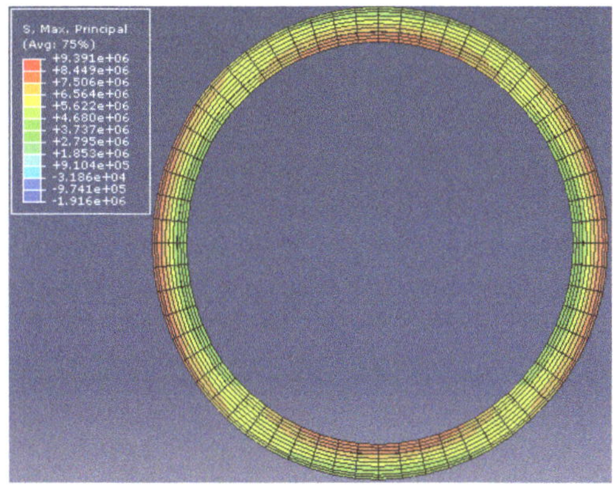

Figure 36. Maximum Principal stress in the middle section of PCCP (Case 6)

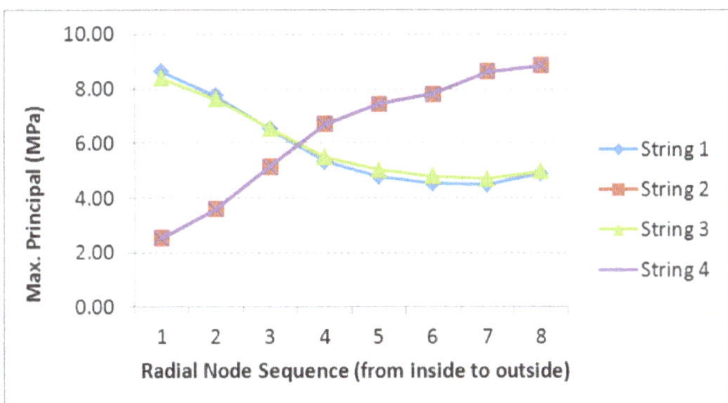

Figure 37. Maximum Principal stress at nodes in four strings in the middle section of PCCP (Case 6)

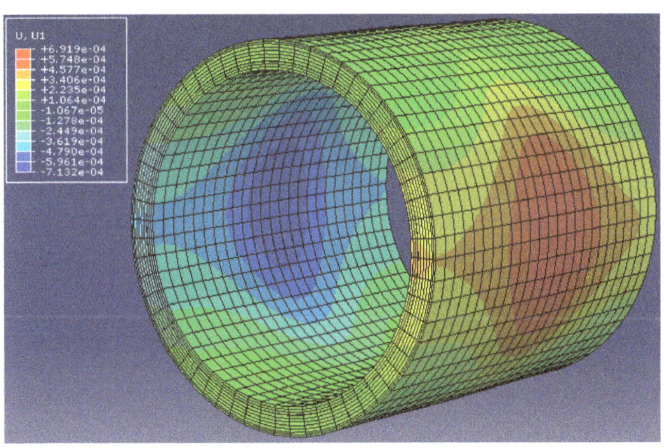

Figure 38. Horizontal deformation of PCCP (Case 6)

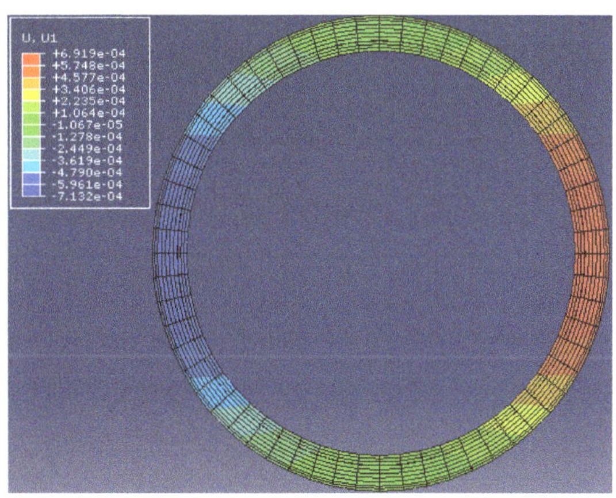

Figure 39. Horizontal deformation in the middle section of PCCP (Case 6)

The deformation of the whole model and in middle section is shown in Figure 38 and Figure 39. The maximum horizontal deformation is 1.37 mm, much higher compared with previous cases. From this case, it can be seen that assuming all prestress in broken wires is lost is too conservative. This is not consistent with the actual failure mode of PCCP. In practice, most failures fail at 4 or 8 o'clock in circumferential position. In the investigation of actual failures of PCCP, most pipes burst at these positions. However, according to this conservative assumption, after removing all broken wires from the model, the maximum tensile stress occurs in the invert of the pipe and springline.

4. Summary of PCCPs with Various Bond Quality

The maximum principal stress of nodes in four strings of each case is summarized to understand the impact of broken wires with different bond quality on the structural integrity of the pipe. The stress at nodes in each string is shown in Figure 40-43. In Figure 40, since the broken wires are cut at around string 1, there is stress concentration in this location. Same is for String 4 in Case 4 shown in Figure 40, the broke wire is cut at around String 4 and stress concentration generates in this location. In general, case 1 and case 2 have very close in almost all String positions. This is because, even though there are broken wires, if the wire bond is very good and no prestress loss in the prestress wires, the pipe is almost not compromised. As the range of broken wires losing prestress increase, the number of nodes with stress higher than tensile strength increases, this is because more load which has been born by the prestressing wires is transferred to the surrounding concrete core and coating. Though in Case 4 and 5 there are locations with very high tensile stress, the total zone with high tensile stress is smaller compared with Case 6 which shows the whole broken wire zone is in danger of cracking. However, if the prestress wires have only a small range without prestress as is shown in Case 2 and Case 3, the pipe is still in good condition with any crack.

The horizontal deformations of the pipe in broken wire zone and non-broken wire zone are compared in Figure 44, which shows that the horizontal deformation in the broken wire zone increases as the range of broken wire without prestress increases. Similarly, the difference of the horizontal deformation between broken wire zone and non-broken wire zone increases, which is consistent with the increase of maximum principal stress in the pipe.

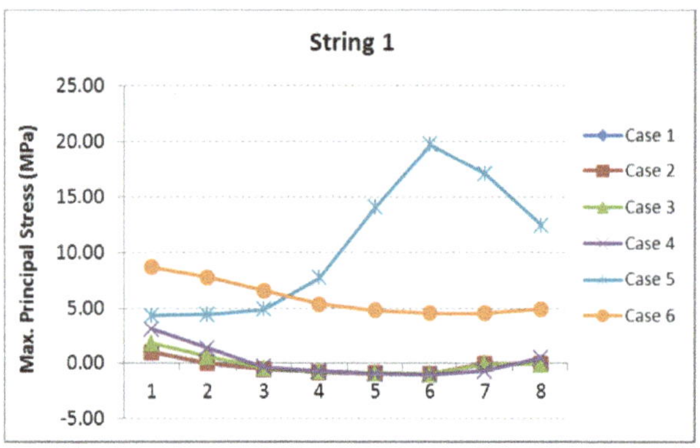

Figure 40. Maximum Principal stress at nodes in Strings 1

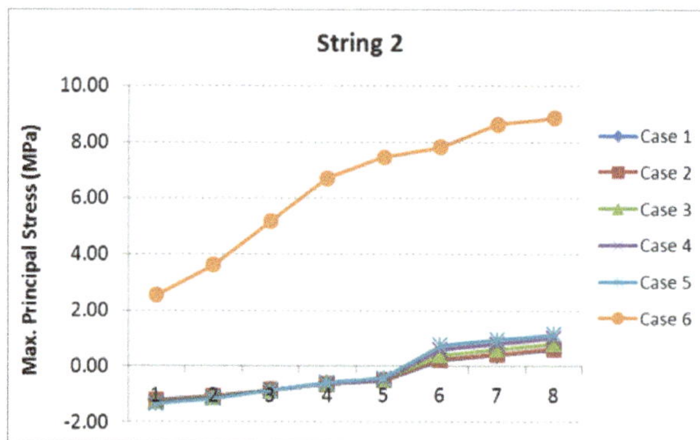

Figure 41: Maximum Principal stress at nodes in Strings 2

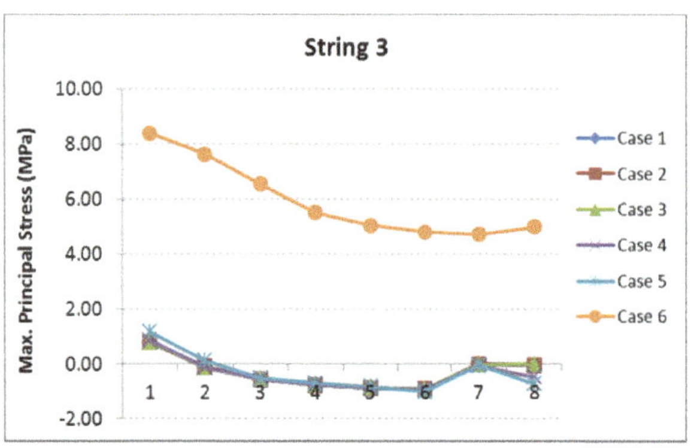

Figure 42. Maximum Principal stress at nodes in Strings 3

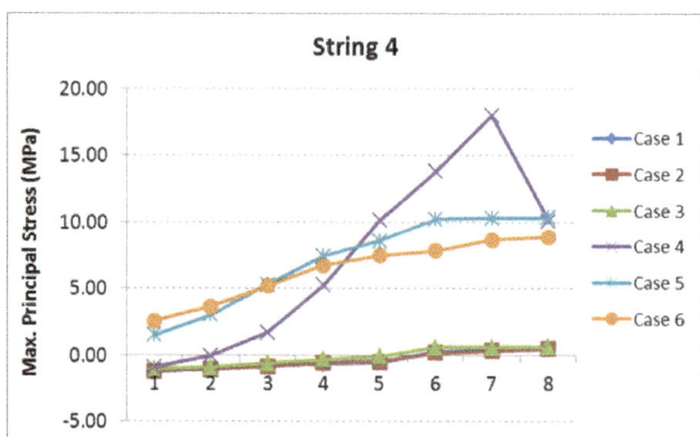

Figure 43. Maximum Principal stress at nodes in Strings 4

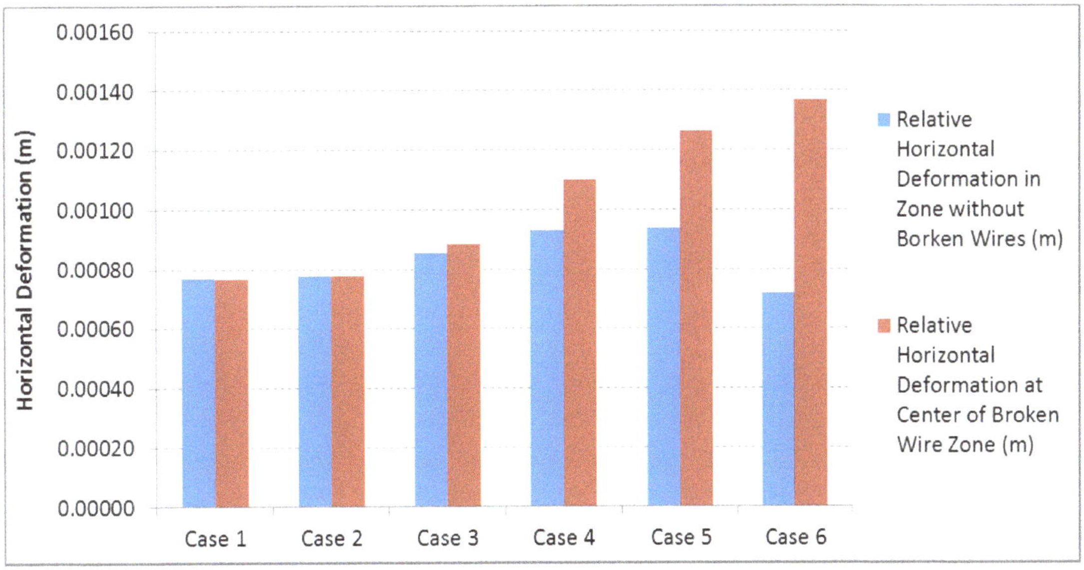

Figure 44. Horizontal deformation in the PCCP

5. Conclusions

The bond quality of mortar coating is a critical factor to consider in the modeling of PCCP using numerical models. Current models have very conservative assumption and remove all broken wires from the model. This could be over-conservative in cases where the mortar coating has very good bond. Even though the wire is broken, there still could be prestress remaining in a certain length of the broken wires.

To have a more realistic simulation of PCCP and to understand the effect of mortar coating's quality on the structural integrity of PCCP, this paper proposes a theoretical methodology to determine the length of wires needed to resume prestress beyond the broken point. Based on this methodology, a numerical model is established to analyze scenarios in which the mortar coating has different bond qualities. In this paper, six cases are analyzed and compared. The findings from the results indicate that even though the pipe has broken wires, however, if the mortar has good bond, the pipe is not much compromised and there could be no cracks in coating or concrete core. In contrast, if the mortar coating has poor bond, then there could be very high tensile stress exceeding the tensile strength of the mortar coating or concrete cores, causing cracks in the pipe. Compared with all other cases, the assumption of zero prestress in all broken wires could cause very high tensile stress in the pipe and cause cracks in invert, crown, and external springline of the pipe. This is overly conservative and is not consistent with most of the actual pipe failures modes, in which most pipes fail at 4 or 8 o'clock, not at invert, crown, or springlines.

The research presented in this paper highlights the importance to consider the bond quality of mortar coating and provides guidance for development of numerical models for PCCP. The findings of this research could help utilities have a better understanding of the limitations of current numerical models used for PCCP condition evaluation, and help them better manage PCCP pipelines. In this proposed model, due to the prestress change in broken wires, local stress concentration could be generated at places where the wire is cut. In future research, more work should be done to consider the prestress distribution between the broken wire point and the point where full prestress is resumed.

Acknowledgement

The development of WATER*i*D (WATER Infrastructure Database) is funded by the Water Environment Research Foundation (WERF) under the contract INFR9SG09 and INFR10SG09 through the U.S. Environmental Protection Agency's Aging Water Infrastructure Research Program, and stems from EPA's Sustainable Water Infrastructure Initiative. The authors would like to thank the National Science Foundation (NSF) for the partial funding under the grant NSF-CMMI-0801018. We would also like to thank the support of the Sustainable Water Infrastructure Management (SWIM) Center of Excellence in the Institute of Critical Technology and Applied Science (ICTAS) at Virginia Tech. The authors also appreciate the constructive disucssions with Mr. John Marshall from J.W. Marshall and Associates, and Mr. Mike Woodcock from Washington Suburban Sanitary Commission.

References

Alavinasab, A., Padewski, E., Holley, M., Jha, R., & Ahmadi, G. (2010). Damage Identification Based on Vibration Response of Prestressed Concrete Pipes. In *ASCE Pipelines Conf.: Climbing New Peaks to Infrastructure Reliability-Renew, Rehab, and Reinvest, Keystone, CO* (pp. 909-919).

AWWA C301-07 (2007). *AWWA Standard for Prestressed Concrete Cylinder Pipe, Steel-Cylinder Type.* American Water Works Association, Denver, CO.

ANSI/AWWA C304-07 (2007). *Design of Prestressed Concrete Cylinder Pipe.* American Water Works Association, Denver, CO.

AwwaRF. (2008). *Failure of Prestressed Concrete Cylinder Pipe, C301-07.* AWWA Research Foundation, Denver, CO.

Bell, G. E., Kendall, D. R., & Mulligan, S. B. (2001). An Assessment and Reliability Program for PCCP. *Proc. Advances in Pipeline Engineering and Construction, American Society of Civil Engineers* (pp. 1-10).

Diab, Y. G., & Bonierbale, T. (2001). A numerical modeling and a proposal for rehabilitation of PCCPs. In *Proc. Int. Conf. on Advances in Pipelines Eng. & Constr, ASCE, San Diego, California.*

Erbay, O. O., Zarghamee, M. S., & Ojdrovic, R. P. (2007). Failure risk analysis of lined cylinder pipes with broken wires and corroded cylinder. In *Int. Conf. on Pipeline Eng. and Construction* (pp. 1-10).

Essamin, O., El-Sahli, K., Hovanessian, G., & LeDiouron, T. (2005). Risk Management System for Prestressed Concrete Cylinder Pipeline: Practical Results and Experience on the Great Man Made River. *Pipelines 2005* (241-251).

Galleher Jr, J. J., & Stift, M. T. (1998). Internal inspection and database development of PCCP. In *Pipelines in the Constructed Environment* (pp. 721-730).

Ge, S., & Sinha, S. (2012). Failure Analysis, Condition Assessment Technologies, and Performance Prediction of Prestressed-Concrete Cylinder Pipe: State-of-the-Art Literature Review. *Journal of Performance of Constructed Facilities*, *28*(3), 618-628.

Gomez, R., Muñoz, D., Vera, R., & Escobar, J. A. (2004). Structural model for stress evaluation of prestressed concrete pipe of the Cutzamala system. In *Proc. Int. Conf. on Pipeline Eng. & Constr.: What's on the Horizon, ASCE, San Diego, California.*

Kola, R. (2010). Development of predictability and condition assessability indices for PCCP water mains, Master Thesis, Virginia Polytechnic Institute and State University, Blacksburg, VA.

Lee, Y. (2011). *Analysis of prestressed concrete cylinder pipe for rehabilitation* (Ph.D. Dissertation). University of California, Irvine, California, USA.

Lewis, R. O. (2002). Improving prestressing wire reliability in prestressed concrete cylinder pipe. *Materials performance*, *41*(7), 48-54.

Marshall, J, Fisk, P. (2015). Measuring the Condition of Prestressed Concrete Cylinder Pipe. http://www.ndtcorporation.com/resources_1.html (Visited in January, 2015).

Ojdrovic, R. P., Zarghamee, M. S., Hegarty, J. R., and Westman, T. (2001). "Condition Assessment of a PCCP Line Accessible from Outside Only." *Pipelines 2001-Advances in Pipeline Engineering & Construction*, 1-13.

Parks, R. R., Drager, J. K., and Ojdrovic, R. P. (2001). "Condition Assessment and Rehabilitation of the Windy Gap Pipeline – An Owner's Perspective." *Pipelines 2001-Advances in Pipeline Engineering & Construction*, 1-13.

Price, R. E., Lewis, R. A., & Erlin, B. (1998). Effects of environment on the durability of prestressed concrete cylinder pipe. In *Pipelines in the constructed environment, Proceedings of the 1998 Pipeline Division Conference* (pp. 584-593).

Rauniyar, S. (2013). *Development of finite element model for analysis of prestressed concrete cylinder pipe - embedded cylinder pipe* (M.S. Thesis). University of Texas at Arlington, Texas, USA.

Romer, A., Ellison, D., Bell, G., and Clark, B. (2008). "Failure of Prestressed Concrete Cylinder Pipe." *Pipelines 2007*, 1-17.

Semanuik, S. and Mergelas, B. (2006). "Comparison of indentified distress in CCP pipelines operated by water utilities in North America." *Proceedings of the ASCE Pipeline Division Specialty Conference*, July 30-August 2, Chicago, Illinois, 1-8.

UIM. (2011). Prestressed Pipe Deterioration Prompts Large Utility to Seek Effective, Cost-efficient Approach. Retrieved December 29th, 2013, from http://uimonline.com/index/webapp-stories-action/id.563/archive.yes/Issue.2011-10-01/title.prestressed-pipe-deterioration-prompts-large-utility-to-seek-effective,-cost-efficient-approach

USEPA. (2013). Miami-Dade County Clean Water Act Settlement. Retrieved May 9th, 2014, fromhttp://www2.epa.gov/enforcement/miami-dade-county-clean-water-act-settlement

Villalobos, J. L. (1998). "Effects of Prestressed Concrete Cylinder Pipe in a high chloride environment after 19 years of service." *Pipelines in the Constructed Environment*, ASCE, Reston, VA, 575-583.

Walsh, T. L. and Hodge, D. S. (1998). "Overcoming the challenges of replacing 20 km of defective 1524 mm diameter PCCP." *Pipelines in the Constructed Environment*, ASCE, Reston, VA, 602-611.

Wardany, R. A. (2008). Condition assessment of prestressed concrete cylinder water pipes. *Proceedings of 60th Annual WCWWA Conference and Trade Show* (pp. 1-9).

Weare, R. E. (2007). PCCP sewerage force main structural condition assessment and asset management approach. *Pipelines 2007: Advances and Experiences with Trenchless Pipeline Projects* (pp. 1-11).

Woodcock, M. (2008). *48-inch PCCP Pipeline Break.* Technical report, Washington Suburban Sanitary Commission, Laurel, Maryland.

WSSC (2012). *80 foot setback*. Washington Suburban Sanitary Commission, Laurel, Maryland. Retrieved September 18, 2014, from http://www.wsscwater.com/file/EngAndConst/80footsetbackparts/part2.pdf

Xiong, H., Li, P., & Li, Q. (2010). FE model for simulating wire-wrapping during prestressing of an embedded prestressed concrete cylinder pipe. *Simulation Modelling Practice and Theory*, *18*(5), 624-636.

Zarghamee, M. S., Ojdrovic, R. P., & Dana, W. R. (1993). Preventing coating delamination in prestressed concrete pipe. In *Pipeline Infrastructure II* (pp. 574-594). ASCE.

Zarghamee, M. S., & Ojdrovic, R. P. (2001). Risk assessment and repair priority of PCCP with broken wires. In *Pipelines 2001@ sAdvances in Pipelines Engineering and Construction* (pp. 1-8). ASCE.

Zarghamee, M. S., Eggers, D. W., & Ojdrovic, R. P. (2002). Finite-element modeling of failure of PCCP with broken wires subjected to combined loads. In *Pipelines 2002@ sBeneath Our Feet: Challenges and Solutions* (pp. 1-17). ASCE.

Zarghamee, M. S., Eggers, D. W., Ojdrovic, R. P., & Rose B. (2003). Risk analysis of prestressed concrete cylinder pipe with broken wires. *Pipelines 2003* (pp. 599-609).

2

On the Effect of SiC Content and Processing Temperature on Relative Density and Hardness of Hot Compacted Aluminum AA6061 Composite – Mathematical Empirical and Response Surface Approach

W. H. El-Garaihy[1], S. E. Oraby[2], El-Sayed M. A. Rassoul[3] & H. G. Salem[4]

[1] Mechanical Engineering Department, Faculty of Engineering, Suez Canal University, Egypt

[2] Department of Manufacturing Engineering Technology, College of Technological Studies, PAAET, Kuwait

[3] Production Engineering Department, Faculty of Engineering, Mansoura University, Egypt

[4] Mechanical Engineering Department, Yousef Jamil Science and Technology Research Center (YJSTRC), American University in Cairo, Egypt

Correspondence: W.H. El-Garaihy, Mechanical Engineering Department, Faculty of Engineering, Suez Canal University, Ismailia, Egypt. E-mail: wgaraihy@eng.suez.edu.eg.

Abstract

The AA6061 is reinforced by adding SiC at various volume fractions and, the mixture is hot compacted at different processing temperatures. The influences of such parameters are investigated on the product relative density along with its relevant Vickers hardness using quantitative and qualitative formulation approach. Empirical relationships are established to relate each of the controlling (independent) parameters (SiC% and hot compaction (HC) temperature), to the composites relative density and the hardness, as dependent variables. The developed models are examined for its adequacy and significance using several statistical criteria. Response surface and contour graphs are established to reflect the relevant function interrelations and, to provide a data base source for the prior design stage.

Within the specified experimental domain, first order and nonlinear models are found independently adequate and significant to grasp the functional dependence between the relative density and both SiC and HC temperature. However, second order multiple model with quadratic components of SiC percent is found to best suit the hardness-SiC%-temperature functional relationship.

Increasing SiC content is found to reduce the relative density of the composites regardless the hot compaction temperature while, up to about 18 vol.% SiCp relative ratio, it enormously and nonlinearly increases the composite hardness. Further increase in SiC% addition seems not to affect the composite hardness. Relative density of the resulting composite is decreased by increasing HC temperature.

Keywords: AA6061, hot compact powder technology, relative density, vickers hardness, linear and nonlinear regression procedures, response surface methodology

1. Introduction

Aluminum alloys are notable by vast diversity in industrial application thanks to their many advantages regarding specific strength, corrosion resistance, thermal conductivity, low density, and good workability. On the other hand, their use is limited due to their relatively low yield strength and poor tribological characteristics (Min, 2009; O'Donnel et al., 2001). Therefore, applicability was affected negatively in many circumstances as a result of their reduced hardness and wear resistance (Huang et al., 2004). Recently, the interest to increase aluminum strength has risen and the study of metal matrix composites (MMCs) has been motivated in particular applications such as the aerospace and aeronautic industries. MMCs are considered as outstanding materials to obtain properties that are superior to those of the constituent phases and also to satisfy the above requirements. Aluminum is the most common metal used in MMCs; in particular, particles reinforced Aluminum-based MMCs are focal composites grasping an increasing attention recently thanks to their lightness, higher specific strength, and wear resistance (Senapati et al., 2014; Zakaria, 2014; Mohanakumara et al., 2014).

Aluminium-Silicon alloys, as a matrix material, are frequently selected for their good wear resistance, strength-to-weight ratio, thermal conductivity, ease of fabrication at reasonable cost, high strength at elevated temperature, as well as excellent corrosion resistance. As a result of the previous mentioned properties, the suitability of these alloys for aerospace, automotive and military applications has been evident (Senapati et al., 2014; Rao & Das, 2011). The strengthening of Aluminum and its alloys can be done by dispersing hard particles such as carbides, oxides, or nitrides into the aluminum matrix by using various techniques in the solid or liquid state (Pawar & Utpat, 2014). Al_2O_3 and SiC reinforcements are two widely used types of reinforcing agents in aluminum metal matrix composites (AMCs). Their use is focused mainly in automotive and aircraft industries due to the importance of material tribological properties in these applications (Senapati et al., 2014; Mazahery & Shabani, 2013). SiC is a covalent material of huge technological attention thanks to its excellent overall properties. It has good thermal shock behavior and mechanical resistance with has high thermal conductivity, oxidation, and erosion resistances (Liu et al., 2010). The SiC as whisker or particle reinforced Al matrix composite (Al-SiC) is perhaps the most successful class of MMCs produced ever (Mandal & Viswanathan, 2013). As a result, AMCs reinforced with SiC particles offer higher modulus, wear resistance, and better dimensional stability than conventional aluminum alloys (Mohanakumara et al., 2014).

Powder Metallurgy (PM) can be used to prepare aluminum composites in the solid state. The process consists of mixing reinforcement particles with the metallic powder, followed by consolidation and sintering processes. Other methods that could be also adopted include mechanical alloying (MA) and mechanical milling (MM), which renders composites with fine and homogeneous distributions of the particles (Mazahery & Shabani, 2013; Showaiter & Youssefi, 2008; Kim et al., 2001). The best properties of PM for fabrication of composites can be obtained when the reinforcement is homogeneously dispersed in the matrix (Ravindran et al., 2013).

MMCs can be reinforced using alternatives such as continuous fibres, discontinuous particles, or whiskers (Yan et al., 2008). Particle-reinforced MMCs possess distinct advantages over fibre reinforced composites regarding low cost and isotropic mechanical properties considerations. Therefore, they are relatively easier to process via powder compared to AMCs reinforced with ceramic whiskers and fibers (Mazahery & Shabani, 2012). The mechanical properties of a composite under loading are typically controlled by the interfaces formed between matrix and reinforcement particles. Generally, a good interface bonding with coherency or semi-coherency is advantageous for better mechanical properties. Conversely, interface with in-coherency degrades its properties, especially with the presence of brittle intermetallic phases (Mandal & Viswanathan, 2013). Discontinuous particles reinforced MMCs have gained much interest recently due to their promising mechanical properties regarding matrix-reinforcement coherency (Mohanakumara et al., 2014; Mazahery & Shabani, 2013; Zhanwei et al., 2014). Additionally, the common problems accompanying the fabrication of continuous reinforced MMCs such as fiber-damage, microstructural heterogeneity, fiber mismatch, and inter-facial reactions are minimized by the use of discontinuous reinforcements (Kalkanl & Yilmaz, 2008).

During the last several decades, optimizing the mechanical properties of the SiC reinforced aluminum alloy composites has been a main point of interest for many researchesrs, for instance (Min, 2009). The improvement of mechanical properties of produced composites could be reached by having a well performed homogenization which would enable uniform distribution of reinforcement particles in the composite matrix. The effectiveness of the powder and the performance of produced components are typically determined by the appropriate relative mixture contents and constituents (Bozic et al., 2010).

In the current study, a quantitative procedure is adopted to explore the effect of the SiC volume fraction and the HC temperature on the physical and mechanical properties of AMCs, and to develop a general approach describing the dependence of the relative density and hardness of the AA6061-SiC composite on SiC content and hot compaction temperature. Based on a previous study (El_Garaihy, 2012), experimental data are used to establish mathematical models of the functional relationship of the aforementioned parameters. Response surface in terms of three dimensions and contours representations are introduced as database reference to help in the design stages.

2. Materials and Experimental Procedures

2.1 Materials

The investigated aluminum alloy AA6061 (supplied by Powders Company Limited) was received in the form of powders 30 μm in average size, Figure 1a. AA6061 particles were characterized by irregularity in shape with variation in size from 10-to-75 μm. The as-received SiC powders (supplied by American Elements Company) were used as reinforcement. As shown in Fig. 1b, SiC powder was characterized by non-uniformity in shape with particles size ranging from 1-to-5 μm with an average size of 2 μm.

2.2 Experimental Procedures

To fabricate theAA 6061- SiC metal matrix composite, the required fractions of AA6061 and SiC powders were mixed under vacuum in a glove box followed by mixing in a turbula mixer operating at a rotation speed of 96 rpm. Four different volume fractions of SiC (5, 10, 15, and 20%) were selected to reinforce AA6061 matrices. Both AA6061, AA6061-SiC composites powders were subjected to single sided uniaxial HC into cylindrical sample 10 mm in diameter and 10 mm in height. The hot compaction process was carried out at compaction pressure of 525 MPa and temperature of 400, 450, and 500°C for 30 min.

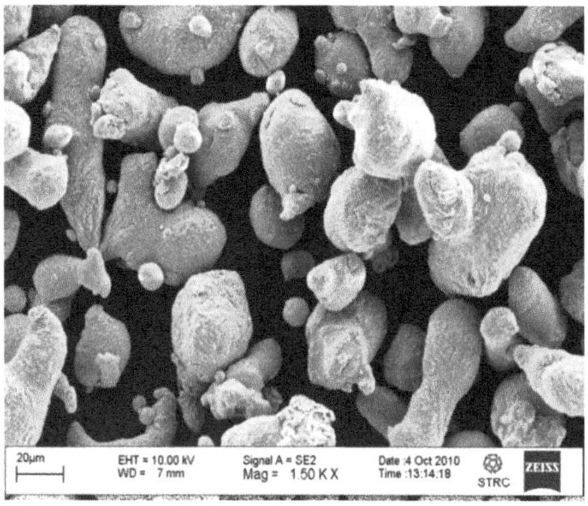

Figure 1a. SEM micrograph of AA6061 powders

Figure 1b. SEM micrograph of SiC powders

Following HC, the disks were polished to a mirror-like finish. Density measurements were carried out on the base alloy and reinforced samples using a digital densitometer that employs the Archimedean principle for density measurements. Vickers Hardness (Hv) values were measured on the disk sections cut parallel to the loading direction using digital metallic Vickers's hardness tester. Testing was carried out on the surfaces under applied load of 10 N and a dwell time 15 sec for each separate measurement. The microstructure of the monolithic alloy and the composites were characterized by LICKA inverted metallurgical OM and a LEO field emission scanning electron microscopy (FESEM) operating at 200KV at 1 nm resolution of ground polished and etched samples using "keller" solution.

3. Data Processing Modeling Methodology

3.1 LinearModeling

To establish an empirical functional interrelationship between a dependent variable, response (R), and some dependent variables (ξ's), a first-order model is usually postulated to take the general form (Douglas et al., 2012; Myres et al., 2012):

$$R = bo + \sum_{j=1}^{p} b_j \xi_j + \varepsilon_n, \tag{1}$$

where b's are the model estimates (coefficients) and ε n is the error absolute value using linear non-transformed model. However, in terms of real parameters, the model takes the final first order multiple linear model:

$$R = b_o + b_1(x_1) + b_2(x_2) + \ldots\ldots + b_n(x_n), \tag{2}$$

where x's are the independent variables.

To take into account the possible interaction and quadratic effect among the independent variables, a second order structure is usually accounted for. For two independent variables x_1 and x_2, the second order model usually takes the form (Myres et al., 2012):

$$R = b_o + b_1(x_1) + b_2(x_2) + b_{11}\,(x_1^2) + b_{22}\,(x_2^2) + b_{12}\,(x_1.x_2), \tag{3}$$

where b_{ii} and b_{ij} are the quadratic and the interaction effects to be determined by fitting the experimental data using the appropriate mathematical technique.

3.2 NonlinearModeling

However, to detect the possible nonlinear nature of the independent variables, a nonlinear model is usually proposed in the form (Myres et al., 2012; Kowalski, 1977):

$$R = ao\ x_1^{a1}\ x_2^{a2} \ldots\ldots\ldots x_n^{an}, \tag{4}$$

where a's are the model coefficients to be determined by the nonlinear regression procedures using the experimental data.

Regression statistical routine is used together with the experimental data to reach the most adequate and significant relationship between each of the dependent variable; relative density (RD) and Vicker's hardness (H_V), and the independent variables; AA6061 percentage in the alloy (Al%) and compact temperature (T_{HC}).

Regression routine is one of the most familiar statistical tool to detect the influence of more than one independent variables are involved. This is a necessity if a response surface is required. Using regression analysis asseses and estimates the effect of each individual variable on the measured response together with the possible interaction effect among all involved variables. As in many modeling and forecasting techniques, such as neural network, genetic algorithm, etc, regression analysis uses the most widely used ordinary least squares analysis. Such a fitting technique is included as a toolbox facility in many commercial programs such as MathWork and MatLab. Least squares method creats the best fit line or surface through all of the available data points so as to minimize error sum of squares. Fitting a regression model requires several assumptions of which the assumption that the errors are uncorrelated random variables with mean zero and constant variance. Also, tests of hypotheses and interval estimation require that the errors are normally distributed. There are a number of advanced statistical tests that can be used to examine whether or not these assumptions are true for any given regression equation (Myres et al., 2012; Kowalski, 1977).

Through the current study, a fitting strategy is followed so as to begin with the first-order model of form (2) then; interaction among dependent variable and their possible quadratic and nonlinear trend are examined using model structures (3) and (4). Fitting procedures are terminated once the best model is detected. Model adequacy and significant is judged through many criteria which can be defined as follow:

- $Correlation\ Factor\ (R^2) = [1 - (Residuals\ Sum\ Squares/Total\ Sum\ Squares)]$
- $t_{statistics}\ value = [Predictor\ Value/Predictor\ Standard\ Error]$
- $F_{ratio} = [Regression\ Mean\ Squares\ /Residuals\ Mean\ Squares]$

Correlation factor R^2, sometimes denoted as coefficient of determination, measures the percentage of variation in the response variable R explained by the explanatory variable x. Thus, it is an important measure of how well the regression model fits the data. The value of R^2 is always between zero and one. R^2 of unity or, 100%, means all variability are grasped while 50% or below, indicates the the predictionmay be poor. Student $t_{statistics}$ value usually measures the influence strength (weight) of an estimated coefficient for a specivic independent variable x_i through comparing its estimated value to its calculated standard error. For a variable to be significant, $t_{statistics}$ absolute value must not be less than 2.0. As indicated above, F_{ratio} (F-test) indicates the ratio of regression mean squares to the residuals mean squares for a set of independent variables and number of data points (degrees of freedom). This has a preset threshold value from special statistical tables where greater value is not always a judgment of model superiority.

Further statistical judgment (hypothesis tests) is performed by observing the trends of each of the residuals pattern and the normal distribution of the predicted values (Douglas et al., 2012). Based on the selected best model, a surface response representation is introduced in terms of 3-dimensional graphs and contours which will be demonstrated in the following section.

4. Effect of SiC Content and Compact Temperature on the Relative Density

The aforementioned fitting strategy is used to establish a relationship between the relative density (RD), as a dependent variable, and AA6061 percent in the alloy and compact temperature (T_{HC}) as independent variables using the experimental data listed in Table 1. To get rid of computational error arises during fitting process at zero SiC%, AA6061 content (Al%) is transformed and coded so that the value is always greater than zero (Al% = 100% – SiC%).

A first-order linear model of form (2) was obtained taking the form:

$$RD = 90.903 + 0.132\ (Al\%) - 0.13\ (T_{HC}) \tag{5}$$

with R^2 of 90% along with $t_{statistics}$ for b_o, b_1 and, b_2 are 51, 8.7 and, -5.1 respectively together with F_{ratio} value of 52, the model satisfies the adequacy requirements. Model significance is justified as its residuals are with error zero mean, Figure 2a, error normal distribution, Figure 2b, and random distribution, Figures 2c.

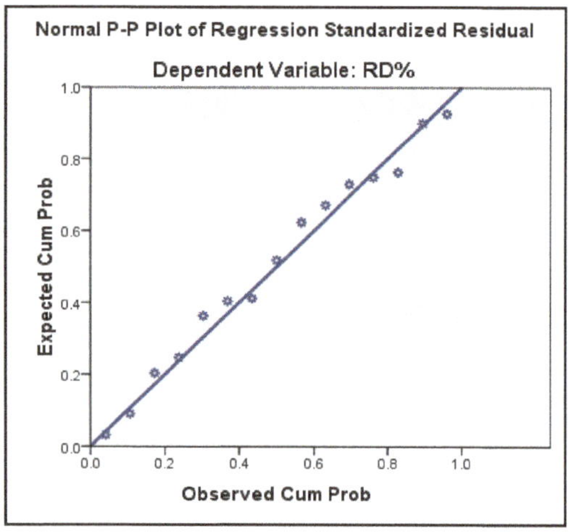

a) Residuals zero-mean hypothesis

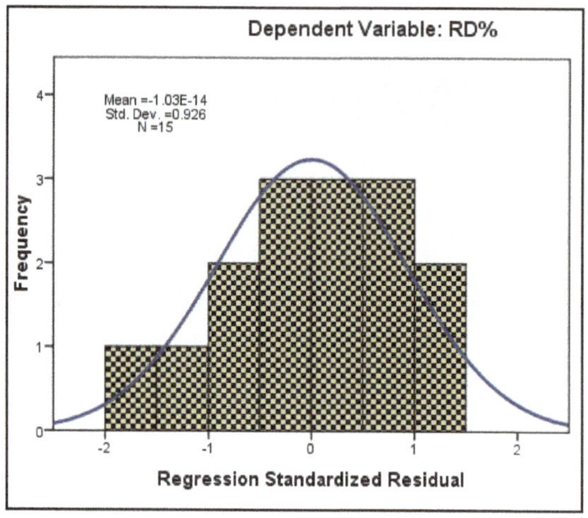

b) Residuals normal probability hypothesis

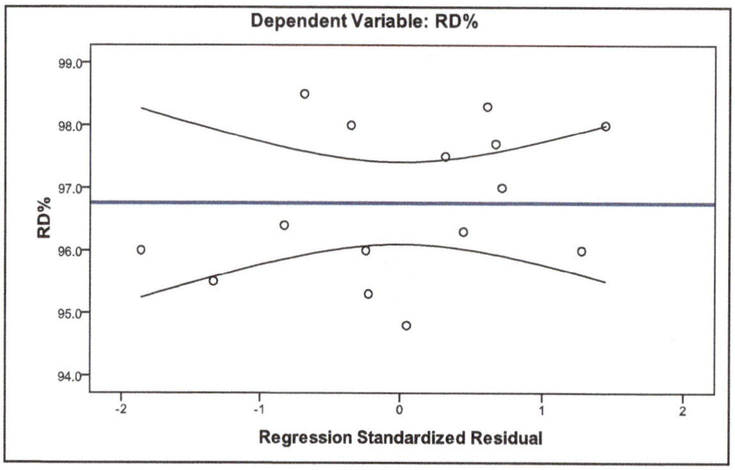

c) Random error distribution hypothesis

Figure 2. Residuals hypothesis tests for the developed model (5)

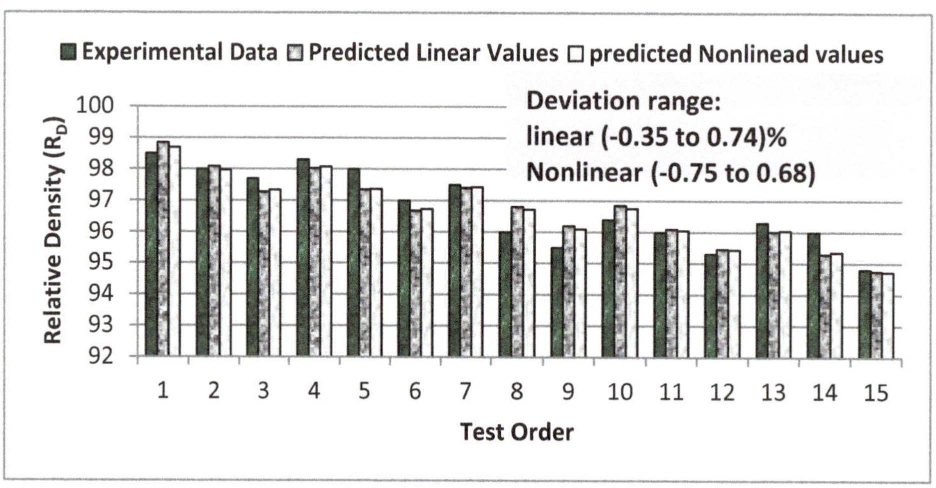

Figure 3. Comparison of experimental and predicted data for both linear and nonlinear models

For the same set of experimental data, a better model predicatability is seeked through the establishment of the counterpart nonlinear model of form (4). The nonconventional nonlinear iterative regression routine in SPSS program is used according to the default convergenece criteria. For each iteration, the model derivatives are numerically calculated and examined against the specified criterion values. Process continues until it reaches the best model with minimum squares of residuals relevant to the specified criteria. This process has led to the model (6).

For practical predictability comparison between the linear and nonlinear models, Figure 3 is dedicated to indicate how predicted (estimated) value from each structure is close to its counterpart experimental value, Table 1.

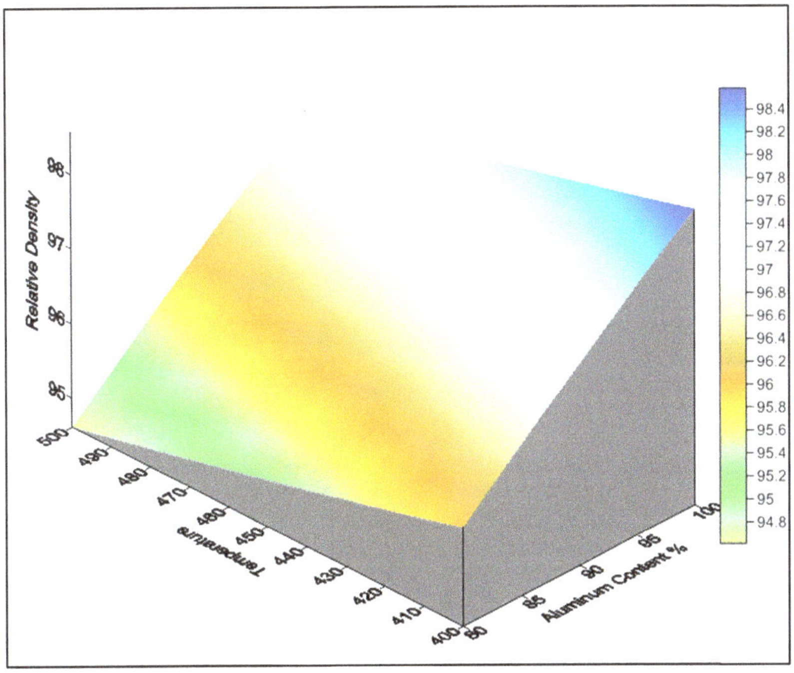

a) Al%-T_{HC}-RD% response surface

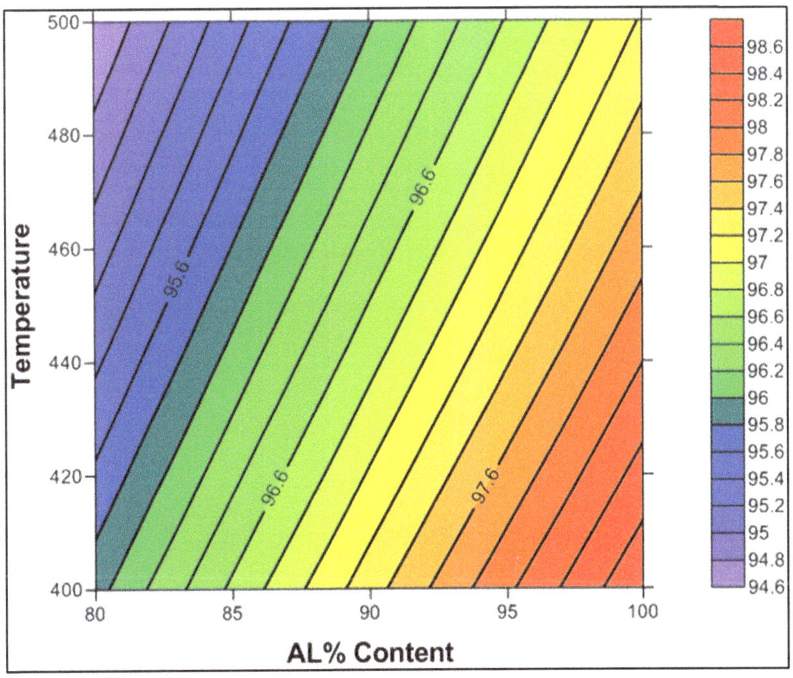

b) Al%-T_{HC}-RD% Contour graph

Figure 4. Surface response of the influence of AA6061% and temperature on the relative density

While a deviation range of (-0.35 to 0.74) % is obtained considering the linear model (5), a wider deviation is noticed for the nonlinear one (6). Furthermore, no interaction or second order trends of the independent variables, Al% and T_{HC}, prove credibility to be included in the model.

Therefore, it is concluded that a first order linear model is adequate and significant enough to represent the set of experimental data for which it was developed.

$$RD = 81.589\,(Al\%)^{0.122}\,(T_{HC})^{-0.062} \tag{6}$$

Figure 4 shows the surface response with contour graph of the influence of AA6061 vol fraction and HC temperature on the RD. As numerically explained by model (5), the relative density slightly decreased as HC temperature increased. This is thought, on the first hand, due to the formation of voids in the cooling stage that is associated with variation in the coefficient of thermal expansion (CTE) of the AA6061 matrices (α = 23.6 *10-6 /°C) and the SiC (α = 2.77 *10-6 /°C). As HC temperature increases, both voids volume fraction and residual stresses increase leading to deterioration in density (Salem et al., 2009). On the other hand, model coefficients and graphs embody the intuitive relationship where lower SiC content improves the HCs density at any HC temperatures. Monolithic aluminum alloy powder possesses good compressibility that is reduced as hard SiC particles (which possess low compressibility) are added and, consequently, the relative density decreases. Adding brittle SiC into ductile matrix of AA6061 leads to the segregation between the reinforcement and matrix at the triple junctions and along grain boundaries and, this widespread as SiC reinforcement content increases as shown in Figure 5. This could be verified by the microstructure, where increasing the amount of voids is proportional to the increase in amount of SiC content, as shown in Figure 6. SiC resulted in the formation of isolated voids which coalesced into larger voids and cavities when SiC content increased.

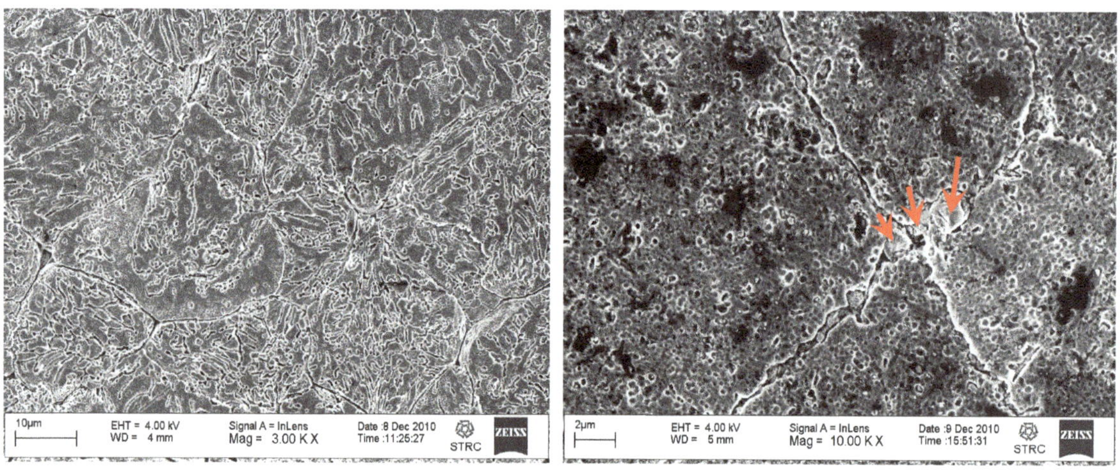

(a) (b)

Figure 5. SEM micrographs of (a) AA6061 and (b) AA6061+15vol.% SiC_p HC at 400 °C, arrows point at SiC particles

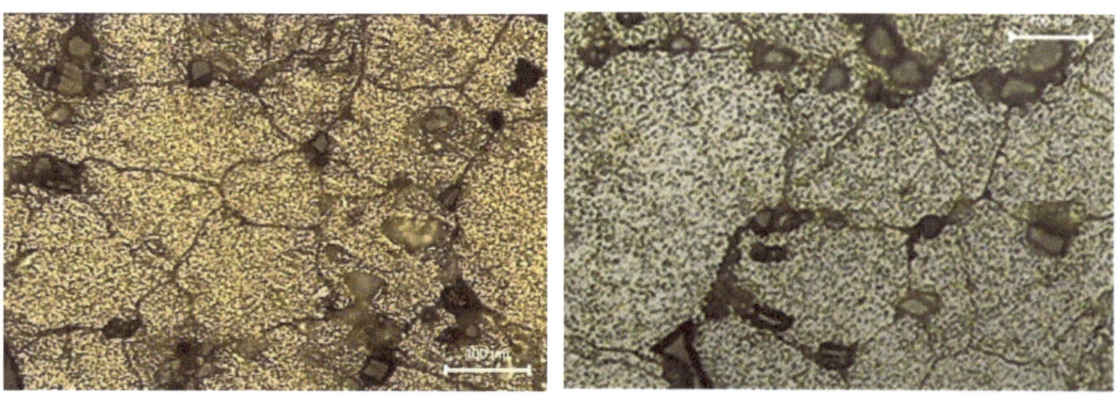

(a) (b)

Figure 6. OM micrographs of AA6061-15vol% SiC_p hot compacted at (a) 400 °C, and (b) 500°C

4. Effect of SiC Content and Temperature on the Hardness

Same fitting procedures and strategy, which has been explained in the last section, are followed to develop a functional interrelationship between the alloy hardness (Vickers Hv) as a dependent variable and both of the AA6061 content (Al%) and HC temperature (T_{HC}) as influential independent variables.

Applying a first order linear multiple form (2) didn't lead to satisfactory general statistical outcome. To resolve such an emerged problem, the partial plots are considered to reveal the individual real natural dependence of Hv on each of Al% and T_{HC}, Figure 7. Hardness-AA6061 content, Figure 7a, exhibits a strong indication of second-order (quadratic) trend while an almost linear trend is noticed regarding hardness-temperature dependency, Figure 7b. Consequently, a second-order multiple regression of form (3) is proposed. Using Stepwise procedure in linear regression routine, which determines the maturity of each individual independent variable to be included into the final equation, the following second order linear model is obtained:

$$HV = 424.17 - 0.081(Al\%)^2 - 0.082(T_{HC}) + 13.325(Al\%) \tag{7}$$

with R2 = 93.7%, t_values= -2.4, -4.3, -4.3 and 3.94 and, F_ratio = 54, the developed model is considered adequate and significant to represent the relevant functional interrelationship.

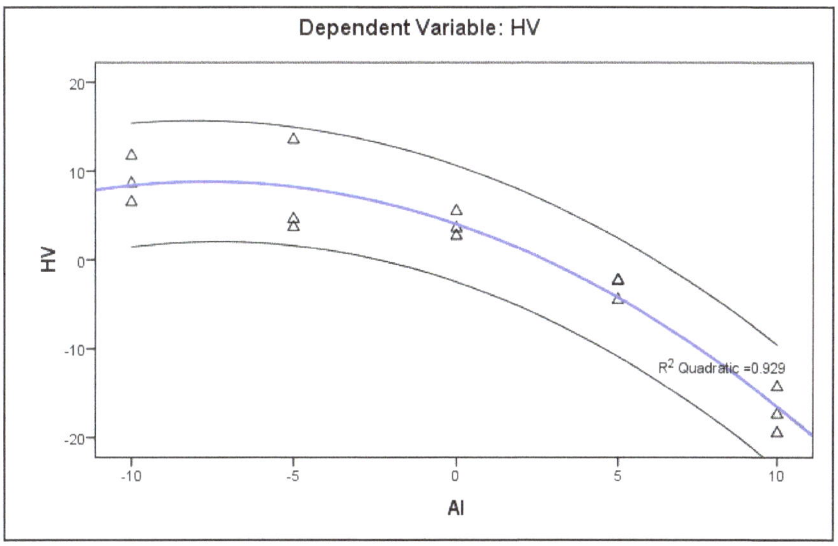

a) Quadratic component in the Hv-Al% real relationship

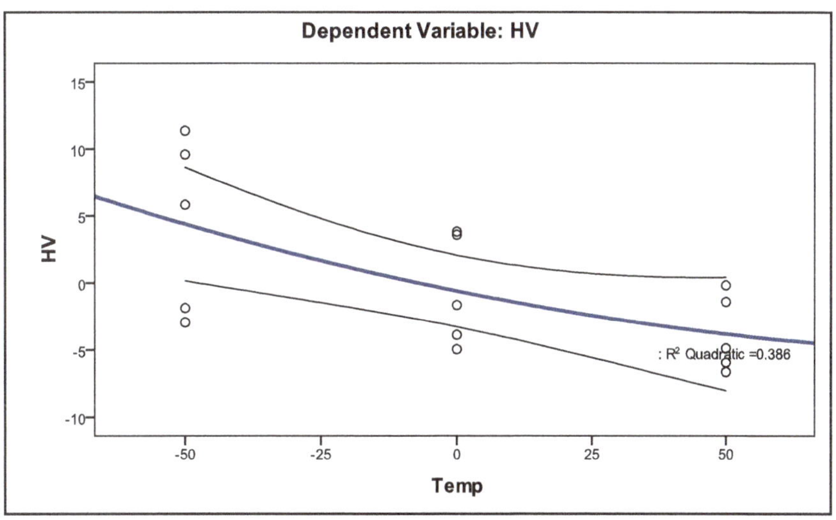

b) Hv-T_{HC} relationship

Figure 7. Partial relation between hardness and each of AA6061 content and temperature

Residuals examination shows nothing against its zero mean, normal probability and, randomness distribution, Figure 8.

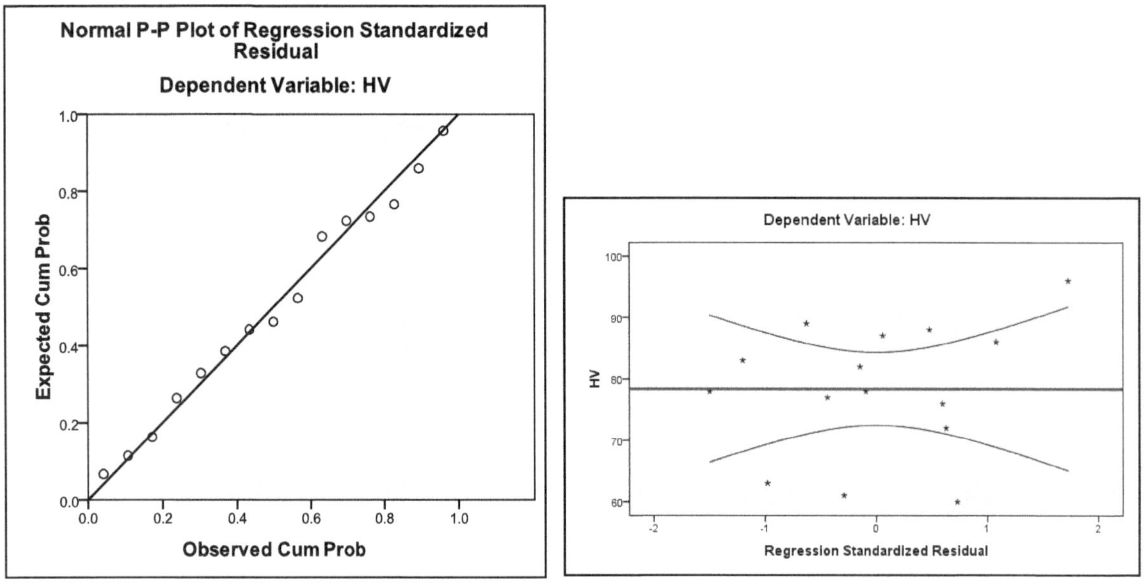

a) Residuals zero-mean hypothesis b) Random error distribution hypothesis

Figure 8. Residuals zero mean, normality, and random distribution of multiple regression model (7)

An equivalent iterative no-linear model is found to have the following structures:

$$HV = 602469.152\ (Al\%)^{-1.348}\ (T_{HC})^{-0.474} \tag{8}$$

with less correlation factor R2 of 87.7 and, therefore, it is of less predictability than model (7) to predict values of Hv as a relation to both AA6061 content (Al%) and HC temperature (T_{HC}).This may be observed from Figure 9 where both models are examined against the experimental values. As shown by Figure 9.b, nonlinear model exhibits larger deviations from the experimental values especially for experiments Nos. 1, 2, 3, 13 and 14 with overestimation trend. For the rest of cases, an approximate trend is observed for both models credibility. However, both models tend to represent the effect of reducing AA6061% or, increasing SiC%, on the hardness of the composite. It is obvious that a reinforced disc has a Hv-value higher than the monolithic alloy. However compact temperature seems to have a limited influence on the hardness. At higher HC temperature, the grain coarsening increases thus lower HV-values result.

Figure 10 shows a 3-D surface response along with a contour graph of the functional Hv-Al%-T_{HC} relationship. The effect of HC temperature has a higher impact of the hardness of monolithic material than on the composite. As SiC percent increases up to 18 %, hardness increases reaching its ultimate value. Further increase in SiC content seems not to affect composite hardness which is slightly and linearly decreases according to a temperature-dependent pattern.

The hardness distribution results showed that good sintering was achieved in the case of AA6061 with or without SiC reinforcement even at the 400°C sintering temperature and adhesion between particles increase as the sintering temperature increase. When temperature rises, the resulting heat leads to the expansion of an aluminum particle making a wider contact with the neighbor particle. As a result, voids between particles are reduced. Furthermore, increasing HC temperature up to 500°C usually leads to strain softening due to grain coarsening as shown in Figure 6.

The presence of the SiC increased the composite hardness as they carry some of the load applied to the material (load transfer from the matrix to the reinforcement due to the difference in the elastic constant). The increase in the Hv-values of the composite can be attributed to the high hardness of the reinforcement. So mainly the interaction is either aluminum with aluminum particles which adhere properly or aluminum with SiC in which adhesion is enhanced by the presence of SiC particle.

It is important to emphasize the fact that the grain size of the processed AA6061 was reduced with SiC addition, where the SiC particles assisted in hindering the grain growth by pinning the grain boundaries and inhibiting grain boundary migration and they also locked the grain boundaries which elevate its strength. The presence of SiC particles at the grain boundaries and at the triple junctions promoted good sites for hindering the dislocation motion and grain growth. Dislocation piling at the boundaries resulted in increasing resistance to deformation and hence higher strength and hardness. SiC particles being hard and brittle led to dispersion hardening of the matrix. The resistance to slip was increased with increasing SiC particles content because the overall lattice strain must increase if a dislocation is torn away from them. The excess dislocations within grains and near grain or subgrain boundaries which increased by increasing the SiC content, made dislocation glide more difficult. Thus, the imposed stress necessary to deform a metal increased, and hence the hardness of the composite revealed to be higher than of the alloy.

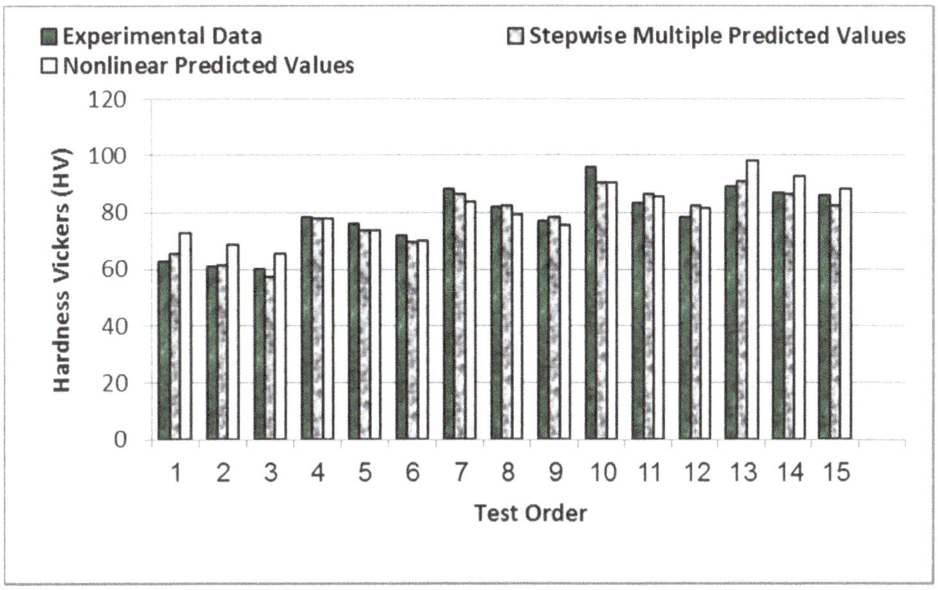

(a) Experimental-second order linearc-nonlinear comparison of predicted values

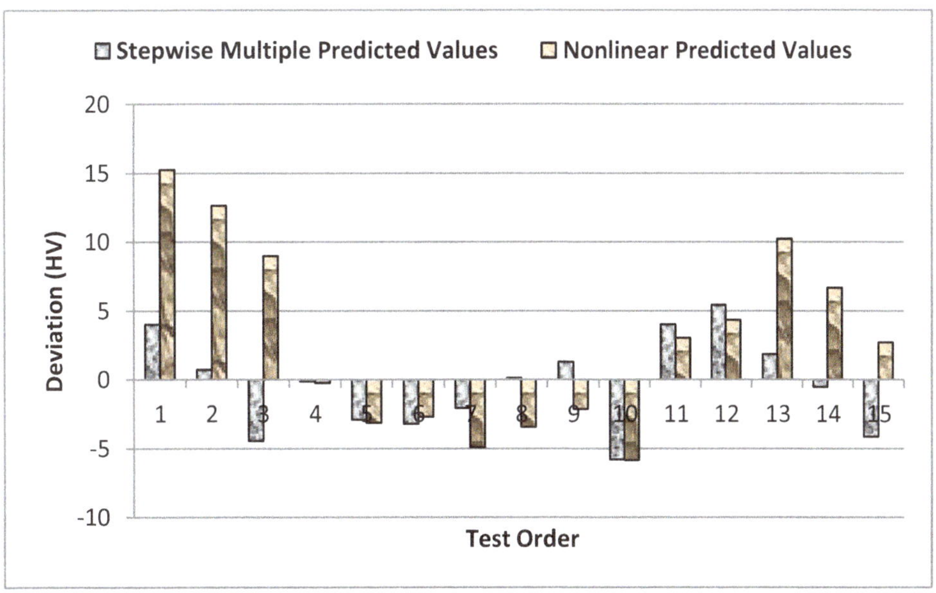

(b) Predictability comparison between second order linear stepwise and nonlinear regression models

Figure 9. Stepwise multiple second order linear vs nonlinear models

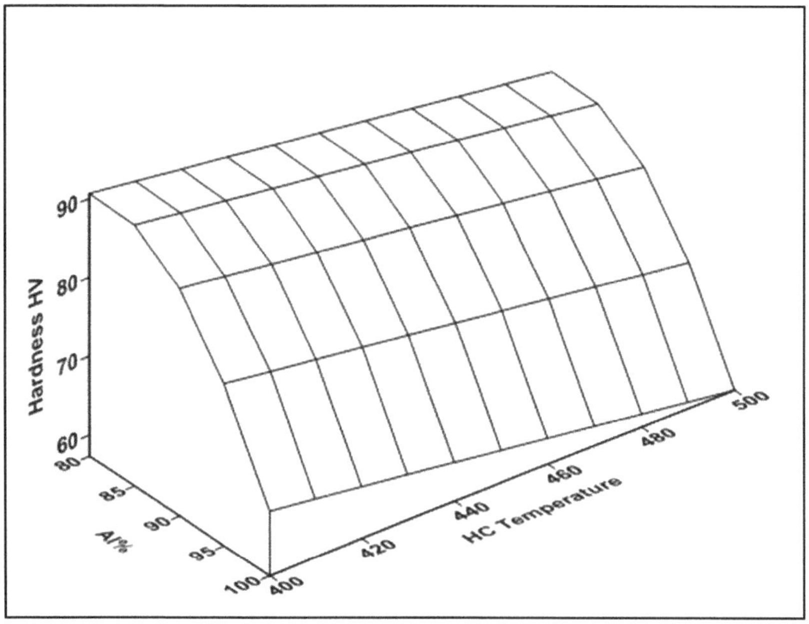

a) HV-Al%-T_{HC} response surface

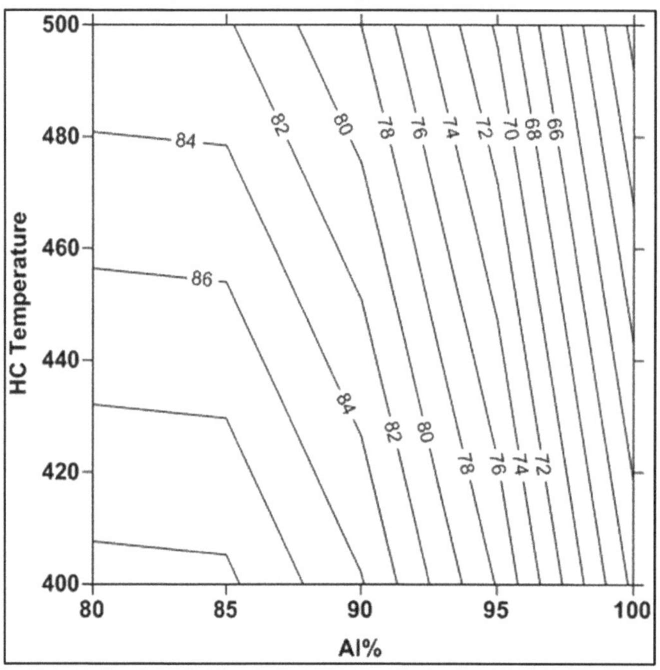

b) HV-Al%-T_{HC} contour graph

Figure 10. Surface response and contour graph of the influence of AA6061 content and temperature on the hardness

As shown by Figure 10a, Hv decreases at 100% Al content while such a trend gradually dimenshes as SiC% increases until it almost vanishes at about 85% Al content (15% SiC). This is graphically evident as area of surface that represents hardness values becomes almost parallel to the horizontal plane suggesting that there is no further change in the hardness values. Such observation is evident in the contour graph, Figure 10.b at given T_{HC}=450, where Hv increases from 62 to 82 (about 32% increase) within the interval 90 to 100 Al% while a corresponding increase within the interval 85 to 90 Al% is only 3.7 % (from 82 to 85 HV increase). As SiC% increases after this point, Hv seems not to be affected. Similar observation may be noticed at any HC temperature within the considered domain (from 400 to 500, Fig. 10.b. This indicates that increasing SiC content more than about 15% does not affect the hardness. Increasing SiC content usually motivies the SiC particles contact and also increases the clustering of SiC particles chances which dramatically affect the material properties. This agrees with the experimatal observations, Figure 11, where at 20% SiC composite, the hardness drops due to the weak contact among SiC particles and, as aresult, higher probability of more than two particle cluster together occurs, Figure 11. Additionally, such decrease in hardness of the AA6061- 20% SiC_p composite can be attributed to the decreasing of the composite compressibility which resulted in increasing voids. The porosity volume fraction dominated the behavior of the composite which resulted in deterioration of the overall hardness of the discs. This probably reflects mixing difficulties in obtaining a uniform particle distribution at high volume fraction for this process, there being a higher number of nucleation sites present for the cracks to form as a result of the particle clusters.

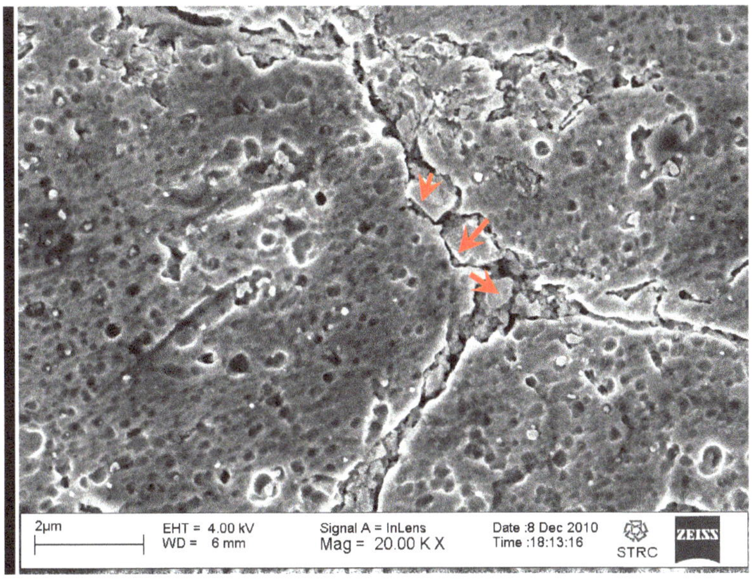

Figure 11 SEM Micrograph for AA6061-20% SiC showing the segregation of SiC particles along boundaries and triple points, arrows point at SiC particles

5. Conclusions

Different volume fractions of SiC are added to the monolithic aluminum AA6061. Mixture is hot compacted at various HC temperatures and both the relative product density and Vickers hardness are recorded. Linear and nonlinear model structures are proposed with sequential preference strategy and robust statistical criterion judgment measures of the model predictability and residuals pattern examination. Although, in many current cases, linear model structure is sufficient for predictability purposes, the need for nonlinear forms is still required to obtain better and more representative respose surface regarding the functional interrelationship considered.

Increasing SiC content is found to reduce the relative density and to increase the hardness of the composites. However, the surface response and the contour graphs interestingly explained that adding greater than (15 to 20%) SiC% to the composite does not benefit increasing of its hardness. HC temperature showed a negative trend (decrease) in each of the composite relative density and its hardness.

Acknowledgements

The authors are grateful to Youssef Jamil Science and Technology Research Center (YJSTRC), and the Mechanical Engineering Department, The American University in Cairo- Egypt, for financial support and continuous assistance of their lab engineers and research assistances.

References

Bozic, D., Dimcic, B., Dimcic, O., Stasic, J., & Rajkovic, V. (2010). Influence of SiC particles distribution on mechanical properties and fracture of DRA alloys. *Materials and Design, 31*(1), 134-141. http://dx.doi.org/10.1016/j.matdes.2009.06.047

Douglas, C. M., Elizabeth, A. P., & Geoffrey, G. V. (2012). Introduction to Linear Regression Analysis. 5th Edition, Publisher: John Wiley & Sons. Retrieved from http://eu.wiley.com/WileyCDA/WileyTitle/productCd-0470542810.html

El_Garaihy, W. H. (2012). *Production of Ultra Hard Nanostructured Materials*. Ph.D Thesis, Man. Un., Egypt. Retrieved from http://www.eulc.edu.eg/eulc_v5/Libraries/Thesis/BrowseThesisPages.aspx?fn=PublicDrawThesis&BibID=11842944.

Huang, J. C., Shin, C. S., & Chan, S. L. I. (2004). Effect of temper, specimen orientation and test temperature on tensile and fatigue properties of wrought and PM AA6061-alloys. *International Journal of Fatigue, 26*, 691-703. http://dx.doi.org/10.1016/j.ijfatigue.2003.11.006

Kalkanl, A., & Yılmaz, S. (2008). Synthesis and characterization of aluminum alloy 7075 reinforced with silicon carbide particulates. *Materials and Design, 29*, 775-780. http://dx.doi.org/10.1016/j.matdes.2007.01.007

Kim, W. J., Hong, S. H., & Lee, J. H. (2001). Superplasticity in PM 6061 Al alloy and elimination of strengthening effect by reinforcement in super plastic PM aluminium composites. *Mat. Sc. and Engineering A, 298*(1-2), 166-173. http://dx.doi.org/10.1016/S0921-5093(00)01341-1

Kowalski, B. R. (1977). Chemometrics: Theory and Application. ACS Publications. http://dx.doi.org/10.1021/bk-1977-0052

Liu, G., Muolo, W., Valenza, F., & Passerone, A. (2010). Survey on wetting of SiC by molten metals. *Ceramics International, 36*(4), 1177-1188. http://dx.doi.org/10.1016/j.ceramint.2010.01.001

Mandal, D., & Viswanathan, S. (2013). Effect of heat treatment on microstructure and interface of SiC particle reinforced 2124 Almatrix composite. *Materials Characterization, 85*, 73-81. http://dx.doi.org/10.1016/j.matchar.2013.08.014

Mazahery, A., & Shabani, M. O. (2013). Microstructural and abrasive wear properties of SiC reinforced aluminum-based composite produced by compocasting. Trans. *Nonferrous Met. Soc., 23*, 1905-1914. http://dx.doi.org/10.1016/S1003-6326(13)62676-X

Mazahery, A., & Shabani, M. O. (2012). Nano-sized silicon carbide reinforced commercial casting aluminum alloy matrix: Experimental and novel modeling evaluation. *Powder Technology, 217*, 558-565. http://dx.doi.org/10.1016/j.powtec.2011.11.020

Min, S. (2009). Effect of volume fraction of Sic particles on mechanical properties of SiC/Al composites. *Transactions of Nonferrous Metals Society of China, 19*(6), 1400-1404. http://dx.doi.org/10.1016/S1003-6326(09)60040-6

Mohanakumara, K. C. M., Rajasheker, H., Ghanara, S., & Ajitprasad, S. L. (2014). Development of Mechanical Properties of SiC Reinforced Cast and Extruded Al Based MMC. *Procedia Materials Science, 5*, 934-943. http://dx.doi.org/10.1016/j.mspro.2014.07.381

Myers, R. H, Montgomery, D. C., Vining, G. G., & Robinson, T. J. (2012). *Generalized Linear Models: With Applications in Engineering and the Sciences* (2nd ed.). John Wiley & Sons. http://dx.doi.org/10.1002/9780470556986

O'Donnell, G., & Looney, L. (2001). Production of aluminium matrix composite components using conventional PM technology. *Materials Science and Eng. A, 303*(1-2), 292-301. http://dx.doi.org/10.1016/S0921-5093(00)01942-0

Pawar, P. B., & Utpat, A. A. (2014). Development of Aluminium Based Silicon Carbides Particulate MMC for Spur Gear. *Procedia Materials Science, 6*, 1150-1156. http://dx.doi.org/10.1016/j.mspro.2014.07.187

Rao, R. N., & Das, S. (2011). Effect of SiC content and sliding speed on the wear behavior of aluminum matrix composites. *Materials and Design, 32*(2), 1066-1071. http://dx.doi.org/10.1016/j.matdes.2010.06.047

Ravindran, P., Manisekar, K., Kumar, S. V., & Rathika, P. (2013). Investigation of microstructure and mechanical properties of aluminumhybrid nano-composites with the additions of solid lubricant. *Materials and Design, 51*, 448-456. http://dx.doi.org/10.1016/j.matdes.2013.04.015

Salem, H. G., El-Eskandarany, S., Kandil, A., & Abdul-Fatah, H. (2009). Bulk Behavior of Ball Milled AA2124 Nanostructured Powders Reinforced with SiC. *Journal of Nanomaterials,* 1-12. http://dx.doi.org/10.1155/2009/479185

Senapati, A. K. , Ganguly, R. I., Dash, R. R., Mishra, P. C., & Routra, B. C. (2014). Production, characterization and analysis of mechanical properties of a newly developed novel aluminium-silicon alloy based metal matrix composites. *Procedia Materials Science, 5*, 472-481. http://dx.doi.org/10.1016/j.mspro.2014.07.290

Showaiter, N., & Youssefi, M. (2008). Compaction, sintering and mechanical properties of elemental 6061 Al powder with and without sintering aids. *Material and Design, 29*(4), 752-762. http://dx.doi.org/10.1016/j.matdes.2007.01.027

Yan, C., Lifeng, W., & Jianyue, R. (2008). Multi-functional SiC/Al composites for aerospace applications. Chinese *Journal of Aerospace, 21*, 578-584. http://dx.doi.org/10.1016/S1000-9361(08)60177-6

Zakaria, H. M. (2014). Microstructural and corrosion behavior of Al/SiC metal matrix composites. *Ain Shams Engineering Journal, 5*(3), 831-838. http://dx.doi.org/10.1016/j.asej.2014.03.003

Zhanwei, Y., Fuguo, L., Peng, Z., Bo, C., & Fengmei, X. (2014). Mechanical properties study of particles reinforced aluminum matrix composites by micro-indentation experiments. *Chinese J. of Aer., 27*(2), 397-406. http://dx.doi.org/10.1016/j.cja.2014.02.010

Appendix A

Table 1. Experimental Date used in the analysis

No	SiC %	Temperature ºC	Relative Density RD%	Vicker's Hardness Hv	AA6061%
1	0	400	98.5	63	100
2	0	450	98	61	100
3	0	500	97.7	60	100
4	5	400	98.3	78	95
5	5	450	98	76	95
6	5	500	97	72	95
7	10	400	97.5	88	90
8	10	450	96	82	90
9	10	500	95.5	77	90
10	15	400	96.4	96	85
11	15	450	96	83	85
12	15	500	95.3	78	85
13	20	400	96.3	89	80
14	20	450	96	87	80
15	20	500	94.8	86	80

3

Advanced Joining Technologies for Load and Fibre Adjusted FRP-Metal Hybrid Structures

M. Klein[1], F. Podlesak[2], K. Höfer[2], H. Seidlitz[1], C. Gerstenberger[1], P. Mayr[2] & L. Kroll[1]

[1] Lightweight Structures and Polymer Technology, TU Chemnitz, Chemnitz, Germany

[2] Welding Engineering, TU Chemnitz, Chemnitz, Germany

Correspondence: M. Klein, Lightweight Structures and Polymer Technology, Chemnitz University of Technology, Chemnitz, Germany. E-mail: mario.klein@mb.tu-chemnitz.de

Abstract

Multi-material-design (MMD) is commonly realized through the combination of thin sheet metal and fibre reinforced plastics (FRP). To maximize the high lightweight potential of the material groups within a multi-material system as good as possible, a material-adapted and particularly fibre adjusted joining technology must be applied. The present paper focuses on two novel joining technologies, the Flow Drill Joining (FDJ) method and Spin-Blind-Riveting (SBR), which were developed for joining heavy-duty metal/composite hybrids. Tests were carried out with material combinations which are significant for lightweight constructions such as aluminium (AA5083) and carbon fibre-reinforced polyamide in sheet thickness of 1.8 mm. The mechanical testing and manufacturing of those multi-material joints was investigated.

Keywords: multi-material design, fibre reinforced plastics, lightweight automotive structures, joining

1. Introduction

Load-path adapted multi material design initiates significant mass reduction in automotive industry and contributes to resource savings. The combined use of metal sheets and fibre-reinforced plastic (FRP) offers a very high lightweight potential (Klein, 1997; Rosato, 2005; Goede et al., 2010) through all the material groups. For a uniform initiation of concentrated loads usually metal components are applied. In contrast to that FRPs are more suitable for load transmission along large distances (Schürmann, 2005; Kroll, 2009).

Due to comparatively short cycle times when manufacturing components and the use of inexpensive matrix systems such as polypropylene, but also higher-quality plastics such as polyamide or polyether ether ketone, the use of thermoplastic FRPs with high strength and stiffness properties has raised significantly through the last years.

The demand on car bodies to become lighter and more efficient lead to massive development in sheet metal production. Not only steel manufacturers presented high strength steel qualities, but also aluminium and magnesium gained influence in automotive applications. Therefore, the task for advanced joining technologies is to offer a wide range of processable materials in both categories, FRP and metal. Through the combined use of thermoplastic FRPs and thin-walled metal sheets or extrusions, the modern body in white consists of weight-optimized components especially developed for vehicle construction. To solve this, the required joining technology takes a central role within the manufacturing process.

The joining technology for thin-walled structural components, made of FRP and metal sheets, is considered to be extremely challenging. Primarily mechanical methods such as screwing, blind and punch riveting as well as hybrid techniques are applicable (Simon, 2005) due to the strongly different thermo-mechanical behaviour and the divergent material composition and morphology.

Furthermore the conventional placing of joining elements leads to a sectioning of load bearing fibres at the load initiation zone of the FRP. As a consequence, high notch stresses are induced, which weaken the FRP component significantly and prevent the optimal exploitation of the high anisotropic lightweight potential (Colins, 2006; Seidlitz et al., 2011). Therefore additional reinforcements like cones and patches are applied at critical notch stress areas, which consequently lead to a strong over-sizing of the joint, an increased component weight and most of all space problems. To fully exploit the high anisotropic material behaviour of the FRP, the joining process must be performed very gentle as well as load path adapted (Rotheiser, 2009).

This paper focuses on the qualification of the new flow-drill joining (FDJ) concept and the Spin-Blind-Riveting (SBR) concept – showing potentials and references for automotive application. The technologies are currently used to join plane metal sheets and thermoplastic FRPs within wall thickness range of 1.0 to 4.0 mm (Seidlitz, 2013b). In analogy to bionic, the fibres can be aligned along the force flux lines at the joint of the FRP component. With these new technologies, it is possible to join different materials without the need of predrilling. Therefore, a minimum preparation effort and less destruction of fibres can be guaranteed.

2. Methods

2.1 Flow Drill Joining Concept

The flow drill joining (FDJ) concept was developed at the Institute of Lightweight Construction at Chemnitz University of Technology. It primarily serves to join comparatively thicker thermoplastic FRPs with metal sheets. The process is characterized by short processing times and high-strength joining properties appropriate for the force fluxes. The high degree of lightweight construction and can be utilized in various industrial sectors, not only automotive, but also space and aviation. The essential advantage towards typical joining technologies relates to the fact that no additional auxiliary joining elements are needed for joining the components.

In order to manufacture an FDJ joint, the plastic flow properties of the metal component are reduced due to heat influence. With the use of a rotating mandrel, a bushing is thermo-mechanically shaped from the metallic base material. While doing so, this bushing is pushed directly through the also locally plasticized thermoplastic FRP component during the shaping operation and is subsequently pulled over in a positive-locking form when the mandrel is retracted or in an additional working step (Figure 1). The rotation of the mandrel and the associated friction serve to release thermal energy which is locally introduced into the FRP component as a result of the process and thus initiates the partial melting of the thermoplastic matrix. Therefore, the continuous fibres arranged in the plastic become "moveable" and can be aligned towards the force fluxes during the shaping of the bushing (Seidlitz et al., 2011).

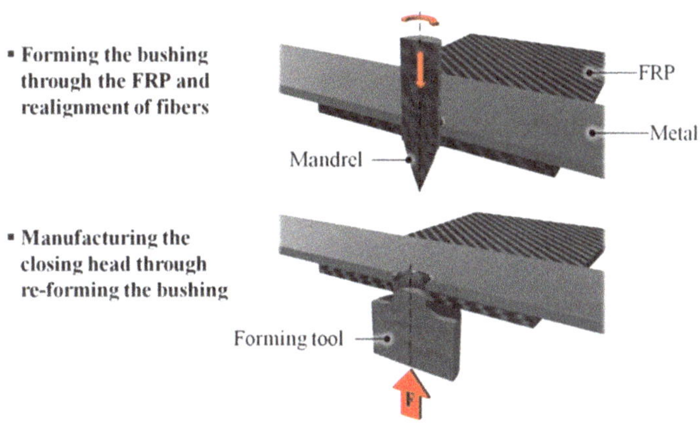

Figure 1. Process scheme for manufacturing FDJ joints (Seidlitz et al., 2011)

2.2 Spin Blind Riveting (SBR)

Spin-Blind-Riveting (SBR) has its origin in conventional blind riveting, which is a common solution to join different materials (Nestler, 2013). Additionally, SBR uses the effects of flow drilling as described in (Biermann, 2014) and the effects of hole moulding (Hufenbach, 2009). Main difference is that SBR uses a modified blind rivet as tool and joint element at the same time. In contrast to conventional riveting, no predrilling is necessary and the process is completely free of chip formation. SBR principle is illustrated in Figure 1.

A metal and a FRP sheet are placed as lap joint with the metal sheet on the upside. A blind rivet with a conical mandrel head is rotated and pressed onto the sheets with a force. Due to rotation and pressure, the metal sheet begins to locally heat up and plastify under the rivet. Thus the rivet penetrates into the metal sheet under formation of a sleeve similar to flow drilling. The formed sleeve is pushed into the FRP sheet and because of the frictional heat between rivet and metal sheet the thermoplastic matrix of the FRP is also heated up and locally melted by conduction. When the rivet is fully penetrated through both sheets, the mandrel is pulled back and the rivet head forms into the blind-head of the rivet body. With this formation of joint the sheets are additionally pressed together and the melted matrix of the FRP is cooling down and adheres partially at the metal sheet.

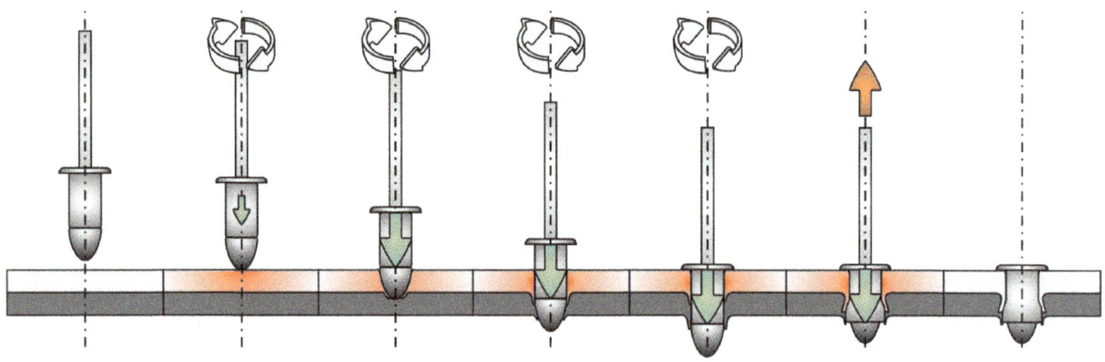

Figure 2. Process scheme for manufacturing SBR joints

Main process parameters are:

- Rivet geometry
- Rotation speed of rivet
- Penetration force

One specific property of the joint is the formation of a sleeve due to material displacement in the upper sheet. This sleeve can transmit significantly higher shear loads than conventional rivet joints. Furthermore SBR causes less damage at the fibres of reinforced plastic sheets because the material is heated up above the glass transition temperature of the thermoplastic matrix and fibres can be displaced instead of destroying them like as with drilling

3. Experimental Setup

3.1 Materials Studied

Organic sheets are flat semi-finished products made of thermoplastic FRP's, which can be prepared by stacking of unidirectional (UD) pre-impregnated layers (prepregs). Subsequent pressing at matrix melting temperature (for polyamide 6: 220°C) produces the final product. These thermo formable unidirectional and multidirectional composite structures with defined mechanical properties can be produced within short cycle times. Similar processing chains as for metal sheet forming are aplicable. Furthermore, in addition to the high specific strength and stiffness properties, thermoplastic FRP's possess comparatively high damping characteristics, good crash resistance as well as material immanent recycling properties. Subject matter of the investigations are FRPs with orthotropic carbon fibre reinforcement, embedded in a polyamide 6 (PA 6) matrix system. Table 2 gives an overview referring to the applied unidirectional (UD) prepreg systems, laminate lay-up and further specific properties.

Table 1. Properties of metallic joining partners

material	young's modulus	tensile strength	yield strength	thickness
Aluminium AlMg4.5Mn (AW5083)	71 GPa	300 MPa	125 MPa	1,5 mm

Table 2. Properties of CFRP joining partners

FRP type	prepreg	lay up	young's modulus $E_1 = E_2$	tensile strength $\sigma_1 = \sigma_2$	fibre volume ratio	thickness
CFRP	UD carbon fibre /PA 6 (Ticona)	$[(0/90)_3]_s$	49.5 GPa	860 MPa	0.6	1.8 mm

3.2 Experimental Setup

Material combinations were joined by both processes, FDJ and SBR according to DIN EN ISO 14273 as tensile shear test specimens with a length of 175 mm, width of 50 mm and an overlapping length of 35 mm.

The described FDJ process was carried out by an automated joining jig (Figure 3 left). Joining of the comparatively thick walled CFRP components was supported by an infrared heating system, which plasticizes the matrix component efficiently within a short processing time.

The SBR riveting process is based on a conventional blind riveting tool but due to reasons of simplicity, the specimen rotates instead of the rivet. The joining force is induced by a pneumatic cylinder. Parameter settings are modified by a programmable control unit. In further work a rivet-setting tool will be developed, which combines the rotation of the blind rivet to penetrate the sleeves and afterwards the forming of the rivet by backwards movement. Both experimental setups are shown in Figure 3 (right).

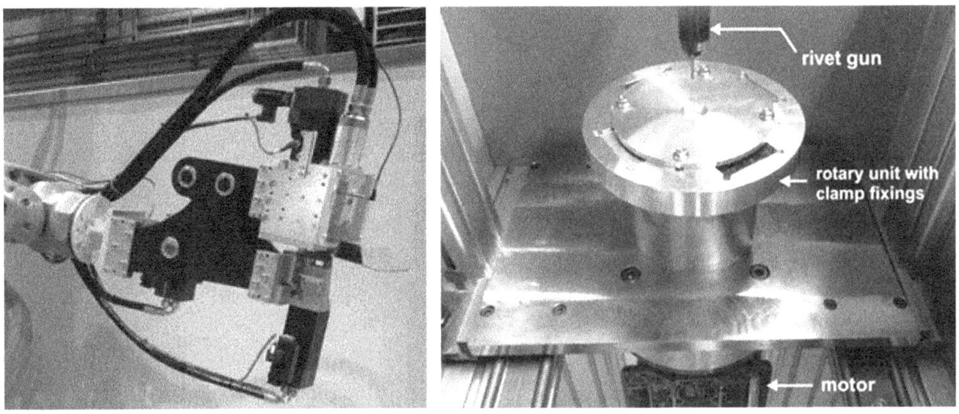

Figure 3. Left: Joining jig for FDJ process; Right: SBR experimental

In Figure 4 (right) a macro section of the used rivet developed by Gesipa Blindnietgeräte GmbH is shown. The mandrel head consists of steel and has a conical shape. In the left image of Figure 4 the geometric properties of mandrel for the FDJ process are shown. The applied joining point diameter of 5.3 mm for FDJ and 4.8 mm for SBR were chosen, since most of the mechanical joints in actual body in white structures show similar geometric parameters. Table 3 summarizes the processing parameters and shows exemplarily samples of the manufactured joints.

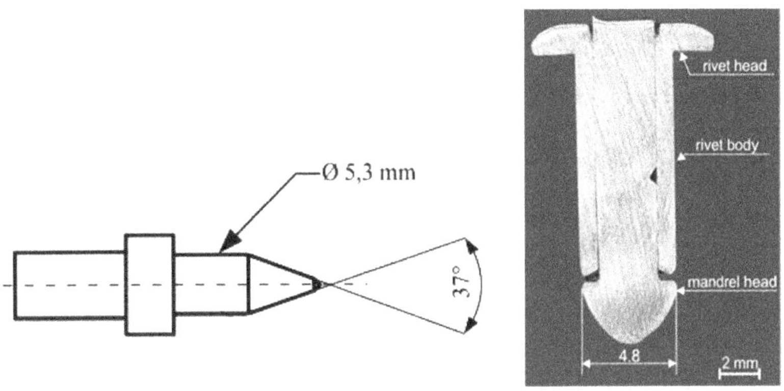

Figure 4. Left: Mandrel for the FDJ process; Right: Rivet used for SBR process

Following experimental investigations were carried out:

- Evaluation of process characteristics
- Visual evaluation
- Mechanical testing

Table 3. Manufactured FDJ / SBR joints and parameters

Tensile shear specimen	Processing parameters
AlMg4.5Mn/CFRP (FDJ) (Number of specimens: 5)	Joining point diameter: 5.3 mm Joining force: 20 kN Flow drill speed: 3000/min Heating time: 14 s
AlMg4.5Mn/CFRP (SBR) (Number of specimens: 5)	Joining point diameter: 4.8 mm Joining force: 3 kN Rotation speed: 4200/min Penetration time: 2 s

At first, specimens were joined with the parameters given in Tab. 1 for visual evaluation. For this, the specimens were viewed and macro sections were made.

At least five specimens per process were joined for shear load tests. Shear tension test were done with an elongation speed of 1 mm/min with a Zwick Z050.

Beneath the comparison of the load values, the fracture behaviour was analysed. Especially failure mode and location indicates the quality of the joint.

4. Results and Discussion

The obtained material combinations were joined successfully by both processes, FDJ and SBR. Images of joint samples are shown in Fig. 5. While the upper side of the SBR joint looks similar to a conventional blind rivet joint, the down side shows the formation of a sleeve and the mandrel head has annealing colours because of the frictional heat. The FDJ joint shows a zone with melted CFRP and a sleeve formation in the metal sheet which is completely penetrated through the CFRP. It also can be seen, that at both processes, some of the fibres of the CFRP are deflected and some are destroyed due to high deformation degrees. Altogether both processes lead to joints with a high optical appearance.

Figure 5. Left: Joint made by SBR process; right: Joint made by FDJ process, upper photos show the metal side, bottom photos show the FRP side of the joint

In Figure 6, macro sections are shown. It can be seen that both processes form a sleeve in joining direction without any chip formation. They have almost the same dimensions in terms of diameter, wall thickness and length. Most of the metal material is displaced to the downside of the sheet while only a minor part is displaced to the upside. The sleeves, formed by FDJ and SBR are pushed completely through the CFRP. While the form lock is established by the rivet at the SBR process, it is made by the sleeve itself at the FDJ process through the forming tool (Figure 1). The macro sections also show that the CFRP is locally melted by the processes und the fibres are deflected at the joining zone without signs of failures like delamination.

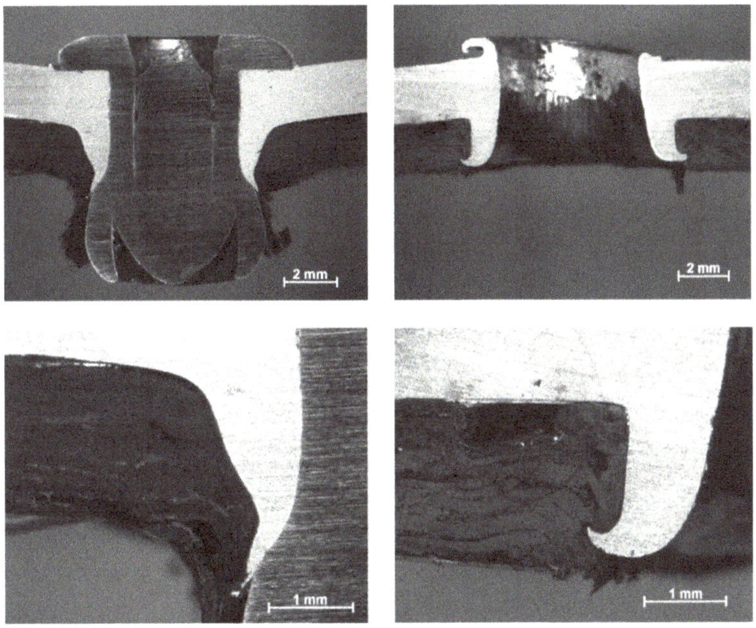

Figure 6. Macro sections; Left: SBR process; right: FDJ process

Characteristic force-displacement curves under shear tension for both processes, FDJ and SBR are shown in Figure 7. As it can be seen shear load curves of all specimens have a linear characteristic at lower load values and a peak value at the end of this linear characteristic at about 2.3 kN. The subsequent values vary greatly and peak loads between 2.3 and 3.1 kN are achieved. The failure mode is a mixture of bearing and pull-out failure. By reaching a load level of 2.3 kN the metal sheet deforms and instead of a pure shear load, a combined load appears. Under these conditions, the SBR joints show slightly higher values than the FDJ joints because the rivet can take up the combined load better than the sleeve of the FDJ joint. All together both kinds of joint processes show good results and a low standard deviation under pure shear load.

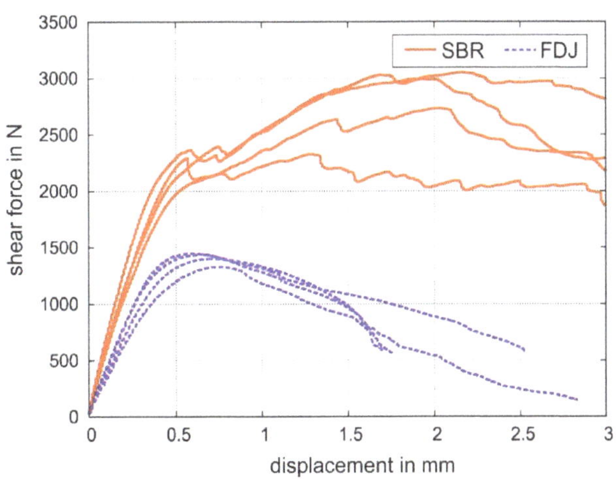

Figure 7. Force-displacement curves for SBR and FDJ joined specimens

Additionally attention should be paid on the weight of every joint with regard to applications in lightweight constructions. While the FDJ joint does not cause any extra weight, the SBR joint causes an extra weight of about 2.2 g per joint due to the inserted rivet.

5. Conclusions

- Two high promising processes for easy and high strength joining of metals to fibre reinforced plastics were developed. A combination of aluminium sheets and CFRP sheets were joined successfully.
- Both processes show short cycle times and a good repeatability.
- Both processes form a uniform sleeve in the metal sheet which can support shear loads. The fibres in the CFRP are partly deflected and only a minor part is destroyed which causes less damage than comparable processes with predrilling.
- The joints show good strength properties so this process becomes attractive for use in applications with high demands in reliability and strength.
- SBR joints show slightly higher shear load values than FDJ joints while these causes less extra weight.

Acknowledgements

This research project was supported by the European Social Fund and Sächsische Aufbaubank (Antragsnummer 100110153).

The publication costs of this article were funded by the German Research Foundation/DFG (Geschäftszeichen INST 270/219-1) and the Chemnitz University of Technology in the funding programme Open Access Publishing.

References

Biermann, D., & Liu, Y. (2014). Innovative Flow Drilling on Magnesium Wrought Alloy AZ31. *Proceedings of the International Conference on Manufacturing of Lightweight Components - ManuLight 2014, 18*, 209-214. http://dx.doi.org/10.1016/j.procir.2014.06.133

Collins, M. W., & Brebbia, C. A. (2006). *Design and Nature II: Comparing Design in Nature with Science and Engineering* (6th ed.). Southampton: WIT Press. http://dx.doi.org/10.1086/431108

Döhler, C., Hälsig, A., Podlesak, H., Czech, A., Nestler, D., Wielage, B., … Kroll, L. (2014). Energy-Efficient Joining Technologies to Realise Dissimilar Joints of Metal and Fibre-Reinforced Plastics. In R. Neugebauer & W. G. Drossel (Ed.), *3rd ICMC 2014 - International Chemnitz Manufacturing Colloquium/3rd International Colloquium of the Cluster of Excellence eniPROD Part 1 Innovations of Sustainable Production for Green Mobility* (Vol. 80, pp. 447-459). Chemnitz, Germany: Fraunhofer-Institut für Werkzeugmaschinen und Umformtechnik IWU. http://dx.doi.org/10.1002/mawe.201000629

Goede, M., Stehlin, M., Rafflenbeul, L., Kopp, G., & Beeh, E. (2009). Super Light Car—lightweight construction thanks to a multi-material design and function integration. *European Transport Research Review, 1*, 5-10. http://dx.doi.org/10.1007/s12544-008-0001-2

Hufenbach, W., Gottwald, R., & Kupfer, R. (2009). Bolded joints with moulded holes for textile thermoplastic composites. *18th International Conference on Composite Materials*.

Ind.Exp. (2013). *Carbon Fibres & Carbon Reinforced Plastics (CFRP) – A Global Market Overview*. London: Industry Experts.

Kroll, L. (2009). Textilverstärkte Kunststoffbauteile in funktionsintegrierender Leichtbauweise. In E. Wintermantel, S.-W. Ha (Ed.), *Medizintechnik: Life Science Engineering* (Vol. 5, pp. 343 - 356). Berlin Heidelberg, Germany: Springer. http://dx.doi.org/10.1007/978-3-540-93936-8_15

Malikova, V., Ossenbrink, R., Viehweger, B., & Michailov, V. (2013). Experimental study of the change of stiffness properties during deep drawing of structured sheet metal. *Journal of Materials Processing Technology, 11*, 1811-1817. http://dx.doi.org/doi:10.1016/j.jmatprotec.2013.05.005

Rosato, D. V. (2005). *Reinforced Plastics Handbook* (3rd Ed.). Oxford: Elsevier Advanced Technology. http://dx.doi.org/10.1016/b978-185617450-3/50005-0

Rotheiser, J. (2009). *Joining of Plastics—Handbook for Designers and Engineers* (3rd Ed.). Munich: Carl Hanser Verlag, http://dx.doi.org/10.3139/9783446445956

Seidlitz, H., Kroll, L., & Ulke, L (2009). Load adjusted joining technology for Composite-Metal Hybrids. In T. Burczyński, W. Cholewa, & W. Moczulski (Ed.), *Methods of Artificial Intelligence* (pp. 51-52). Gliwice, Poland: Silesian University of Technology. Faculty of Mechanical Engineering. Department of Fundamentals of Machinery Design.

Seidlitz, H., Ulke, L., & Kroll, L. (2013, January 31). *DE 102009013265B4:* Methods and tools for producing a mixed module. Chemnitz University of Technology.

Seidlitz, H., Ulke, L., Gerstenberger, C., & Kroll, L. (2013). *Dimensioning of Punctiform Metal-Composite Joints: A Section-Force Related Failure Criterion.* Munich: Open Journal of Composite Materials. http://dx.doi.org/10.4236/ojcm.2014.43018

Viehweger, B., Simon, S., Wagner, L., & Berg, A. (2002). Aluminium - Stahl - Verbundblech. In *VDI-Gesellschaft Fahrzeug- und Verkehrstechnik, Entwicklungen im Karosseriebau - Development in Chassis Engineering (VDI-Berichte Nr. 1674)* (pp. 377 - 384). Düsseldorf, Germany: VDI Verlag.

4

Oxygen Consumption of Laccase-Mediator-Systems (LMS)

Alexander Kirsch[1], Markus Euring[1] & Kolja Ostendorf[1]

[1] Faculty of Forest Science and Forest Ecology, Buesgen-Institute, Department of Molecular Wood Biotechnology and Technical Mycology, Georg-August-University of Goettingen, Göttingen, Germany

Correspondence: Markus Euring, Faculty of Forest Science and Forest Ecology, Buesgen-Institute, Department of Molecular Wood Biotechnology and Technical Mycology, Georg-August-University of Goettingen, Büsgenweg 2, Göttingen 37077, Germany. E-mail: meuring@gwdg.de

Abstract

Crude oil as a non-renewable resource is presenting challenges for the future in many industrial sectors. Due to the dwindling of these resources, costs are increasing negatively affecting the wood based panels industry, which uses mainly petrochemical resins as components for binding agents. In addition harmful formaldehyde emissions arising from conventionally produced wood composites are demonstrable. In order to substitute these conventional binding agents for wood based panels, in particular medium density fiberboards (MDF), Laccase-Mediator-Systems (LMSs) were investigated in several researches. A popular and reliable method used to test the suitability of LMSs is the measurement of their oxygen consumption on wood fibers in aqueous suspension. This enzymatic catalyzed oxidation was the object of this study. The mediators 4-hydroxybenzoic acid (HBA), vanillic acid (VanA), vanillic alcohol (VAl), ethylvanillin (EVan), acetovanillone (AVan), ferulic acid (FA), caffeic acid (CA) and guajacol (Gu) were tested as possible components of the LMSs. The study showed that all of the LMSs have oxidized wood fibers more efficiently than laccase on its own. Among the different mediators, vanillic alcohol, guajacol and caffeic acid in LMSs have shown the fastest O_2 consumption.

Keywords: enzymatic binder, laccase, Laccase-Mediator-System, LMS, MDF, mediators, medium density fiberboard

1. Introduction

Wood's stability arising from its main components lignin, cellulose and hemi-celluloses is a paradigm for the biotechnological production of wood based panels. In the search for natural binding agents for wood composites such as medium density fiberboard (MDF), lignin plays an important role. By activating the lignin on wood fiber surfaces with oxidative enzymes for example, adhesive properties can be induced (Kües et al., 2007). The most important oxidative enzyme in the present case is laccase which is naturally involved in each biosynthesis and degradation of lignin (Kharazipour & Hüttermann, 1998). It belongs to the group of the polyphenol oxidases and is able to catalyze the oxidation of phenolics and other phenolic compounds by utilizing oxygen (Baldrian, 2006; Widsten, Hummer, Heathcote, & Kandelbauer, 2009). In this process oxygen reacts to water under formation of phenoxy radicals (Felby, Nielsen, Olesen, & und Skibsted, 1997). The major disadvantage of laccase is their lower redox potential in comparison to other oxidative enzymes such as peroxidases (Li, Feng, & Eriksson, 1999). In order to improve this, specific redox-molecules acting between laccase and lignin can be added to the reaction (Kües et al., 2007). These redox-molecules are called mediators.

The mediator is oxidized by laccase and following this, it is able to oxidize lignin and even non-phenolic compounds of lignin by itself (Figure 1) (Kües et al., 2007; Bourbonnais & Paice, 1990).

This Laccase-Mediator-System (LMS) enables a faster reaction time due to its increased redox-potential and a higher spectrum of substrates (Kües et al., 2007). The generating of wood fibers by using thermo-mechanical pulping is also beneficial to the reaction. Due to the high temperature of this process of beyond 170 °C, the lignin of middle lamellae plasticizes and encrusts after cooling down on the fiber surfaces (Kharazipour & Hüttermann, 1998; Euring, 2008). In this state the fiber is completely surrounded by lignin which is able to be activated in further steps (Euring, 2008; Euring, Rühl, Ritter, Kües, & Kharazipour, 2011). The formation of radicals during the incubation with LMS and the following hot pressing of incubated fibers to fiberboards, generates a natural adhesive effect. Euring (2008), Euring, Rühl, Ritter, Kües, and Kharazipour (2011), Euring, Trojanowski, and

Kharazipour (2013) achieved with LMS bonded MDF, most of the required European standards for wood based panels. It can be concluded that LMSs are suitable alternatives for conventional binding systems such as urea-formaldehyde resins. For this reason, various mediators were tested for this current paper by measuring their oxygen consumption on wood fibers as a method to evaluate their reactivity and thus their suitability as mediator for a potential industrial application.

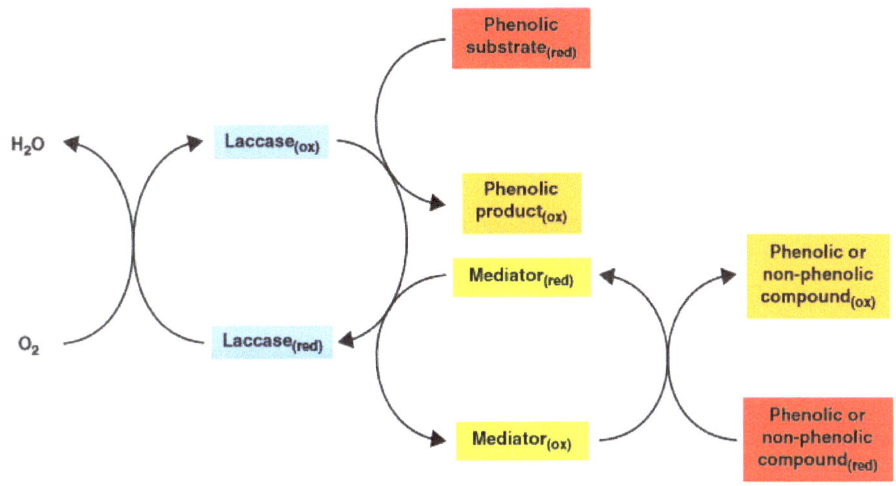

Figure 1. Simplified representation of laccase reactions with phenolic enzyme substrates and Laccase-Mediator-System (LMS) in conformity with Kües (2015)

2. Material and Methods

2.1 Material

2.1.1 Wood Fibers

Manufactured from pine wood (*Pinus sylvestris* L.) by using thermo-mechanical pulping (TMP). Here the arithmetic average fiber length of pine wood is declared as 2.2 mm (Krug, 2010).

2.1.2 Laccase

Novozym 51003 (Novozymes company, Bagsvaerd, Denmark) fermented from *Trametes vilosa* in recombination with *Aspergillus oryzae.* This was at hand in liquid form and a stabilizer was added to increase the durability. Due to this stabilizer the solution got a brown coloration. The product Novozym 51003 was fermented and mixed according to an in-house recipe.

2.1.3 Mediators

Acetovanillone (AVan): 4-hydroxy-3-methoxybenzyl alcohol, Alfa Aesar company (Karlsruhe, Germany), with a chemical purity of 98%.

Caffeic acid (CA): from Sigma-Aldrich company (Seelze, Germany) with a chemical purity of 95%.

Ethylvanillin (EVan): 3-ethoxy-4-hydroxy-benzaldehyde from Sigma-Aldrich (Seelze, Germany) with a chemical purity of 99%.

Ferulic acid (FA): 4-hydroxy-3-methoxy-cinnamic acid from Merck company (Darmstadt, Germany) with a chemical purity of 99%.

Guajacol (Gu): 2-methoxyphenol, Alfa Aesar (Karlsruhe, Germany), chemical purity 98+%.

4-hydroxybenzoic acid (HBA): from Alfa Aesar (Karlsruhe, Germany) with a chemical purity of 99%.

Vanillic alcohol (VAl): 4-hydroxy-3-methoxybenzyl alcohol, Merck (Darmstadt, Germany), chemical purity of ≥ 98%.

Vanillic acid (VanA): from Alfa Aesar company (Seelze, Germany) with a chemical purity of 98%.

2.1.4 McIlvaine Buffer (McIlv)

Composed of 0.2 M di-potassium hydrogen phosphate (K_2HPO_4) and 0.1 M citric acid ($C_6H_8O_7$), buffered to pH 6.0. Both components were from AppliChem, Darmstadt, Germany.

2.2 Methods

The consumption of oxygen is a parameter for the quality of the enzymatic activity related to the oxidation. This oxygen consumption was measured with an O_2 electrode (InPro 6800® series 12/220 by Mettler-Toledo, Urdorf, Switzerland) following the methods of Widsten (2002) and Grönqvist et al. (2005) (Figure 2). Reactions took place in a room with a constant temperature of 20 °C under normal air pressure.

First of all the mediator was dissolved in one liter of buffer while stirring and heating at 80°C. The heating treatment enables a better solution of the mediator but benefits a partial degasification causing a decreased saturation. The amount of the mediator is applied to a defined amount of wood fibers. For reactions in the flask with a maximum volume of 123 ml, 1 g of absolutely dry (atro) fibers was added. The fibers were treated with 10 mM of mediator. Afterwards the buffer-mediator-mixture was ventilated for 0.5 h to achieve a 100-percent saturation regaining the initial state, depending on the specific oxygen capacity of water. In order to verify this, the oxygen content of the buffer-mediator-mixture was measured.

After that all components (mediator dissolved in buffer and wood fibers) were added to the flask. To achieve precise results the flask was filled up to the brim, ousting last deposits of entrapped air. In order to get stable oxygen starting values a calibrating phase of the O_2 electrode was required. Then laccase with an activity of 100 U/ml was added per cannula to start the redox reaction and measurement. The current enzymatic activity of laccase was determined by using the ABTS-test developed by Yamamoto, Numata, Kawano, Shin, and Murao (1986). The oxygen content after calibrating was set as 100% saturated and the following O_2 uptake was calculated relatively to this content. A magnetic stirrer (DRAGONLAB, Bejing, China) kept the suspension with average stirring rates in motion. In intervals of 10 seconds the consumption data was recorded by the software LabView 7.1 (National Instruments, Austin, USA) involving an O_2 Transmitter 4100 e (Mettler-Toldeo, Urdorf, Switzerland) over an incubation time of 180 minutes. Every sample was repeated three times. Below an overview is shown for different samples:

- buffer (123 ml)
- buffer + fibers (1g atro)
- buffer + mediator (10 mM)
- buffer + laccase (100 U/ml)
- buffer + mediator + fibers
- buffer + laccase + mediator
- buffer + laccase + fibers
- buffer + laccase + mediator + fibers

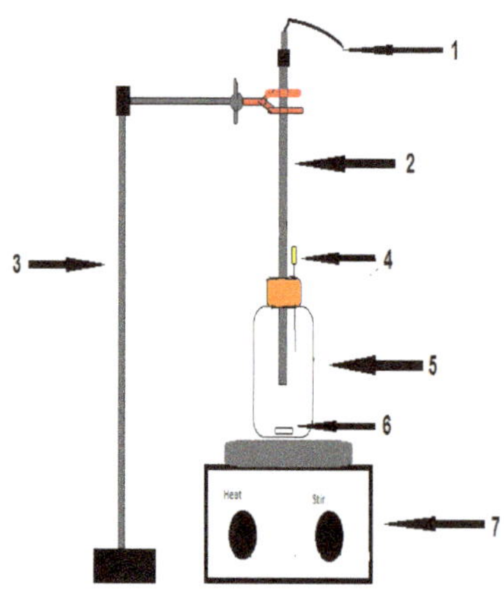

Figure 2. Experimental setup for oxygen consumption tests with 1= access to O_2 transmitter; 2= O_2 electrode; 3= tripod; 4= cannula for adding laccase; 5= flask; 6= magnetic stirring rod; 7= magnetic stirrer

3. Results and Discussion

3.1 Acetovanillone

With acetovanillone acting in the LMS the highest volume of 98% of O_2 was taken up after 180 minutes. In the same time period 90% was consumed by the tests with laccase and fibers alone. Acetovanillone and laccase consumed 15% of O_2. In the case of the references, no sample consumed more than 20 %. The grey shading above the curve represents the positive standard deviation. The curve itself is calculated as an average value of three repetitions (Figure 3).

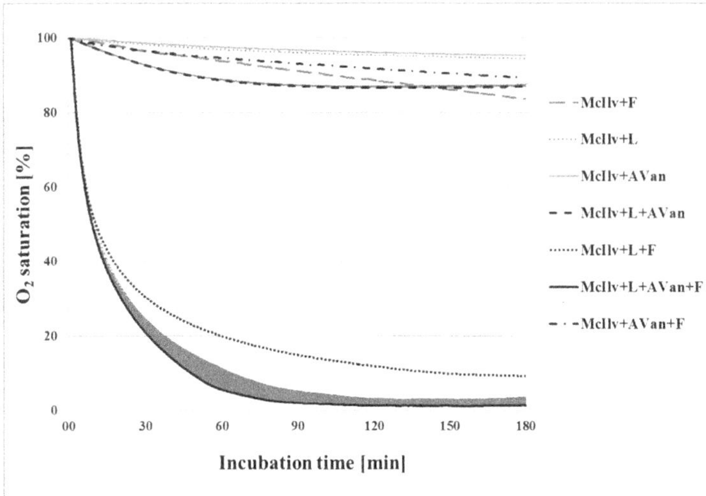

Figure 3. Oxygen consumption versus time for laccase action determined by reduction of oxygen saturation. Comparison of the samples treated with acetovanillone and references; McIlv+F = buffer and fibers McIlv+L = buffer and laccase; McIlv+AVan = buffer and acetovanillone; McIlv+L+AVan = buffer, laccase and acetovanillone; McIlv+L+F = buffer, laccase and fibers; McIlv+L+AVan+F = buffer, laccase, acetovanillone and fibers; McIlv+AVan+F = buffer, acetovanillone and fibers. The grey shading above the curve represents the positive standard deviation. The curve itself is calculated as an average value of three repetitions.

3.2 Caffeic Acid

The LMS with caffeic acid consumed all of the O_2 content within the first 30 minutes. 43% of O_2 was used when caffeic acid and laccase were tested together. 90% was consumed by the tests with laccase and fibers after 180 minutes (Figure 4).

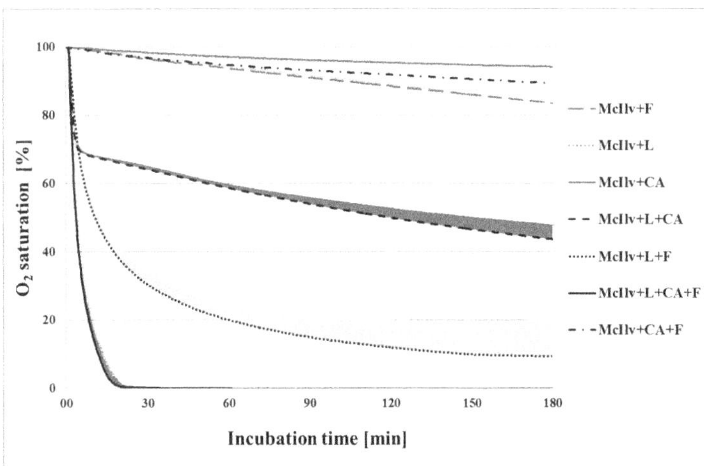

Figure 4. Oxygen consumption versus time for laccase action determined by reduction of oxygen saturation. Comparison of the samples treated with caffeic acid and references; McIlv+F = buffer and fibers McIlv+L = buffer and laccase; McIlv+CA = buffer and caffeic acid; McIlv+L+CA = buffer, laccase and caffeic acid; McIlv+L+F = buffer, laccase and fibers; McIlv+L+CA+F = buffer, laccase, caffeic acid and fibers; McIlv+CA+F = buffer, caffeic acid and fibers. The grey shading above the curve represents the positive standard deviation. The curve itself is calculated as an average value of three repetitions.

3.3 Ethylvanillin

After 180 minutes the LMS with ethylvanillin used up all O_2 in the suspension, more than samples with laccase and fibers. 90% was consumed by the tests with laccase and fibers. 15% was consumed when mediator and laccase were tested. The references consumed less than 20% (Figure 5).

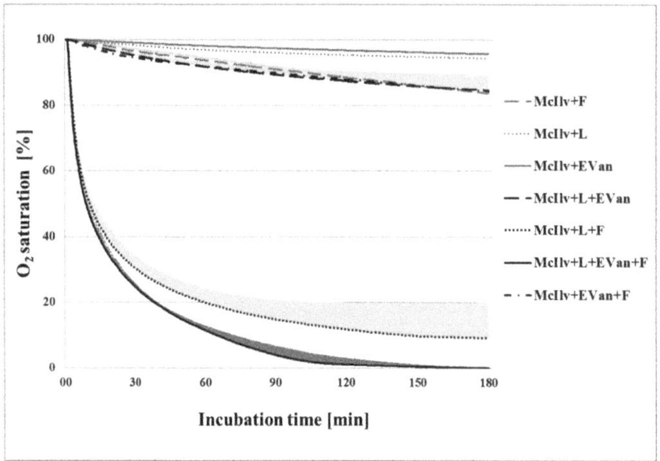

Figure 5. Oxygen consumption versus time for laccase action determined by reduction of oxygen saturation. Comparison of the samples treated with ethylvanillin (EVan) and references; McIlv+F = buffer and fibers McIlv+L = buffer and laccase; McIlv+EVan = buffer and ethylvanillin; McIlv+L+EVan = buffer, laccase and ethylvanillin; McIlv+L+F = buffer, laccase and fibers; McIlv+L+EVan+F = buffer, laccase, ethylvanillin and fibers; McIlv+EVan+F = buffer, ethylvanillin and fibers. The grey shading above the curve represents the positive standard deviation. The curve itself is calculated as an average value of three repetitions.

3.4 Ferulic Acid

An O_2 consumption of nearly 100% within 180 minutes was monitored when ferulic acid was acting in a LMS. 90% was consumed by the tests with laccase and fibers. 34% was taken up with sample ferulic acid and laccase. The references can be associated with the results of acetovanillone or caffeic acid (Figure 6).

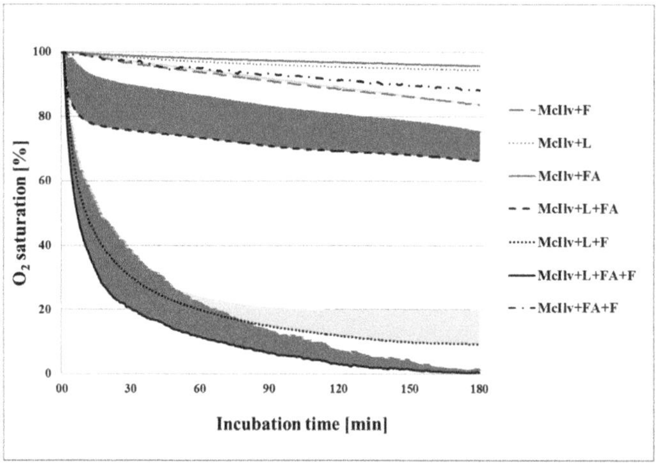

Figure 6. Oxygen consumption versus time for laccase action determined by reduction of oxygen saturation. Comparison of the samples treated with ferulic acid (FA) and references; McIlv+F = buffer and fibers McIlv+L = buffer and laccase; McIlv+FA = buffer and ferulic acid; McIlv+L+FA = buffer, laccase and ferulic acid; McIlv+L+F = buffer, laccase and fibers; McIlv+L+FA+F = buffer, laccase, ferulic acid and fibers; McIlv+FA+F = buffer, ferulic acid and fibers The grey shading above the curve represents the positive standard deviation. The curve itself is calculated as an average value of three repetitions.

3.5 Guajacol

Like caffeic acid all of the O_2 content was consumed by the sample LMS with guajacol within the first 30 minutes. 90% was consumed by the tests with laccase and fibers. A consumption of 35% was monitored when guajacol was tested with laccase. The references can be seen with again O_2 usages of less than 20% (Figure 7).

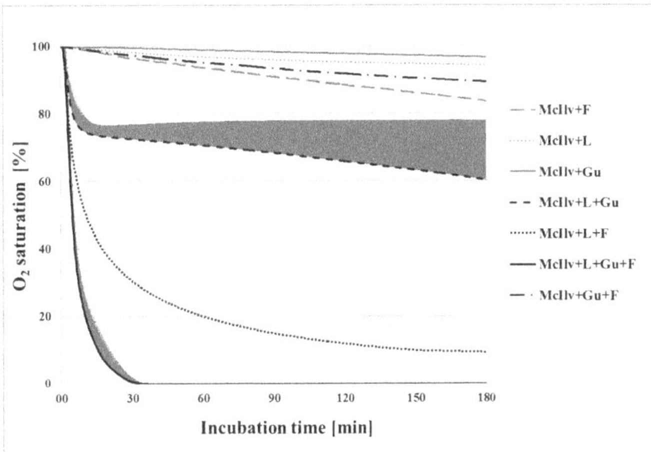

Figure 7. Oxygen consumption versus time for laccase action determined by reduction of oxygen saturation. Comparison of the samples treated with guajacol (Gu) and references; McIlv+F = buffer and fibers McIlv+L = buffer and laccase; McIlv+Gu = buffer and guajacol; McIlv+L+Gu = buffer, laccase and guajacol; McIlv+L+F = buffer, laccase and fibers; McIlv+L+Gu+F = buffer, laccase, guajacol and fibers; McIlv+Gu+F = buffer, guajacol and fibers. The grey shading above the curve represents the positive standard deviation. The curve itself is calculated as an average value of three repetitions.

3.6 4-hydroxybenzoic acid (HBA)

The LMS with HBA with O_2 consumption of 95% uses slightly more O_2 in a direct comparison with sample laccase and fibers (90%). HBA and laccase together had taken up less than 10% ranking in the same area with samples of the references. HBA is, as a consequence, like the combination of buffer, laccase and HBA shows, the least effective of all tested mediators. Nevertheless the combination of buffer, laccase, HBA and fibers shows in comparison to the combination buffer, laccase and fibers an increasing effectivity (Figure 8).

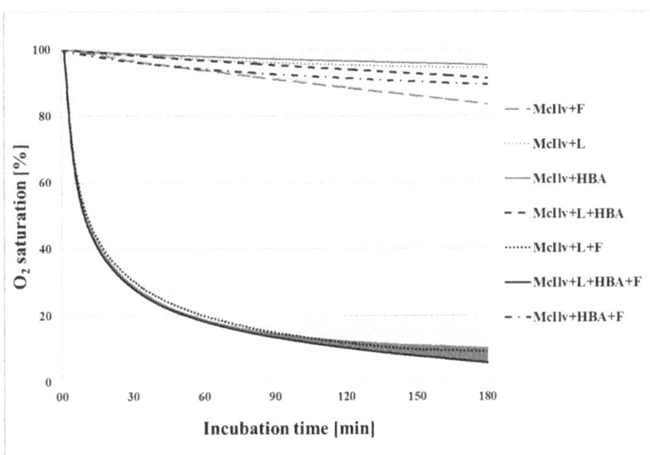

Figure 8. Oxygen consumption versus time for laccase action determined by reduction of oxygen saturation. Comparison of the samples treated with 4-hydroxybenzoic acid (HBA) and references; McIlv+F = buffer and fibers McIlv+L = buffer and laccase; McIlv+HBA = buffer and 4-hydroxybenzoic acid; McIlv+L+HBA = buffer, laccase and 4-hydroxybenzoic acid; McIlv+L+F = buffer, laccase and fibers; McIlv+L+HBA+F = buffer, laccase, 4-hydroxybenzoic acid and fibers; McIlv+HBA+F = buffer, 4-hydroxybenzoic acid and fibers. The grey shading above the curve presents the positive standard deviation. The curve itself is calculated as an average value of three repetitions.

3.7 Vanillic Acid

In the sample LMS with vanillic acid, 100% of O_2 was taken up after 60 minutes. 90% was consumed by the tests with laccase and fibers. Vanillic acid and laccase consumed 37%. The references were localized in the same range in comparison to the other mediators (Figure 9).

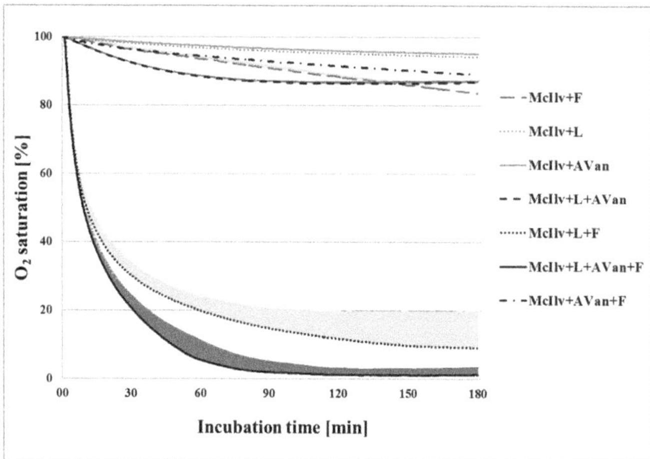

Figure 9. Oxygen consumption versus time for laccase action determined by reduction of oxygen saturation. Comparison of the samples treated with vanillic acid (VanA) and references; McIlv+F = buffer and fibers McIlv+L = buffer and laccase; McIlv+VanA = buffer and vanillic acid; McIlv+L+VanA = buffer, laccase and vanillic acid; McIlv+L+F = buffer, laccase and fibers; McIlv+L+VanA+F = buffer, laccase, vanillic acid and fibers; McIlv+VanA+F = buffer, vanillic acid and fibers. The grey shading above the curve represents the positive standard deviation. The curve itself is calculated as an average value of three repetitions

3.8 Vanillic Alcohol

100% O_2 consumption was monitored within the first 40 minutes with vanillic alcohol as component in an LMS. 90% was consumed by the tests with laccase and fibers. Vanillic alcohol and laccase had taken up 40% O_2 (Figure 10).

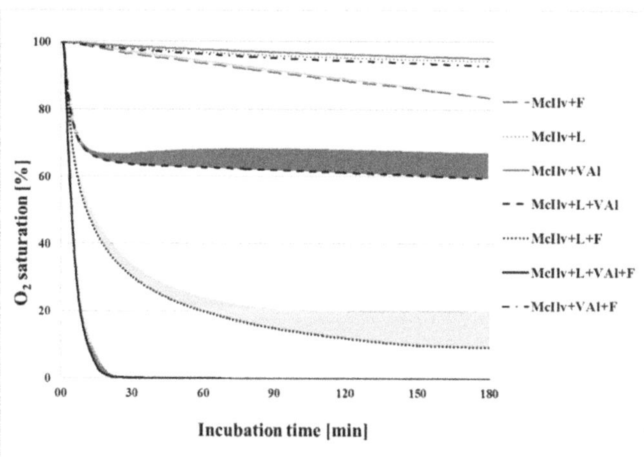

Figure 10. Oxygen consumption versus time for laccase action determined by reduction of oxygen saturation. Comparison of the samples treated with vanillic alcohol (VAl) and references; McIlv+F = buffer and fibers McIlv+L = buffer and laccase; McIlv+VAl = buffer and vanillic alcohol; McIlv+L+VAl = buffer, laccase and vanillic alcohol; McIlv+L+F = buffer, laccase and fibers; McIlv+L+VAl+F = buffer, laccase, vanillic alcohol; McIlv+VAl+F = buffer, vanillic alcohol and fibers. The grey shading above the curve represents the positive standard deviation. The curve itself is calculated as an average value of three repetitions.

To allow an easier comparison, the following table shows in comparison the oxygen consumption of the different mediators concerning saturation and time. These figures are calculated as an average value of three repetitions (Table 1).

Table 1. Oxygen consumption versus time for laccase action determined by reduction of oxygen saturation. Comparison of the samples with buffer, fibers, laccase and different mediators. McIlv+L+AVan+F = buffer, laccase, acetovanillone and fibers; McIlv+L+CA+F = buffer, laccase, caffeic acid and fibers; McIlv+L+EVan+F = buffer, laccase, ethylvanillin and fibers; McIlv+L+FA+F = buffer, laccase, ferulic acid and fibers; McIlv+L+Gu+F = buffer, laccase, guajacol and fibers; McIlv+L+HBA+F = buffer, laccase, 4-hydroxybenzoic acid and fibers; McIlv+L+VanA+F = buffer, laccase, vanillic acid and fibers; McIlv+L+VAl+F = buffer, laccase, vanillic alcohol

Sample	**O_2 saturation [%]**							
time [min]	0	5	10	15	30	60	120	180
McIlv+L+AVan+F	100,00	62,74	45,75	36,35	20,42	5,41	1,26	1,31
McIlv+L+CA+F	100,00	35,82	13,77	3,39	0,07	0,00	0,00	0,00
McIlv+L+EVan+F	100,00	63,81	47,06	38,67	24,65	11,51	1,11	0,01
McIlv+L+FA+F	100,00	57,95	40,74	31,97	20,46	11,62	3,02	0,48
McIlv+L+Gu+F	100,00	39,19	18,19	9,50	0,28	0,00	0,00	0,00
McIlv+L+HBA+F	100,00	63,52	47,64	39,88	27,92	18,11	10,23	5,79
McIlv+L+VanA+F	100,00	55,22	32,56	19,62	2,50	0,00	0,00	0,00
McIlv+L+Val+F	100,00	40,48	12,38	3,39	0,14	0,00	0,00	0,00

One noticeable observation is that reactions were most intense immediately after adding laccase except for the test buffer with laccase. The curves illustrate highest reactions within the first 20 to 30 minutes. The more time passes by the more flat the curves become. This should relate to the high initial availability of substrates such as lignin or lignin and mediator and the high initial value of co-substrate O_2 in the suspension. The temperature-considering the Q_{10} temperature coefficient and pH of the suspension are also important. The pH of McIlvaine buffer is therefore applied to the pH optimum of laccase used for this study which was reported to be best at pH 6.0 (Kües et al., 2007).

In most of the cases, lignin and mediator were apparently oxidized by laccase. Only HBA showed a contradictory result when tested together with laccase consuming less than 10% O_2.

Overall, tests with LMSs showed a higher consumption of oxygen than with laccase on fibers on its own. Most of the LMSs (vanillic acid, vanillic alcohol, ethylvanillin, caffeic acid, ferulic acid and guajacol) even consumed the entire available oxygen content of the suspension. This is attributable to the increased spectrum of substrates and higher redox-potential of LMS (Kües et al., 2007; Bourbonnais & Paice, 1990). Similar results were obtained in the study of Euring, Rühl, Ritter, Kües, and Kharazipour (2011) with HBA acting in a LMS. Potthast, Rosenau, Chen, and Gratzl (1995) found out that LMSs can transfer two electrons at the same time due to the mediators` ability as additional electron donor which adequately describes the better redox-potential of LMS.

Among the different mediators in LMSs, vanillic alcohol, guajacol and caffeic acid have shown most rapid O_2 consumption. It can be concluded that these LMSs catalyzed the oxidation of phenolic and non-phenolic compounds in lignin more efficiently than other LMSs. These results were underlined by the results of preparations with laccase and mediator without wood fibers. Here, uptake rates of up to 40% O_2 were monitored. These three mediators should be focused upon when being tested for an application as a binding system for MDF.

4. Conclusion

In exploring the effectiveness and suitability of LMSs, various mediators were tested in this study. A reliable method for this is the measurement of their oxygen consumption on phenolic (laccase) and non-phenolic (mediator) compounds. This work has shown that mediators such as vanillic acid, vanillic alcohol, ethylvanillin, caffeic acid, ferulic acid and guajacol, which have had the highest oxygen consumption of the tests, are possible alternatives to increase effectiveness of LMS. An important aspect for future studies will be to use the best working mediators to produce MDF in pilot-scale. This is a popular method to determine the effects regarding to technical properties such as modulus of rupture, internal bond and thickness swelling.

Acknowledgements

This work was supported by the Deutsche Forschungsgemeinschaft (DFG), grant EU 124/2-2 to Markus Euring and Alexander Kirsch.

References

Baldrian, P. (2006). Fungal Laccases - Occurrence and Properties. *FEMS Microbiology Reviews,* 30(2), 215-42. http://dx.doi.org/10.1111/j.1574-4976.2005.00010.x

Bourbonnais, R., & Paice, M. G. (1990). Oxidation of non-phenolic substrates: An expanded role for laccase in lignin biodegradation. *FEBS Letters, 267*(1), 99-102. http://dx.doi.org/10.1016/0014-5793(90)80298-W

Euring, M. (2008). Einsatz von Mediatoren bei der enzymatischen Aktivierung der fasereigenen Bindekräfte zur Herstellung von enzymgebundenen, bindemittelfreien Holzwerkstoffen (Dissertation). Georg-August-University of Göttingen, Göttingen, Germany. Retrieved from https://ediss.uni-goettingen.de/bitstream/handle/11858/00-1735-0000-0006-B111-9/euring.pdf?sequence=1

Euring, M., Rühl, M., Ritter, N., Kües, U., & Kharazipour, A. (2011). Laccase Mediator Systems for Eco-Friendly Production of Medium-Density Fiberboard (MDF) on a Pilot Scale: Physicochemical Analysis of the Reaction Mechanism. *Biotechnology Journal, 6*(10), 1253-1261. http://dx.doi.org/10.1002/biot. 201100119

Euring, M., Trojanowski, J., & Kharazipour, A. (2013). Laccase-Mediator Catalyzed Modification of Wood Fibers: Studies on the Reaction Mechanism and Making of Medium-Density Fiberboard, Göttingen.

Felby, C., Nielsen, B. R., Olesen, P. O., & und Skibsted, L. H. (1997). Identification and Quantification of Radical Reaction Intermediates by Electron Spin Resonance Spectrometry of Laccase-Catalyzed Oxidation of Wood Fibers from Beech (Fagus Sylvatica). *Applied Microbiology and Biotechnology, 48*(4), 459-64. http://dx.doi.org/10.1007/s002530051080

Grönqvist, S., Viikari, L., Niku-Paavola, M.-L., Orlandi, M., Canevali, C., & und Buchert, J. (2005), Oxidation of Milled Wood Lignin with Laccase, Tyrosinase and Horseradish Peroxidase. *Applied Microbiology and Biotechnology,* 67(4), 489-494. http://dx.doi.org/10.1007/s00253-004-1800-6

Kharazipour, A., & Hüttermann, A. (1998). Biotechnological production of wood composites. In A. Bruce, & J. W. Palfreyman (Eds.), *Forest Products Biotechnology* (chapter 9, pp. 141-150). Taylor & Francis.

Krug, D. (2010). Einfluss der Faserstoff-Aufschlussbedingungen und des Bindemittels auf die Eigenschaften von mitteldichten Faserplatten (MDF) für eine Verwendung im Feucht- und Außenbereich (Dissertation). Hamburg University, Hamburg, Germany. Retrieved from http://ediss.sub.uni-hamburg.de/volltexte/2010/4679/pdf/Detlef-Krug_Dissertation.pdf

Kües, U. (2015). Fungal enzymes for environmental management. *Current Opinion in Biotechnology, 33,* 268-278. http://dx.doi.org/10.1016/j.copbio.2015.03.006

Kües, U., Bohn, C., Euring M., Müller C., Polle, A., & Kharazipour, A. (2007). Enzymatically Modified Wood in Panel Board Production. In U. Kües (Ed.), *Wood Production, Wood Technology, and Biotechnological Impacts* (pp. 433-467). Universitätsverlag Göttingen.

Li, K., Feng, X., & Eriksson, K. E. L. (1999). Comparison of Fungal Laccases and Redox Mediators in Oxidation of a Nonphenolic Lignin Model Compound. *Applied and Environmental Microbiology, 65*(6), 2654-2660

Potthast, A., Rosenau, T., Chen, C.-L., & Gratzl, J. S. (1995), Selective Enzymic Oxidation of Aromatic Methyl Groups to Aldehydes. *The Journal of Organic Chemistry, 60*(14), 4320-4321. http://dx.doi.org/10.1021/jo00119a006

Widsten, P. (2002). Oxidative Activation of Wood Fibers for the Manufacture of Medium-Density Fiberboard (MDF) (Dissertation). Helsinki University of Technology, Helsinkim Finland. Retrieved from https://aaltodoc.aalto.fi:443/handle/123456789/2239

Widsten, P., Hummer, A., Heathcote, C., & Kandelbauer, A. (2009). A preliminary study of green production of fiberboard bonded with tannin and laccase in a wet process. *Holzforschung, 63*, 545-550. http://dx.doi.org/10.1515/HF.2009.090

Yamamoto, E., Numata, A., Kawano, T., Shin, T., & Murao, S. (1986). Structures of the Laccase-catalysed Oxidation Products of Hydroxybenzoic Acids in the Presence of ABTS. Japan Society for Bioscience,

Biotechnology and Agrochemnistry. Agric. Biol. Chem., 50(5), 1355-1357. http://dx.doi.org/10.1271/bbb 1961.50.1355

5

Comparative Relationship of Fiber Strength and Yarn Tenacity in Four Cotton Cultivars

Yongliang Liu[1], B. Todd Campbell[2], Chris Delhom[1] & Vikki Martin[3]

[1] USDA, ARS, Cotton Structure & Quality Research Unit, Southern Regional Research Center (SRRC), New Orleans, LA, USA

[2] USDA, ARS, Coastal Plain Soil, Water and Plant Conservation Research, Florence, SC, USA

[3] Cotton Incorporated, Cary, NC, USA

Correspondence: Yongliang Liu, USDA, ARS, Cotton Structure & Quality Research Unit, Southern Regional Research Center (SRRC), New Orleans, LA, USA. E-mail: Yongliang.liu@ars.usda.gov

Abstract

High volume instrumentation (HVI^{TM}) measurement is a primary and routine tool of providing fiber properties to cotton researchers. There have been considerable studies designed to derive yarn quality from acquired fiber quality data by various means, including HVI. There is also of desired information about the comparison of yarn quality within a cotton cultivar or among the cultivars, as such knowledge could be informative in attempts to understand the selection of cotton cultivars. The purpose of this preliminary study was to characterize the fiber HVI strength and yarn skein tenacity of four cotton cultivar harvested from three locations in different crop years. Instead of developing linear regression models from acquired fiber property parameters to predict yarn tenacity, this study applied a simple ratio method (i.e., correct fiber strength or yarn tenacity with fiber micronaire component) to relate fiber strength with yarn tenacity. The results indicate that three cultivars (DP 393, Phytogen 72, and FM 958) show stronger correlation between micronaire corrected yarn tenacity and micronaire corrected fiber HVI strength. It implies the feasibility of utilizing HVI fiber micronaire and strength property data, as a semi-quantitative and fast tool, to compare the yarn tenacity performance within a cotton cultivar or between cultivars.

Keywords: cotton fiber, fiber quality, yarn quality, HVI, cotton cultivar

1. Introduction

Cotton is one of the most important and widely grown crops in the world. It is a well-traded agricultural commodity mostly for textile fiber purpose, but it also yields a high grade vegetable oil from cottonseed for human consumption as well as multiple cellulosic byproducts and whole seeds used as a primary source of fiber and protein in animal feed (Wakelyn & Chaudhry, 2010).

Cotton production is determined by at least three main factors and the interactions among them. These include genotype, environment, and production practices. These three factors and their interactions influence both yield and fiber quality potential and ultimately determine the growers and processors' profitability (Bradow & Davidonis, 2000). Thus, the desired cotton cultivars would provide high fiber yield, good fiber quality, and efficient processing at the ginning site and textile mill.

Cotton fiber, as a raw and starting material, could impact its intermediate yarn production and quality. Since yarn manufacturing requires specific equipment and uses a large amount of cotton fibers, substantial research has been done to optimize the methods of predicting yarn properties, such as tenacity (strength), uniformity, hairiness, and elongation, from available properties of raw fibers (Cai et al., 2013; Faulkner et al., 2012; Frydrych, 1992; Kelly & Hequet, 2013; Kelly et al., 2013; Long et al., 2013; Pan et al., 2001; Ramey et al., 1977; Thibodeaux et al., 2008; Üreyen & Kadoglu, 2006). In general, two approaches (theoretical vs. statistical) were explored in these studies. The theoretical approach is based on certain assumptions and the generated models provide good information about interactions among different fiber properties and yarn characteristics (Frydrych, 1992; Pan et al., 2001). The statistical approach, being preferred more frequently, bridges a relationship between yarn and fiber quality characteristics through multiple linear regression methods (Cai et al., 2013; Faulkner et al., 2012; Kelly & Hequet, 2013; Kelly et al., 2013; Long et al., 2013; Ramey et al., 1977; Thibodeaux et al., 2008; Üreyen & Kadoglu, 2006).

This approach has become more preferred as a result of voluminous fiber quality attributes available from routine, standard, and improved fiber quality measurements, including high volume instrument (HVI™), advanced fiber information system (AFIS), Favimat, Fineness Maturity Tester (FMT), and fiber cross-sectional image analysis.

The tenacity property of a spun yarn is an important index in determining the quality of the yarn, as it directly affects the winding and knitting efficiency as well as warp and weft breakages during weaving. Therefore, understanding the yarn tenacity potential from raw cotton fibers directly is advantageous to fiber researchers in the periods of next-generation genotype development and crop field management and also to fiber spinners during mill preparation. As such, great efforts have been made to estimate yarn tenacity from diversified fiber properties (Cai et al., 2013; Faulkner et al., 2012; Frydrych, 1992; Kelly & Hequet, 2013; Kelly et al., 2013; Long et al., 2013; Pan et al., 2001; Ramey et al., 1977; Thibodeaux et al., 2008; Üreyen & Kadoglu, 2006). Üreyen and Kadoglu (2006) observed a high positive correlation between fiber and yarn strength and also found that fiber strength was the most important parameter for yarn tenacity, while fiber elongation, fiber length, uniformity index, fiber fineness, yarn count, yarn twist, roving count and unevenness of roving are other parameters that can have a significant influence on yarn tenacity. Thibodeaux et al. (2008) examined the effect of short fiber content in raw cotton on the yarn quality and found that the yarn strength model developed using the four basic HVI properties (strength, micronaire, short fiber content, and uniformity index) alone was nearly as good as those using all 23 fiber properties from the AFIS, HVI, and Suter-Webb (SW) Array method. By analyzing the impacts of fiber length parameters on yarn properties, Cai et al. (2013) reported the effectiveness of these length parameters and their combinations in predicting yarn properties such as strength and irregularity. In a more recent investigation, Long et al. (2013) assessed alternative cotton fiber quality attributes and their relationship with yarn strength, and revealed that the substitution of alternative fiber fineness variables for micronaire or single fiber strength for bundle strength improved the prediction of yarn strength in their models. Hequet and collaborators (Faulkner et al., 2012; Kelly & Hequet, 2013; Kelly et al., 2013) have taken great attention to predict yarn quality from combined HVI and AFIS data, along with the consideration of harvest method and cultivar information.

As a different strategy of better understanding the relationship between cotton fiber strength and yarn tenacity, we present the concept of dividing fiber HVI strength and yarn skein tenacity by fiber HVI micronaire component, instead of modeling yarn tenacity from a number fiber properties using multiple linear regression. The main objective of this study was not to develop a set of universal equations to predict yarn tenacity; rather, it was intended to provide a relatively simple and semi-qualitative screening method for a rapid comparison of yarn tenacity performance either within or between the cultivars.

2. Method

2.1 Cotton Samples

In crop years of 2011, 2012, and 2014, four commercial cultivars Deltapine 393 (DP 393, PVP #200400206), Fibermax 958 (FM 958, PVP #200100208), Phytogen 72 (PVP #200100115), and UA 48 (PVP #201100041), along with additional elite breeding lines, were grown in four replicated field tests at the Clemson University Pee Dee Research and Education Center near Florence, SC (Florence), the Clemson University Edisto Research and Education Center near Blackville, SC (Blackville), and the North Carolina State University Sandhills Research Station near Jackson Springs, NC (Sandhills). It is a Norfolk loamy sand soil in Florence, a Barnwell loamy sand soil in Blackville, and a Candor sand soil in Sandhills. Each trial was arranged in a randomized complete block design with four replications. Each entry was planted in a two-row plot 10.7 m long with 96.5 cm spacing between rows. Plots were managed conventionally and followed the established local practices.

From each plot in each trial, 50 bolls were picked by hand. These boll samples were subsequently ginned on a 10-saw laboratory gin and lint fibers were collected. Cotton lint fibers were conditioned at a constant relative humidity of 65 ± 2% and temperature of 21 ± 1 °C for at least 24 hours, prior to routine fiber and yarn quality measurement. Table 1 summarizes the fiber distributions of each cotton cultivars grown at three locations over three crop years.

2.2 Fiber Quality Measurement

Average micronaire and strength values were obtained from five replicates on each sample by an Uster® HVI™ 900A (Uster Technologies Inc., Knoxville, TN). All measurements were performed at the Southern Regional Research Center of USDA's Agricultural Research Service (USDA-ARS-SRRC). The same instrument was used for all fibers throughout the study.

Table 1. Fiber distributions of four cotton cultivars grown at three locations and three crop years

		Florence	Blackville	Sandhills
	2011	4	4	4
DP 393 (total: 32)	2012		4	4
	2014	4	4	4
	2011	4	4	4
FM 958 (total: 20)	2012		4	4
	2014			
	2011	4	4	4
Phytogen 72 (total: 31)	2012		3	4
	2014	4	4	4
	2011	4	4	4
UA 48 (total : 32)	2012		4	4
	2014	4	4	4

2.3 Yarn Quality Measurement

With the limited quantity of lint sample available, a mini-spinning protocol was applied by carding approximately 60 g per sample on a modified Saco Lowell Model 100 card. The carded web was drawn into sliver on a modified Saco Lowell DF 11 draw frame. Two bobbins of ring spun yarn were spun to a nominal count of Ne 30/1 per sample. A 54.9 m (or 109.8 m for the 2012 cottons) mini-skein was produced from each bobbin and tested on an Instron tensile tester (ASTM, 2012). Additionally, each bobbin was tested for 1 min at 91.4 m/min on an Uster Tester 4 (UT4) and 20 single end breaks on an Uster Tensorapid 4. The skein strength was reported in single-end tenacity equivalent (g/tex), which is a normalization process to account for yarn size variations.

3. Results and Discussion

3.1 Relationship between Fiber HVI Strength and Micronaire

Figure 1 shows the plot of HVI strength against micronaire for a total of 115 cotton fibers from 4 known varieties grown in 3 crop years. HVI fiber micronaire is determined by both maturity (degree of secondary cell wall development) and fineness (weight per unit length) of the fibers (Lord, 1956), while HVI strength reflects the normalized external force required to break a bundle of parallel fibers. Overall, fiber HVI strength did not change along with fiber micronaire, although the scatter becomes more apparent as micronaire elevates. As anticipated, Pearson correlation or univariate correlation coefficient (r) between HVI strength and micronaire within these samples was extremely low ($r = 0.12$). In this work, we generally considered a significant correlation between two variables when absolute r value is greater than 0.50, a moderate correlation between two variables when absolute r value is between 0.20-0.50, and an insignificant correlation between two variables when absolute r value is less than 0.20. In Figure 1, mean HVI strengths were 30.2, 31.0, 31.5, and 33.6 gm/tex for DP 393, FM 958, Phytogen 72, and UA 48, corresponding to respective micronaire units of 4.84, 4.63, 4.61, and 4.99.

Characteristics of regression lines for individual varieties are also inserted in Figure 1. Regardless of growing locations and crop years, HVI strengths among DP 393 fibers showed no correlation with fiber micronaire ($r = -0.08$), strengths in FM 958 fibers had negative and moderate correlation with micronaire ($r = -0.46$), while those within Phytogen 72 and UA 48 fibers showed positive and moderate correlations with micronaire ($r = 0.44$ and 0.21, respectively).

In a previous study (Liu et al., 2011), fiber HVI strength readings were corrected by respective micronaire values in both quotient (HVI_{str} / HVI_{mic}) and product ($HVI_{str} * HVI_{mic}$) forms. Following, the effect of modified HVI strength indices on NIR model performance were examined through partial least squares (PLS) regression using large data sets. Two modified strength indices were observed to have better correlations than raw HVI strength index with NIR spectra, and the use of the quotient form resulted in an improved model performance than that of the product one. When plotting micronaire corrected HVI strength against HVI micronaire for those samples in Figure 1, significant and linear relationships were expected for each cultivar (Figure 2). Relatively lower correlation ($r = 0.72$) among Phytogen 72 fibers than the other three cultivars ($r > 0.85$) suggests either great differences in strength readings for fibers having the similar micronaire values or great variations in micronaire readings for samples having the similar strength values.

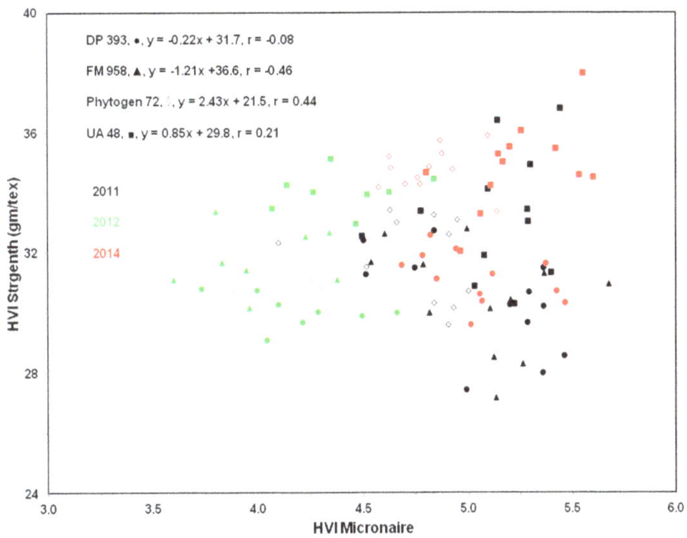

Figure 1. Plot of fiber HVI strength against HVI micronaire for four commercial cultivars grown in crop years of 2011, 2012, and 2014. Regressions for each cultivar were inserted for comparison. Pearson correlation r was 0.12 among all samples

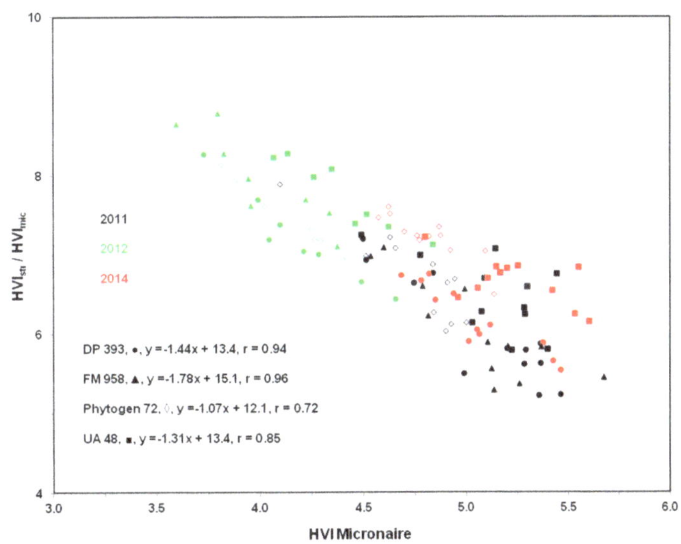

Figure 2. Plot of corrected fiber HVI strength against HVI micronaire for four commercial cultivars grown in crop years of 2011, 2012, and 2014. Regressions for each cultivar were inserted for comparison. Pearson correlation r was 0.84 among all samples

The difference in magnitude of slopes in Figure 2 might reflect differing response of HVI_{str} / HVI_{mic} ratio to HVI micronaire parameter among the four cultivars. An example of both calculated HVI_{str} / HVI_{mic} ratio from regression lines in Figure 2 and converted HVI strength is tabulated in Table 2, given the fiber micronaire (HVI_{mic}) = 4.0 and 5.0, respectively. From low to high micronaire, the four cultivars showed a decrease in modified strength. Also, variations in either HVI_{str} / HVI_{mic} ratio or HVI strength among 4 cultivars within lower micronaire fibers were smaller than those among higher micronaire fibers.

Table 2. Comparison of calculated HVI_{str}/HVI_{mic} from regression lines in Figure 2. Converted HVI strength was included in parentheses

Cultivar	$HVI_{mic} = 4.0$	$HVI_{mic} = 5.0$
DP 393	7.7 (30.8)	6.2 (31.0)
FM 958	8.0 (32.0)	6.2 (31.0)
Phytogen 72	7.8 (31.2)	6.7 (33.5)
UA 48	8.2 (32.8)	6.9 (34.5)

3.2 Relationship between Fiber HVI Strength and Yarn Skein Tenacity

The plots of yarn tenacity against fiber strength for four commercial cultivars are given in Figure 3. Considering data combined across cultivars and years, a reasonable trend with a scattered pattern was evident (r = 0.55). The average fiber strength and yarn tenacity were 30.2 gm/tex and 45.7 g/tex for DP 393 fibers, 31.0 gm/tex and 45.4 g/tex for FM 958 fibers, 31.5 gm/tex and 50.1 g/tex for Phytogen 72 fibers, and 33.6 gm/tex and 50.6 g/tex for UA 48 fibers. Overall, yarn tenacity displayed a strong relationship with fiber strength for DP 393 fibers (r = 0.60), followed by moderate correlations for FM 958 (r = 0.49), Phytogen 72 (r = 0.47), and UA 48 fibers (r = 0.24). Meanwhile, the slope decreased in the order of DP 393, Phytogen 72, FM 958, and UA48 cultivars. Both higher r and slope values reflected stronger correlation between fiber strength and yarn tenacity for DP 393 as opposed to FM 958 and Phytogen 72. The lowest correlation between fiber strength and yarn tenacity was found for UA 48 fibers.

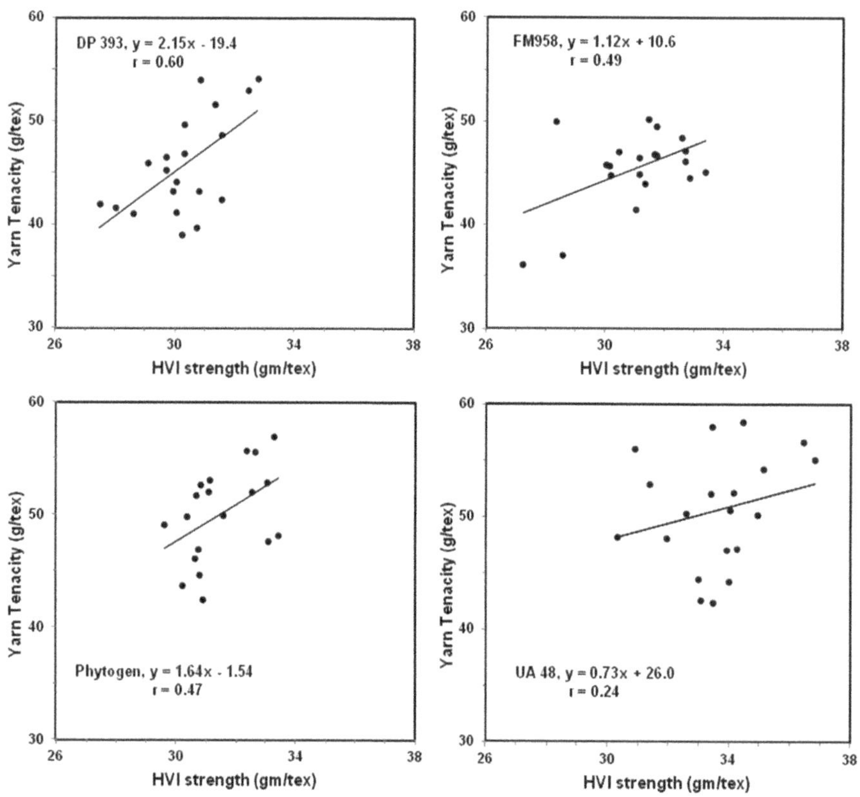

Figure 3. Plot of yarn tenacity against fiber strength for four commercial cultivar grown in crop years of 2011 and 2012 (Yarn data for 2014 was not available this time). Regressions for each cultivar were inserted for comparison. Pearson correlation r was 0.55 among all samples

When both fiber strength and yarn tenacity values were modified by fiber micronaire, the resulting plot in Figure 4 suggests a much stronger linear correlation for each cotton cultivar. Improvement of r from the 0.24 to 0.60 range in Figure 3 to the 0.54 to 0.93 range in Figure 4 implies the sensitivity of fiber / yarn breaking property to the inherent fiber chemical and physical structure. Especially, DP 393, Phytogen 72, and FM 958 fibers showed a larger r (> 0.85) than UA 48 fibers (~0.54), indicating that micronaire corrected yarn tenacities could be strongly related with micronaire corrected fiber HVI strength for these three cultivars. Meanwhile, the slope in Figure 4 decreased in the order of Phytogen 72 (2.10), DP 393 (1.99), FM 958 (1.50) and UA 48 (0.71) fibers. Concurrently, the UA 48 fibers displayed the lowest fiber strength-yarn tenacity correlation in Figure 3.

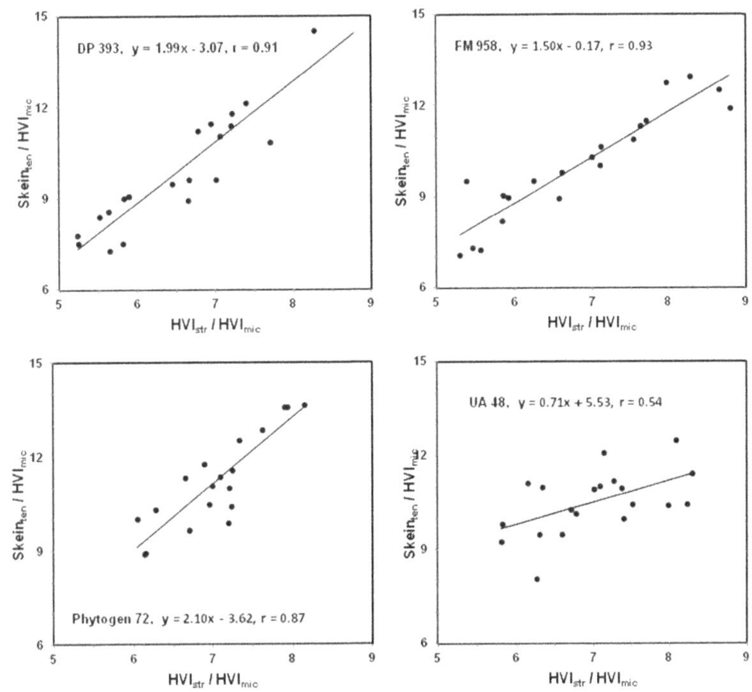

Figure 4. Plot of corrected yarn tenacity against corrected fiber strength for four commercial cultivars grown in crop years of 2011 and 2012. Regressions for each cultivar were inserted for comparison

Potentially, relationships in Figure 4 could be utilized to compare the yarn tenacity property either between the same or differing cultivars on the basis of HVI micronaire and strength data only. For instance, Table 3 presents the predicted $Skein_{ten}/HVI_{mic}$ values from the regressions in Figure 4, assuming the $HVI_{str} / HVI_{mic} = 6.0$ and 8.0, respectively. Interestingly, DP 393, FM 958, and Phytogen 72 cultivars had a close $Skein_{ten}/HVI_{mic}$ when $HVI_{str} / HVI_{mic} = 6.0$, and the difference between any two of the three became apparent when $HVI_{str} / HVI_{mic} = 8.0$. UA 48 fibers showed the largest $Skein_{ten}/HVI_{mic}$ among the four cultivars when $HVI_{str} / HVI_{mic} = 6.0$, in turn, it has the greatest ratio of yarn skein tenacity to HVI micronaire among the four cultivar. As a comparison, UA 48 fibers had smaller $Skein_{ten}/HVI_{mic}$ than the other three cultivars when $HVI_{str} / HVI_{mic} = 8.0$, leading to a weak ratio between yarn tenacity and fiber micronaire.

Table 3. Calculated $Skein_{ten}/HVI_{mic}$ from relationships in Figure 4

Cultivar	$HVI_{str}/HVI_{mic} = 6.0$	$HVI_{str}/HVI_{mic} = 8.0$
DP 393	8.9	12.9
FM 958	8.8	11.8
Phytogen 72	9.0	13.2
UA 48	9.7	11.1

Compared to the earlier strategy of multiple fiber property derived regression models for estimating yarn tenacity, it might be desirable to consider only 2 fiber quality components, micronaire and strength, as it avoids the need to develop multiple regression equations from a number of fiber quality input traits. Undoubtedly, introducing more fiber quality parameters into regression equations has improved the accuracy in predicting yarn tenacity. However, there are concerns about over-fitting the model and how to validate the equations by an independent set of fibers or cultivars. As observed in the present study, one cultivar (UA 48) could perform quite differently from other cultivars when simply relating micronaire corrected yarn tenacity to fiber strength, possibly leading to a compromised model if it was compiled into calibration, validation or independent set. Nevertheless, the ratio values of micronaire corrected fiber strength might be utilized as a semi-quantitative and fast approach to compare cultivar performance for yarn tenacity.

4. Conclusions

Numerous fiber quality parameters are accessible from routine and standard fiber testing procedures, including HVI measurement. This investigation was carried out to relate fiber strength with yarn tenacity on four cotton cultivars grown in three locations and two crop years. Instead of developing linear regression models to predict yarn tenacity with the consideration of multiple fiber property parameters, this study utilized simple micronaire corrected strengths to compare the relationship between fiber strength and yarn tenacity. The results indicate that, reasonably, the relations between two types of qualities were influenced by cotton cultivars. For example, DP 393, Phytogen 72, and FM 958 fibers showed a larger slope and r (> 0.80) than the UA 48 fibers, indicating that micronaire corrected yarn tenacities could be strongly related with micronaire corrected fiber HVI strength for these three cultivars. The observation suggests the feasibility of applying fiber micronaire and strength properties, as a semi-quantitative and fast tool, to compare cultivar performance for yarn tenacity.

Acknowledgments

The authors would like to acknowledge all technical supporters at both ARS locations (New Orleans, LA and Florence, SC) for their diligent work in field, mill, and laboratory operations.

Mention of a product or specific equipment does not constitute a guarantee or warranty by the U.S. Department of Agriculture and does not imply its approval to the exclusion of other products that may also be suitable.

References

American Society for Testing and Materials (ASTM). (2012). *Standard test method for breaking strength of yarn in skein form* (D1578). West Conshohocken, PA: ASTM International.

Bradow, J. & Davidonis, G. (2000). Quantitation of fiber quality and the cotton production-processing interface: a physiologist's perspective. *J Cotton Sci.,* 4, 34–64.

Cai, Y., Cui, X., Rodgers III, J. E., Thibodeaux, D. P., Martin, V., Watson, M., & Pang, S. (2013). A comparative study on cotton fiber length parameters' effects on modeling yarn property. *Text Res J.,* 83, 961–970. http://dx.doi.org/10.1177/0040517512468821.

Faulkner, W. B., Hequet, E. F., Wanjura, J., & Boman, R. (2012). Relationships of cotton fiber properties to ring-spun yarn quality on selected high plains cottons. *Text Res. J.,* 82, 400–414. http://dx.doi.org/10.1177/0040517511426613.

Frydrych, I. (1992). A new approach for predicting strength properties of yarns. *Text Res J.,* 62, 340–348. http://dx.doi.org/10.1177/004051759206200606.

Kelly, B., & Hequet, E. F. (2013). Breeding for improved yarn quality: importance of non-HVI fiber properties. *ICAC Recorder, 31*, 13–20.

Kelly, C. M., Hequet, E. F, & Dever, J. K. (2013). Breeding for improved yarn quality: Modifying fiber length distribution. *Industrial Crops and Products, 42*, 386– 396. http://dx.doi.org/10.1016/j.indcrop.2012.06.018.

Liu, Y., Gamble, G., & Thibodeaux, D. (2011). Potential of near infrared spectroscopy in the prediction of cotton fiber strength indices. *Applied Engineering in Agriculture, 27*, 839-843. http://dx.doi.org/10.13031/2013.39563.

Long, R. L., Bange, M. P., Delhom, C. D., Church, J. A., & Constable, G. A. (2013). An assessment of alternative cotton fibre quality attributes and their relationship with yarn strength. *Crop and Pasture Science, 64*, 750-762. http://dx.doi.org/10.1071/CP12382.

Lord, E. (1956). Air flow through plugs of textile fibers part II. The micronaire test for cotton. *J. Text Inst., 47*, T16-T47.

Pan, N., Hua, T., & Qiu, Y. (2001). Relationship between fiber and yarn strength. *Text Res J., 71*, 960-964. http://dx.doi.org/10.1177/004051750107101105.

Ramey, H. H., Jr., Lawson, R., & Worley, S., Jr. (1977). Relationship of cotton fiber properties to yarn tenacity. *Text Res J., 47*, 685-691. http://dx.doi.org/10.1177/004051757704701008.

Thibodeaux, D., Senter, H., Knowlton, J. L., McAlister, D., & Cui, X. (2008). The impact of short fiber content on the quality of cotton ring spun yarn. *J Cotton Sci., 12*, 368–377.

Üreyen, M. E., & Kadoglu, H. (2006). Regressional estimation of ring cotton yarn properties from HVI fiber properties. *Text Res J., 76*, 360–366. http://dx.doi.org/10.1177/0040517506062262.

Wakelyn, P. J., & Chaudhry, M. R. (Eds). (2010). *Cotton: Technology for the 21st Century*. Washington, D.C., USA: International Cotton Advisory Committee.

6

Preparation of Activated Carbon from Desiccated Coconut Residue by Chemical Activation with NaOH

Mohd Adib Yahya[1], C.W. Zanariah C. W. Ngah[1] M. A. Hashim[1] & Z. Al-Qodah[2]

[1] Faculty of Science and Technology, Universiti Sains Islam Malaysia (USIM), Malaysia

[2]Department of Chemical Engineering, Taibah University, Saudi Arabia

Correspondence: C. W. Zanariah C. W. Ngah, Faculty of Science and Technology, Universiti Sains Islam Malaysia (USIM), Malaysia. E-mail: cw.zanariah@usim.edu.my

Abstract

This study investigates the effect of temperature and impregnation ratio on the physicochemical properties of activated carbon prepared from agricultural waste; desiccated coconut residue (DCR) by chemical activation using sodium hydroxide (NaOH). DCR sample was first carbonized at three different temperatures for 1 hour at 400°C, 500°C, and 600°C respectively. The resulting chars were impregnated with NaOH at three different impregnation ratio; 1:1, 1:2, and 1:3 respectively and activated under nitrogen atmosphere for 1 hour at three different temperatures based on its carbonization temperature. The specific surface area was strongly affected by impregnation ratio in which increased with impregnation ratio. The specific surface area also increased with temperature but then decreased at highest desired temperature.

Keywords: desiccated coconut residue, activated carbon, sodium hydroxide

1. Introduction

Activated carbon is a kind of amorphous carbonaceous material with high porosity and surface area. The production of activated carbon can be divided into two types: physical activation and chemical activation. Physical activation is a two step process which involves carbonization and followed by gasification by an oxidizing agent (typically CO_2 and steam). Meanwhile, chemical activation refers to a single step process in which the carbonization and activation process occurs simultaneously. The most common chemical agents used are $ZnCl_2$, H_3PO_4, KOH and NaOH (ElShafei et al., 2014). The huge cost of the production of activated carbon has been the most challenging parts for commercial manufacturers. Therefore, the use of cheap raw materials with high contents of carbon as well as low inorganic compounds has been a focus of research efforts in the last decades (Yahya et al., 2015). Agricultural by-products and waste materials such as coconut shell (Olafadehan et al., 2012), wood (Choy et al., 2005), and rice husk (Yalcin & Sevinc, 2000) are among the low-cost precursors for activated carbon production. The key criteria in selecting a carbonaceous material as activated carbon precursor includes low inorganic content, low cost, workability of the material, minimum impact upon environmental, and its storage life (Choy et al., 2005). A good precursor for activated carbon should also have high carbon content (Hiremath et al., 2012). Waste materials can be divided into two categories; conventional wastes and non-conventional waste. Conventional wastes refer to waste from agricultural and wood industry, such as coconuts and olives whereas, non-conventional waste refer to waste from municipal and industrial activities such as polyethyleneterephthalate (PET) and polyvinyl chloride (PVC) (Dias et al., 2007). With this respect, our precursor known as desiccated coconut residue (DCR); a by-product of coconut industry that obtained from the extraction of milk is categorized as conventional wastes which fulfill the criteria desirable for activated carbon precursor. It is believed that the use of chemical activation in activated carbon preparation has been more frequent for conventional waste (Dias et al., 2007).

DCR was selected in this study to prepare activated carbon as it is inexpensive, even no cost required to obtain the material. This conversion of DCR into activated carbon would increase its economic value, reducing the disposal cost of the material and provide cheap material for activated carbon preparation (Prahas et al., 2008). Preparation of activated carbon by alkaline hydroxide activation basically involve two steps; carbonization of the precursor under an inert atmosphere followed by activation of the char impregnated with activating agents (Xu et al., 2010). Consequently, in this study the raw DCR was first carbonized into the char prior to activation with NaOH for activated carbon productions. It suggests that the use of NaOH is very effective for activation of the char of the materials (Prahas et al., 2008).

Activation mechanisms for NaOH comprise the reaction of carbon material with the hydroxides. Carbon molecules depletion from char network gives rise to development of porosity. The activation mechanisms are commonly complex. The reaction mechanisms have been proposed as follows (Hu et al., 2013):

$$6MOH + 2C \rightarrow 2M_2CO_3 + 2M + 3H_2 \quad (1)$$

$$2MOH \rightarrow M_2O + H_2O \quad (2)$$

$$C + H_2O \rightarrow CO + H_2 \quad (3)$$

$$M_2CO_3 + 2C \rightarrow 2M + 3CO \quad (4)$$

$$M_2CO_3 \rightarrow M_2O + CO_2 \quad (5)$$

$$M_2CO_3 + C \rightarrow M_2O + 2CO \quad (6)$$

$$2M + CO_2 \rightarrow M_2O + CO \quad (7)$$

$$M_2O + C \rightarrow 2M + CO \quad (8)$$

Where, $M = Na^+$

Our previous study (Yahya et al., 2015b) has shown that DCR was successfully utilized for activated carbon preparation using potassium hydroxide (KOH) as an activating agent. In this study we used sodium hydroxide (NaOH) as an activating agent instead of KOH. It is hoped that the research would provide a cheap material for the preparation of activated carbon.

2. Experimental

2.1 Activated Carbon Preparation

The preparation of the activated carbon was done following our previous study (Yahya et al., 2015b) except for NaOH. The DCR was collected from local market in Nilai and was dried for 24 hours and then ground to powder with kitchen grinder to attain particle size of 1 to 2 mm. DCR was first carbonized for 1 hour at three different temperatures (400°C, 500°C, and 600°C) under N_2 flow (300 ml min^{-1}) and was cooled to room temperature. The DCR-char was then impregnated with NaOH (Merck, Germany) at three different ratios (1:1, 1:2, and 1:3) at room temperature for 1 hour duration. The impregnated char was then dried in an oven overnight at 105°C, followed by heating for 1 hour at three different temperatures which based on its char production temperature, respectively. The resulting activated carbon was then washed with 1 M HCl and followed by plenty of deionized water to remove any residual organic and mineral matter. The DCR-AC was then dried at 110°C overnight, before being cooled and stored for further studies. The samples were abbreviated as AC 1:400, AC 2:400, AC 3:400, AC 1:500, AC 2:500, AC 3:500, AC 1:600, AC 2:600, and AC 3:600 (1, 2, 3 indicated ratio and 400, 500 and 600 indicated desired temperature).

2.2 Proximate and Ultimate Analysis

Proximate analysis was done by using a thermogravimetric analyser (Mettler Toledo TGA-DSC STAR^e system). The ultimate analysis was done to determine the existence of the carbon, hydrogen, nitrogen, and oxygen was determined by CHNS analyzer (Thermo Finnigan, USA).

2.3 Bulk Density Determination

The bulk density was determined following Al-Qodah & Shawabkah (2009). The bulk density was calculated on the dry basis as in Equation (9):

$$\text{Bulk density} = \text{weight of sample (g)/volume of sample (cm}^3\text{)} \quad (9)$$

2.4 Physical Properties of DCR-AC

The morphology of the resulting activated carbon was studied using a Scanning Electron Microscope (Phenom, G2 pro) operated at 15 kV. The magnifications were selected at x 1000.

2.5 Nitrogen Adsorption Measurements

The specific surface area (SSA) of the DCR-AC, micropore volume and mesopore volume was studied by N_2 adsorption at -196°C (77 K), using an automatic volumetric sorption analyzer (Quantachrome, Autosorb 2). The instrument was equipped with Dewar flasks to hold the coolant (liquid nitrogen), analysis stations to hold the sample cell during analysis, sample cells with glass filler rods to put the sample inside the glass, cold trap to prevent the diffusion of sample degassing products into the manifold and heating mantles to outgas the sample.

DCR-AC was first degasified at 350°C for a certain period (more than 5h) prior to analysis. In the analysis period, the adsorbate was introduced to the manifold. The valve of the sample cell was opened to allow the adsorbate to interact with DCR-AC samples. The pressure was repeatedly measured until a stable reading was achieved.

The Brunauer, Emmett and Teller (BET) method was used to measure the SSA of DCR-AC. The equation of BET equilibrium adsorption was based on the assumption of multi-layer adsorption, as following Equation (10) (Devnarain et al., 2002):

$$\frac{1}{W[(\frac{P_o}{P})-1]} = \frac{1}{CW_m} + \frac{C-1}{CW_m}\left(\frac{P}{P_o}\right) \quad (10)$$

Where, W represents weight of nitrogen adsorbed at a relative pressure, P/P_o, W_m is the weight of adsorbate that constitutes a monolayer of surface coverage, and C is a constant related to adsorption energy in the first adsorbed layer in which its value indicates the magnitude of the adsorbent/adsorbate interactions.

The volume adsorbed was based on the Equation (11):

$$\frac{V}{V_m} = \frac{C(P/P_O)}{[1-P/P_O][1+(C-1)(P/P_O]} \quad (11)$$

The quantity of N_2 gas adsorbed in monolayer was determined by plotting P/P_o vs $1/V[P/P_o)-1]$. The SSA was determined by following Equation (12):

$$S = \frac{[\sigma N_A]\rho V_m}{MW} \quad (12)$$

Where, σ denotes the adsorbate molecule area, ρ is the liquid density of adsorbate, N_A is the Avogadro's number (6.022×10^{23} number/mole) and MW signifies molecular weight of adsorbate molecule (nitrogen). The values of BET surface area (S_{BET}) (also known as SSA) were calculated from the adsorption isotherms at the range of $0.05 \leq P/P_o \leq 0.30$ (Okada et al., 2003). Micropore volume was obtained by *t*-plot method. Total pore volume was obtained at 0.99 of relative pressure (Hu et al., 2001).

3. Results and Discussion

3.1 Proximate and Ultimate Analysis

The proximate analysis of raw DCR has been reported by our previous study (Yahya et al., 2015b) which showed that the contents of the moisture, volatile matter, ash, and fixed carbon were 3.0 %, 84.20 %, 5.51 %, and 7.29 %, respectively. The high value of volatile matter and low value of ash content in DCR indicated that it was suitable for the biomass to be converted into activated carbon. The ultimate analysis of the DCR and DCR-AC was shown in Table 1. The carbon contents of the activated carbon were generally higher than that of raw materials. It was observed that the carbon contents were decreased with impregnation ratio for all types of temperature which may be attributed to the carbon monoxide that was produced during activation (Choy et al., 2005).

Table 1. Ultimate analysis of DCR and DCR-AC

Ultimate analysis (%)	DCR	AC 1:400	AC 2:400	AC 3:400	AC 1:500	AC 2:500	AC 3:500	AC 1:600	AC 2:600	AC 3:600
Carbon	49.48	63.24	53.07	46.00	74.93	65.80	53.34	75.25	72.55	65.70
Hydrogen	12.80	1.83	2.11	1.82	1.70	1.83	3.35	1.21	1.64	2.70
Nitrogen	0.19	3.23	2.48	2.42	3.19	2.12	0.06	2.24	2.28	0.20
Oxygen*	37.53	31.70	42.34	50.24	20.18	30.25	43.25	21.30	23.83	31.40

* Oxygen was calculated by difference.

3.2 Yield (%) of DCR-AC

Table 2 shows yield percentage of the resulting DCR-AC. It was observed that the yield of the DCR-AC in the range 11.20 to 14.63 (%). The low yield of the activated carbon was expected due to low lignin contents in the raw DCR hence resulted in low yield (Yahya et al., 2015b).

3.3 Bulk Density

Bulk density is one of the important characteristics of AC as it is a measure of the amount of the adsorbate that the carbon can hold per unit volume (Abechi et al., 2013). Table 2 shows the bulk density of the DCR-AC. Bulk density of raw DCR was found to be 0.24 g/cm^3. It was found that the bulk density of the DCR-AC was found between 0.16 and 0.21 g/cm^3. The lower bulk densities of DCR-AC compared to the raw DCR were attributed to the porosity development during activation process (Lima & Marshall, 2005).

3.4 Surface Area, Pore Size Distribution and Porosity of DCR-AC

Figure 1 shows the N_2 adsorption isotherms of DCR-AC prepared. All isotherms adsorption exhibited Type I isotherms with horizontal plateau at higher relative pressures. The isotherms showed that all DCR-AC indicated highly microporous materials with a narrow pore size distribution except for AC 1:400, AC 2:400 and AC 3:400. Some of the isotherms indicated an increasing slope at values of $P/P^o > 0.9$. This shows there was a development of the wider pores (Demiral et al., 2008). From Table 2, the SSA obtained from N_2 adsorption ranged from 27.15 to 1394.79 m^2/g. The total pore volume was between 0.05 to 0.78 cm^3/g. The SSA was in the order: AC3:500 > AC3:600 > AC2:500 > AC2:600 > AC1:500 > AC1:600 > AC2: 400 > AC3:400 > AC1:400. The total pore volumes were ranged from 0.05 to 0.78 cm^3/g. The SSA values corresponding to all samples were plotted in Figure 2.

Table 2. Physicochemical properties of DCR-AC

Property	AC 1:400	AC 2:400	AC 3:400	AC 1:500	AC 2:500	AC 3:500	AC 1:600	AC 2:600	AC 3:600
Yield (%)	13.05	15.28	14.54	14.63	12.31	11.20	12.48	11.60	11.52
Bulk density (g/cm^3)	0.17	0.20	0.16	0.17	0.19	0.18	0.20	0.21	0.20
S_{BET} (m^2/g)	27.15	112.94	65.53	244.20	702.98	1394.79	189.68	563.71	1199.73
V_{mi} (cm^3/g)	0.00	0.03	0.02	0.09	0.31	0.60	0.08	0.26	0.52
V_{me} (cm^3/g)	0.05	0.10	0.04	0.05	0.10	0.18	0.04	0.06	0.16
V_t (cm^3/g)	0.05	0.13	0.06	0.14	0.41	0.78	0.12	0.32	0.68
Porosity (%)	0.00	23.08	33.33	64.29	75.60	76.92	66.67	81.25	76.47

S_{BET} = BET surface area, V_{mic} = micopore volume, V_{mes} = mesopore volume, V_t = total pore volume.

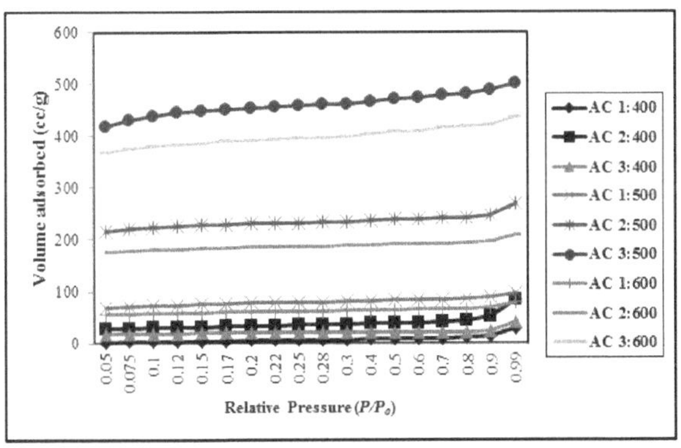

Figure 1. N_2 adsorption isotherms of DCR-AC

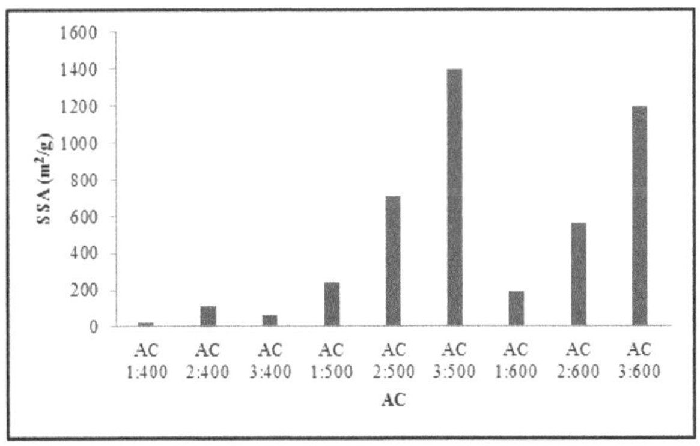

Figure 2. SSA corresponding to all activated carbon prepared

3.5 Effect of Temperature

Figure 3 shows the effects of temperature on the SSA of DCR-AC. It was observed that the SSA for DCR-AC was significantly increased with the increased of temperature from 400 °C to 500 °C and then decreased at 600 °C for impregnation ratio from 1 to 3. It was reported that the initial increased of the SSA was attributed to the increased in the reaction between DCR-char with the chemical agent (NaOH) during activation process hence developing new pores. However, the increased of temperature from 500°C to 600°C caused widening of the micropores thus the SSA to be decreased. Xu et al. (2010) reported this situation was due to mechanism that involved in NaOH activation. Some of active sites in the char were selectively etched by NaOH at high temperature to generate porous structure. During carbonization, the carbon contents in the char were increased and the porous structure was built by eliminating non-carbon atoms. The crystal structure in the char was rearranged resulted in well-ordered structure. These rearrangements obstructed the subsequent activation reactions of the char with NaOH to generate porous structure. Therefore, the higher the temperature resulted in more ordered of the crystal structure of the char and became more difficult to be etched by NaOH. This caused the SSA and pore volume of NAC to decrease with increasing temperature. Results also showed that at temperature of 400°C, NAC had shown very low SSA. This finding was in agreement with a study on preparation of AC from rice husk by Isoda et al. (2014) in which the author also reported that this phenomenon was attributed to the mechanism involved in NaOH activation. NaOH was first reduced to Na^+ and introduced into the structure of the char and expanded it. With the increased in temperature, Na^+ was moved forcefully separating the grapheme type-layers hence creating the porosity. The reaction was severely a function of temperature, therefore at low temperature (400°C) it was not effective for the formation of AC.

3.6 Effect of Impregnation Ratio

Figure 4 shows the effects of impregnation ratio on the SSA of the DCR-AC. It was observed that the SSA was significantly increased with impregnation ratio from 1:1 to 1:3 except for all temperatures except for 400°C. The pore volume was also increased with the increased of the impregnation ratio. It has been reported that the particle size of the resulting activated carbon could change depending on the impregnation ratio, hence increased the SSA and pore volume of the DCR-AC. In this study, activating agent (NaOH) may react as templates for porosity development in activated carbon (Molino-Sabio & F. Rodriguez-Reinoso, 2004). Our previous study using KOH had shown quite different findings in which SSA was increased from impregnation ratio of 1:1 to 1:2 and finally decreased at impregnation ratio of 1:3. We have reported that this situation was attributed to the initial template reaction by KOH and widening of micropores that caused the increased and decreased of SSA, respectively. These differences indicate that different activating agents could promote different mechanisms which could probably attributed to the ionic radii differences and concentration of both types of activating agents.

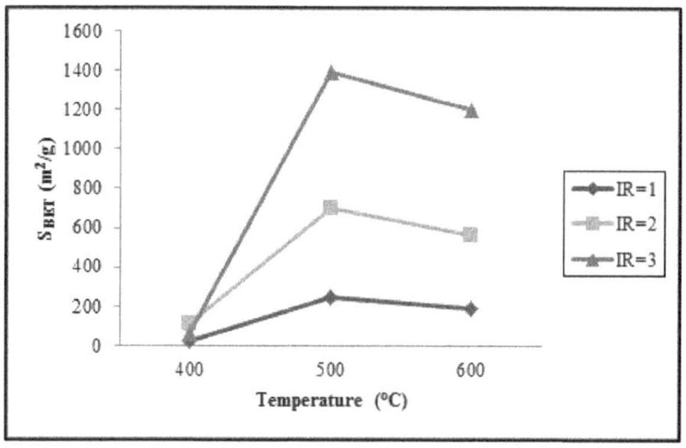

Figure 3. SSA value as a function of temperature

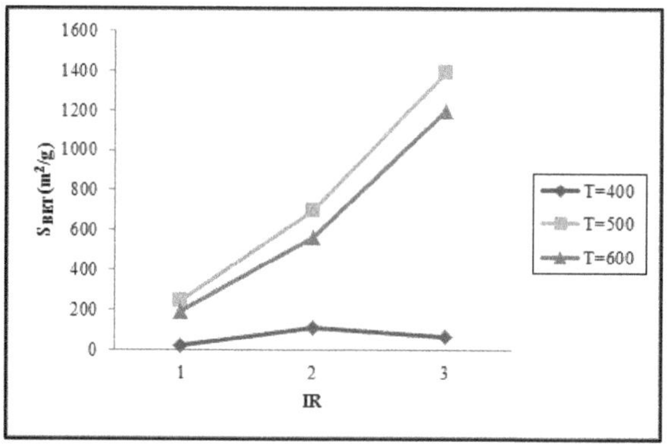

Figure 4. S_{BET} value as a function of impregnation ratio

3.7 Effect of Temperature on Porosity

Figure 5 shows the effect of temperature on the porosity of the DCR-AC. In general, porosity of the DCR-AC was increased from 400°C to 600°C for all types of IR. Porosity is defined as the ratio of micropore volume to total per volume (Boonpoke et al., 2011). Temperature has been identified as a determinant factor in the structure of the activated carbon. Carbonization of the starting material (DCR) which resulted in char production basically had low porosity and disordered structure. Subsequently, activation would further develop the porosity. Carbonization removed the volatile matter in the DCR, whereas activation with chemical agents (NaOH) produced a well developed porosity of DCR-AC (Ketcha et al., 2012). As temperature increased, the diameter of the sample also decreased and resulted in the development of the micropore volume (Mohamed et al., 2010).

3.8 SEM Morphology of DCR-AC

Figure 6 shows the SEM morphology of the selected DCR-AC for all temperatures studied. It was observed that DCR-AC had a wide variety of pores with honey-comb look like structure. There were cavities and regular at the external surface of the DCR-AC. Results also showed that NaOH played an important role in porosity developments. It was also found that all selected DCR-AC had homogenous types of pore structure distribution. In summary, the activation process with NaOH was effective in creating well developed pores on the DCR-AC surfaces.

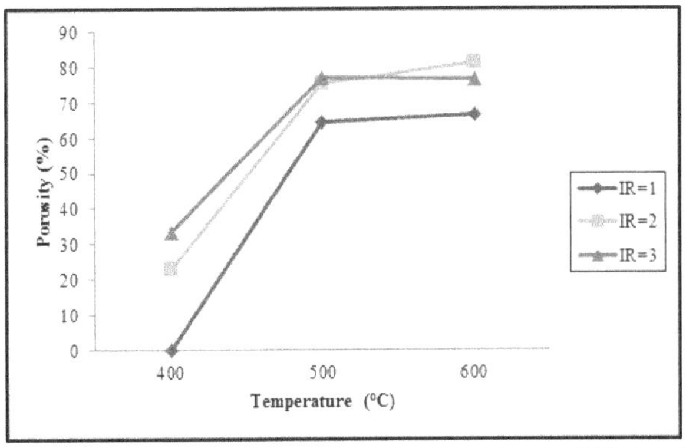

Figure 5. Effect of temperature on particle porosity at different IR value

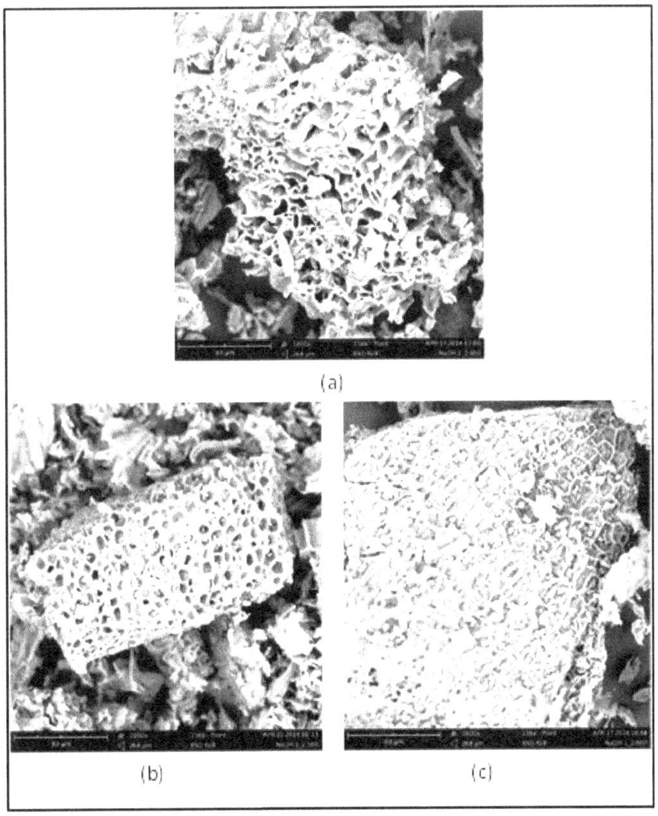

Figure 6. SEM micrographs of DCR-AC, (a) A2:400, (b) A2:500, (c) A2:600

3.9 Micropore Size Distribution

The evolution of the micropore size distributions (MPSD) was observed by plotting the normalized nitrogen adsorption isotherms. The curves were acquired by dividing the amount of N_2 adsorbed at a given relative pressure by the amount of N_2 adsorbed at a relative pressure near to saturation pressure (0.99) (Lozano-Castello et al., 2003). Figure 7 shows the plot of normalized N_2 adsorption for all DCR-AC. It was observed that the narrowest MPSD for NAC correspond to AC 3:500, AC 3:600 and AC 2:500. These results generally were in agreement with results found in Table 2 associated with the micropore volumes. Figure 8 shows the effects of micropore contributions for DCR-AC as a function of impregnation ratio and temperature studied. These micropore contributions were calculated by dividing the amount of N_2 adsorbed at relative pressure of 0.1 to amount of N_2 adsorbed at relative pressure of 0.95 (Ahmadpour & Do 1997). Results showed that at temperatures of 500°C and 600°C for all types of impregnation ratio, the microporosity contributions had higher values as compared to temperature of 400°C. As previously mentioned in this study it showed that temperature has been important factors that could determine microporosity development in DCR-AC.

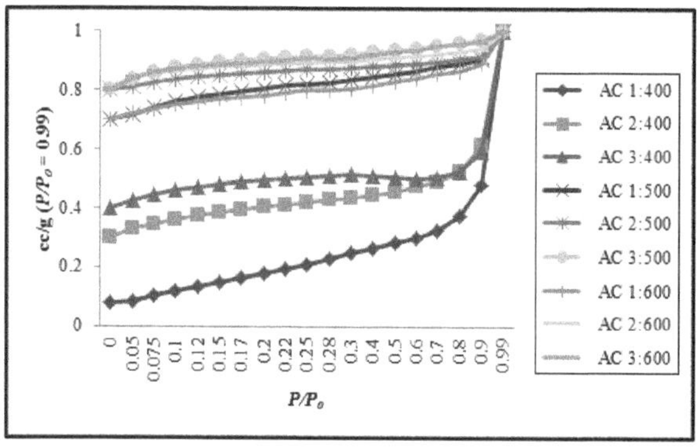

Figure 7. Normalized N_2 adsorption isotherms at 77 K for NAC

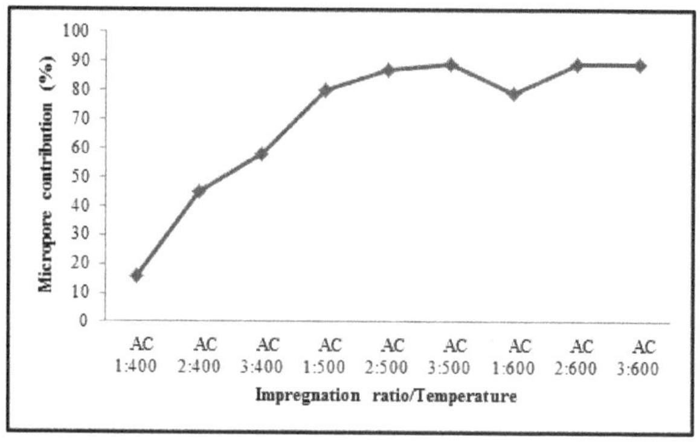

Figure 8. Effects of impregnation ratio and temperature on the relative micropore contribution of DCR-AC

4. Conclusion

A low cost activated carbon from waste material can be converted into activated carbon which a value added product. Chemical activation by using sodium hydroxide produces activated carbon with high porosity.

Acknowledgements

The authors are grateful to University Sains Islam Malaysia (USIM) and Ministry of Higher Education (MyBrain15) for findings of this project.

References

Abechi, S. E., Gimba, C. E, Uzairu, A., & Dallatu, Y. A. (2013). Preparation and characterization of activated carbon from Palm Kernel shell by chemical activation. *Research Journal of Chemical Sciences, 3*(7), 54-61.

Ahmadpour, A., & Do, D. D. (1997). The preparation of activated carbon from Macadamia nutshell by chemical activation. *Carbon, 35*(12), 1723-1732. http://dx.doi.org/10.1016/S0008-6223(97)00127-9

Al-Qodah, Z., & Shawabkah, R. (2009). Production and characterization of granular activated carbon from activated sludge. *Brazilian Journal of Chemical Engineering, 26*(1), 127-136. http://dx.doi.org/10.1590/S0104-66322009000100012

Boonpoke, A., Chiarakorn, S., Laosiripojana, N., Towprayoon, S., & Chidthaisong, A. (2011). Synthesis of activated carbon and MCM-41 from Bagasse and Rice Husk and their carbon dioxide adsorption capacity. *Journal of Sustainable Energy and Environment, 2*, 77-81.

Choy, K. K. H., Barford, J. P., & McKay, G. (2005). Production of activated carbon from bamboo scaffolding waste-process design, evaluation and sensitivity analysis. *Chemical Engineering Journal, 109*, 147-165. http://dx.doi.org/10.1016/j.cej.2005.02.030

Demiral, H., Demiral, I., Tumsek, F., & Karabacakoglu, B. (2008). Pore structure of activated carbon prepared from hazelnut bagasse by chemical activation. *Surface and Interface Analysis, 40*, 616-619. http://dx.doi.org/10.1002/sia.2631

Devnarain, P. B., Arnold, D. R., & Davis, S. B. (2002). Production of activated carbon from South African sugarcane bagasse. *Proceedings of the South African Sugar Technologists' Association, 76*, 477-489.

Dias, J. M., Alvim-Ferraza, M. C. M., Almeida, M. F., Rivera-Utrilla, J., & Sanchez-Polo, M. (2007). Waste materials for activated carbon preparation and its use in aqueous-phase treatment: A review. *Journal of Environmental Management, 85*, 833-846. http://dx.doi.org/10.1016/j.jenvman.2007.07.031

ElShafei, G. M. S., ElSherbiny, I. M. A., Darwish, A. S., & Philip, C. S. (2014). Silkworms' feces-based activated carbons as cheap adsorbents for removal of cadmium and methylene blue from aqueous solution. *Chemical Engineering Research and Design, 92*, 461-470. http://dx.doi.org/10.1016/j.cherd.2013.09.004

Hiremath, M. N., Shivayogimath, C. B., & Shivalingappa, S. N. (2012). Preparation and characterization of granular activated carbon from corn cob by KOH activation. *International Journal of Research in Chemistry and Environment, 2*(3), 84-87.

Hu, Z., Srinivasan, M. P., & Nia, Y. (2001). Novel activation process for preparing highly microporous and mesoporous activated carbons. *Carbon, 39*, 877-886. http://dx.doi.org/10.1016/S0008-6223(00)00198-6

Hu, Y., Wang, H., Yang, L., Liu, X., Zhang, B., Liu, Y., … Fu, H. (2013). Preparation of chitosan-based activated carbon and its electrochemical performance for EDLC. *Journal of The Electrochemical Society, 160*(6), H321-H326. http://dx.doi.org/10.1149/2.062306jes

Isoda, N., Rodrigues, R., Silva, A., Goncalves, M., Mandelli, D., Figueiredo, F. C. A., & Carvalho, W. A. (2014). Optimization of preparation conditions of activated carbon from agricultural waste utilizing factorial design. *Powder Technology, 256*, 175-181. http://dx.doi.org/10.1016/j.powtec.2014.02.029

Ketcha, J. M., Dina, D. J. D., Ngomo, H. M., & Ndi, N. J. (2012). Preparation and characterization of activated carbons obtained from maize cobs by zinc chloride activation. *American Chemical Science Journal, 2*(4), 136-16. http://dx.doi.org/10.9734/ACSJ/2012/1806

Lima, I. M., & Marshall, W. E. (2005). Granular activated carbons from broiler manure: physical, chemical and adsorptive properties. *Bioresource Technology, 96*, 699-706. http://dx.doi.org/10.1016/j.biortech.2004.06.021

Lozano-Castello, D., Cazorla-Amoros, D., Linares-Solano, A., Shiraishi, S., Kuriara, H., & Oya, A. (2003). Influence of pore structure and surface chemistry on electric double layer capacitance in non-aqueous electrolyte. *Carbon, 41*, 1765-1775. http://dx.doi.org/10.1016/S0008-6223(03)00141-6

Mohamed, A. R., Mohammadi, M., & Darzi, G. N. (2010). Preparation of carbon molecular sieve from lignocellulosic biomass: A review. *Renewable and Sustainable Energy Reviews, 14*, 1591–1599. http://dx.doi.org/10.1016/j.rser.2010.01.024

Molino-Sabio, M., & Rodriguez-Reinoso, F. (2004). Role of chemical activation in the development of carbon porosity. *Colloids and Surfaces A: Physicochemistry and Engineering Aspects, 241*, 15-25. http://dx.doi.org/10.1016/j.colsurfa.2004.04.007

Ng, S. P., Tan, C. P., Lai, O. M., Long, K., & Mirhosseini, H. (2010). Extraction and characterization of dietary fiber from coconut residue. *Journal of Food, Agriculture and Environment, 8*(2), 172-177.

Okada, K., Yamamoto, N., Kameshima, Y., & Yasumori, A. (2003). Porous properties of activated carbons from waste newspaper prepared by chemical and physical activation. *Journal of Colloid and Interface Science, 262*, 179-193. http://dx.doi.org/10.1016/S0021-9797(03)00107-3

Olafadehan, O. A., Jinadu, O. W., Jinadu, S. L., & Popoola, O. T. (2012). Treatment of Brewery wastewater effluent using activated carbon prepared from coconut shell. *International Journal of Applied Science and Technology, 2*(1), 165-178.

Prahas, D., Kartika, Y., Indraswati, N., & Ismadji, S. (2008). Activated carbon from jackfruit peel waste by H_3PO_4 chemical activation: Pore structure and surface chemistry characterization. *Chemical Engineering Journal, 140*, 32-42. http://dx.doi.org/10.1016/j.cej.2007.08.032

Xu, B., Chen, Y., Wei, G., Cao, G., Zhang, H., & Yang, Y. (2010). Activated carbon with high capacitance prepared by NaOH activation for supercapacitors. *Materials Chemistry and Physics, 124*, 504-509. http://dx.doi.org/10.1016/j.matchemphys.2010.07.002

Yahya, M. A., Al-Qodah, Z., & Ngah, C. W. Z. (2015). Agricultural bio-waste materials as potential sustainable precursors used for activated carbon production. A review. *Renewable and Sustainable Energy Reviews, 46*, 218-235. http://dx.doi.org/10.1016/j.rser.2015.02.051

Yahya, M. A., Al-Qodah, Z., Ngah, C. W. Z. C. W., & Hashim, M. A. (2015b). Preparation and characterization of activated carbon from desiccated coconut residue by potassium hydroxide. *Asian Journal of Chemistry, 27*(6), 2331-2336. http://dx.doi.org/10.14233/ajchem.2015.18804

Yalcin, N., & Sevinc, V. (2000). Studies of the surface area and porosity of activated carbons prepared from rice husks. *Carbon, 38*, 1943-1945. http://dx.doi.org/10.1016/S0008-6223(00)00029-4

7

Freezing Thawing Resistance of Concrete Incorporating Glass Waste

Raed Abendeh[1], Khaled Ramadan[2], Rami Haddad[3] & Mousa Bani Baker[1]

[1] Al-Zaytoonah university of Jordan, Amman, Jordan

[2] Applied Science University, Amman, Jordan

[3] Jordan University of science and Technology, Irbid, Jordan

Correspondence: Raed Abendeh, Al-Zaytoonah university of Jordan, Amman, Jordan. E-mail: r.abendeh@zuj.edu.jo

Abstract

Glass waste, which is dumped around the world in huge amounts, can be used as a partial replacement of mineral aggregate in concrete industry. This would not only contribute to reducing pollution extent but also enhancing physical properties and durability of concrete. This article reports the mechanical performance of concrete with different replacement percentages of crushed glass waste of cement or fine aggregates before and after being subjected to standard cycles of freezing and thawing. Mechanical performance was evaluated in terms of compressive and flexural strengths. Furthermore, internal damage extent was evaluated using ultrasonic pulse velocity and dynamic modulus of elasticity. The results (average of six specimens for each test) revealed the feasibility of incorporating glass waste in concrete mixtures for the purpose of improving strength and durability; especially in environments where concrete is being exposed to an effective freezing and thawing cycles.

Keywords: glass waste, freeze-thaw cycles, non-destructive tests

1. Introduction

Glass waste is one among many waste materials used in concrete industry. Since glass is a non-organic material made by sintering selected raw materials, thus cannot be incinerated or decomposed. Glass recycling can reduce the complete reliance on natural aggregate resources and decrease the generated environmental waste (green solution).

Glass waste can be used in construction industry as a partial replacement of fine aggregate or/and as a partial replacement of Portland cement (glass powder). Pattengill and shutt (1973) were the first who used the grounded powdered glass as pozzolanic material to replace the cement partially (Pattengil & Shutt, 1973).Various research works have tackled the topic of incorporating glass waste in concrete mixtures to establish a ground for its usage in concrete industry (Kim et al., 2014; Bazant & Mayer, 1998; Saikia & de Brito, 2014; Madandoust & Ghavidel, 2013; Tuan, Hwang, Lin, Chen, & Young, 2013; Saribiyik, Piskin,& Saribiyik, 2013; Ling, Poon, & Wong, 2013; Soroushian, 2012; Ismail & Al-Hashmi, 2009; Federico & Chidiac, 2009; Taha & Nounu, 2008; Shi & Zheng, 2007; Topcu & Canbaz, 2004; Metwally, 2007; Schmidt & Saia, 1963; Carpenter & Cramer, 1999; American Psychological Association, 1972). Expansive reactions between amorphous silica (from glass) and alkalis from cement have been noticed to cause disintegration of concrete structures. Bazant and Mayer (1998) found that the glass particle sizes to have a great an impact on glass-alkalis reactivity ; stipulating that using particles of sizes of around 1.5 mm would lead to excessive expansion whereas using sizes less than 0.25 mm would have negligible negative impact on concrete behavior; agreeing with the findings by Carpenter and Cramer (Carpenter & Cramer, 1999).

Freeze-thaw cycles (F/T) may cause deterioration of concrete; especially in concrete structures exposed continuously to water such as water tanks and seashore structures. The accumulative effects of freeze-thaw action throughout successive winter seasonsmay disruption both paste and aggregate leading to their expansion and crackingleading ultimately to scaling and crumbling of the concrete.One of the few studies thattackled durability aspect of glass waste concrete was that of Kim et al (2014) which revealed that utilizing waste glass sludge as a partial replacement of cement can improve the resistance of concrete to (F/T) cyclesin absence orpresence of deicing salts.

A preliminary study has been overtaken to assess the potential of replacingcement or fine aggregate by glass waste in concrete. Different percentages of crushed and powder glass of up to 20% with an increment of 5% were incorporated into concrete. Concrete specimens without and with glass waste replacements were subjected to F/T action of up to 230 cycles using ASTM test method C666 (plan A). Compressive and flexural strengths measurements were acquired before and after exposure to different F/T cycles along with resonance frequency and ultrasonic. Obtained data were analyzed to determine optimal replacement ratios.

2. Materials

Natural crushed fine and coarse limestone aggregates and ordinary Portland cement (OPC) were used in preparing different concrete mixes. Coarse aggregate at a maximum aggregate size of 12.5 mm was used along with fine aggregates having a fineness modulus of 2.8. The physical properties for coarse and fine aggregates and mineral filler were determined according to ASTM specifications: results are summarized in Table 1.

Table 1. The aggregate properties used in the research

Aggregate type (limestone)	ASTM test designation	Bulk specific gravity	Apparent specific gravity	Absorption percentage
Coarse	C 127	2.5	2.52	4.5
Fine	C 128	2.47	2.485	7
Mineral filler	C 128	2.515	2.534	

The waste glass was obtained from crushing and milling bottles and window glass which is referred toas Soda-lime glass type; the gradation of which is shown in Table 2. The sizes of crushed glass used as a partial replacement of fine aggregate were between 4.75 mm and 0.075 mm, whereas, the sizes of those used as a partial replacement of cement wereless than 0.075 mm.

Table 2. Sieve analysis gradations of the crushed waste glass

Sieve size	% passing (mid point)	% retained	% cumulative retained
3/4" (19.5mm)	100	0	0
½"(12.5mm)	95	5	5
3/8" (9.5mm)	88	7	12
# 4 (4.76mm)	51	37	49
# 8 (2.36mm)	31	20	69
0.85 mm	11	20	89
0.425 mm	6	5	94
0.075 mm	2	4	98
pan		2	100

3. Experimental Work and Test Methods

3.1 Mixture Proportions

To achieve the main objective of this research, a total of eight concrete mixtures were prepared to determine the impact of replacing fine aggregates and cement by crushed and milled glass wastes, respectively, on mechanical properties and durability of concrete subjected to freezing-thawing cycles: glass waste replaced fine aggregates at percentages of 5%, 10%, 15% and 20 % and cement at percentages of 5%, 10 %, and 15%. The basic concrete mixture (without waste glass) had a water-to-cement ratio (w/c) of 0.5, and a proportion by weight for fine and coarse aggregates of 1.75 and 3.25 respectively. A superplasticizer was used whenever needed to maintain a slump value above 70 mm.

3.2 Test Methods

3.2.1 Specimens Preparation and Testing

For each type of test and type of mixture, six specimens were prepared. Control mixtures and those with glass replacements were cast in molds consisting of 75-mm cubes and prisms (75 x 75x 300 mm) before de-molded then cured under standard conditions for 28 days. The cubes were designated for determining compressive according to

(BS 1881 Part 108). The prisms, on the other hand, where tested for flexural strength according to ASTM test method C78, resonance frequency according to ASTM test method C215- 14, and ultrasonic pulse velocity according to ASTM test method C 597 -09. These measurements were carried out prior and after exposing concrete to varying cycling of F/T up to 350. The letter F have been used to designate concrete mixtures that incorporated glass as replacements of fine aggregate whereas the letter P designated those prepared with glass waste powder replacements of the cement.

3.2.2 Freezing and Thawing Test

Specimens were subjected to repeated freezing and thawing cycles according to ASTM C666 (procedure A) during a total period of 2.5 hours according to the scheme of schematic of Figure 1.

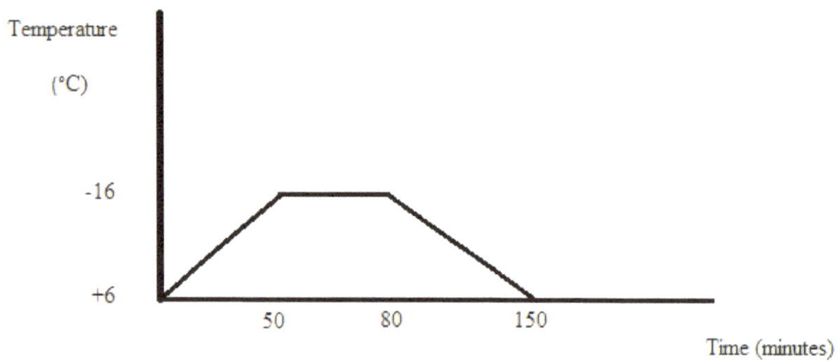

Figure 1. Freezing and thawing repeated cycle used in the experimental work 3.6.2 Adverse Events

4. Results and Discussion

The results and discussed in different subsections. The results (average of six specimens) related to loss in weight of concrete specimens upon F/T cycles are reported first. After that, measurements of compressive and flexural strengths are reported and discussed in sections 4.2 and 4.3 respectively. Finally, damage induced in different concrete specimens was evaluated nondestructively as reported in section 4.4.

4.1 Weight Loss in Concrete Prisms under F/T Effect

Figure 2 shows the weight loss percentage of concrete specimens due to surface scaling by cycles of freezing and thawing. The results showed that the use of glass waste contributed significantly to decreasing surface scaling of concrete, whether that was through replacement of fine aggregate or cement. The decrease in surface scaling was more pronounced in mixtures with cement replacements by glass waste powder; suggesting that some pozzolanic activity between the glass and hydrated cement might have taken place; confirming the findings by Kim et al. (2014).

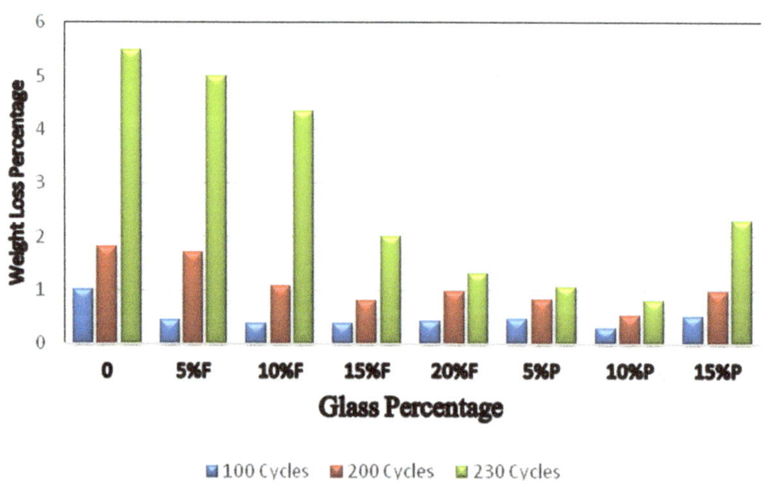

Figure 2. Weight loss percentage of concrete prisms containing different glass percentages at 100, 200 and 230 F/T cycles (results are average of six specimens at each percentage)

4.2 Compressive and Flexural Strengths

Figure 3 also shows compressive strength results of different concrete prisms containing different glass waste percentages and subjected to freezing and thawing cycles up to 230 cycles, as compared to control specimens. The results showed that incorporation of glass waste in concrete had contributed to enhancing compressive strength; owning to possible improvement of pore system by transformation of calcium hydroxide to calcium silica hydrate as a result of pozzolanic nature of the glass waster. Concrete mixtures, incorporated waster glass at percentages greater than 5%, showed better resistance than that of those prepared without glass waste replacement of sand or cement. The residual compressive strength after 230 cycles reached as high as 48% for mixtures with glass waster replacement as compared to about 34% to control mixture. As may be expected, the performance of the concrete mixtures under F/T action was better for specimens with glass waste replacement of fine aggregate as compared to that of those with glass waste replacement of cement.

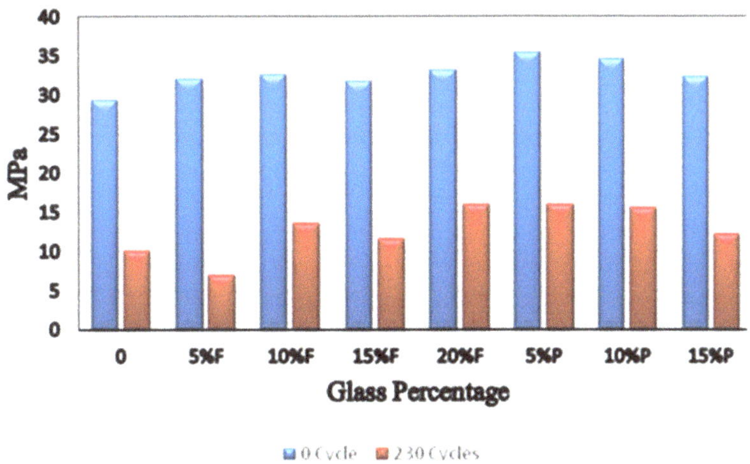

Figure 3. Compressive strength of concrete cubes containing glass waste at different replacement percentages of fine aggregate or cement; before and after exposure to 230 F/T cycles (results are average of six specimens at each percentage)

Figure 4 shows the flexural strength for mixtures containing 0%, 5%, 10%, and 15% of crushed glass waste as a partial replacement of fine aggregate and 5%, 10% and 15% of milled glass as a partial replacement of cement content. It can be seen that the flexural strength of the concrete increased slightly in almost all mixtures containing glass waste as compared to that without replacements. After exposure to 230 F/T cycles, concrete prisms, containing glass waste powder at 5-10% as partial replacement of cement content contributed the most to maintaining flexural strength. This can be related to the pozzolanic characteristics of the powder glass, hence the lower permeability of concrete containing glass waste (Pattengil & Shutt, 1973; Kim et al., 2014; Bazant & Mayer, 1998).

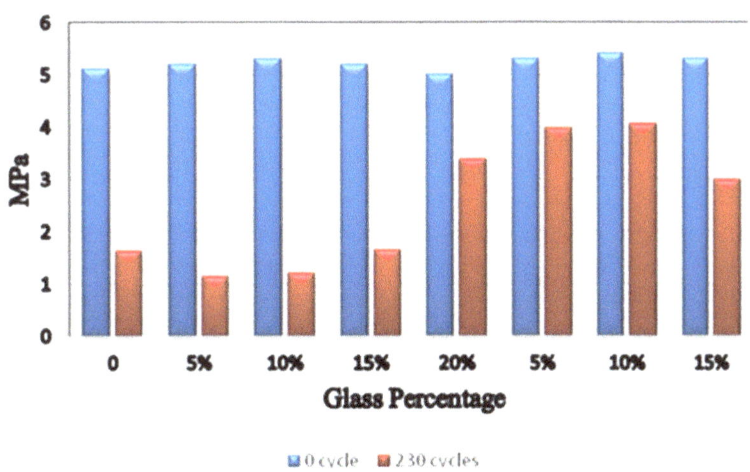

Figure 4. Flexural strength of concrete prisms containing glass waste at different replacement percentages of fine aggregate or cement; before and after exposure to 230 F/T cycles(results are average of six specimensat each percentage)

4.3 Nondestructive Tests Results

The damage induced in various concrete mixtures under different F/T cycles was evaluated using two nondestructive techniques, namely ultrasonic pulse velocity and resonance frequency. Based on these measurements, certain damage indices were computed and used for evaluating the impact of replacing cement or fine aggregate with glass waste in concrete on its resistance to freezing and thawing.

4.3.1 Ultrasonic Velocity Method

The damage index in terms of UPV, was computed using Equation (1) as follows:

$$DI_{UPV} = 1 - \left(\frac{UPV^{dam}}{UPV^{o}} \right)^2 \tag{1}$$

Where, UPV^{o} is the initial ultrasonic pulse velocity and UPV^{dam} is the ultrasonic pulse velocity after certain number of F/T cycles.

Figure 5 depicts damage index in terms of ultrasonic pulse velocity versus number of freezing thawing cycles for concrete prisms, prepared with varying glass waste replacement of cement and fine aggregates. The results showed that exposing concrete to up to 100 F/T cycles had imparted limited cracking, hence damage in various mixtures. However, the damage index in various concrete mixtures showed a drastic increase upon exposure to F/T cycles equal to or greater than 200. It is rational to conclude that mixtures incorporated glass waste as replacement of cement at percentages of 10 and 15% were the least to be damaged by 200 F/T cycles. In general, various prisms were significantly damaged at 230 F/T cycles without positive contribution of glass replacement of either cement or fine aggregates.

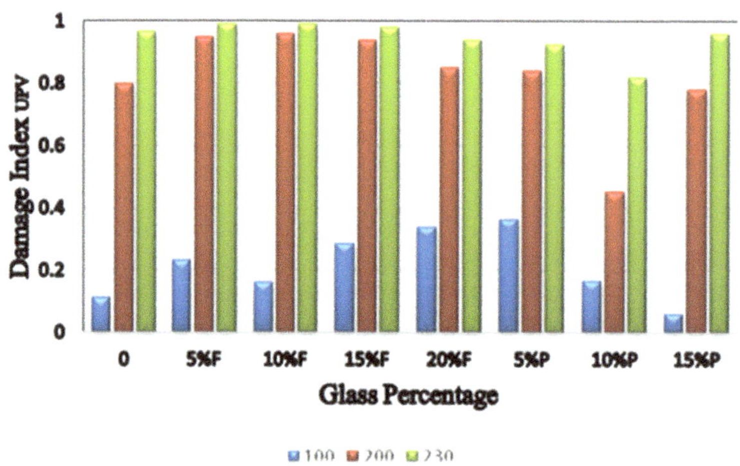

Figure 5. Damage index in terms of UPV versus percentage replacement of glass waster in concrete (results are average of six specimens at each percentage)

4.3.2 Resonance Frequency

As well stipulated, resonance frequency is a nondestructive test that enables computing the dynamic modulus of elasticity for concrete. Accordingly, a damage index in terms of the initial dynamic modulus of elasticity, DI_{DME}, is written as:

$$DI_{DME} = 1 - \frac{(E_d)_{dam}}{(E_d)_0} \tag{2}$$

Where, $(E_d)_o$ are the initial dynamic elastic modulus whereas $(E_d)_{dam}$ is the dynamic elastic modulus for prisms exposed to F/T cycles.

Figure 6 depicts the damage index, as defined by Equation (2), versus glass waste replacement percentage of cement and fine aggregates after prisms exposure to 230 F/T cycles. The least damage index was obtained in concrete mixture containing 10% of glass waste powder as a partial replacement of cement content with mixtures incorporating 5-15% of glass waste as replacement of cement content showing higher damage indices than that of control ones.

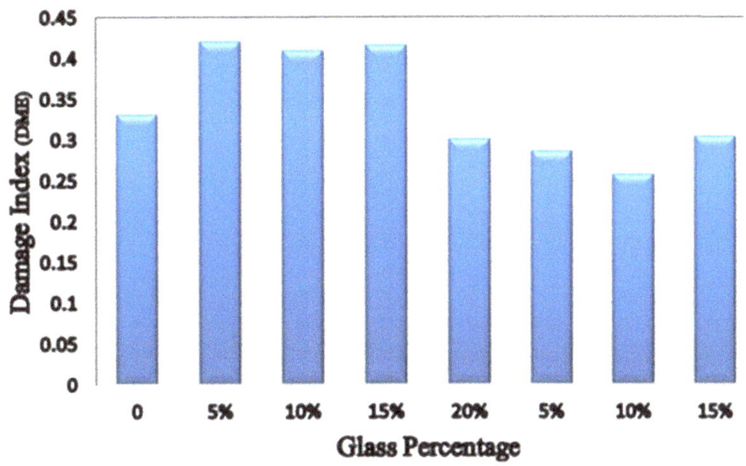

Figure 6. Damage Index of concrete prismscontaining different glass percentages which were exposed to the action of freezing and thawing cycles using the frequency resonance method (results are average of six specimens at each percentage)

5. Conclusions

The findings of this research indicate that using of glass waste in concrete can not only reduce problems encountered with environmental pollution, but also it may enhance the durability and mechanical properties of the concrete subjected to an environmental condition of freezing and thawing cycles. The following conclusions were drawn from the present work:

1) The compressive and flexural strengths of concrete were improved upon cement or fine aggregate replacement by glass waste at percentages of 10-20.
2) Incorporation of the glass waste partially in concrete improved its resistance against surface scaling upon exposure to freezing and thawing cycles. The best resistance was achieved when glass waste powder was used as a partial replacement of cement at a percentage of 10.
3) The damage extent as evaluated by two different nondestructive techniques was decreased in concrete mixtures prepared at 10-20 percentages of glass waste replacement of cement or fine aggregates as compared to that of conventional concrete.
4) In general, the addition of crushed and milled glass wastes to the concrete mixtures can be considered not only a green solution, but also a method to enhance the mechanical properties and the durability of concrete mixtures.

Acknowledgments

The writers would like to thank the Applied Science University for their supporting and funding the research work.

References

American Psychological Association. (1972). Ethical standards of psychologists. Washington, DC: American Psychological Association.

Bazant, Z., & Mayer, C. (1998). Fracture mechanics of concrete structures. Proceedings of fracture mechanics: part 3 (pp. 1687-1693).

Carpenter, A., & Cramer, S. (1999). Mitigation of alkali-silica reaction in pavement patch concrete that incorporates highly reactive fine aggregate. Transportation Research Record: Journal of the Transportation Research Board, (1668), 60-67.

Federico, L. M., & Chidiac, S. E. (2009). Waste glass as a supplementary cementitious material in concrete–critical review of treatment methods. Cement and concrete composites, 31(8), 606-610. http://dx.doi.org/10.1016/j.cemconcomp.2009.02.001

Ismail, Z. Z., & Al-Hashmi, E. A. (2009). Recycling of waste glass as a partial replacement for fine aggregate in concrete. Waste management, 29(2), 655-659.

Kim, J., Moon, J. H., Shim, J. W., Sim, J., Lee, H. G., & Zi, G. (2014). Durability properties of a concrete with waste glass sludge exposed to freeze-and-thaw condition and de-icing salt. Construction and Building Materials, 66, 398-402. http://dx.doi.org/10.1016/j.conbuildmat.2014.05.081

Ling, T. C., Poon, C. S., & Wong, H. W. (2013). Management and recycling of waste glass in concrete products: Current situations in Hong Kong. Resources, Conservation and Recycling, 70, 25-31. http://dx.doi.org/10.1016/j.resconrec.2012.10.006

Madandoust, R., & Ghavidel, R. (2013). Mechanical properties of concrete containing waste glass powder and rice husk ash. Biosystems Engineering, 116(2), 113-119. http://dx.doi.org/10.1016/j.biosystemseng.2013.07.006

Metwally, I. (2007). Investigations on the performance of concrete made with blended finely milled waste glass. Advances in Structural Engineering, 10(1), 47-53. http://dx.doi.org/10.1260/136943307780150823

Pattengil, M., & Shutt, T. C. (1973). Use of ground glass as a pozzolan. In Albuquerque symposium on utilization of waste glass in secondary products. New Mexico, USA.

Saikia, N., & de Brito, J. (2014). Mechanical properties and abrasion behaviour of concrete containing shredded PET bottle waste as a partial substitution of natural aggregate. Construction and Building Materials, 52, 236-244. http://dx.doi.org/10.1016/j.conbuildmat.2013.11.049

Saribiyik, M., Piskin, A., & Saribiyik, A. (2013). The effects of waste glass powder usage on polymer concrete properties. Construction and Building Materials, 47, 840-844.

Schmidt, A., & Saia, W. H. F. (1963). Alkali-aggregate reaction tests on glass used for exposed aggregate wall panel work. ACI Mater J, 60, 1235-1236.

Shi, C., & Zheng, K. (2007). A review on the use of waste glasses in the production of cement and concrete. Resources, Conservation and Recycling, 52(2), 234-247. http://dx.doi.org/10.1016/j.resconrec.2007.01.013

Soroushian, P. (2012). Strength and durability of recycled aggregate concrete containing milled glass as partial replacement for cement. Construction and Building Materials, 29, 368-377. http://dx.doi.org/10.1016/j.conbuildmat.2011.10.061

Taha, B., & Nounu, G. (2008). Properties of concrete contains mixed colour waste recycled glass as sand and cement replacement. Construction and Building Materials, 22(5), 713-720. http://dx.doi.org/10.1016/j.conbuildmat.2007.01.019

Topcu, I. B., & Canbaz, M. (2004). Properties of concrete containing waste glass. Cement and Concrete Research, 34(2), 267-274. http://dx.doi.org/10.1016/j.cemconres.2003.07.003

Tuan, B. L. A., Hwang, C. L., Lin, K. L., Chen, Y. Y., & Young, M. P. (2013). Development of lightweight aggregate from sewage sludge and waste glass powder for concrete. Construction and Building Materials, 47, 334-339. http://dx.doi.org/10.1016/j.conbuildmat.2013.05.039

8

Temperature Controlled of Self-Oscillating for Polymer Chains Induced by BZ Reaction with $Fe(phen)_3$ Catalyst

Jie Ren[1,2], Mengqi Yao[2], Guangcheng Zhang[1], Jinfen Gu[2], Li Tao[2] & Wu Yang[2]

[1] School of Natural and Applied Science, Northwestern Polytechnical University, Xi'an 710072, PR China

[2] Chemistry & Chemical Engineering College, Northwest Normal University, Key Lab of Bioelectrochemtry & Environmental Analysis of Gansu, Lanzhou 730070, PR China

Correspondence: Jie Ren, School of Natural and Applied Science, Northwestern Polytechnical University, Xi'an 710072, PR China. E-mail: jieren@nwnu.edu.cn

Abstract

The polymer chain with $Fe(phen)_3$-catalyzed was demonstrated to occur self-oscillation phenomenon in BZ reaction system. The self-oscillation polymer composed of N-isopropylacrylamide and $Fe(phen)_3$ catalyst moiety, which were synthesized by radical polymerization. It was found that the self-oscillation parameters (induction period, T_i, oscillation period, T_p and amplitude) could be regulated by subtle changes of temperature. It can be well explained by the FKN mechanism.

Keywords: self-oscillation, poly(NIPAAm-*co*-$Fe(phen)_3$), temperature control

1. Introduction

Conventional stimuli-responsive hydrogels, which change their properties in response to external stimulus, such as temperature, pH, electric field and magnetic field, have been studied extensively for applications in drug delivery systems, biosensors, and actuators, *etc* (Tanaka et al., 1980; Annaka & Tanaka, 1992; Chen & Hoffman, 1995; Hirokawa & Tanaka, 1984; Bae, Okano, & Kim, 1990; Yoshida et al., 1995). Yoshida's group reported a novel polymeric gel which autonomously swelled and de-swelled periodically in a closed homogeneous solution without any external stimuli, similar to autonomic phenomena in life such as heartbeat, brain waves, periodic hormone secretion, *etc*. The mechanical oscillation is achieved by coupling nonlinear chemical dynamics with polymers, transferring the dissipating chemical energy of a self-oscillating reaction to the driving force for the conformation changing of the polymers. The coupling of oscillating reactions with polymerization processes may provide methods for construction of novel materials capable of transmitting chemical energy. One area of research has been the construction of pulsating gels for periodic delivery of drugs, with oscillating chemical reactions supplying the necessary energy.

In general, the Belousov-Zhabotinsky (BZ) reaction (Noszticzius & Bódiss, 1979; Field, Körös, & Noyes, 1972) was employed for the simple model for a spatiotemporal structure in an unstirred solution and with periodicity in a stirred solution. Yoshida's group prepared a series of copolymer gels in which ruthenium(II) tris-(2,2′-bipyridine) ($Ru(bpy)_3^{2+}$), a catalyst for the BZ reaction, was covalently bonded to the polymer chains (Yoshida, Takahashi, Yamaguchi, & Ichijo, 1996; Yoshihiro, Masayuki, & Yoshida, 2003; Suzuki, Taniguchi, & Yoshida, 2009; Hidaka & Yoshida, 2011; Yoshida, Sakai, Ito, & Yamaguchi, 2002; Hara et al., 2005; Suzuki & Yoshida, 2008). There the catalyst acts as a transducer to convert its periodical redox changes into the mechanical oscillation of the polymer network through the hydrophilicity changes of the polymer chain.

Up to now, the polymer system has been studied for application to self-walking gel robot, actuator and mass transport system, *etc* (Murase, Maeda, Hashimoto, & Yoshida, 2009; Mitsumata, Ikeda, Gong, & Osada, 2000; Tateyama, Shibuta, & Yoshida, 2008; Maeda, Hara, & Yoshida, 2008). Besides, the self-oscillating macrogels would have potential applications such as pumps with autonomous beating, oscillatory drug release synchronized with cell cycles, *etc* (Weissman, Sunkara, Tse, & Asher, 1996; Suzuki, McGrath, Kawaguchi, & Lyon, 2007; Lyon et al., 2004). Moreover, self-oscillating microgel particles (less than 1 μm in diameter) have been synthesized by surfactant free aqueous precipitation polymerization (Suzuki, Sakai, & Yoshida, 2008; Suzuki & Yoshida, 2008). Due to the properties of colloidal dispersions (Pelton, 2000; Nayak & Lyon, 2005; Saunders &

Vincent, 1999; Kawaguchi, 2000), microgel particles are used as microreactors, emulsifiers, and photonic crystals (Suzuki, McGrath, Kawaguchi, & Lyon, 2007; Suzuki & Kawaguchi, 2006; Suzuki, Tsuji, & Kawaguchi, 2007). Therefore, the self-oscillating microgels may be useful not only as artificial oscillators, but also as rheological modifiers, and so on.

However, this kind of self-oscillating polymer system is too expensive for application in many fields owing to the use of expensive rare metal Ru as the catalyst moiety required for BZ reaction. Recently, Hara developed a novel self-oscillating polymer by covalently bonding the $Fe(bpy)_3$ moiety to the polymer chain, which acted as the metal catalyst for the BZ reaction. However, the amplitude of the soluble-insoluble self-oscillation of poly(NIPAAm-co-$Fe(bpy)_3$) was significantly smaller than that of the corresponding Ru-based self-oscillating polymer system.

To achieve the applications of self-oscillating polymers, we developed a novel self-oscillating polymer by covalently bonding a $Fe(phen)_3$ moiety, which acted as the metal catalyst for the BZ reaction. It was demonstrated that the polymer underwent soluble-insoluble self-oscillation that originated from the periodic change in the solubility of the $Fe(phen)_3$ catalyst moiety (Ren et al., 2015). In order to clarify the effect of temperature on the polymer chains oscillation, and to understand how self-oscillation of polymer can be controlled, the influence of temperature on the self-oscillation of the polymer chain was investigated in the present study.

2 Experimental

2.1 Materials

N-isopropylacrylamide (NIPAAm), azo-bis-isobutyronitrile (AIBN) were brought from Shanghai Zhongqin chemical industry and used without further purification. Ethanol, dichloromethane and methanol (Tianjin Guangfu fine chemical industry research institute, Tianjin, China), and $FeSO_4 \cdot 7H_2O$, malonic acid (MA), $KBrO_3$, H_2SO_4, $Ce_2(SO_4)_3$ and $Ce(SO_4)_2$ were analytical regents and used as received. 5-acrylamide-phenanthroline (5-AM-phen) was synthesized by the method in references. The deionized water was used throughout the experiments.

2.2 Preparation of Poly(NIPAAm-co-Fe(phen)$_3$)

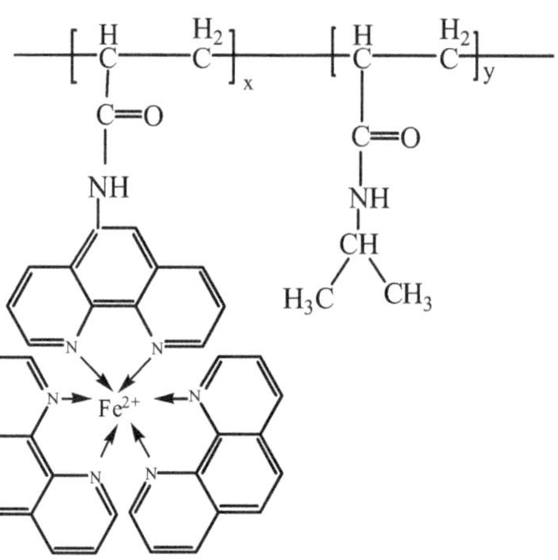

Poly(NIPAAm-co-Fe(phen)$_3$

Figure 1. The chemical structure of the self-oscillating polymer chain

The self-oscillating polymer chain with the Fe catalyst moiety(as shown Figure. 1) was synthesized via free-radical polymerization as previously reported (Ren et al., 2015). The process is as following: N-isopropylacrylamide (NIPAAm, 5.0 g), 5-acrylamide-phenanthroline (5-AM-phen, 0.20 g), and azo-bis-isobutyronitrile (AIBN, 0.05 g) were dissolved in ethanol (25 g). The mixture was poured into a three-neck flask equipped with a stirrer, a condenser, and a nitrogen gas inlet. The polymerization reaction was carried at 80 °C for 6 h with N_2 bubbling. The obtained polymer was dissolved in dichloromethane. The solution of $FeSO_4$ in methanol was added dropwise into the polymer solution with stirring. Finally, the solution of phenanthroline in ethanol was added with stirring at room temperature. The resulting sample was dialyzed with ethanol for a week while changing ethanol frequently, and then dialyzed with water for a week, freeze-dried finally.

2.3 Measurement of Redox Potential Self-Oscillation

The experiments were performed in a glass reactor (*ca.* 50 mL) coupled with a SY-601 thermostat and a Model ML-902 magnetic stirrer. A mixture containing malonic acid, KBO_3, H_2SO_4 and poly(NIPAAm-co-Fe(phen)$_3$) was placed in the reactor. A CHI-832 electrochemistry analyzer (Shanghai Chenhua Instrument Company, China) was directly connected to the reactor through two Pt-electrodes (Rex, 213, China) as the working electrode and the auxiliary electrode and a Hg_2SO_4 reference electrode to record the potential changes.

2.4 Measurement of Transmittance Self-Oscillation

The solution was mixed as the measurement of redox potential self-oscillation and stirred until its color began to change between red and light blue, then the transmittance self-oscillation was measured. A wavelength of 389 nm was also used because of being the isosbestic point of the reduced and oxidized states of the polymer chain. The time course of the transmittance at 389 nm was monitored using an UV-Vis spectrophotometer.

3. Results and Discussion

A wide variety of metal ions and metal complex catalysts have been used in the BZ reaction: Ce^{3+}, Mn^{2+}, $[Fe(phen)_3]^{2+}$, $[Fe(bpy)_3]^{2+}$, $[Ru(bpy)_3]^{2+}$, and so on. These catalysts fill the requirement that they are capable of a single electron transfer and have two stable oxidation states separated by a potential of 1.0-1.5 V/SCE. Despite having some features in common, these catalysts can sometimes have remarkably different effects on the observed behavior of the BZ reaction. There is a pronounced color change from red to blue as ferroin ($Fe(phen)_3^{2+}$) is oxidized to ferriin ($Fe(phen)_3^{3+}$). Smoes emphasized the fact that ferroin does not react in the same manner as cerium in the BZ reaction (Smoes, 1979). In fact, the ferriin/ferroin couple has a lower standard reduction potential (1.06 V) than that of Ce^{4+}/Ce^{3+} (1.44V). Keki, Magyar, Beck and Gaspar (1992) demonstrate that direct oxidation of ferroin by bromine species other than BrO_2• radical species is possible due to the low redox potential of the catalyst couple. Therefore, the behavior of the ferroin-catalysed BZ can be explained by the following reactions.

$$2Fe(phen)_3^{2+} + BrO_3^- + 3H^+ \rightarrow 2Fe(phen)_3^{3+} + HBrO_2 + H_2O \quad \text{(R1)}$$

$$2Fe(phen)_3^{2+} + HBrO_2 + 2H^+ \rightarrow 2Fe(phen)_3^{3+} + HOBr + H_2O \quad \text{(R2)}$$

$$2Fe(phen)_3^{2+} + HOBr + H^+ \rightarrow 2Fe(phen)_3^{3+} + Br^- + H_2O \quad \text{(R3)}$$

$$2Fe(phen)_3^{2+} + Br_2 \rightarrow 2Fe(phen)_3^{3+} + 2Br \quad \text{(R4)}$$

Körös performed calorimetric studies on the BZ reaction and determined that the heat released per mole bromated consumed (650kJ) in the same and is independent of the catalyst used. Most of heat released during FKN process C, *i.e.* the oxidation of MA or BrMA by Ce^{4+}. Although the rate of heat production is oscillatory, however, the temperature of the system does not oscillate in the BZ system. Therefore, the effect of temperature on the self-oscillation is completely from the surrounding temperature changing, rather than the BZ system itself.

The self-oscillating mechanism of the polymer is due to the change in the hydrophilic-hydrophobic properties of polymer chains with oxidation and reduction. With the development of BZ reaction, $Fe(phen)_3$ moiety hanging on the polymer chain is oxidized or reduced. When Fe(III) is reduced to Fe(II), the polymer chains stretch and the solubility of the polymer increases due to the good hydrophilicity and solubility of $Fe(phen)_3^{2+}$. However, when Fe(II) is oxidized to Fe(III) in BZ medium, the polymer chains curl and twist each other and the solubility of $Fe(phen)_3^{3+}$ decreases correspondingly. Therefore, the polymer undergoes soluble-insoluble self-oscillation along with the periodic change of $Fe(phen)_3$ catalyst moiety between the oxidation and reduction states in BZ reaction. The temperature plays an important part in the dynamics of the BZ system. Figure 2 shows the redox potential self-oscillation of the poly(NIPAAm-co-Fe(phen)$_3$) solution at different temperature against fixed concentrations of the BZ system and Figure 3 shows the oscillation profiles of optical transmittance (T%) with the same solution. Note that, when the temperature is lower than 19 °C, both the periodic redox potential and transmittance changes could no longer be observed due to the reaction rate being too slow. However, when the temperature is higher than 29 °C, there is some flocculent substance in the system which adheres onto the electrodes due to the temperature sensitivity of N-isopropylacrylamide. Therefore, the self-oscillation parameters were compared when the temperature was in the range of 22- 28 °C.

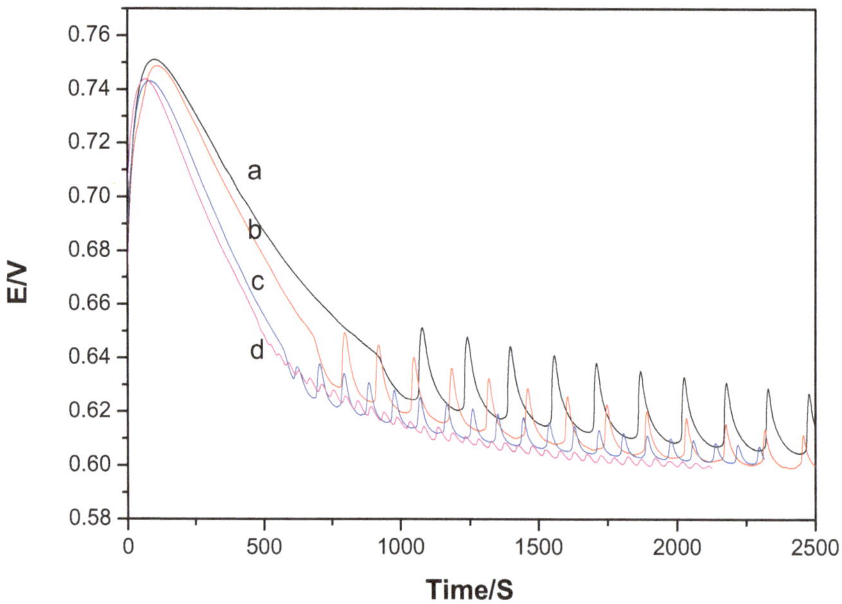

Figure 2. Typical oscillation waveforms of redox potential with different temperature. a: 22°C, b: 24°C, c: 26°C, d: 28°C. The common conditions: $[polymer]_0$=0.20%, $[KBrO_3]_0$=0.0625mol/L, $[MA]_0$=0.135 mol/L, $[H_2SO_4]_0$=0.175 mol/L

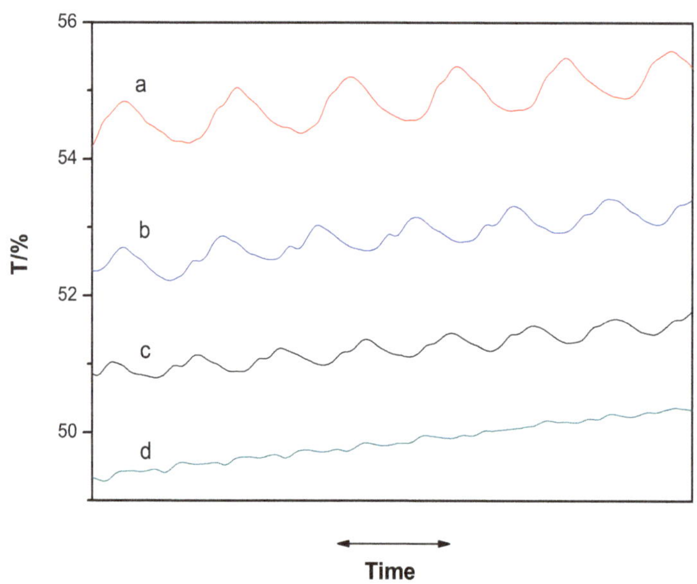

Figure 3. Typical oscillation waveforms of transmittance with different temperature. The experimental conditions are the same as which in Figure 2

In the induction period, BrMA is accumulating. A crucial amount of BrMA is necessary to induce regular oscillations. The rate for production of BrMA increases with increasing temperature, and the induction period decreases correspondingly as shown in Figure 4(a). We also compared the oscillation periods with different temperature. In this study, the average value of the periods between the second and the sixth oscillation were regarded as the oscillation period. The frequency of oscillation is mainly determined by the rate of removal of bromide. In the FKN mechanism, bromide ion is consumed by reaction with bromated ion mainly as in R5.

$$BrO_3^- + 3H^+ + Br^- \rightarrow 3Br_2 + 3H_2O \qquad \text{(R5)}$$

The period of oscillation decreases for all reactions rate involved increasing at high temperature. So as shown in Figure 4(a), the oscillation period (T_p) decreased with increasing temperature.

Figure 4(b) shows the dependence of the amplitudes of the oscillation on temperature. It can be seen from the figure, the oscillation amplitudes both of redox potential and transmittance decreased with the increasing of temperature. The main component of the polymer is N-isopropylacrylamide with temperature sensitivity. The lower critical solution temperature of N-isopropylacrylamide is 32 °C. Suzuki and Yoshida (2008) suggested that the dramatic change in oscillation waveform and amplitude is related to the difference in colloidal stability between in the Ru(II) and the Ru(III) states for self-oscillation polymer containing $Ru(bpy)_3$ moiety. At low temperature, the polymer chain is stretch and the $Fe(phen)_3$ moiety hanging on the chain catalyzes the BZ reaction effectively and changes between the oxidation and reduction states freely. The $Fe(phen)_3$ moiety can exist only in its oxidation state at a certain time and in its reduction state at another certain time. That is to say, $Fe(phen)_3$ moiety can transfer from one state to the other completely. The concentration difference between the two states is the maximum at oxidation state or reduction state. Therefore, both the redox potential and transmittance amplitudes are big at low temperature. However, at high temperature and close to the lower critical solution temperature of N-isopropylacrylamide, the phase separation is coming, and the polymer chains curls and twists each other. Some of the moiety is embedded and can't contact with BZ medium adequately. Both the reduction of $Fe(phen)_3^{3+}$and the oxidation of $Fe(phen)_3^{2+}$ are incompletely, the concentration difference between the two states become smaller on raising the temperature and the amplitudes smaller correspondingly.

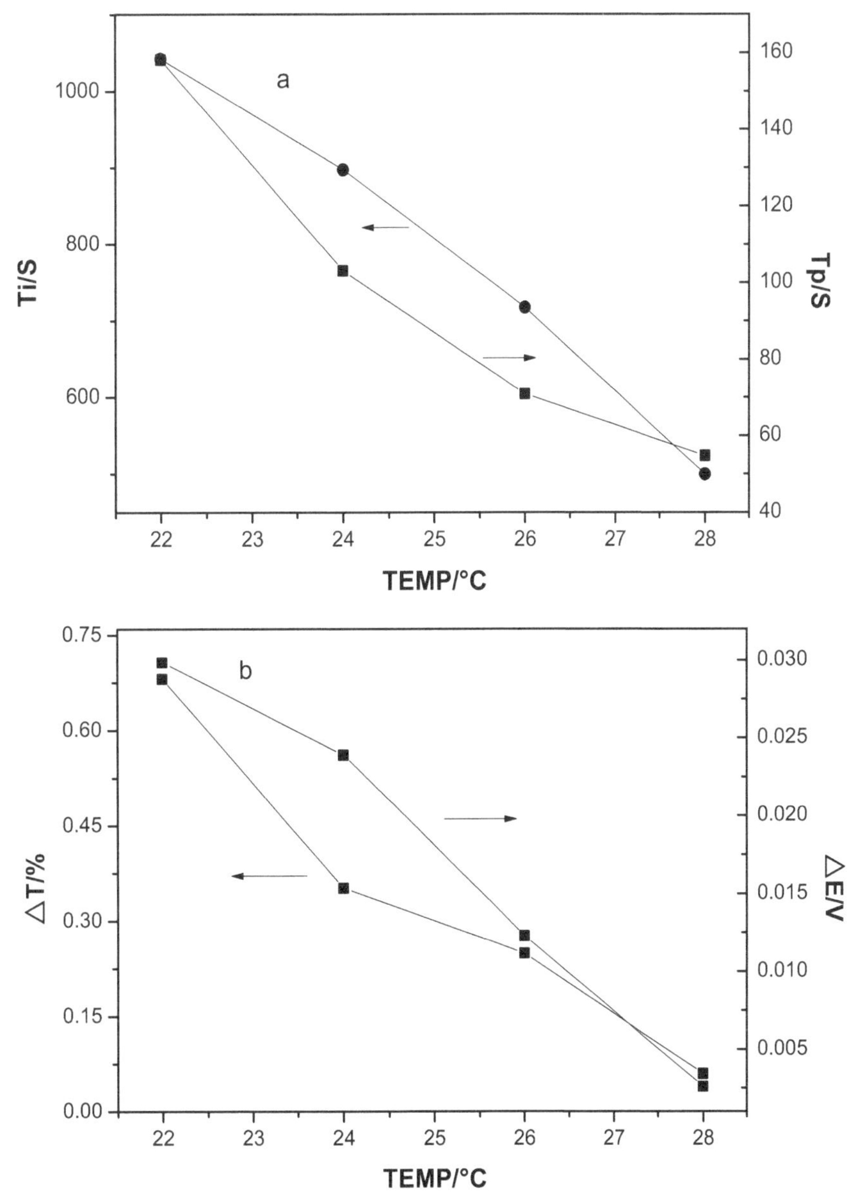

Figure 4. Effect of temperature on induction period and oscillation period (a), amplitude (b) of the self-oscillation

4. Conclusions

The polymer chain containing $Ru(bpy)_3$ catalyst is well-known as a self-oscillation phenomenon. Photo-irradiation and temperature have been employed to switch the self-oscillation on and off. In the present work, for the first time we found that the self-oscillation could be controlled by temperature. When the temperature increased, the oscillation frequency increased and the amplitudes both of redox potential and transmittance decreased correspondingly.

Acknowledgments

This work was supported in part by the Basic Project of Science and Research of Colleges and Universities of Gansu Province (5001-109), the Project for Young Teacher of Northwest Normal University (NWNU-LKQN-13-6).

Referrences

Annaka, M., & Tanaka, T. (1992). Multiple phases of polymer gels. *Nature, 355,* 430-432. http://dx.doi.org/ 10.1038/355430a0

Bae, Y. H., Okano, T., & Kim, S. W. (1990). Temperature dependence of swelling of crosslinked poly (N, N'-alkyl substituted acrylamides) in water. *Journal of Polymer Science Part B, 28,* 923-936. http://dx.doi.org/ 10.1002/polb.1990.090280609

Chen, G., & Hoffman, A. S. (1995). Graft copolymers that exhibit temperature-induced phase transitions over a wide range of pH. *Nature, 373*, 49-52. http://dx.doi.org/ 10.1038/373049a0

Field, R. J., Körös, E., & Noyes, R. M. (1972). Oscillations in chemical systems. II. Thorough analysis of temporal oscillation in the bromate-cerium-malonic acid system. *Journal of the American Chemical Society, 94*(25), 8649-8664. http://dx.doi.org/ 10.1021/ja00780a001

Hara, Y., Sakai, T., Maeda, S., Hashimoto, S., & Yoshida, R. (2005) Self-oscillating soluble-insoluble changes of a polymer chain including an oxidizing agent induced by the Belousov-Zhabtinsky reaction. *Journal of Physical Chemistry B, 109*, 23316-23319. http://dx.doi.org/10.1021/jp055157y

Hidaka, M., & Yoshida, R. (2011). Self-oscillating gel composed of thermosensitive polymer exhibiting higher LCST. *Journal of Controlled Release, 150*, 171-176. http://dx.doi.org/ 10.1016/j.jconrel.2010.11.026

Hirokawa, Y., & Tanaka, T. (1984). Volume phase transition in a nonionic gel. *The Journal of Chemical Physics, 81*, 6379-6380. http://dx.doi.org/10.1063/1.447548

Kawaguchi, H. (2000). Functional polymer microspheres. *Progress in Polymer Science*, *25*, 1171-1210. http://dx.doi.org/10.1016/S0079-6700(00)00024-1

Keki, S., Magyar, I., Beck, M.T., & Gaspar, V. (1992). Modeling the oscillatory bromate oxidation of ferroin in open systems. *The Journal of Chemical Physics, 96*(4), 1725-1729. http://dx.doi.org/ 10.1021/j100183a042

Lyon, L. A., Debord, J. D., Debord, S. B., Jones, C. D., McGrath, J. G., Serpe, & M. J. (2004). Microgel colloidal crystals. *Journal of Physical Chemistry B, 108,* 19099-19108. http://dx.doi.org/ 10.1021/jp048486j

Maeda, S, Hara, Y, & Yoshida, R. (2008). Control of the dynamic motion of a gel actuator driven by the Belousov-Zhabotinsky reaction. *Macromolecular Rapid Communications, 29*, 401-405. http://dx.doi.org/ 10.1002/marc.200700717

Mitsumata, T., Ikeda, K., Gong, J. P., & Osada, Y. (2000). Controlled motion of solvent-driven gel motor and its application as a generator. *Langmuir, 16*, 307-312. http://dx.doi.org/ 10.1021/la990483o

Murase, Y., Maeda, S., Hashimoto, S., & Yoshida, R. (2009). Design of a mass transport surface utilizing peristaltic motion of a self-oscillating gel. *Langmuir, 25*, 483-489. http://dx.doi.org/ 10.1021/la8029006

Nayak, S., & Lyon, L. A. (2005). Soft nanotechnology with soft nanoparticles. *Angewandte Chemie, International Edition, 44*, 7686-7708. http://dx.doi.org/ 10.1002/anie.200501321

Noszticzius, Z., & Bódiss, J. (1979). A heterogeneous chemical oscillator. The Belousov-Zhabotinskii-type reaction of oxalic acid. *Journal of the American Chemical Society, 101*, 3177-3182. http://dx.doi.org/10. 1021/ja00506a007

Pelton, R. (2000). Temperature-sensitive aqueous microgels. *Advances in Colloid and Interface Science, 85,* 1-33. http://dx.doi.org/ 10.1016/S0001-8686(99)00023-8

Ren, J., Yao, M. Q., Zhang, G. C., Yang, X. C., Gu, J. F., & Yang, W. (2015). Self-oscillating soluble-insoluble changes of polymer chain. *Journal of Polymer Research, 22*(44), 1-6. http://dx.doi.org/10.1007/s10965-015-0686-6.

Saunders, B. R., & Vincent, B. (1999). Microgel particles as model colloids: theory, properties and applications. *Advances in Colloid and Interface Science, 80*, 1-25. http://dx.doi.org/10.1016/S0001-8686(98)00071-2

Smoes, M. L. (1979). Period of homogeneous oscillations in the ferroin - catalyzed Zhabotinskii system. *The Journal of Chemical Physics, 71*(11), 4669-4670. http://dx.doi.org/ 10.1063/1.438250

Suzuki, D., & Kawaguchi, H. (2006). Hybrid microgels with reversibly changeable multiple brilliant Color. *Langmuir, 22*(8), 3818-3822. http://dx.doi.org/ 10.1021/la052999f

Suzuki, D., & Yoshida, R. (2008). Effect of initial substrate concentration of the Belousov-Zhabotinsky reaction on self-oscillation for microgel system. *Journal of Physical Chemistry B, 112*, 12618-12624. http://dx.doi.org/10.1021/jp8037973

Suzuki, D., & Yoshida, R. (2008). Temporal control of self-oscillation for microgels by cross-linking network structure. *Macromolecules, 41*(15), 5830-5838. http://dx.doi.org/ 10.1021/ma800684d

Suzuki, D., McGrath, J. G., Kawaguchi, H., & Lyon, L. A. (2007). Colloidal crystals of thermosensitive, core/shell hybrid microgels. *Journal of Physical Chemistry C, 111*, 5667-5672. http://dx.doi.org/ 10.1021/jp068535n

Suzuki, D., Sakai, T., & Yoshida, R. (2008). Self-flocculating/self-dispersing oscillation of microgels. *Angewandte Chemie, International Edition, 47*, 917-920. http://dx.doi.org/ 10.1002/ange.200703953

Suzuki, D., Taniguchi, H., & Yoshida, R. (2009). Autonomously oscillating viscosity in microgel dispersions. *Journal of the American Chemical Society, 131*(34), 12058-12059. http://dx.doi.org/ 10.1021/ja904677g

Suzuki, D., Tsuji, S., & Kawaguchi, H. (2007). Janus microgels prepared by surfactant-free pickering emulsion-based modification and their self-assembly. *Journal of the American Chemical Society, 129*, 8088-8089. http://dx.doi.org/ 10.1021/ja072258w

Tanaka, T., Fillmore, D., Sun, S. T., Nishio, I., Swislow, G., & Shah, A. (1980). Phase transitions in ionic gels. *Physical Review Letters, 45*, 1636-1639. http://dx.doi.org/10.1103/PhysRevLett.45.1636

Tateyama, S., Shibuta, Y., & Yoshida, R. (2008). Direction control of chemical wave propagation in self-oscillating gel array. *Journal of Physical Chemistry B, 112*, 1777-1782. http://dx.doi.org/10.1021/jp709882h

Weissman, J. M., Sunkara, H. B., Tse, A. S., & Asher, S. A. (1996). Thermally switchable periodicities and diffraction from mesoscopically ordered materials. *Science, 274*, 959-960. http://dx.doi.org/ 10.1126/science.274.5289.959

Yoshida, R., Sakai, T., Ito, S., & Yamaguchi, T. (2002). Self-oscillation of polymer chains with rhythmical soluble-insoluble changes. *Journal of the American Chemical Society, 124*, 8095-8098. http://dx.doi.org/ 10.1021/ja012584q

Yoshida, R., Takahashi, T., Yamaguchi, T., & Ichijo, H. (1996). Self-oscillating gel. *Journal of the American Chemical Society, 118*(21), 5134-5135. http://dx.doi.org/ 10.1021/ja9602511

Yoshida, R., Uchida, K., Kaneko, Y., Sakai, K., Kikuchi, A., Sakurai, Y., & Okano, T. (1995) .Comb-type grafted hydrogels with rapid deswelling response to temperature changes. *Nature, 374*, 240-242. http://dx.doi.org/ 10.1038/374240a0

Yoshihiro, I., Masayuki, N., & Yoshida, R. (2003). Temperature control of the Belousov-Zhabotinsky reaction using a thermoresponsive polymer. *Langmuir, 19*, 9577-9579. http://dx.doi.org/ 10.1021/la0355242

9

Two Challenges to the System of Periclase Quality Evaluation

L. M. Axelrod[1], I. G. Maryasev[1], A. A. Platonov[1] & D. R. Melnikova[1]

[1] Magnezit Group, Russia

Correspondence: I. G. Maryasev, Magnezit Group, Russia. E-mail: imaryasev@magnezit.com

Abstract

A new method of estimating the fused periclase quality – the revision of international standards. Quality and service life of refractories on the basis of fused periclase depend to a large extent upon the size of periclase crystals and its mineralogical composition. During fusion of periclase it is impossible to obtain completely homogeneous material with identical crystals size. That is why it is generally accepted to evaluate quality of fused periclase by the average crystals size. Measurements of this item are usually done with the help of generally adopted method of chords, which was developed in the last century and has a number of drawbacks. Magnezit Group developed a new objective method of digital analysis during microscopic examination of structural elements of fused periclase. It allows to considerably improve objectivity of obtained data. The main feature of the new method is application of the system of images analysis, which allows to carry out in automatic mode measurements on the preliminary created digital model of the whole area of the polished section. With the help of the digital model it is possible to calculate average size of fused periclase crystals taking into account number of crystals as well as percentage of the area occupied by them.

Does CaO/SiO_2 ratio influence service life of refractories? New view of old rules. Till today it was considered that one of the characteristics of fused periclase quality is coefficient of basicity – CaO/SiO_2, ratio, which should be more than 2. Magnezit Group carried out investigations of coarse-crystalline fused periclase with MgO >97.5 % content and with various CaO/SiO_2 ratios. We present in this report the main study results: if the impurities content is low in the fused periclase with MgO >97.5 %, then coefficient of basicity exerts limited influence onto the service life of periclase-carbon bricks.

Keywords: Fused periclase, linear method of determination of crystals size, method of random chords, digital method of determination, periclase-carbon bricks, lining, CaO/SiO_2 ratio

1. Introduction

Quality of refractory materials, produced with application of fused and sintered periclase, depends upon a number of its characteristics. It was found out that the size of periclase crystals – in addition to its chemical composition, mineral composition of phases, located between crystals, apparent density – plays considerable role. As was proved on practice, the size of crystals exerts considerable influence upon wear resistance of bricks, produced with application of periclase, first of all of periclase – carbon refractories, which are widely used in steel making ((Riepl & Barthel, 1991; Steinwender & Bugajski, 1991; Mosser & Riepl, 1988). Maximally exact evaluation of crystals size in quantitative analysis is necessary not only for selecting periclase of this or that quality for production of refractory materials, but also for control of technological factors during production of fused periclase.

Determination of chemical composition and bulk density are standardized while determination of crystal size is done with the help of various procedures, that is why precision of results, obtained in different laboratories, is often not identical (Riepl & Barthel, 1991; Mosser & Riepl, 1988). It is related both to the difficulty of selecting a representative sample for investigation and to discrepancies in applied methodologies. There is no universally accepted classification of fused periclase by crystals size.

It is known, that for the products of melt crystallization including periclase there is a logarithmic normal law of crystals diameters distribution (under the term “periclase crystal diameter” we understand average crystal size, i.e. average value between the length and the width of a crystal, analogously to the term “periclase crystal size”)

Quantitative evaluation of crystals size distribution is related to obtaining and analysis of considerable statistical material, permitting to establish a distribution law (Sivash, Perepelitsin, & Mityushov, 2001).

There are a number of methods for quantitative assessment of periclase crystal size in metallographic sections on the basis of random sampling. As other quantitative methods they are based on presentation of limitlessly large number of intersections of grains with parallel planes or on overlaying onto images of calibrated nets of certain size. Theoretical basics of the methods are given in detail in (Saltykov, 1976). But stereometric methods (developed in the last century) for microscopic analysis of crystals require much time and are rather labour intensive. Quantitative assessments and calculations of microstructure elements on polished sections or metallographic sections were previously done visually by way of viewing directly into the microscope eyeglasses.

In the century of digital technologies, when spaceships are ploughing the space, the time has come to improve the quality of optical-microscopic measurements using modern digital equipment and programs of images analysis.

2. Linear Method of Assessment

Linear sizes were first measured in divisions of eyeglass scale. Then obtained results were multiplied for the division value. It was necessary to use during analysis various magnifications from 10× up to 500×, as inadequate magnification in quantitative analysis often led to erroneous results in assessment of two-dimensional structure parameters as a consequence of detection of new small crystals (grains) under high magnification (Sivash, Perepelitsin, & Mityushov, 2001). Technology of the most widely used «manual» linear method of crystal size evaluation by way of random chords consists of the following:

– Selection and preparation of a sample for investigation:

- visual evaluation of the total sample and classification of lumpy periclase by colour, structure, texture and cryslalline particles size; evaluation of share of each periclase class;
- preparation of sample for investigation. If periclase is in the form of large lumps, it is sawn along the chosen plane, finished and polished. Lumps of periclase less than 10 mm in size are preliminary coated with epoxy resin, then finished and polished;

– Methodology of microscopic investigation:

- adjustment of microscope for 100× magnification with the help of eyeglass.
- installation of polished sample into holder on microscope table. The first site in the periclase sample is chosen and image is focused.
- displacement of the sample to the edge of the line of one crystal from the edge of the net.
- determination of crystals size, crossing the line (Figure 1). If a crystal possesses isometrically rounded shape, then one of its middle diameter is measured. If the shape of a crystal is elongated, then its average size is determined, i.e. average value between the maximal length and width of the grain.
- after finishing analysis of a certain fragment the microscope table is displaced lower on the line parallel to the previously investigated site in order to measure another part of the sample and to repeat the process up to the end of the sample edge (Figure 2).
- for determination of average crystal size it is necessary to assess the size of not less than 250 crystals.
- for calculation of final average crystal size the following equation is used: $L_{av.} = s/n$, where $L_{av.}$ –average crystal size (microns); s – sum of crystals size (microns); n – number of examined crystals (pcs.).

Methodology of grid application (ASTM Norm E 112-96/2004) and a number of other methods can be considered as more or less perfected variants of linear method.

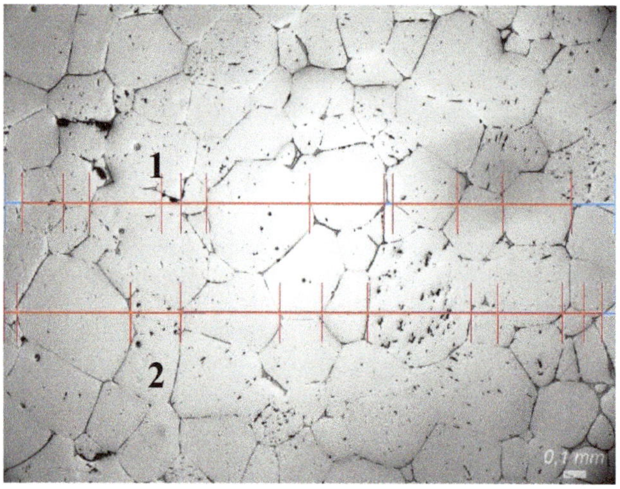

Figure 1. Evaluation of dimensions of separate periclase crystals, found out in the field of microscope objective by method of overlapping lines (chords). In reflected light. Magnification: 25×. Plane lines 1 and 2 were chosen for measurements

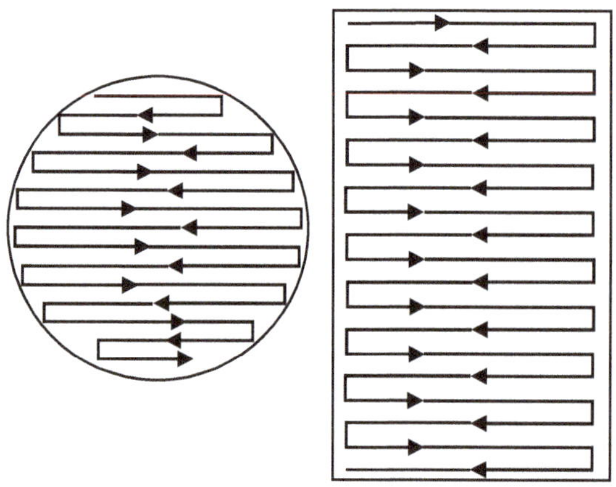

Figure 2. Schemes of moving of sample during measurement of crystals size of periclase by line method

3. Digital Method of Crystals Assessment

Modern methods of microscopic investigation allow to obtain more reliable and exact results in shorter time. At Magnezit Group equipping of the existing optical microscope Axioplan with the system of digital analysis allowed to modify and to supplement methods of determination of microscopic parameters of refractory materials and raw materials. The essence of the suggested method consists in modeling of a digital image of a large area of polished section 1200-1500 mm^2 under any necessary magnification, embracing its whole surface Figure 3).

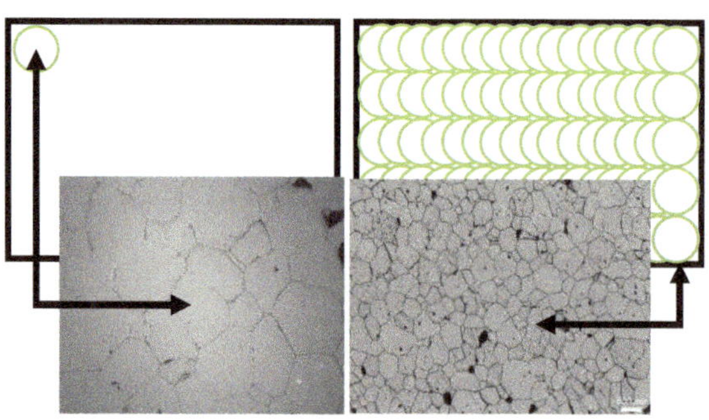

Figure 3. Scheme of image capture single (left) and crosslinked (right)

After formation of an image is finished, evaluation of geometric parameters of two-dimensional structure is done with the help of special software, in particular of the software Image-Pro Premier by the company Media Cybernetics (Figure 4).

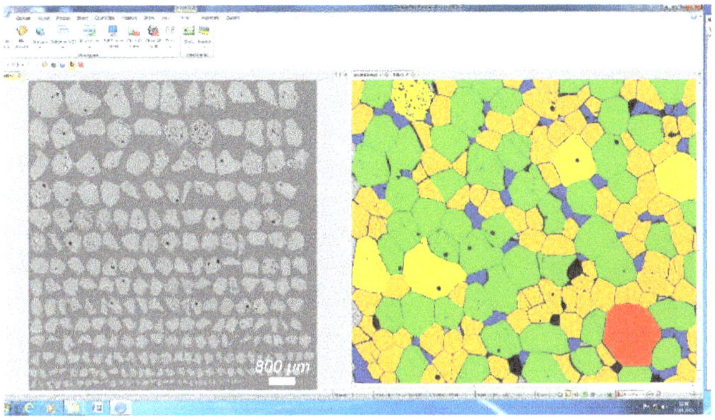

Figure 4. Example of automatic crystal size distribution

Zonal formation of crystals of different sizes occurs during melting of a block of fused periclase from a few microns to several centimeters. During crushing pieces of periclase crystals with different sizes (from different zones) are mixed. It becomes necessary to get an objective picture about crystals size and get the average data. With the help of image analysis system it is possible to carry out automatic measurements of material of any grain composition.

During investigation of fused periclase we obtain distinct notion about crystals shape in typical polished sections and determine size and area of all the crystals, which are found in the image (Figure 5a and 5b).

When the borders of crystals are poorly distinguishable it is necessary to etch the surface of the polished section in acid. A typical number of measured crystals for fused periclase is from 600 to 1200 depending upon crystals size.

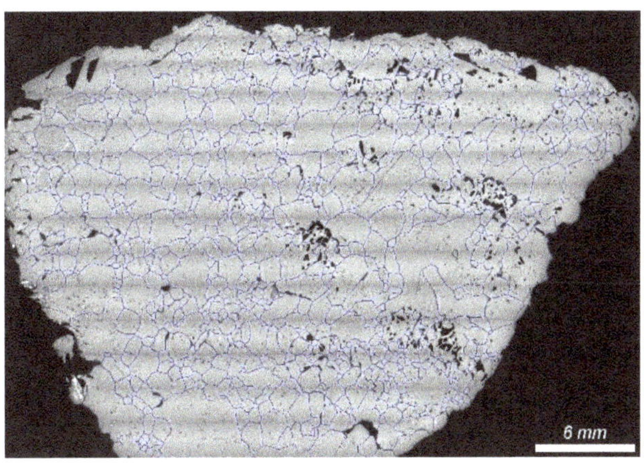

Figure 5a. Example of evaluation of periclase crystal area over the whole surface of the polished section. In reflected light. Magnification: 4×

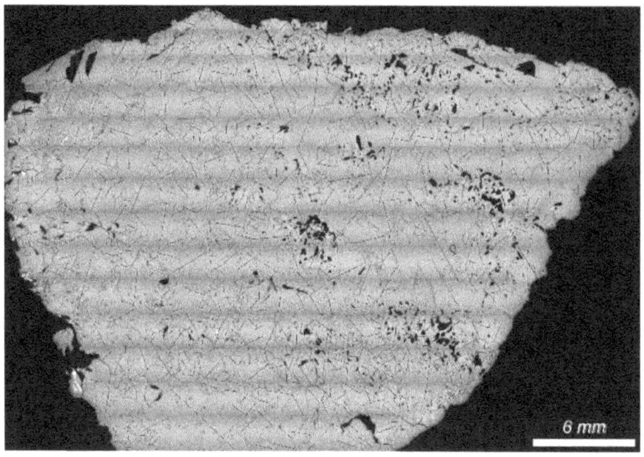

Figure 5b. Example of evaluation of periclase crystals average size over the whole surface of the polished section. In reflected light. Magnification: 4×

After measurements are finished, statistical treatment of data takes place and the maximal, minimal and average crystal size are determined, as well as graphs of sizes distribution are drawn with evaluation of parameters L_{10}, L_{50} and L_{90} (subscripts mean crystal size corresponding to 10, 50 and 90 % share of crystals, which are smaller than a certain diameter).

Often fused periclase has different crystalline structure with areas, where aggregation of small crystals is observed. These small crystals occupy in the total volume of the sample a small share, but nevertheless, exert strong influence onto the parameter of average crystal size based on the total number of crystals (Figure 6).

That is why it is reasonable to determine and compare average area crystal size (La average), which takes into account the share of the area occupied by crystals. It is calculated by the formula:

$$La_{av}.\ equals = \sum(Ai \times Li) / \sum Ai,$$

where Ai – area of crystal, Li – average crystal size.

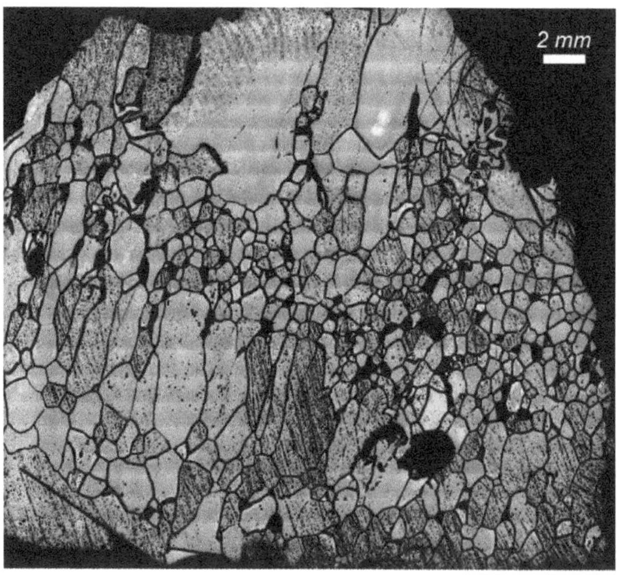

Figure 6. Fused periclase with areas of different crystalline structure

For comparison we presented in Table 1 and on Figure 7a, 7b data on the character of crystal size distribution in fused periclase, produced by Magnezit Group. The size was determined by two methods: the «old» one and the «new» one.

Table 1. Example of parameters values, characterizing size of fused periclase crystals

Parameters	characteristics	
	new method	old method
A, mm^2	651	–
N, pieces	767	250
L_{min} - L_{Max}, microns	50-5090	50-3910
$L_{av.}$, microns	824	1115
$La_{av.}$, microns	1792	–
Size of crystals by number, microns:		
L_{10}	180	180
L_{50}	575	720
L_{90}	1320	2170
Size of crystals by area, microns:		
La_{10}	560	–
La_{50}	1400	–
La_{90}	3270	–
Mass fraction of crystals by number, %		
M_{500}	57	65
M_{1000}	22	37
Mass fraction of crystals by area, %		
Ma_{500}	93	–
Ma_{1000}	67	–

A – area of surface of the polished section, mm^2

N – total number of crystals in the polished section, pcs.

L_{min} - L_{Max} – minimal and maximal size of periclase crystals in the polished section, microns.

$L_{av.}$ – arithmetic mean value by parameter of average crystal size, microns.

$La_{av.}$ – average area of crystal size, microns.

L_{10} – 10 % of crystals smaller in size than the size in question;

L_{50} – 50 % of crystals of the size in question;

L_{90} – 90 % of crystals smaller in size than the size in question.

La_{10} – 10 % of the area of the polished section consist of crystals smaller in size than the size in question;

La_{50} – 50 % of the area of the polished section consist of crystals of the size in question;

La_{90} – 90 %. of the area of the polished section consist of crystals smaller in size than the size in question.

M_{500} and M_{1000} – share of crystals in % larger than 500 microns and 1000 microns depending upon the number of crystals.

Ma_{500} and Ma_{1000} – percentage of crystals larger than 500 microns and 1000 microns depending upon the area occupied by them.

500 microns and 1000 microns depending upon the area occupied by them.

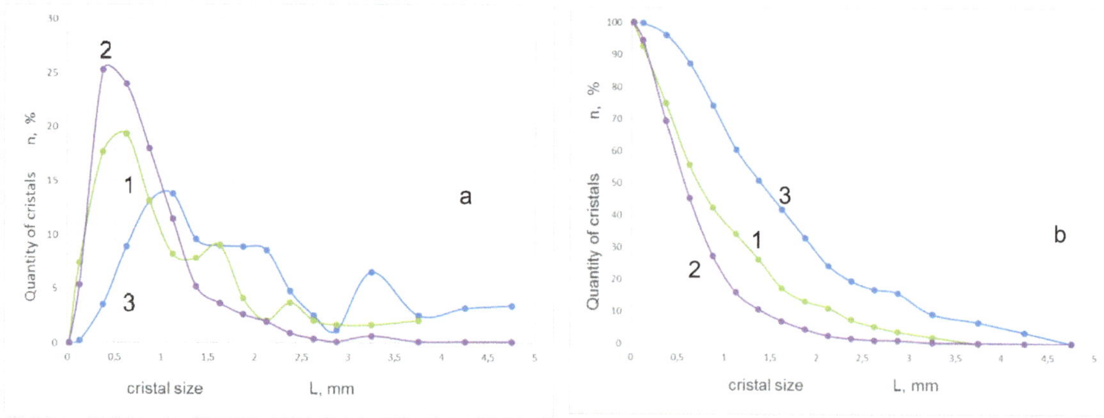

Figure 7. Curves of crystal size distribution in fused periclase

a – differential curve of crystal size distribution, b – integral curve of crystal size distribution.

1 – old method of chords, 2 – new method of global analysis; distribution by number of crystals.

3 – new method of global analysis; distribution by area of crystals.

From the Table 1 we can see, that method of microscopic analysis with the help of chords (the old method) gives overstated values of crystals size (especially by the item L_{90}), which differ from the real size of all the crystals. Positive results after introduction of the new method consist in the following:

- visualization of obtained results;
- possibility of control and repeated checks;
- organization of electronic data bases and accumulation of statistics;
- possibility to do on one and the same image measurements of not only crystals size, but of any basic geometric parameters of plane structure, such as, for example, surface of a separate particle and total surface of contacts, diameter of a spherical particle, dihedral angle between facets, number of microparticles in a unit of volume, number of contacts and other;
- measurements of crystals size directly on one and the same image allow to obtain precise data on size of very large crystals, which don't fit entirely in the microscope eyeglass field. Previously it was necessary to move the microscope table several times during measurement of such crystals. Now there appeared possibility to obtain precise data on the elements with absolutely different linear sizes, while previously it was necessary in such a case to change microscope magnification during measurements of small crystals;
- reliability and objectivity of obtained results increased. Good repeatability of data is ensured and influence of human factor is lowered. Such results are achieved thanks to increase of amount of data retrieval as well as thanks to retention in real time of all the measurement results. It is possible to assess the obtained result at any moment and if necessary to introduce corrections into determined structure parameters.

All the grades of tested fused periclase are standard products, manufactured by Magnezit Group. They were studied with the help of the new method of crystals size determination. Mineral and chemical compositions were studied as well. Obtained results for some of grades, which are part of the new product line Russian Magnezit™, are presented in Table 2.

New product line Russian Magnezit™ consists of fused periclase and dead-burned clinker grades with MgO content not less than 97 %, manufactured in Russia on the basis of magnesite from Siberian deposits. We tried to suggest a number of indicators, which in our opinion, describe in fullest measure distribution of fused periclase crystals by size.

Often fused periclase has different crystalline structure with areas, where aggregation of small crystals is observed, These small crystals occupy in the total volume of the sample a small share, but nevertheless, exert strong influence onto the parameter of average crystal size based on the total number of crystals. That is why it is feasible to determine and compare average area crystal size, which takes into account the share of area occupied by crystals.

Obtained results for some of grades are presented in Table 2. We have developed techniques, allowing to change within certain limits average size of crystals and, accordingly, average areal size and area of the whole number of crystals by way of changing fusion mode parameters during production of fused periclase of one and the same composition.

Table 2. Characteristics of different grades of fused periclase, which are part of Magnezit Group's product line Russian Magnezit™. ρ – bulk density, g/cm^3

Parameters		Imported FM 98CS	DTMF 97CS	DTMF 98CS	DTMF 98	DTMF 98,5CS
$L_{av.}$, microns		768	869	1077	901	801
$La_{av.}$,microns		3288	3458	2779	2137	2168
L_{min} -L_{Max}, microns		48-8200	80-10200	70-7150	110-7360	60-5280
L_{10}	μm	160	180	200	240	170
L_{50}	μm	550	560	900	700	550
L_{90}	μm	1450	1300	1750	1340	1120
La_{10}	μm	910	620	900	640	540
La_{50}	μm	2340	2650	1910	2240	2520
La_{90}	μm	5620	6100	5150	5140	5320
M_{500}	%	72	60	73	70	55
M_{1000}	%	32	17	35	25	14
Ma_{500}	%	94	98	97	95	93
M_{a1000}	%	72	75	83	67	62
Mineral composition, % (volume fraction):						
periclase MgO		97-98	97-98	97-98	97-98	>98
Monticellite $CaMgSiO_4$		–	–	–	1-2	–
Mervenite $Ca_3Mg(SiO_4)_2$		≤1	≤1	<1	1-2	≤1
Belite Ca_2SiO_4		1-2	2-3	1-2	–	1-2
Chemical composition, % (mass fractions):						
MgO		98,26	97,44	98,09	97,90	98,66
Al_2O_3		0,05	0,11	0,17	0,15	0,13
SiO_2		0,42	0,60	0,44	0,66	0,34
CaO		0,89	1,45	1,07	1.02	0,72
Fe_2O_3		0,38	0,41	0,24	0,27	0,12
sum, Σ		100,0	100,00	100,00	100,00	100,00
CaO/SiO_2		2,12	2,60	2,43	1,55	2,12
ρ, g/cm^3		3,52	3,48	3,52	3,51	3,52

4. Evaluation of influence of CaO/SiO_2 ratio

With the aim of evaluation of influence of fused periclase quality with various CaO and SiO_2 ratio we tested fused periclase of DTMF 98 grade, which is a part of the new product line Russian Magnezit™, produced by Magnezit Group (97,6-98,1% MgO, CaO/SiO_2 1,4-1,6, bulk density 3,48-3,52 g/cm^3) and imported fused periclase of FM 98CS grade (more than 97,7% MgO, CaO/SiO_2 ≥2,2, apparent density 3,48-3,52 g/cm^3) which are used as main components in production of periclase-carbon refractories for slag line of 130-tons steel-teeming ladles.

Monticellite (t_m = 1500 °C) and mervenite (t_m = 1575 °C) are present as intergranular phase in periclase of DTMF 98 grade in the form of layers 10-20 microns thick (Figure 8).

A part of monticellite and mervenite are distributed in crystals of periclase. Belite (t_m = 2130 °C) prevails in the imported periclase of FM 98CS grade (Figure 9). Mervenite (see Table 2) is in subordinate amount.

Periclase-carbon bricks have the following composition: basis – fused periclase of fraction 0-6 mm, 9±1% of graphite, antioxidant – metallic Al, Carbores P and phenol-formaldehyde binder. Physical and chemical properties of bricks are identical both after thermal treatment and after coking (see Table 3).

Table 3. Physico-mechanical parameters of periclase-carbon bricks after and before coking

Physical and mechanical properties		PC-bricks on the basis of fused periclas	
		DTMF 98	FM 98 CS
Ration CaO / SiO_2		1.55	2.12
Apparent (open) porosity, П, %	before coking	3,8	3,5
	after coking	9,6	8,4
Bulk density, ρ, g/см^3	before coking	3,05	3,07
	after coking	3,00	3,02
Ultimate compressive strength, δ, N/mm^2	before coking	42,8	50,4
	after coking	35,4	45,6

Bricks made from imported periclase of FM 98CS grade and from domestic DTMF 98 grade were tested at a number of plants in slag line of different vessels. Operating conditions were approximately the same. Share of vacuumized metal which influences corrosion resistance of bricks in slag line was also approximately the same at all the plants.

At the plant №1 with EAF-130 and at the plant №2 with EAF-150, steel teeming ladles operate in the cycle: ladle-furnace (100% of heats) – vacuum degasser RH (50-60% of heats) – casting at continuous casting machine (CCM); at the plant №3 with EAF-150 the ladles operate in the following cycle: ladle-furnace (100% of heats) – vacuum degasser VD/VOD (50-60% of heats). Basicity of slag ($CaO+MgO/SiO_2+Al_2O_3$) is in the range of 1,9-2,4.

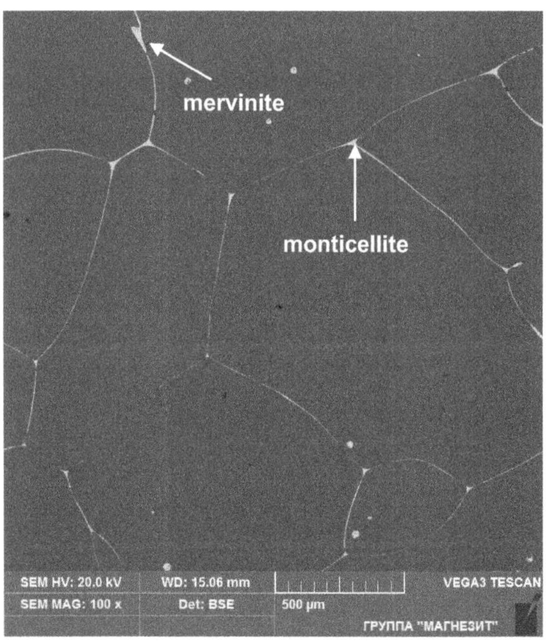

Figure 8. Microstructure of the sample of fused periclase of DTMF 98 grade. SEM. Detector BSE. Magnification: 100×

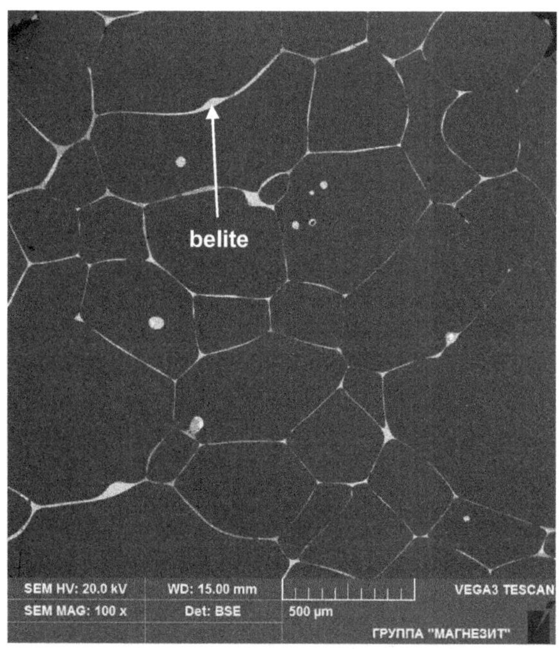

Figure 9. Microstructure of the sample of fused periclase of FM 98CS grade. SEM. Detector BSE. Magnification: 100×

At the plant №1 (EAF-130) service life of the slag line of steel teeming ladle amounted averagely to 40 heats, no matter what grade of fused periclase was used in composition of periclase-carbon bricks.

At the plant №2 (EAF-150) they use technology of hot repair of some areas of steel ladle slag line. Final service life is 110 heats and consumption of periclase-carbon bricks (on the basis of DTMF 98 grade periclase) for repair after one campaign amounted to 4,5 tons, instead of 8,7 tons when bricks made of imported periclase FM 98CS were used. And it should be pointed out that repair works were begun 10-12 heats later.

At the plant №3 service life of bricks in the slag line of steel teeming ladle (before substitution of bricks) amounted to 43-45 heats in both cases.

At the plant №4, where alloyed corrosion resistant steel is produced, melting is done in EAF-120 and then metal is treated in the vessel ASEA or in the degasser VD/VOD. In severe conditions of aggressive action of slag during treatment of metal in the ladle for up to three hours service life of slag lines made of bricks with periclase of DTMF 98 grade achieves averagely 37 heats, while service life of bricks with periclase of FM 98CS grade amounted averagely to 34 heats.

Resistance of slag lines of ladles lined periclase-carbon bricks based on different periclase were approximately the same. Parameters of thermal vessels, operation conditions and service life are shown in Table 4.

Table 4. Parameters of thermal vessels, operation conditions and service life

Technical characteristics and parameters		Steel ladles of the companies			
		№ 1	№ 2	№ 3	№ 4
Steelmaking unit EAF, t		130	150	150	120
The proportion of metal processed in the unit ladle furnace, %		100	100	100	-
Vacuumizing of metal,%	Unit RH	50-60	50-60	-	-
	VD/VOD, ASEA	-	-	50-60	100
	Processing time, min	18-25	18-30	18-30	150-180
Basicity of the slag $(CaO+MgO) / (SiO_2+Al_2O_3)$		2,4	2,0	2,1	1,7-2,3
Resistance of PC-bricks in slag line, number of heats	with DTMF 98	> 40	≥ 110	43-45	av. 37
	with FM 98 CS	> 38	≥ 110	43-45	av. 34
Residual thickness of refractory	%	25-35	30-40	30-40	30-40

Process of periclase-carbon lining wear in metallurgical vessels takes place according to the well known scheme:

– oxidation of carbon component of the refractory with enlargement of pores. Appearance of large remified pores connected with each other. Oxidizing and reduction processes taking place inside refractory also add their share as well as processes providing for removal of gaseous products from reaction zone of the refractory (from the pores of the refractory) during out-of-furnace treatment of metal in the ladle;

– slag goes along formed pores – channels, in which there is no carbon, and comes into contact with coarse – gariend periclase and with fine particles;

– grains of periclase are subjected to corrosion along the surface of contact with formation of secondary silicates, spinellides and others. At the same time washing out of grains from the structure takes place. It is confirmed by presence of periclase grains in the slag contacting the working surface of periclase-carbon refractory (Figure 10).

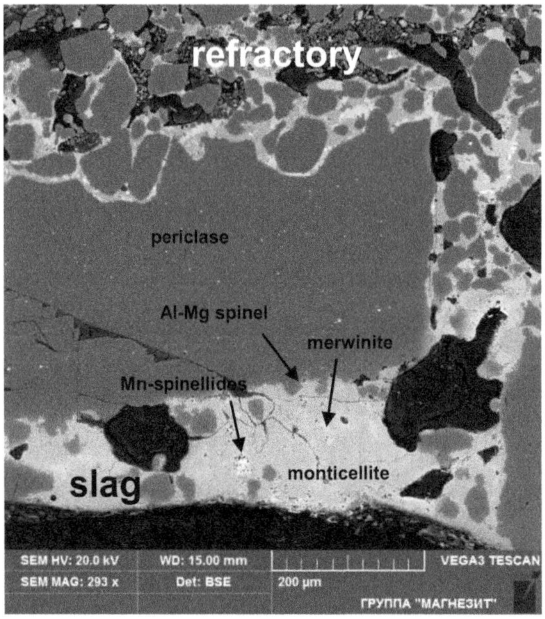

Figure 10. Microstructure of the working zone of a sample of periclase – carbon brick after service. SEM. Detector BSE. Magnification: 293×. Washing out of grains of fused periclase from the refractory with formation of new phases – silicates, Al-Mg spinel (AMS) and complex spinelids

One can suppose, that the final stage of refractories wear has mechanical character and the speed of washing out of periclase grains from the oxidized refractory prevails over the process of slag interaction with the components in intercrystalline space. This version has been checked by comparison of periclase-carbon bricks made of fused periclase of different quality after service in slag lines.

In case of application of fused periclase and content of low-melt monticellite and mervinite totaling to 2,5% we have not discovered considerable influence onto the process of interaction of metallurgical slag (of basic composition) with periclase-carbon refractories in comparison with the case, when more high-melting belite occurs in intercrystalline space of periclase. The share of silicates in periclase DTMF 98 grade is so small that there is no continuous layer of silicates between crystals along which interaction with the slag may be possible, and in this case their quality isn't so important.

In the case of increasing share of silicates (from 2.5 to 4.0 percent) interlayer between the crystals become thicker and at them begins erosion, destruction grains (Figure 11).

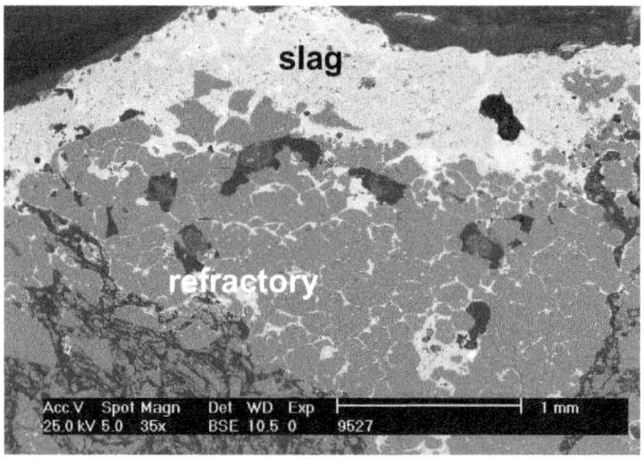

Figure 11. Microstructure of the working zone of a sample of periclase – carbon brick after service. SEM. Detector BSE. Magnification: 35×. Erosion, destruction grains of fused periclase

Identical result was obtained on EAFs at two plants. When carbon containing refractories on the basis of coarse-crystalline fused periclase of grades DTMF 98 and FM 98CS were used, service life of bricks in slag lines was found out to be equal: on the level of 800 heats.

At present periclase-carbon refractories for steel-teeming ladles and electric arc furnaces are manufactured at Magnezit Group with application of own coarse crystalline periclase. CaO/SiO_2 ratio in fused periclase with MgO content 97,5% and more is predetermined, but it has no decisive influence.

We are of the opinion that much greater influence onto the service life is exerted by such a characteristic as average size of periclase crystals and average areal size of crystals. Critical values of these items are correspondingly on the level of 300-500 microns and 1200-1500 microns. The larger is the size of periclase crystals, the higher is service life thanks to lowering the speed of the process of mechanical washout of decarbonized area.

In the case, considered in our report, difference between the average and average areal size of crystals of fused periclase of grades DTMF 98 and FM 98CS didn't influence service life of refractories as these characteristics were higher than 500 microns and 2000 microns correspondingly.

5. Conclusions

We are of the opinion that it is time to develop general universal requirements for determination of periclase crystal size and other characteristics of its quality with the help of novel equipment and software with the aim of development in the future of generally accepted methods and classification of fused periclase by size of crystals and standardization of measurements.

References

Riepl, K., & Barthel, H. (1991). *Large-crystal magnesia clinker for advanced refractories, an update and overview*. UNITECR (pp. 97-101).

Mosser, J., & Riepl, K. (1988). *Method of determining the periclase crystal size of sintered magnesia*. cfi/Ber (pp. 212-216).

Steinwender, W., & Bugajski, J. (1991). *Periclas Cristal Size – Importance, Determination, Propagation*. UNITECR (pp. 91-97).

Sivash, V. G., Perepelitsin, V. A., & Mityushov, N. A. (2001). *Fused Periclase*. Yekaterinburg: Urals' Worker (p. 578).

Saltykov, S. A. (1976). *Stereometric Metallography*. M.: Metallurgy (p.270).

10

Tribological Investigations of PVD Coated Multi-Layer Constructions

Marcus Schulze[1], Holger Seidlitz[1], Franziska König[2] & Sabine Weiß[2]

[1] Chair of Lightweight Construction, Brandenburg University of Technology Cottbus – Senftenberg, Cottbus, Germany

[2] Department of Physical Metallurgy and Materials Technology, Brandenburg University of Technology Cottbus – Senftenberg, Cottbus, Germany

Correspondence: Marcus Schulze, Chair of Lightweight Construction, Brandenburg University of Technology Cottbus – Senftenberg, Cottbus, 03046, Germany. E-mail: marcus.schulze@b-tu.de

Abstract

Multi-layer constructions become more and more relevant in lightweight applications due to their high strength to weight ratio. They offer excellent crash, damping and recycling properties. Also, the morphology of thermoplastic carbon fibre reinforced plastics (CFRP) render them interesting for large scale manufacturing processes. Nevertheless, a major disadvantage results in a poor resistance against wear and tear, e.g. erosion, which is attributed to weak hardness properties. Hence, this work deals with tribological investigations on orthotropic carbon fibre reinforced polymers (PA 6) either with protective ceramic coating or without. The chosen coating system is a well-known protective covering of metal components, e.g. metal cutting tools, produced by physical vapor deposition (PVD). To characterize the coating system on thermoplastic CFRP, standard analyzing methods are utilized, like optical and scanning electron microscopy (SEM). The tribological investigations are conducted by the tribological ball on disk method to generate wear tracks on the sample surfaces and hence to calculate the wear rates. These results are compared to literature findings with respect to a certain protective coating system (TiN) and a second nano-structured gel coating system, where both systems are deposited on a thermosetting material, i.e. carbon fibre reinforced epoxy resin, respectively. For this purpose the feasibility of depositing a protective ceramic layer on thermoplastic CFRP is demonstrated. First results on suitable surface pre-treatments have shown a significant influence on the coating quality. The improved performance regarding the wear behavior with respect to tribology compared to the poor substrate and existing technologies is shown additionally.

Keywords: thermoplastic composites, protective coating systems, TiAlN – ceramic coating, tribology, wear, physical vapor deposition, scanning electron microscopy

1. Introduction

In the last decade the application of fibre reinforced plastics (FRP) steadily increased in transportation industries. Recent examples are the BMW MegaCity Vehicle Project with the i3 and i8 as well as the BMW i7. For the first time FRP are used in large scale serial production in the automotive industry. In the aerospace industry the Airbus A350 and the Boing 787 have an overall quantity of composite materials of approximately 50 % (Marsh, 2015). Up until now FRP are used for lightweight structures. But increasingly they come into to focus for the application as complex technological systems. For example train couplings in FRP design or hydraulic cylinders manufactured with the filament winding process (Scholz & Kroll, 2014). Polymer coatings are used to enhance the erosion resistance of composite materials (Hertel & Watson, 1996) in aerospace industry. Nevertheless, the life time of such coatings is limited (Tiong & Clark, 2011) and therefore alternative solutions are needed. In a different study the deposition of zinc by pulsed gas dynamic spraying process was used to create wear resistant coatings (Metallic coating of aerospace carbon/epoxy composites by the pulsed gas dynamic spraying process, Surface & Coatings Technology 2009. However the hardness of the zinc coatings is lacking and therefore other solutions are necessary. Other studies have demonstrated the feasibility of the deposition of Ti and TiN layers on CFRP and examined the erosion resistance of these coatings (Maurer & Schulz, 2013, 2014). In this study the feasibility of the application of TiAlN coatings is investigated. TiAlN is one of the most studied hard coating systems. It is used due to its excellent properties such as high hardness, good wear behavior and chemical stability when applied on metal surfaces (Sproul, 1996; PalDey & Deevi, 2003; Jehn, 2000). A lot of former studies examined the deposition of TiAlN on metallic and ceramic substrates (Sveen, Andersson, M'Saoubi, & Olsson, 2013; Ibrahim, Rahmat,

Oskouei, & Singh Raman, 2015; Dobrzański, Pakuła, Křiž, Soković, & Kopač, 2006; Kottfer, Ferdinandy, Hvizdoš & Kaczmarek, 2012). The goal of this work is to prove the possibility to deposit TiAlN hard coatings on FRP and therefore lay the foundation for further investigations into the applications for TiAlN coatings on composite materials. Therefore TiAlN coatings are deposited on thermoplastic composites by means of magnetron sputter deposition. Two different coating thicknesses (s1 = 6 µm, s2 = 20 µm) are prepared. The surface roughness before and after the coating is analyzed. By means of tribological ball on disk tests the coefficient of friction is measured. Scanning electron microscope (SEM) is used to analyze the wear traces.

2. Experimental Setup

2.1 Sample Preparation

To ensure high quality sample surfaces the CFRP specimen (Ø = 30 mm) with a Polyamide 6 matrix and orthotropic reinforced continuous carbon fibres (Table 1) are prepared using a mechanical grinding process with SiC wet sandpapers up to 2500 grit. After the grinding process the samples are cleaned with ethanol. With this pretreatment well prepared sample surfaces are obtained, on which a sufficient coating deposition can be executed. Therefore, the CC 800/9 HPPMs PVD coating machine from CemeCon (Würselen, Germany) is utilized. The conducted coating process is described in detail by (Naveed, Obrosov & Weiß, 2015). In spite of, Figures 1a), 1b) and 1c) give an insight on the possible sample positioning within in the working chamber (Figure 1a)). In addition, Figure 1b) depicts the clamping of the sample and Figure 1c) the directly positioned CFRP sample in front of the combined TiAl-target. The necessary nitrogen to obtain a ceramic TiAlN coating is added in the gaseous state. To compare the influence of varying coating thicknesses among each other, the overall coating time of selected samples has been changed. Therewith, three different coating thicknesses are prepared. A first variation is represented by specimen I with a coating thickness of t = 6 µm. A second variation leads to a coating thickness of t = 20 µm represented by specimen II and III. The third variation regarding specimen III is an additional plasma etching pretreatment prior to the coating deposition. Here, specimen III is pretreated to probably increase the adhesion between TiAlN coating and CFRP surface (Maurer, 2014). Table 2 lists all necessary process parameters regarding the conducted deposition process.

2.2 Measurement of Surface Roughness

The evaluation of the surface roughness allows conclusions regarding the quality of the deposition process as well as the influence of the sample pretreatment. Hence, within the present work detailed investigations on the measurement of the surface roughness value Ra are presented and discussed, respectively. To realize independent conclusions, different measurement techniques were in use, like the Surftest 501 series 178 from Mitutoyo (Illinois, USA), a Keyence VHX-5000 3D microscope (Itasca, USA) and common optical microscopy. Even the influence of the fibre orientation is considered. Therefore Ra is determined parallel as well as perpendicular to the fibres of the top layer. Suitable statistics are generated by conducting a sufficient number of measurements, i.e. 5 points over the surface, with respect to the measurement direction. Finally, the resulting average value (Ra) is calculated and compared among each other.

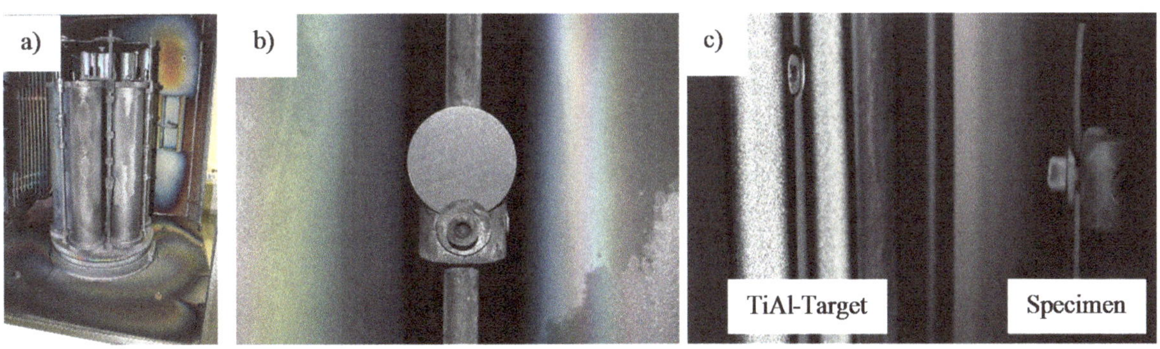

Figure 1. a): Working chamber with specimen holder b): clamping of the sample; c): sample positioned in front of the TiAl – Target

Table 1. Properties of the utilized CFRP

FRP type	Prepreg	Lay up	Young´s modulus $E_1 = E_2$	Tensile strength $\sigma_1 = \sigma_2$	Fibre volume ratio	Thickness	Glass transition temperature
CFRP	UD carbon fibre/PA 6 (Ticona)	$[(0/90)_3]_s$	49.5 GPa	860 MPa	0.6	1.8 mm	220 °C

Note. UD = Unidirectional.

Table 2. Process parameters for deposition process using CC 800/9 HPPMs

Coating process			Unit
Test Chamber	Process temperature	150 ± 5	°C
	Working pressure	< 8.0	mPa
	Argon	200	mln
	Nitrogen	20	mln
Cathode	TiAl Target	2000	W
	Voltage	90	V
	Frequency	280	kHz
Plasma etching pre-treatment			
Argon		160	mln
Table Voltage		650	V
Frequency		240	kHz
Time		300	s

Note. mln = milliliter normal.

2.3 Tribology Testing

To treat the influence of protective coatings deposited on CFRP with respect to the wear and tear behavior in comparison to uncoated CFRP and two other existing coating systems on composites, the tribological ball-on-disc test method is applied (DIN 50324). Therefore, a high temperature tribometer from CSM Instruments (Reseux, Switzerland) is used. In order to realize comparable results the present experimental boundary conditions are in accordance with the investigations of nanocomposite reinforced CFRP (Scholz, Kroll & Schettler, 2014; Scholz & Kroll, 2014). Here, attention should be paid to one limiting condition, which is the material availability. Thus, it was necessary to conduct two tribological tests on each specimen. Consequently, two different working radii (rw = 4 mm; rw = 7 mm) are selected to avoid interferences. Together with the predefined boundary conditions, like wear tracking distance and wear tracking velocity, the tests conducted at a working radius of 4 mm are limited to a certain number of rotations (n = 12,000), which corresponds to a wear tracking distance of 300 m. For the second working radius of 7 mm the limiting requirements result in n = 25,000, which corresponds to a wear tracking distance of 1,100 m, respectively. According to (Scholz, Kroll & Schettler, 2014; Scholz & Kroll, 2014) a load of 10 N is used. The wear partner (counterpart) for all tests is a 100Cr6 steel ball with a diameter of 6 mm applied to the system without lubrication and at room temperature.

3. Results & Discussion

3.1 Coating Quality

The coating quality is an important property, which has to be discussed prior to the evaluation of the surface roughness and the wear behavior. Therefore, all coated samples are investigated by means of optical microscopy (Figure 1). The first results show that on specimen I (Figure 1a)) small cracks along the left side of the sample occur. This peculiarity can be linked to small electric discharges on the sample surface, which were observed at the beginning of the coating process. Larger defects are observed at the surface of specimen II (Figure 1 b)). Here, deformations as well as delaminations took place. Such a delamination was only observed after the start of the coating process. Therefore, it is supposed that low pressure (< 8 mPa) and relatively low temperature (T ≈ 150 °C) boundary conditions in combination with the coating process cause these structural changes. Certain irregularities in the composite matrix and trapped air can be possible reasons as well. The third prepared sample (Figure 1c)) shows merely a small area of delamination on the left side, whereas the remaining specimen surface shows a satisfying structure. At this phase of the work it is particularly noticeable, that all the samples have defects on the left side. This observation can be lessened by the fact, that all specimens were randomly oriented during installation with respect to the fibre orientation of the top layer. Thus, the reason for this defect phenomenon has to be the coating process. Within an industrial coating process multiple targets are used and the sample stage is additionally rotated to improve the coating process quality. In this work, only one TiAl–target was available. Subjected to this condition, a table rotation would have had the consequence of an increased testing time and considerable consumption of the target, respectively. To prevent such a high degradation of the target and for an optimized utilization of the limited conditions, the samples are positioned in static mode right in front of the TiAl- target during the coating process. Besides, a static magnetic field accelerates electrons and Argon ions, wherewith slightly different conditions on the sample surface are generated and possibly produce the observed defects. Nevertheless, all observations regarding the quality of ceramic coating on CFRP confirm the feasibility of protective coatings on composite materials, which will be a great feature in many technical applications.

3.2 Surface Roughness

The surface roughness of either the substrate material or the coated samples has a great influence on the resulting properties. Thus, a suitable surface roughness of the substrate material is decisive for an improved clamping of the deposited protective coating. On the other hand, this positive effect can turn negative, if the surface roughness is too high due to an insufficient bonding between substrate and coating with respect to resulting cavities (Maurer, 2014). It is also necessary to consider, that the utilized PVD deposition process reproduces the surface of the substrate material (Maurer, 2014). This is confirmed by the observations due to optical microscopy. To get first information about the nature of the surface, representative images produced by means of optical microscopy are shown within Figure 2. Here, Figure 2a) depicts the surface morphology of the uncoated and mechanically grinded CFRP substrate. It is peculiar, that the surface shows a rough behavior, i.e. edges and grooves, which are aligned parallel to the top fibre layer. The mechanical pretreatment even exposed the top fibre layer. This rough surface morphology is in accordance with the measured surface roughness values (Ra,0° = 0.5 μm; Ra,90° = 0.7 μm), which are presented by Figure 3. Here, only a slight deviation occurred with respect to the fibre orientation. Comparable results are discussed within the work of Maurer (Maurer, 2014), who measured the surface roughness of a thermoplastic CFRP, i.e. polyether ether ketone (PEEK), as well. The stated value for the surface roughness is Ra = 0.4 μm. As obvious from Figure 3 the surface of specimen I (t = 6 μm) is less rough compared to the substrate material, and no corresponding anisotropy with respect to the fibre orientation is observed. This effect is also representative in Figure 1a) and Figure 2a). To get a more detailed insight on the surface morphology a 3D map of the sample surface is generated using a Keyence VHX-5000 3D microscope (Figure 4). Here, the surface irregularities, like grooves and edges, are obvious. Specimen II (t = 20 μm), without plasma etching pretreatment, exhibits the highest surface roughness value for the 90° fibre orientation. Note that all measurements are taken in the non-delaminated area. Thus, the high Ra values cannot be explained with the delamination occurred during the coating process. From the literature it is well known, that an increasing coating thickness causes increasing residual stresses within the coating. This may be an explanation for the higher surface roughness. Also the coating growth, which can be different for very thin coatings, can increase the roughness of the coated surfaces. In contrast to these results are the resulting surface roughness values of specimen III (Figure 3). The roughness value Ra is smaller compared to specimen II but with the same coating thickness, and even the fibre orientation has only a slight influence. This can be caused by reduces residual stresses and a more homogeneous coating growth.

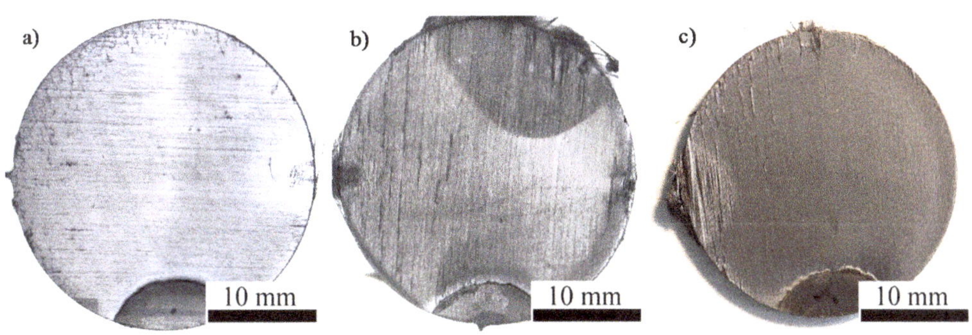

Figure 2. a) specimen I with 6 μm TiAlN layer; b) specimen II with 20 μm TiAlN layer; c) specimen III with 20 μm TiAlN layer with plasma etching pretreatment; d) cross section detail of b

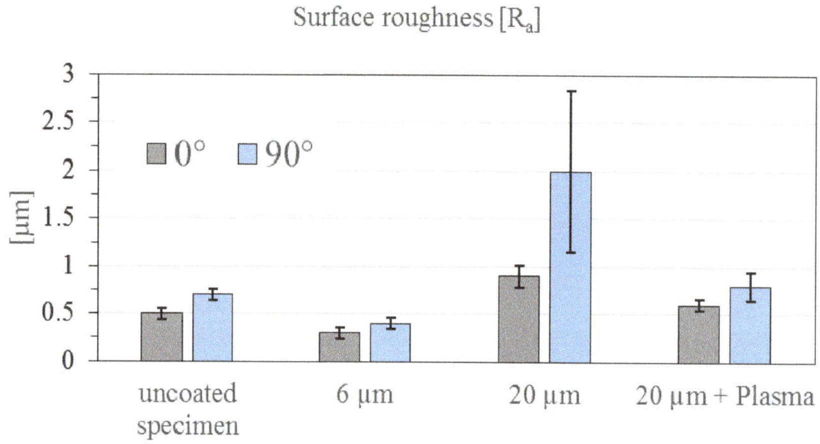

Figure 1. Representation of measured surface roughness values

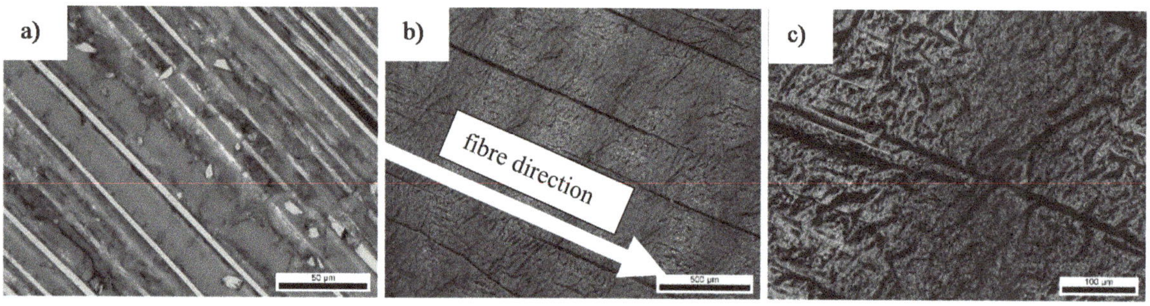

Figure 4. left: Optical microscopic image of an uncoated CFRP specimen produced by mechanical grinding up to 2500 grit; middle: Optical microscopic image of specimen I; right: specimen I with higher magnification

3.3 Tribology Testing

The friction coefficient as a function of the sliding distance is presented in Figure 6. Compared are the specimen I - III and a reference sample without coating (Table 3). The curves exhibit two states, the transient state with a rising coefficient of friction and the steady state with constant or slowly rising friction. From the graphs can be seen, the coating with plasma etching shows a similar friction behavior as the uncoated specimen, albeit at a higher coefficient of friction. Overviews of each wear track and representative detailed magnifications as well an exemplary abrasion of tribological counterpart are given in the Figures 7 - 12.

Specimen I

The first test (normal load 10 N, $r_w = 4$ mm) aborted before the intended 12000 rotations (300 m sliding distance) could be reached. This was caused by the exceeding of the maximum allowed shear force. The specimen shows an exposure of the substrate material and the fibres in two areas (marked in Figure 7). The damaged areas are at points, where the tribological counterpart moved with 90° to the fibre direction. In areas where the counterpart moved with 0° to the fibre layer, no damage was observed. To avoid the premature termination of the experiments the load is reduced to 5 N for further tests. In the second test (normal load 5 N, $r_w = 7$ mm) on specimen I, fibre exposure still occurred and the test aborted at 909 m. However, most of the test was completed before termination (909 m of 1100 m). An overview of the wear track is given in Figure 8 and an area of fibre exposure is shown. The area of fibre exposure is drastically reduced, on the whole wear track only a single coating failure is found (Figure 8 middle). At a closer examination the grooves shown in Figure 5 can be seen. The coating failed exactly at one of these groves. Furthermore the parallelism of grooves and fibres is clearly discernible.

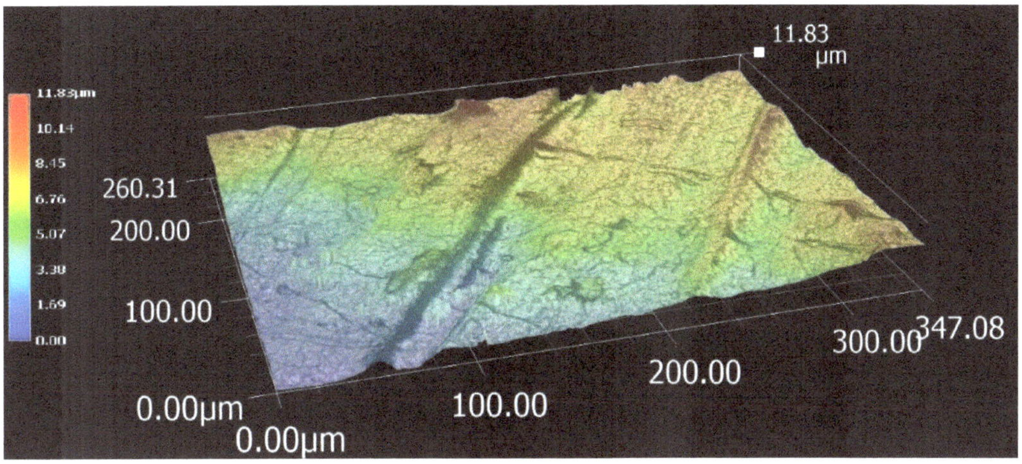

Figure 5. 3D map of the specimen surface with the 6 µm coating (Figure 2a) showing intrusions and extrusions of the TiAlN coating

Specimen II

Specimen II shows no exposure of matrix material or fibres. The test with 4 mm working radius was completed. But the test with 7 mm working radius terminated prematurely similarly to specimen I. The wear of the counterpart in the test with 7 mm working radius is larger than the one with 4 mm working radius, even with a smaller covered distance (Table 3). A possible reason may be the delaminated area (Figure 2b). In this delaminated area the surface roughness is increased and the specimen is no longer flat, but shows a wavy surface. That leads to a greater wear of the counterpart. It should be noted, that even in this delaminated area no fibre exposure occurred. The wear tracks for both test (normal load 5 N, rw = 4 mm & rw = 7 mm) are shown in Figure 9 and 10. On specimen I a difference in the wear could be observed. If the counterpart moved 90° or 0° to the fibre direction. For the 20 µm coating no such effect is observed.

Specimen III

For specimen III both tests (normal load 5 N, rw = 4 mm & rw = 7 mm) where completed. Figure 11 and 12 show the two tribological tests, no exposure of fibres or matrix material was found.

In all tests the wear of the counterpart was conspicuously high. A hardness of 849 ± 29 HV 0.025 for the 100Cr6 balls was measured using micro hardness testing with a load of 25 g. Because of the much higher wear rate of the balls compared to the one of the coating it can be assumed that the deposited TiAlN coating is harder than the 100Cr6 balls. In Figure 13 are given the wear tracks of an uncoated specimen. In the upper part of the wear tracks the deformed matrix material can be seen. Furthermore fibre damage can be observed. In comparison to the results of the coated specimen this shows the suitability of the TiAlN system for the application as a wear resistant coatings. With a high coefficient of friction a higher wear can be expected. When comparing the results of the tests with rw = 4 mm of the specimen II and III, which both are subjected to the same sliding distance and load. A considerable smaller wear was observed on the counterpart of the specimen with the smaller coefficient of friction. The same can be observed in the results of for the tests with rw = 7 mm for specimen I and III. Specimen I shows a higher coefficient of friction and, even with a smaller sliding distance, a greater wear of the counterpart (Figure 6, Table 3).

Table 3. Diameter of Abrasion of tribological counterpart

specimen	Radius r_w [mm]	Load [N]	rotations	Sliding distance [m]	Diameter abrasion 100Cr6 balls [µm]	test
I – 6 µm coating	4	10	7569	190	1470.98	Aborted
	7	**5**	**20 560**	**909**	**1764.57**	**Aborted**
II – 20 µm coating	**4**	**5**	**12 013**	**302**	**1463.87**	**Finished**
	7	5	4247	187	1851.94	Aborted
III – 20 µm coating with plasma etching	**4**	**5**	**12 018**	**302**	**1181.46**	**Finished**
	7	**5**	**25 012**	**1113**	**1673.14**	**Finished**

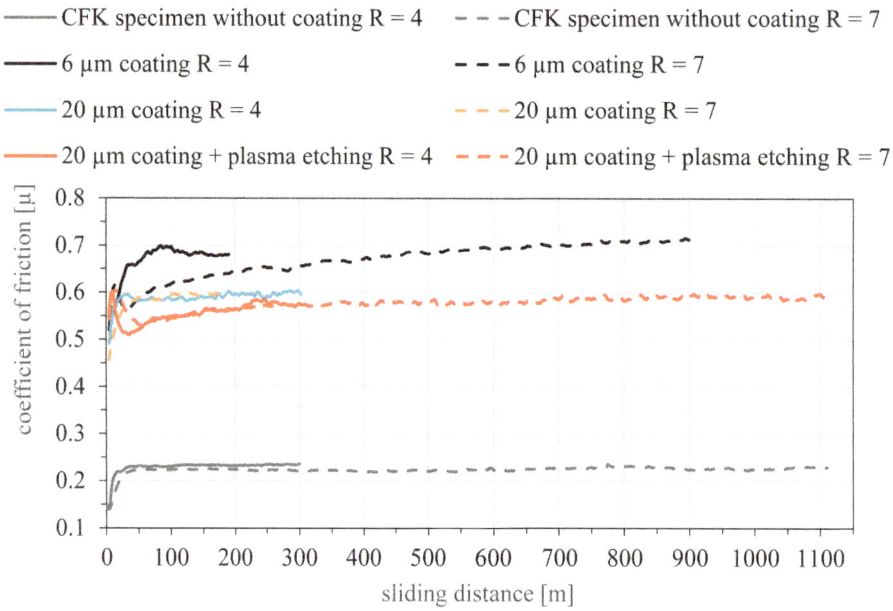

Figure 6. Friction coefficient plotted in relation to the sliding distance

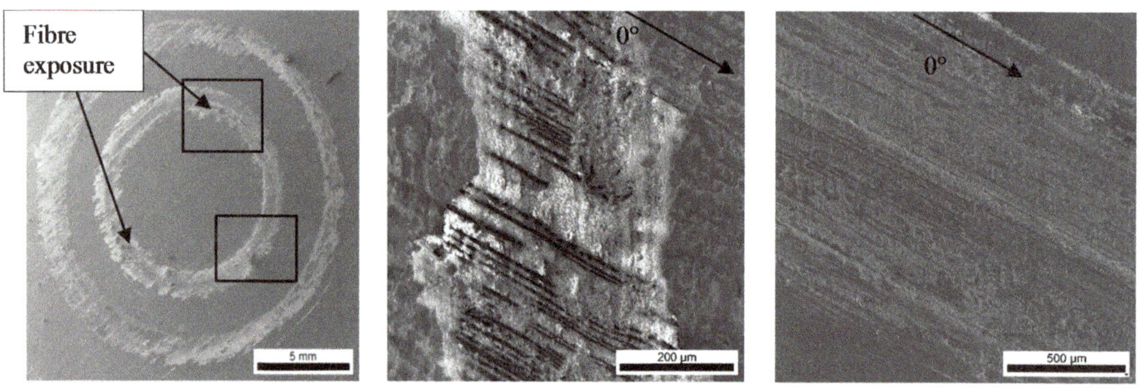

Figure 7. left: overview of Specimen I r_w = 4 mm, middle: exposed fibres, right: undamaged coating

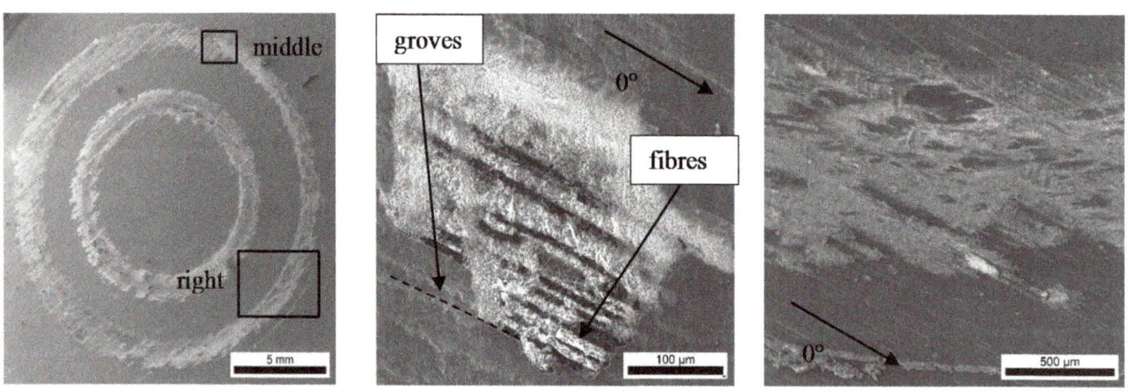

Figure 8. left: overview of Specimen I r_w = 7 mm, middle: exposed fibres, right: undamaged coating

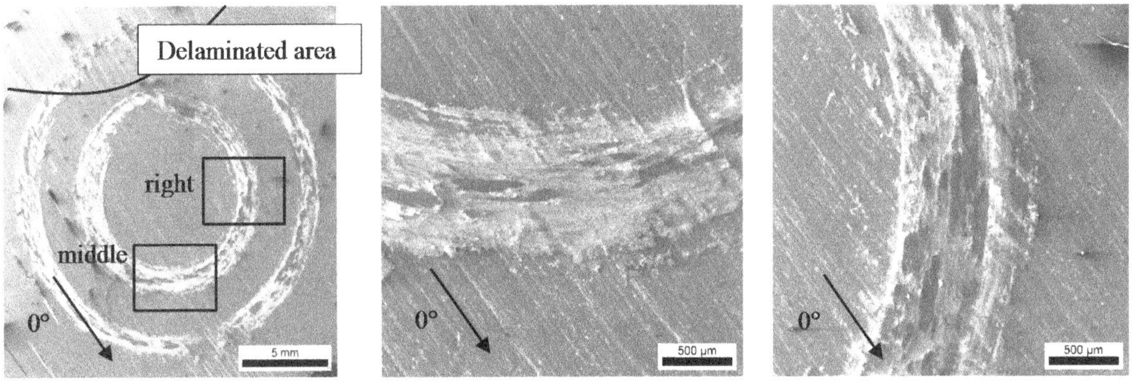

Figure 9. left: overview of Specimen I $r_w = 4$ mm with delaminated area, middle: undamaged coating, right: undamaged coating

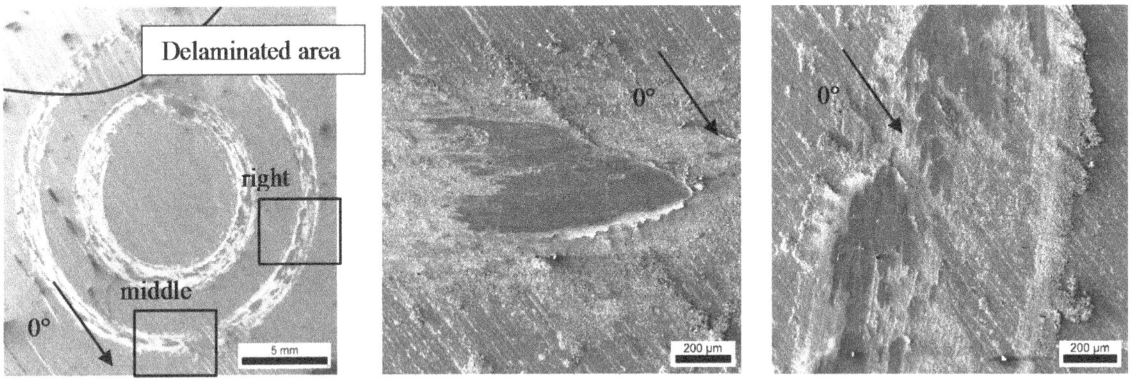

Figure 10. left: overview of Specimen II $r_w = 7$ mm with delaminated area, middle: undamaged coating, right: undamaged coating

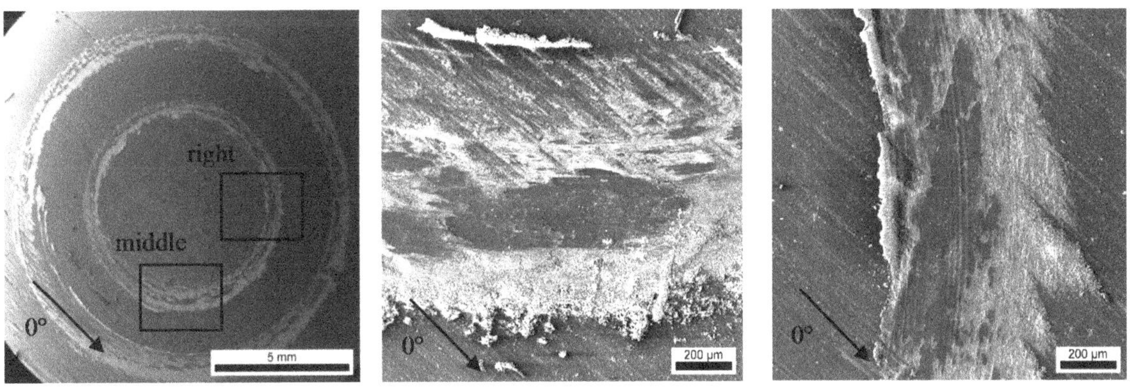

Figure 11. left: overview of Specimen III $r_w = 4$ mm, middle: undamaged coating, right: undamaged coating

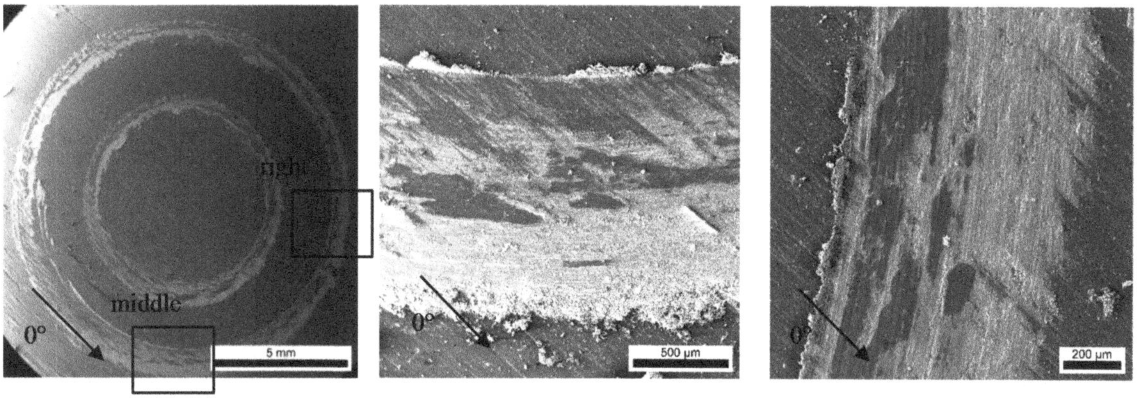

Figure 12. left: overview of Specimen III $r_w = 7$ mm, middle: undamaged coating, right: undamaged coating

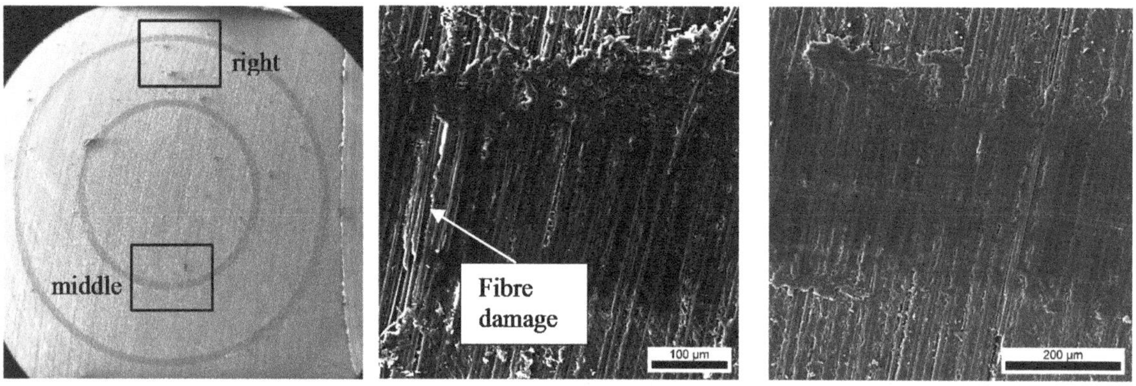

Figure 13. left: overview uncoated specimen, middle: wear track 300 m sliding distance, right: wear track 1100 m sliding distance

4. Conclusions

In this study the deposition of a ceramic TiAlN layer by means of PVD coating process on thermoplastic CRFP is evaluated. The investigations reveal that TiAlN coatings can be applied on CFRP at least up to a coating thickness of t = 20 µm. However, the process parameters have to be improved, to realize high quality protective coating systems, which can be utilized to cover components and parts in many technical applications. Additionally, the presented results show, that a growing coating thickness is directly related to an increase in surface roughness. This effect can be mitigated by applying a suitable plasma etching pretreatment prior to the coating deposition. Nevertheless, further investigations on this topic are necessary. In addition, the presented TiAlN-coating system offer considerable wear resistance compared to uncoated CFRP. This statement is supported due to the absence of discernible effects regarding the conducted tribological investigations for specimen II and III (t = 20 µm). Thus, the presented ceramic coating is well suited for tribological applications, e.g. for substituting sleeves in hydraulic actuators. The wear of the tribological counterpart (849 ± 29 HV 0.025) is considerably higher than that of the coating. Therefore it can be assumed, that the hardness of the coating surpasses that of the counterpart. Here, it has to be considered that hardness testing in areas of thin film technology becomes more complex and has to be treated carefully. From these primary results it is even recommended to utilize ceramic counterparts instead of steel material for further investigations. Since the coating quality is strongly influenced by exposed fibers of the CFRP surface, this has to be treated with great attention. Therefore investigations of applied initial layers, like metallic interface layer, can provide useful results for better adhesion. Future results will be related to these topics.

Acknowledgments

This work was performed within the InnoProfile-Transfer project Lightweight with structured materials, which is supported by the German research foundation BMBF. Financial support is gratefully acknowledged.

References

Dobrzański, L. A., Pakuła, D., Křiž, A., Soković, M. & Kopač, J. (2006). Tribological properties of the PVD and CVD coatings deposited onto the nitride tool ceramics. *Journal of Materials Processing Technology, 175(1-3)*, 179–185. http://dx.doi.org/10.1016/j.jmatprotec.2005.04.032

Hertel, C. J., & Watson, C.R. (1996). Erosion resistant surface protection. *U.S Patent*, Patent Number: US5486096 A

Ibrahim, R. N., Rahmat, M. A., Oskouei, R. H. & Singh Raman, R. K. (2015). Monolayer TiAlN and multilayer TiAlN/CrN PVD coatings as surface modifiers to mitigate fretting fatigue of AISI P20 steel. *Engineering Fracture Mechanics, 137*, 64–78. http://dx.doi.org/10.1016/j.engfracmech.2015.01.009

Jehn, H. A. (2000). Multicomponent and multiphase hard coatings for tribological applications. *Surface and Coatings Technology, 131*(1-3), 433–440. http://dx.doi.org/10.1016/S0257-8972(00)00783-0

Kottfer, D., Ferdinandy, M., Hvizdoš, P., & Kaczmarek, L. (2012). Comperative study of properties of Ti based coatings depositetd by selected PVD techniques.

Marsh, G. (2015). Composites in commercial jets. *Reinforced Plastics, 59*(4), 190-193. http://dx.doi.org/10.1016/j.repl.2015.06.001

Maurer, C., & Schulz, U. (2013). Erosion resistant titanium based PVD coatings on CFRP. *Wear, 302*(1-2), 937–945. http://dx.doi.org/10.1016/j.wear.2013.01.045

Maurer, C., & Schulz, U. (2014). Solid particle erosion of thick PVD coatings on CFRP. *Wear, 317*(1-2), 246–253. http://dx.doi.org/10.1016/j.wear.2014.05.016

Maurer, C. (2014). *Versagensmechanismen von PVD-Beschichtungen auf CFK unter Erosionsverschleiß* (Doctoral dissertation), Rheinisch-Westfälischen Technischen Hochschule Aachen, Germany.

Naveed, M., Obrosov, A., & Weiß, S. (2015). Investigation of the Wear Resistance Properties of Cr/CrN Multilayer Coatings against Sand Erosion. *Conference Papers in Science, 2015*, 1–9. http://dx.doi.org/10.1155/2015/873543

PalDey, S., & Deevi, S. C. (2003). Single layer and multilayer wear resistant coatings of (Ti,Al)N: a review. *Materials Science and Engineering: A, 342*(1-2), 58–79. http://dx.doi.org/10.1016/S0921-5093(02)00259-9

Scholz, S., Kroll, L., & Schettler, F. (2014). Nanoparticle reinforced epoxy gelcoats for fiber-plastic composites under multiple load. *Progress in Organic Coatings, 77*(7), 1129–1136. http://dx.doi.org/10.1016/j.porgcoat.2014.03.012

Scholz, S., & Kroll, L. (2014). Nanocomposite glide surfaces for FRP hydraulic cylinders – Evaluation and test. *Composites Part B: Engineering, 61*, 207–213. http://dx.doi.org/10.1016/j.compositesb.2014.01.044

Sproul, W. D. (1996). Physical vapor deposition tool coatings. *Surface and Coatings Technology, 81*(1), 1–7. http://dx.doi.org/10.1016/0257-8972(95)02616-9

Sveen, S., Andersson, J. M., M'Saoubi, R., & Olsson, M. (2013). Scratch adhesion characteristics of PVD TiAlN deposited on high speed steel, cemented carbide and PCBN substrates. *Wear, 308*(1-2), 133–141. http://dx.doi.org/10.1016/j.wear.2013.08.025

Tiong, U. H., & Clark, G. (2011). Impact of Mechanical Strain Environment on Aircraft Protective Coatings and Corrosion Protection. *Journal of Aircraft, 48*, 1315–1330. http://arc.aiaa.org/doi/abs/10.2514/1.C031270

Woelfle, M. (n.d.). With the A350 XWB Premium AEROTEC pushes in new technological dimensions. Press release.

11

Spectroscopic and Thermal Studies of Polyalkoxysilanes and Silica-Chitosan Hybrid Materials

Antonio Sánchez-Fernández[1], Laura Peña-Parás[2], Elisa M. Mendoza[2], Alejandra Leyva[1], Luis A. Bautista[2], Franz-Xaver Bulach[1], Alejandra Monsivais-Barrón[1], Jaime Bonilla-Ríos[1] & Luis E. Elizalde[3]

[1] Instituto Tecnológico de Estudios Superiores de Monterrey, Eugenio Garza Sada 2501, Col. Tecnológico, Monterrey, N. L., México

[2] Departamento de Ingeniería, Universidad de Monterrey, Ave. Morones Prieto 4500 Pte., Col. Jesús M. Garza, San Pedro Garza García, N. L., México

[3] Centro de Investigación en Química Aplicada, Blvd. Enrique Reyna Hermosillo 140 Saltillo, Coahuila CP25250, México

Correspondence: Laura Peña-Parás, Departamento de Ingeniería, Universidad de Monterrey, Mexico. E-mail: laura.pena@udem.edu

Abstract

Polymers of organoalkoxysilanes (or polyalkoxysilanes) were synthesized through sol-gel process using different organosilanes: Tetraethylorthosilicate (TEOS), (3-Glycidyloxypropyl) trimethoxysilane (GPTMS), Trimethoxy(propyl)silane (TMPS), and/or Triethoxy(octyl)silane (EOS). Different carboxylic acids (acetic, formic, and citric) were used to prepare polyalkoxysilanes, with citric acid acting as a reactant rather than a catalyst for the sol-gel reaction. Silica-chitosan hybrid materials were synthesized reacting polyalkoxysilanes and chitosan, one of the most important and most studied natural polymers, through sol-gel process. The synthesized materials were characterized using Fourier transform infrared spectroscopy (FT-IR), thermogravimetrical analysis (TGA), differential Scanning Calorimetric (DSC) and solid-state Carbon-13 Nuclear Magnetic Resonance (solid-state ^{13}C-NMR) spectroscopy, obtaining information on the structural properties of each modified materials. The results obtained by NMR indicate that the oxirane group of GPTMS remains in all reaction products to form different polyalkoxysilanes.

Keywords: sol-gel process, hybrid materials, polyalkoxysilanes, silica-chitosan

1. Introduction

Polyalkoxysilanes (alkoxysilanes or organofunctional alkoxysilanes) are inorganic/organic hybrid materials at a molecular level, which have been used on different materials in order to modify their surface (Chen & Yakovlev, 2010; Ebnesajjad, 2011) and act at the interface between organic polymers and inorganic substrates as adhesion promoters or coupling agents (Gelest, 2015). They provide a great level of adhesion with resistance to environmental and other destructive forces, such as heat and moisture, without affecting physical or chemical forces at the interface (Ebnesajjad, 2011). Chitosan is the deacetylated form of chitin *K*-glucosamine for utilization in chemical reactions, and it has tree kind of reactive functional groups at the C-2, C-3, and C-6 positions. The chemical modification of C-2, C-3 and C-6 positions can provide numerous materials with different physical properties for applications in biomaterials, anti-bacterial textiles (Ivanova & Philipchenko, 2012), tissue reconstruction (Gingras, Paradis, & Berthod, 2003), artificial skin (Muzzarelli, 1993), drug delivery systems (Rui & San Román, 2005), food additives (Ruiz-Hitzky, Darder, & Aranda, 2008) supports for cells (Khor, 2004), enzymes, and catalysts (Khor, 2004).

1.1 Hybrid Materials

Silica-chitosan hybrid materials have been commonly synthesized by sol-gel process reacting chitosan and alkoxysilanes. For example, Smitha, Shajesh, Mukundan, and Warrier (2008) obtained highly transparent silica-chitosan hybrids through a sol-gel process by in situ hydrolysis–condensation reaction of methyltrimethoxysilane (MTMS) and vinyltrimethoxysilane (VTMS) with chitosan that may be used for optical applications. Similarly,

Budnyak, Pylypchuk, Tertykh, Yanovska, and Kolodynska (2015) synthesized a chitosan-silica material by the sol-gel method through hydrolysis of Tetraethoxysilane (TEOS) in a chitosan solution.

These materials have been synthesized for biomedical applications, such as for bone regeneration (Dorozhkin, 2009). For example, Eun-Jung et al. (2009), fabricated a silica-chitosan xerogel with Tetramethylorthosilane (TMOS) and medium molecular weight chitosan for guided bone regeneration (GBR) through sol-gel process. The osteoblastic cells adhered well and grew actively on the silica-chitosan material higher than on the chitosan material. Also, silica-chitosan hybrid materials have been applied in nanomedicine (Solansk et al., 2009). Moreover, Solanki et al. (2009), developed a cholesterol biosensor through multi-walled carbon nanotubes (MWNTs) silica-chitosan nanobiomaterials by sol-gel process using Tetraethylorthosilicate (TEOS), 86% deacetylated chitosan, and acetic acid. Other applications have been developed in order to aid healing processes (Susuki, & Mizushima, 1997; Knapczyk, Macura, & Pawlik, 1992; Biagini et al., 1991), as well as for separation membranes (Liu, Su, & Lai, 2007).

The properties of hybrid materials do not simply result from the sum of the individual contributions of their components, but also from the strong synergy created by an extensive hybrid interface (Sanchez et al., 2001). For instance, the mineral/organic interface, including the types of interactions present, the surface energy, and the existence of labile bonds play a strong role in modulating a number of properties (optical, mechanical, separation capacity, catalysis, and chemical and thermal stability) (Nicole, Rozes, & Sanchez, 2010). The existence of covalent chemical bonds between organic and mineral components presents a number of advantages: (i) the potential of synthesizing entirely new materials from functionalized alkoxides, (ii) the minimization of phase separation, and (iii) a better definition of the organic-inorganic interface. Such defined control can lead to a better understanding of the material and the relationship between microstructure and properties, and an easier adjustment of the hydrophilic-hydrophobic balance (Nicole, Rozes, & Sanchez, 2010; Ciriminna et al., 2013).

Soft chemistry-based routes including sol–gel chemistry (hydrolytic or non-hydrolytic) may be performed from simple precursors or via the use of specific bridged and polyfunctional precursors. Conventional sol–gel pathways to hybrid networks are obtained through condensation of organically modified metal alkoxides. These materials, exhibiting infinity of microstructures, are amorphous, generally transparent, and easily shaped as films or bulks (Sanchez et al., 2014). The sol-gel process is very versatile as it allows the incorporation of inorganic component like metal alkoxides, namely Si (Ahmad, Sarwar, & Mark, 1997; Ahmad, Sarwar, & Mark, 1997), Ti (Ahmad, Sarwar, & Mark, 1998; Ahmad, & Sarwar, 1997), Zr (Rehma, Sarwar, Ahmad, Krug, & Schmidt, 1997), or even bioactive materials (Juhasz, & Best, 2011) in order to modify final chemical and physical properties of materials at low processing temperatures (Al-Sagheer, & Muslim, 2010).

In this work we prepared organoalkoxysilane [RSi-(OR')3] polymers (polyalkoxysilanes), and hybrids of polyalkoxysilanes with low molecular weight chitosan obtaining silica-chitosan hybrid materials, synthesized by sol-gel method. Spectroscopic and thermal characterization was done by solid-state Carbon-13 Nuclear Magnetic Resonance (^{13}C-NMR), Fourier transform infrared spectroscopy (FT-IR), thermogravimetrical analysis (TGA), and differential Scanning Calorimetric (DSC) to obtain a better understanding of the structure and chemistry of the different reactions.

2. Experimental Section

2.1 Materials

Tetraethylorthosilicate (TEOS), (3-Glycidyloxypropyl)trimethoxysilane (GPTMS), Trimethoxy(propyl)silane (TPMS), Triethoxy(octyl)silane (EOS), formic acid, acetic acid, and citric acid were obtained from Sigma-Aldrich and used without further purification. Low molecular weight chitosan with a degree of deacetylation of ≥75%, distilled water, and ethanol, were also purchased from Sigma-Aldrich.

2.2 Synthesis of Polyalkoxysilanes

In order to prepare polyalkoxysilanes 45 mL of ethanol, 50 mL of distilled water, and 12.5 mL of acetic acid, or formic acid were placed a glass container; and then the reagents TEOS, TMPS, GPTMS, and EOS were added in the concentrations shown in Table 1 (in mol). The reaction mixture was kept under constant stirring, and placed at 50 °C during 24 h; after evaporating the solvent, then the sample was aged for 24 h at room temperature. Polyalkoxysilanes prepared with citric acid were done in a 1 L reactor mixing 250 mL of distilled water, 20 mL of ethanol, and TMPS or GPTMS (Table 1). The temperature was kept at 50 °C, and the reaction time was 2 h. This was done in order to obtain esters of organoalkoxysilanes with the carboxylic groups of citric acid. Thus, if each of the carboxylic groups of the citric acid reacts with an organoalkoxysilane it may lead to a three-dimensional cross-linked polymer, as shown in Figure 1.

Table 1. Reagents and quantities in moles for polyalkoxysilanes and silica-chitosan hybrid materials

Polyalkoxysilanes	Acetic acid (mol)	Formic acid (mol)	Citric acid (mol)	TEOS (mol)	TMPS (mol)	GPTMS (mol)	EOS (mol)
S-T$_1$	0.22				0.275		
S-T$_1$-f	-	0.33	-	-	0.275	-	-
S-T$_1$-c	-	-	0.092	-	0.275	-	-
S-G-c	-	-	0.074	-	-	0.222	-
S-GE	0.22	-	-	-	-	0.057	0.040
S-T$_1$G	0.22	-	-	-	0.049	0.057	-
S-T$_1$E	0.22	-	-	-	0.049	-	-
S-TT$_1$	0.22	-	-	0.056	0.049	-	-
S-TT$_1$G	0.22	-	-	0.038	0.048	0.039	-
S-TGE	0.22	-	-	0.038	-	0.039	0.027
S-T$_1$GE	0.22	-	-	-	0.048	0.039	0.027
S-TT$_1$GE	0.22	-	-	0.006	0.005	0.006	0.004
Silica-chitosan hybrid	Chitosan (g)			TEOS (mol)	TMPS (mol)	GPTMS (mol)	EOS (mol)
SCh-TT$_1$	6			0.11	0.14	-	-
SCh-T$_1$	6			-	0.28	-	-
SCh-GE	6			-	-	0.11	0.08

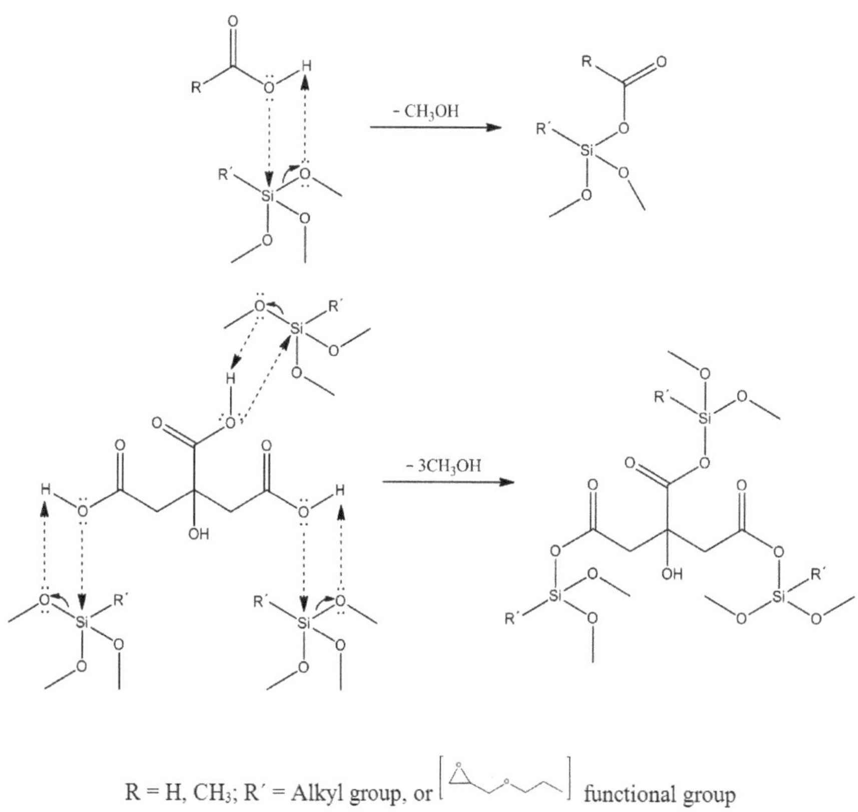

Figure 1. Reaction mechanism of organoalkoxysilanes with carboxylic acids (acetic, formic, or citric)

2.3 Synthesis of Silica-Chitosan Hybrid Materials

Several studies have demonstrated the commercial importance of chitosan as a core component for different applications (Ivanova & Philipchenko, 2012; Gingras, Paradis, & Berthod, 2003; Muzzarelli, 1993; Rui & San Román, 2005; Lee et al., 2009; Solanski et al., 2009; Susuki & Mizushima, 1997; Knapczyk, Macura, & Pawlik, 1992; Biagini et al., 1991; Liu, Su, & Lai, 2007). Though chitosan offers a great alternative to other polysaccharides, it is imperative to understand the effect of the incorporation of suitable functional groups on its surface for higher-end applications

Selected polyalkoxysilanes were chosen to prepare silica-chitosan hybrid materials due to their thermal behavior by DSC, as shown in Table 4; thus, silica-chitosan hybrids SCh-T_I, SCh-TT_I, and SCh-GE were synthetized.

Low weight chitosan (6g) and acetic acid were placed in a glass container in a concentration of 5 % v/v to distilled water and stirred at 90 °C for 1 h. Afterwards 20 mL of ethanol was added to the solution and stirred at 90 °C. Finally, organoalkoxysilanes TEOS, TMPS, GPTMS, and EOS were added to the solution drop by drop using a funnel. The completed solution was aged at room temperature (25 °C) for 48 h. The reagents and concentrations for the preparation of silica-chitosan materials are shown in Table 1.

2.4 Characterization Methods

All samples were analyzed by solid-state ^{13}C-NMR spectroscopy with an 11.7 Tesla Bruker Avance III equipment. Each sample was tested using cross-polarization (CP) and magic-angle-spinning (MAS) with a spinning rate of 125 MHz. A 4 mm inner diameter rotor with a spinning rate of 7 kHz was used. FT-IR characterization was carried out with a Thermo Nicolet 6700 spectrophotometer using the Attenuated Total Reflectance (ATR) technique. Each sample was scanned 42 times at a resolution of 4 cm^{-1} from 4000 - 400 cm^{-1}. TGA was done with a SDT Q600 TA Instruments Thermogravimetric Analyzer with DSC, with a heating rate of 10 °C min^{-1} from room temperature (25°C) to 800°C, and a flow rate of 100 ml/min of nitrogen gas.

3. Results and Discussion

Hybrid organic-inorganic solids represent an important class of engineering materials, usually prepared by sol-gel processes through cross reaction between organic and inorganic precursors. For the reaction under acid catalysis and nucleophilic attack of a silanol oxygen atom at the Si atom of a protonated silanol under displacement of water, the nucleophilic oxygen atom in such a reaction is not restricted to be a central oxygen atom in a siloxane chain, as a silanol or alcohol or water oxygen atom will act in a similar manner. Furthermore, it has long been known that polyalkoxysilanes can be formed by strong Bronsted or Lewis acids. Therefore, while the pKa of formic acid (3.74) is smaller than that of acetic acid (4.76) the resulting product is similar in structure and their infrared spectrum has no appreciable differences for each material synthesized, except for the reaction of citric acid and GPTMS producing the corresponding ester. Thus, the pKa of each of the carboxylic acids of the citric acid molecule does not affect the reaction. Polyalkoxysilane S-T_I-c formed a viscous gel, whereas S-G-c was a crystalline product most likely due to the opening of the epoxy group of GPTMS (Figure 3b) in S-G-c forming C–OH groups. This in turn may result in a further condensation reaction forming Si–OR from a hydroxyl group (where R is alkyl radical), as depicted in Figure 1.

Solid-state NMR is an extraordinarily powerful physical method widely applied in different fields of chemistry. It is particularly valuable at probing complex molecular systems, representing an ideal tool for characterizations of materials with dimensions on a nanoscale. Solid-state NMR experiments at MAS, performed for nuclei 1H, 13C, 15N, 31P, 29Si, and 29Xe, play a central role in various investigations of structures and dynamics in catalysts, polymers, glasses, zeolites, batteries, liquid crystals, natural products, pharmaceutical compounds, membrane proteins, and amyloid fibrils (Lesage, 2009). Figure 2 shows the 13C NMR spectra from samples S-GE, and S-T1G, indicating the obtained signals in ppm.

Table 2. Signals obtained from silica-chitosan materials using ^{13}C-NMR spectroscopy

Sample	Signal (ppm)							
	C1	C2	C3	C4	C5	C6		
SCh-TT_I	103.6	61.1	76.8	84.4	62.5	64.5	-----	17.7
SCh-T_I	105.1	61.1	77.4	86.5	64.5	66.3	18.9	17.9
SCh-GE	102.9	61.0	76.9	84.0	61.0	64.0	18.4	18.0

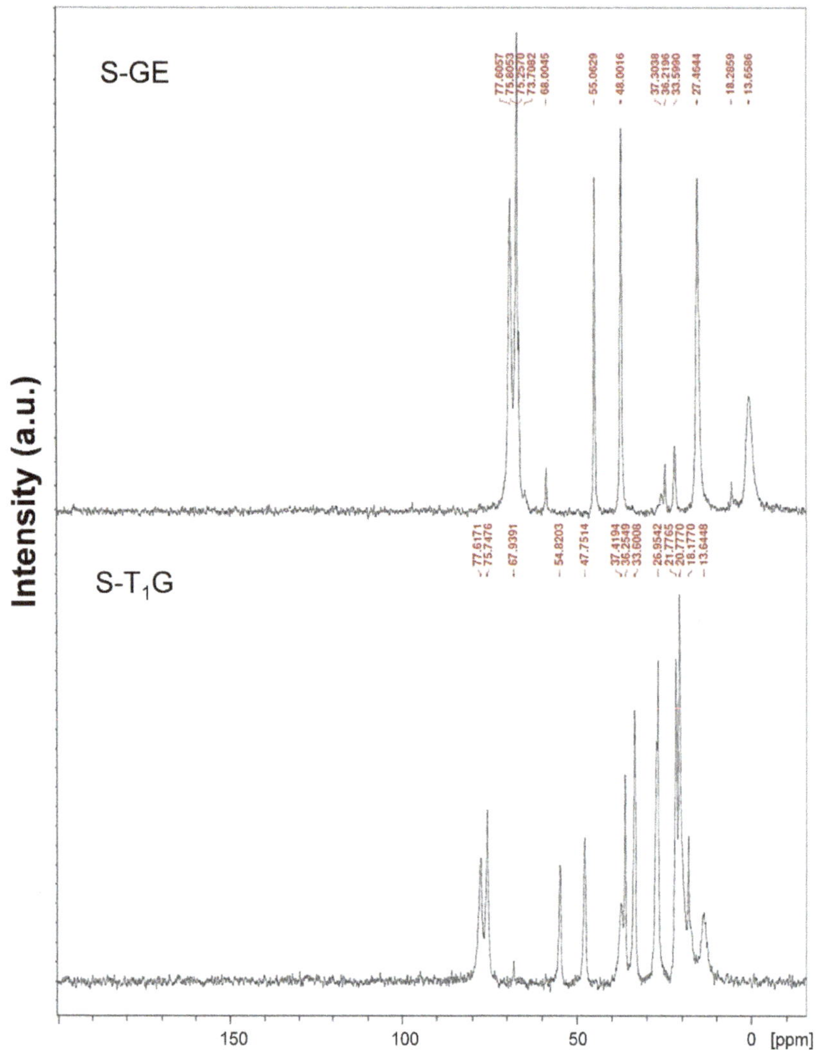

Figure 2. ^{13}C NMR spectra from samples S-GE and S-T_1G

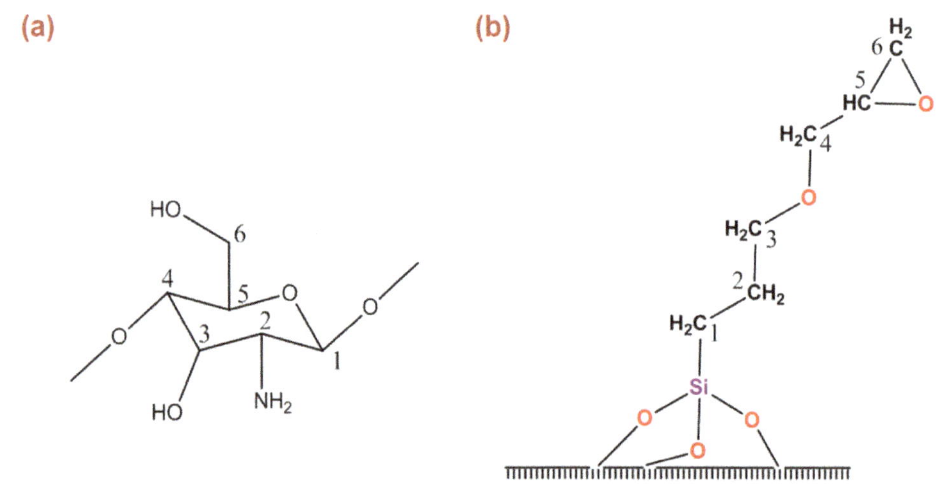

Figure 3. a) Structural unit of chitosan biopolymer indicating carbon numbers to establish the spectroscopic assignments in ^{13}C NMR. b) Schematic representation of a GPTMs unit grafted onto the surface of a polyalkoxysilane showing the spectroscopic designations of each C

Spectral assignments in solid-state ^{13}C NMR of SCh-TT$_1$, SCh-T$_1$, and SCh-GE are shown in Table 2. The carbon numbers are indicated in structure of the Figure 3b. The signals at 18.9 ppm and 17.7 and 17.9 ppm indicate the presence of the corresponding organoalkoxysilanes.

Solid-state ^{13}C-NMR showed that there was no cleavage of Si–C bonds. Likewise, the oxirane functional group remained at the experimental reaction conditions. In solids, because of the absence of high-amplitude molecular motions, resonance shapes detected in static samples, will strongly depend on the symmetry of tensors σ (or δ) and chemical shift anisotropies. Significantly smaller anisotropies can be observed for nuclei ^{1}H, ^{13}C, ^{27}Al, and ^{29}Si, which are also used as target nuclei in studies of different materials. The oxirane groups in GPTMS may undergo epoxy opening when reacted with carboxylic acid, thus leading to the formation of primary and/or secondary alcohols. At the same time GPTMS may crosslink via hydrolysis and condensation of alkoxysilane fragments forming species containing Si–O–Si moieties under the acid conditions provided by carboxylic acid. However, the ^{13}C CP efficiencies are generally different among different carbons depending on their spatial vicinity to protons in the system. As a result, the observed signal intensities of ^{13}C-NMR may not relate directly to the number of ^{13}C nuclei present to the different functional groups. However, in this case, the signals (δ_C) of Si–C at about 10.8 ppm and 13.6 ppm in all polyalkoxysilanes, and about 19.0 ppm and 21.8 ppm when TMPS was in the compounds (Fig 2, bottom), confirmed that cleavage of Si–C bond had not occurred. Additionally, the signals between 73.1 to 75.8 ppm, and 74.9 to 77.6 ppm correspond to C4 and C3, respectively, and Si–O–C. Similarly, the signals at 47.8 ppm (C6) and 54.9 ppm (C5) indicate that the opening of the oxirane function did not occur during the synthesis of polyalkoxysilanes, as shown in Figure 2 and Figure 3b.

The FT-IR spectra shown in Figure 4 and Figure 5 depict differences and similarities between the structure of the obtained polyalkoxysilanes and their respective chitosan hybrids. The absorption IR spectra of chitosan and its hybrids are complex due to complicate and specific network of H-bonds, in which the -OH, C=O and, NH groups are involved, and to typical broadness of the IR bands of natural polymers (Duarte, Ferreira, Marvao, & Rocha, 2002).

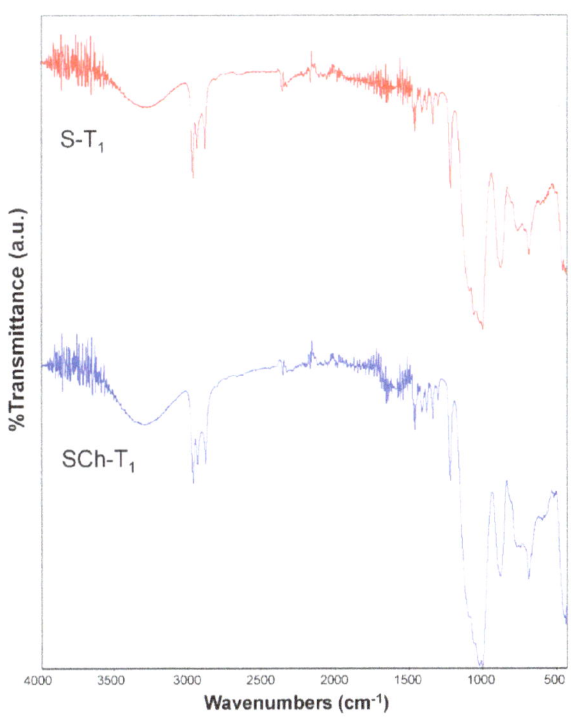

Figure 4. Infrared spectroscopy of samples S-T$_1$ and SCh-T$_1$

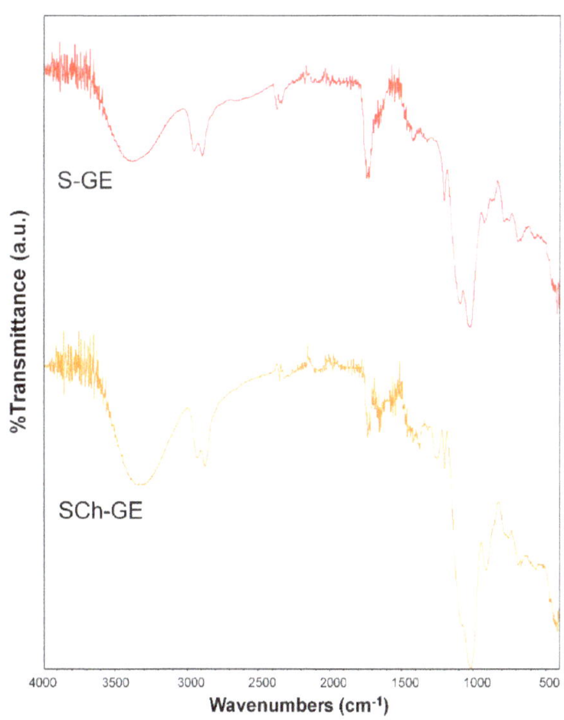

Figure 5. Infrared spectroscopy of samples S-GE and SCh-GE

FT-IR spectroscopy results of silica-chitosan samples depicted in Figure 4 and Figure 5 showed absorption in the domain of -OH, –NH$_2$, and Si–O functional groups as a result of the reaction of carboxylic groups in citric acid with organoalkoxysilanes. This reaction allowed chitosan to react with the free -OH from citric acid. The peaks centered at 2850 and 2920 ῡ (cm^{-1}) are due to C–H_{str} (C–H stretching) and 1450 cm^{-1} for C–H_{bend} (C–H bending). The bands at 1100 and 1000 cm^{-1} can be assigned due to Si–O–Si_{str} from cage-structured, and incompletely condensed linear/network siloxanes, respectely. The characteristic Si–O–Si band at 1100 cm^{-1} for a cage is due to its highly symmetric structure. The peak centered at 905 cm^{-1} confirmed the presence of Si–OH from incompletely condensed cage-structured and incompletely condensed linear/network siloxanes. Moreover, the peak at 3400 cm^{-1} suggests the presence of adsorbed water/alcohol. However, the spectroscopic band at 3400 cm^{-1} in samples of chitosan hybrid is more pronounced due to the free hydroxyl groups present in that molecule, as evident in Figure 4 and Figure 5. For example, the hydroxyl group of the SCh-T_1 hybrid (Figure 4, bottom) is more pronounced than in the S-T_1 polyalkoxysilane (Figure 4, top). The absorption peak at 1730 cm^{-1} in S-GE is characteristic of the ester obtained by reacting citric acid with GPTMS, confirming that the esterification reaction was performed successfully (Figure 5, top), whereas in SCh-GE hybrid (Figure 5, bottom) this peak is not present. This is possibly the result of the reaction of the hydroxyl group in C6 of the chitosan (Figure 3) with the silicon in GPTMS and not with the oxirane group; that can be confirmed by the disappearance of oxirane group in ^{13}C-NMR (Figure 2, top) and the appearance of the signal of a carbonyl group observed in the infrared spectra that does not appear in the silica-chitosan hybrid, as evidenced in Figure 5.

Table 3 shows the weight loss obtained by TGA for polyalkoxysilanes and silica-chitosan hybrids. In general, the main-weight loss of the samples were located in two ranges of temperature.

Table 3. TGA results showing a main-weight loss between 375 °C and 700 °C

<table>
<tr><th>Sample</th><th>375-475 °C (%)</th><th>475-700 °C (%)</th></tr>
<tr><td>S-T_1</td><td colspan="2">31*</td></tr>
<tr><td>S-T_1-f</td><td colspan="2">35*</td></tr>
<tr><td>S-T_1-c</td><td colspan="2">31*</td></tr>
<tr><td>S-G-c</td><td colspan="2">41*</td></tr>
<tr><td>S-GE</td><td>30</td><td>20</td></tr>
<tr><td>S-T_1G</td><td>25</td><td>20</td></tr>
<tr><td>S-T_1E</td><td>5</td><td>45</td></tr>
<tr><td>S-TT_1</td><td colspan="2">18*</td></tr>
<tr><td>S-TT_1G</td><td>20</td><td>15</td></tr>
<tr><td>S-TGE</td><td>30</td><td>20</td></tr>
<tr><td>S-T_1GE</td><td>30</td><td>25</td></tr>
<tr><td>S-TT_1GE</td><td>15</td><td>25</td></tr>
<tr><td>SCh-TT_1</td><td colspan="2">15*</td></tr>
<tr><td>SCh-T_1</td><td>10</td><td>30</td></tr>
<tr><td>SCh-GE</td><td>30</td><td>10</td></tr>
</table>

* Samples have a single transition between 375 and 700 °C.

Thermograms of prepared materials showed a gradual loss of about 4% up to 200 °C, a significant loss above 375 °C, leading to an approximately 40% thermally stable residue at 700 °C. The weight loss taking place above 375 °C is attributed to the degradation of the alkyl chain with production of volatiles, which end up with a thermally stable Si–O–Si residue at 700 °C.

The thermograms of selected polyalkoxysilanes and silica-chitosan materials are shown in Figure 6 and 7. As previously mentioned, due to their DSC behavior some polyalkoxysilanes, such as S-T_1, S-T_1-f, S-T_1-c, were chosen to synthesize the hybrid SCh-T_1, as shown in Figure 6. In the same way, S-TT_1 was used to prepare SCh-TT_1 in order to determine the presence of a similar endotherm (Figure 7c). Also, due to the presence of an exotherm in the polyalkoxysilane S-GE the hybrid SCh-GE was synthesized (Figure 7c).

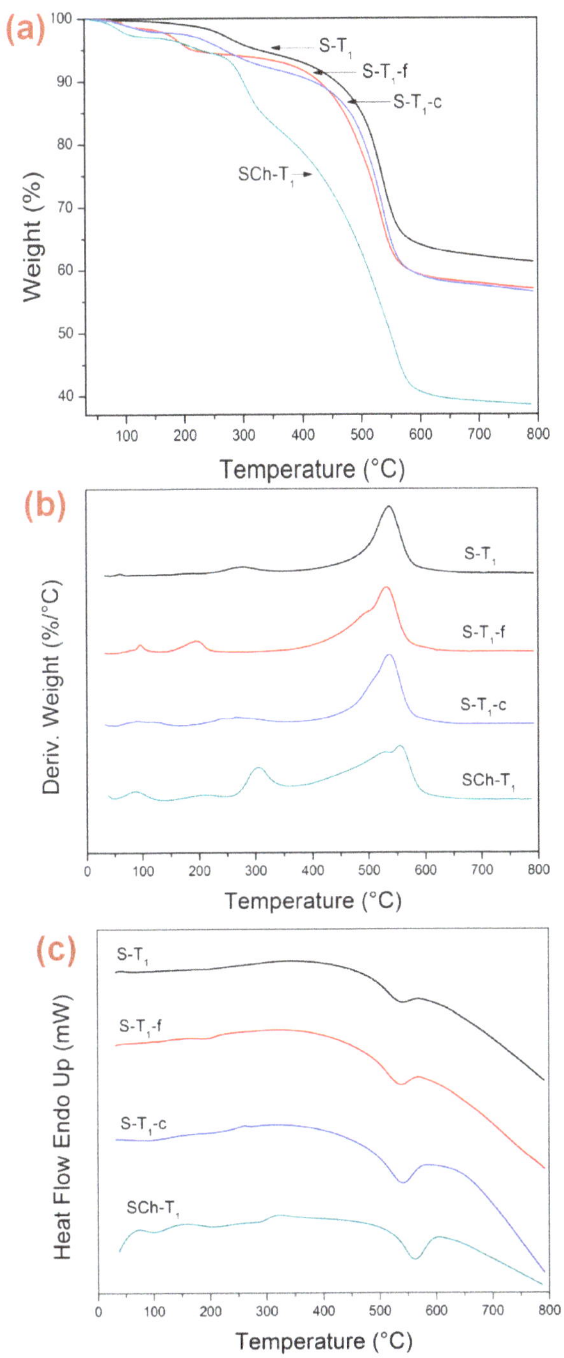

Figure 6. (a) TGA and (b, c) DSC thermograms of S-T_1, S-T_1-f, S-T_1-c, and SCh-T_1

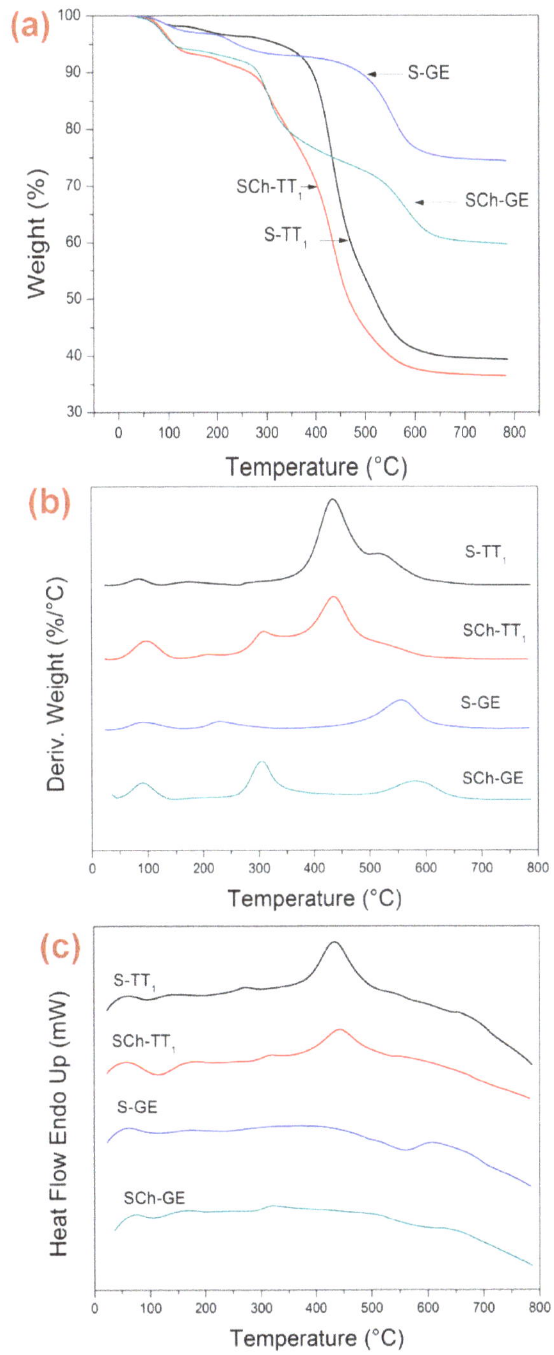

Figure 7. (a) TGA and (b, c) DSC thermograms of S-GE, S-TT_1, SCh-GE, and SCh- TT_1

As shown in Figure 6, differences in thermal behaviour of polyalkoxysilanes are evident according to the media: acetic (S-T_1), formic (S-T_1-f), or citric acid (S-T_1-c), compared to the hybrid SCh-T_1. While S-T_1 and S-T_1-f show only endothermic transitions, S-T_1-c and SCh-T_1 exhibit endothermic, as well as exothermic transitions (Figure 6c). In S-Ch-T_1 a shift of 68°C and 27°C was found in the exotherm and endotherm, respectively, compared to S-T_1-c. More importantly, chitosan formulated materials transitions showed a wider DTGA area (Figure 6b).

TGA results of silica-chitosan hybrids SCh-T_1 (Figure 6a) and SCh-TT_1 (Figure 7a), show that their weight loss is slower, indicating that the degree of polymerization is probably lower than the samples without chitosan. The gradual mass loss below 150 °C shown in Figure 6a and Figure 7a may be due to the evaporation of adsorbed water molecules formed through the condensation of silanol (Si–OH) groups, the vaporization of trapped solvent in the compound, as well as decomposition of low molecule weight polyalkoxysilanes, and other trapped impurities. Also, at 800 °C the hybrid of SCh-T_1 had lost more than 60 % of its weight, indicating that after adding chitosan, the heat-resistant property decreased, compared to S-T_1, S-T_1-f, S-T_1-c (Figure 6a). Nevertheless, this did not occur in SCh-GE (Figure 7a), where after adding chitosan, the sample slightly increased its heat-resistant property, compared to S-GE. Furthermore, the molecular weight distribution for SCh-GE was broader, and the degree of polymerization was probably lower, due to the polymer converting at lower temperatures.

Table 4 shows the maximum transition temperature (T_{peak}) found by DSC of polyalkoxysilanes and silica-chitosan hybrids materials.

Table 4. DSC transitions of polyalkoxysilanes and silica-chitosan hybrids materials. Values in parentheses correspond to the maxima of the derivative weight loss obtained by TGA.

Sample	**Exotherm (°C)**	**Endotherm (°C)**
S-T_1	-	541 (*534*)
S-T_1-f	-	192, 536 (*197, 529*)
S-T_1-c	256 (*261*)	535 (*530*)
S-G-c	418 (*417*)	-
S-GE	434 (*435*)	-
S-T_1G	431 (*426*)	567 (*548*)
S-T_1E	-	557 (*511*)
S-TT_1	-	233, 557 (*232, 558*)
S-TT_1G	435 (*435*)	576 (*546*)
S-TGE	433 (*433*)	-
S-T_1GE	430 (*442*)	554 (*517*)
S-TT_1GE	439 (*450*)	538 (*521*)
SCh-TT_1	345 (*305*)	negligible (*580*)
SCh-T_1	324 (*306*)	562 (*555*)
SCh-GE	325, 450 (*313, 440*)	-

Also shown is a comparison of DSC transitions and the maximum values of the derivative weight loss obtained by TGA. For example, the difference in the exothermic and endothermic transitions found in S-T_1, S-T_1-f, and S-T_1-c are due to the different media used for preparing each polyalkoxysilane. While the exotherms can be easily attributed to the cleavage of Si—Si and Si—C, resulting in a weight loss at the indicated temperature, the endotherm can be attributed to the melting point of polyalkoxysilanes.

4. Conclusions

Polymerization of polyalkoxysilanes of TEOS, TMPS, GPTMS, and EOS with acetic, formic, and citric acid was done through sol-gel process. Through characterization by FT-IR, TGA, DSC, and solid-state NMR it was possible to confirm the successful polymerization of alkoxysilanes and obtain detailed quantitative information on the different kinds of silicon sites. Solid-state ^{13}C-NMR showed that there had been no cleavage of Si–C bonds. Likewise, the oxirane functional group remained at the experimental reaction conditions. However, when reacted to carboxylic acid it lead to epoxy opening forming primary and/or secondary alcohols. The exotherms obtained by TGA, can be easily attributed to the cleavage of Si—Si and Si—C, where the presence of Si—C in polyalkoxysilanes was confirmed by ^{13}C NMR. The endotherms in DSC are attributed to the melting point of polyalkoxysilanes.

FT-IR spectra showed differences and similarities between the structure of the obtained polymers using citric acid and chitosan. Silica-chitosan materials obtained by TMPS, TEOS/TMPS, and GPTMS/EOS presented similar infrared spectroscopic results; this suggests no significant changes regarding chemical composition. On the contrary, for the citric acid and GPTMS reaction, the infrared spectrum indicated the formation of an ester by opening the oxirane functionalization. It was also found that citric acid was too weak to react itself with the organoalkoxysilanes. Therefore, if a polymerization of organoalkoxysilanes with carboxylic groups is intended, a higher concentration of carboxylic acid in water should be used. Since citric acid is solid and has to be dissolved in water first, it is adequate as a reactant but rather as a catalyst for the sol-gel reaction.

Acknowledgments

The authors acknowledge the support from Universidad de Monterrey, Fondo de la Vicerrectoría Académica para Investigación (UDEM VIAC) grant number UIN-14019.

References

Ahmad, Z., & Sarwar, M. I. (1997). Preparation and properties of hybrid organic-inorganic composites prepared from poly (phenylene terephthalamide) and titania. *Polymer, 38*, 4523–4529. http://dx.doi.org/10.1016/S0032-3861(96)01034-8

Ahmad, Z., Sarwar, M. ., & Mark, J. E. (1997). Chemically bonded silica-polymer composites from linear and branched polyamides in a sol-gel process. *J. Mater. Chem., 7*, 259–263. http://dx.doi.org/ 10.1039/A604028D

Ahmad, Z., Sarwar, M. I., & Mark, J. E. (1997). Dynamic-mechanical thermal analysis of aramid-silica hybrid composites prepared in a sol-gel process. *J. Appl. Polym. Sci., 63*, 1345–1352. http://dx.doi.org/0.1002/(SICI)1097-4628(19970307)63:10<1345::AID-APP14>3.0.CO;2-3

Ahmad, Z., Sarwar, M. I., & Mark, J. E. (1998). Thermal and mechanical properties of aramid-based titania hybrid composites. *J. Appl. Polym. Sci., 70*, 297–302. http://dx.doi.org/0.1002/(SICI)1097-4628(19981010)70:2<297::AID-APP9>3.0.CO;2-P

Al-Sagheer, S., & Muslim, S. (2010). Thermal and Mechanical Properties of Chitosan/Hybrid Composites. *J. Nanomater.,* 1-7. http://dx.doi.org/10.1155/2010/490679

Biagini, G., Bertani, A., Muzzarelli, R., Damadei, A., DiBenedetto, G., Belligolli, A., & Riccotti, G. (1991). Wound management with N-carboxybutyl chitosan. *Biomaterials, 12*, 281-291. http://dx.doi.org/10.1016/0142-9612(91)90035-9

Budnyak, T. M., Pylypchuk, I. V., Tertykh, V. A., Yanovska, E. S., & Kolodynska, D. (2015). Synthesis and adsorption properties of chitosan-silica nanocomposite prepared by sol-gel method. *Nanoscale Res. Lett., 10*(87), 1-10. http://dx.doi.org/10.1186/s11671-014-0722-1

Chen, Q., & Yakovlev, N.L. (2010). Adsorption and interaction of organosilanes on TiO_2 nanoparticles. *Appl. Surf. Sci., 257*(5), 1395-1400. http://dx.doi.org/10.1016/j.apsusc.2010.08.036

Ciriminna, R., Fidalgo, A., Pandarus, V., Béland, F., Ilharco, L. M., & Pagliaro, M. (2013). The sol-gel route to advanced silica-based materials and recent applications. *Chem. Rev., 113*, 6592-6620. http://dx.doi.org/10.1021/cr300399c

Dorozhkin, S. V. (2009). Calcium orthophosphate-based biocomposites and hybrid materials. *J. Mater. Sci., 44*, 2343-2387. http://dx.doi.org/10.1007/s10853-008-3124-x

Duarte, M. L., Ferreira, M. C., Marvao, M. R., & Rocha, J. (2002). An optimized method to determine the degree of acetylation of chitin and chitosan by FT-IR spectroscopy. *Int. J. Biol. Macromol., 31*, 1-8. http://dx.doi.org/10.1016/S0141-8130(02)00039-9

Ebnesajjad, S. (2011). Handbook of Adhesives and Surface Preparation: Technology, Applications and Manufacturing (pp. 369-384.) Elsevier: Oxford, UK.

Gelest. (2015). http://www.gelest.de/goods/pdf/couplingagents.pdf Accessed 17 February 2015.

Gingras, M., Paradis, I., & Berthod, F. (2003). Nerve regeneration in a collagen–chitosan tissue-engineered skin transplanted on nude mice. *Biomaterials, 24*, 1653–1661. http://dx.doi.org/10.1016/S0142-9612(02)00572-0

Ivanova, N. A., & Philipchenko, A. B. (2012). *Superhydrophobic chitosan-based coatings for textile processing. Appl. Surf. Sci., 263*, 783-787. http://dx.doi.org/10.1016/j.apsusc.2012.09.173

Juhasz, J. A., & Best, S. M. (2011). Bioactive ceramics: processing, structures and properties. *J. Mater. Sci., 47*, 610-624. http://dx.doi.org/ 10.1007/s10853-011-6063-x

Khor, E. (2004). Chitin-Based Biomaterials. In T. S. Hin (Ed.), *Engineering materials for biomedical applications* (pp. 11-1-11-13). World scientific, Singapore.

Knapczyk, J., Macura, A. B., & Pawlik, B. (1992). Simple tests demonstrating the antimycotic effect of chitosan. *Int. J. Pharm., 80*, 33-38. http://dx.doi.org/10.1016/0378-5173(92)90259-5

Lee, E. J., Shin, D. S., Kim, H. E., Kim, H. W., Koh, Y. H., & Jang, J. H. (2009). Membrane of hybrid chitosan-silica xerogel for guided bone regeneration. *Biomaterials, 30*, 743-750. http://dx.doi.org/10.1016/j.biomaterials.2008.10.025

Lesage, A. (2009). Recent advances in solid-state NMR spectroscopy of spin I = 1/2 nuclei. *Phys. Chem. Chem. Phys., 11*, 6876-6891. http://dx.doi.org/10.1039/B907733M

Liu, Y-L., Su, Y-H., & Lai, J-Y. (2007). In situ crosslinking of chitosan and formation of chitosan-silica hybrid membranes with using γ -glycidoxypropyltrimethoxysilane as a crosslinking agent. *Polymer, 45*, 6831–6837. http://dx.doi.org/10.1016/j.polymer.2004.08.006

Muzzarelli, R. A. A. (1993) Biochemical significance of exogenous chitins and chitosans in animals and patients, Carbohydr. *Polym., 20*, 7-16. http://dx.doi.org/10.1016/0144-8617(93)90027-2

Nicole, L., Rozes, L., & Sanchez, C. (2010). Integrative approaches to hybrid multifunctional materials: From multidisciplinary research to applied technologies. *Adv. Mater., 22,* 3208-3214. http://dx.doi.org/10.1002/adma.201000231

Rehman, H. U., Sarwar, M. I., Ahmad, Z., Krug, H., & Schmidt, H. (1997). Synthesis and characterization of novel aramid-zirconium oxide micro-composites. *J. Non-Cryst. Solids, 211*, 105–111. http://dx.doi.org/10.1016/S0022-3093(96)00614-X

Rui, L. R., & San Román, J. (2005). *Biodegradable Systems in Tissue Engineering and Regenerative Medicine.* CRC Press: Boca Raton, United States of America.

Ruiz-Hitzky, E., Darder, M., & Aranda, P. (2008). An Introduction to Bio-nanohybrid Materials. In E. Ruiz-Hitsky, A. Katsuhiko, Y. Lvov (Eds.), *Bio-inorganic hybrid nanomaterials: Strategies, Synthesis, Characterization and Applications* (pp. 1-32). Wiley-VCH Verlag GmbH: Madrid, Spain.

Sanchez, C., Boissiere, C., Cassaignon, S., Chaneac, C., Durupthy, O., Faustini, M., ... Sassoye, C. (2014). Molecular engineering of functional inorganic and hybrid materials. *Chem. Mater., 26*, 221-238. http://dx.doi.org/10.1021/cm402528b

Sanchez, C., Soler-Illia, G. J. de A. A., Ribot, F., Lalot, L., Mayer, C. R., & Cabuil, V. (2001). Designed hybrid organic-inorganic nanocomposites from functional nanobuilding blocks. *Chem. Mater., 13*, 3061-3083. http://dx.doi.org/10.1021/cm011061e

Smitha, S., Shajesha, P., Mukundana, P., & Warrier, K. G. K. (2008). Sol-gel synthesis of biocompatible silica-chitosan hybrids and hydrophobic coatings. *J. Mater. Re., 23*(8), 2053-2060. http://dx.doi.org/10.1557/JMR.2008.0271

Solanski, P. R., Kaushik. A., Ansari, A. A., Tiwari, A., & Malhotra, B. D. (2009). Multi-walled carbon nanotubes/sol–gel-derived silica/chitosan nanobiocomposite for total cholesterol sensor. *Sens. Actuators B, 137*, 727–735. http://dx.doi.org/10.1016/j.snb.2008.12.044

Susuki, T., & Mizushima, Y. (1997). Characteristics of silica-chitosan complex membrane and their relationships to the characteristics of growth and adhesiveness of L-929 cells cultured on the biomembrane. *J. Ferment. Bioeng, 2*, 128-132. http://dx.doi.org/10.1016/S0922-338X(97)82541-X

12

Effect of addition JORPHOS to Self-Compacting Concrete for Chloride Penetration

Mahmoud B. Alhasanat[1], Arabi N. S. Al Qadi[1], Loai A. Al Tarabulsi[1], & Omar A. Al khashman[2]

[1] Civil Engineering Department, College of Engineering, Al-Hussein Bin Talal University, Ma'an, Jordan

[2] Environmental Engineering Department, College of Engineering, Al Hussien Bin Talal University, Ma'an, Jordan

Correspondence: MahmoudAlhasanat, Civil Engineering Department, College of Engineering, Al-Hussein Bin Talal University, Ma'an, Jordan. E-mail: 20petra07@gmail.com

Abstract

Reinforced concrete structures are exposed to harsh environments, yet they are often expected to last with little or no repair or maintenance for long periods of time (often 100 years or more). For this reason, a durable structure needs to be produced. In addition, bridge reinforced concrete suffering from attacks of chloride penetration that cause corrosion of the steel in the reinforcement concrete that leads to reduction in strength and serviceability cracking, which cause further repair and rehabilitation for the structure and greater maintenance and repair costs. This research focuses on self-compacting concrete with Jordanian Phosphate (JORPHOS) as a filler. The permeability for different percentages of JORPHOS were tested by using Rapid Chloride Permeability Test (RCPT). It was concluded that the addition of different percentages of JORPHOS minimizes chloride penetration in concrete.

Keywords: permeability, corrosion, JORPHOS, durability, chloride

1. Introduction

Concrete is the ideal environment for steel because of its highly alkalinity (pH 12.5-13.8) pore solution contained in the cement paste component, which passivates the steel. However, concrete contains cracks and pores; and is not impervious to the ingress of chloride salts that break down the passive layer of the reinforcing steel (i.e. rebar). Considerable work has been done to examine chloride ingress through the concrete pore network (Sandberg et al., 1998; Dhir & Jones et al., 1996; Jensen et al., 1999; Gjorv & Vennesland, 1979) and its effect on the corrosion of rebar (Mohammed et al., 2004; Hope & Ip, 1987; Beaudoin et al., 2000). It is clear that the quality and depth of the concrete cover is of major importance to determine the durability of the structure (Jaegermann, 1990).

Chloride is considered the major cause of the corrosion of steel reinforcement in the concrete structures. Concrete provides physical and chemical protection to the reinforcing steel from penetrating chlorides, which may cause steel depassivation that leads to increased risk of steel corrosion. Chloride resistance depends on the permeability of concrete and the thickness of the cover to the reinforcement. Additionally, the integrity of the concrete cover under service load in terms of cracking and crack width influences resistance to penetrating chlorides. Corrosion of steel reinforcement is an electrochemical process. Hence, the electrochemical properties of concrete, such as resistivity, are important inherent properties that affect the corrosion rate of reinforcing steel (Guirguis et al., 1989).

Metha (1988) reconfirmed from a review of case studies that it is the permeability of concrete, rather than its chemistry that is the key to overall durability. The causes of high permeability are not limited to poor concrete proportion but poor concreting practices, such as incomplete mixing, inadequate consolidation, curing after placement, insufficient cover to reinforcing steel, and badly constructed joints. In service, concrete may exhibit various forms of cracking for reasons such as settlement, premature loading, overloads, and repeated impact. The control of concrete cracking in service through proper mix proportioning and concreting practice is as important as the control of concrete permeability.

Steel reinforcement embedded in concrete is inherently protected against corrosion by passivation of the steel surface due to the high alkalinity of the concrete. When a sufficient amount of chlorides reach the steel reinforcement, then it permeates the passivating layer and increases the risk of corrosion. The resistivity of concrete can also be reduced, which affects the corrosion rate of the steel.

For use in prestressed or reinforced concrete structures, the chloride concentrations in mixing water, cements, admixtures, and aggregates are carefully controlled; and the maximum allowable concentrations are given by the building standards (AS 1379). According to this standard the chloride concentration must not exceed 0.8 kg/m3 of the fresh concrete.

However, in most cases, good amount of chloride in concrete comes from external sources. The chloride penetration into the concrete occurs by different transport mechanisms that depend on exposure conditions. There are significant amounts of chloride in seawater but chlorides are more limited in groundwater and soil. In many countries, de-icing salts, which are used to combat the build-up of snow and ice on transport infrastructures, are the greatest sources of chlorides. In seawater, chlorides usually pose a greater threat to steel in concrete than sulfates do to concrete because calcium sulphoaluminate or ettringite (the expansive reaction product of sulfate and tricalcium aluminate in the cement) is more soluble in the presence of chloride; hence, it does not cause adisruptive expansion. Portland cement reacts with sodium chloride to form chloroaluminates or Friedel's salt thatimmobilizesthe chloride and reducesthe free chloride ions available to depassivate the steel. The results of a 34-year long-term exposure of plain and reinforced concrete beams in a tidal seawater exposure in Los Angeles Harbor in California in 1959 and 1961 (Stark & David, 1995), found that all plain concrete mixtures (stored at approximately mean tide) display excellent resistance to seawater attack, regardless of cement composition, water-cement ratio, cement content, the use of SCM, and method of curing as well as freezing and thawing do not occur. However, severe cracking due to corrosion of embedded reinforcing steel developed in some beams stored above high tide, while only minor or no cracking developed in companion beams stored in seawater near mean tide levels. The most severe corrosion-induced cracking occurred in concrete with the highest w/c of 0.49 and least cover of 25 mm. The relatively greater degree of steel corrosion in beams stored above high tide is attributed to the greater availability of oxygen at the reinforcing steel surface. Corrosion-related distress was found to be sensitive to concrete cover and water-cement ratio. Prestressed steel appeared to be no more vulnerable to galvanic corrosion than ordinary deformed reinforcing steel bars. The nickel and painted-on epoxy coatings appeared to provide little (if any) additional protection against corrosion (Concrete & Australia, 2009).

2. Method

In this section the number of methods that were applied such as the volume method for determining the mix ratios are elucidated. The test conducted was the Rapid Chloride Permeability Test (RCPT) to determine the charges passing the core samples having different JORPHOS ratios.

2.1 Materials Used

Portland Cement: Ordinary Portland Cement (OPC) conforming to ASTM C150-85A (2006) as available in the local market was used in the investigation. The cement used has a specific gravity of 3.15g/cm^3 and Blaine's fineness (Blaine) of 4500 cm^2 gm^{-1} (Table1).

Jordanian Phosphate: The specific gravity is 2.8 done according to ASTM–C188, while chemical composition of JORPHOS (Table 1) as follows.

Table 1. Chemical composition of JORPHOS and Cement

Type	JORPHOS (%)	Cement (%)
Tri Calcium Phosphate (TCP)	58.44	-
Acid Insoluble Resolution (cilia Air)	11.64	-
Loss of Ignition (LOI)	9.41	1.43
CO2	7.74	-
CaO	46.27	63.21
SiO2	-	19.6
MgO	0.49	3.15
Fe2O3	0.37	3.26
Cl	0.142	-
Al2O3	0.86	5.32
SO3	1.5	2.89
Na2O3	0.68	0.57
K2O	0.075	0.6
Organic Matter	0.18	-

JORPHOS: JORPHOS was sieved using sieve #200.

Potable Tap Water: Water conforming to BS3148 (1981) was used for mixing and curing.

Crushed Aggregate: Crushed angular limestone material of 20 mm nominal size from a local source was used as course and medium aggregate, while fine aggregate crashed silica sand with a combined of aggregate (28% coarse aggregate, 38% medium aggregate, and 34% fine aggregate) all of these aggregate were of oven dry weight. Figure1 shows the sieving analysis according to ASTM C136. Specific gravity for combined aggregate was 2.534 and the absorption value was 1.751% conforming to ASTM C 33-86, (2006) (See Figure 1).

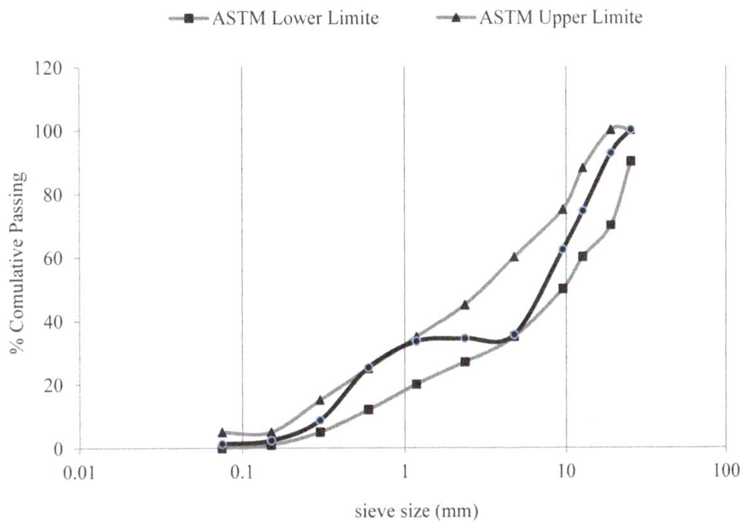

Figure 1. Sieving analysis of combined coarse, medium, and fine aggregate

2.2 The Mixture Design

The control mixture contains only the base components for any mixture without any additives, which are cement, water, and aggregate (Figure 2a). It will be the reference for all other mixtures that contain JORPHOS which is the replacement for the cement (Figure 2b). To achieve a high performance concrete depends on the European Guidelines by weight or volume of components for the related mixtures.

a. The materials after proper preparation (sieving, weighing)

b. the JORPHOS passed through sieve #200

Figure 2. The mixture components

The mix proportions that were used in the investigation and presented in Table 2 along with the response. All concrete mixes were prepared manually.

Table 2. Mix proportions of SCC with different percentages of JORHPOS

Mix No.	C(kg)	W(kg)	CA(kg)	FA(kg)	Sand (kg)	W/P	JORHPOS mass %	JORHPOS. Weight(kg)
PC control	1.15395	0.426	0.5321	2.327	1.933	0.37	0%	0
MJ4	1.10779	0.4337	1.32759	1.8	1.6	0.38	4%	0.041
MJ8	1.06	0.4519	1.315	1.785	1.59	0.40	8%	0.082
MJ12	1.015	0.47	1.303	1.77	1.58	0.41	12%	0.123

2.3 Test Method

The test consists of measuring the amount of electrical current passed through 50mm thick slices of 100mm diameter cylinders during a 6h period. A 60 volt DC is maintained across the ends of the specimen, one of which is sink in a sodium hydroxide solution and the other in a sodium chloride solution.

2.4 Test Specimens

The cylinders cast in the laboratory were prepared according to C 192. Specimen mounting (rubber gasket alternative): Place a 4in. outside diameter by 3in. inside diameter by 1⁄4in. (100mm outside diameter by 75 mm inside diameter by 6 mm) circular vulcanized rubber gasket in each half of the test cell are placed. The Specimen inserted and the two halves of the test cell are together clamped to seal. The side of the cell containing the top surface of the specimen is filled with 3.0 % NaCl. This side of the cell was connecting to the negative terminal of the electrical current. The other side of the cell was connecting to the positive terminal of the electrical current if filled with the 0.3M NaOH solution.

The electrical current 60.0 V connected on, and initial current reading was recorded. Temperatures of the specimen, applied voltage cell, and solutions was around 23°C. During the test, the air temperature around the specimens was maintained in the range of 68–77°F (20–25°C). The current was recorded at every 30 min. Each half of the test cell was filled with the appropriate solution for the entire period of the test. Terminate test was after 6 h.

3. Results

The main test of this study is the rapid chloride permeability test whose name is self-explanatory and determines the amount of chloride ions passing the section of a 4 in concrete core that is subjected to two solutions on both sides (NaOH 0.3M) and (3% NaCl) by applying DC with a 60V power source to induce ionic movement in the concrete core. The general assumption is that the higher the charge passed through the concrete core, the higher the permeability rate of concrete.

3.1 Rapid Chloride Permeability Test (RCPT)

RCPT results for that samples have different mass percentages of JORPHOS in the mix. Table 3 indicates the charge passed coulombs in core sample after 6 h.

Table 3. Charge passed the core sample after 6h

Sample Number	JORPHOS to Cement mass percentage ratio in the mix			
	0%	4%	8%	12%
1st	1690.2 C	681.2 C	312.5 C	126.4 C
2nd	1543.5 C	731.8 C	283.4 C	133.9 C
3rd	1764.1 C	722.7 C	207.5 C	128.7 C
4th	1708.4 C	715.0 C	221.6 C	133.1 C
5th	1793.3 C	620.6 C	308.1 C	125.2 C
6th	1805.7 C	554.4 C	255.9 C	136.4 C
7th	1844.5 C	497.5 C	216.7 C	122.4 C
8th	1674.3 C	656.8 C	311.1 C	130.7 C
Avg.	1728 C	647.5 C	264.6 C	129.6 C

Figure 3 shows the relationship between charges passing in coulombs and the percentages of JORPHOS in the concrete mixture. The graph shows that as the percentage of JORPHOS increases, the charge passing in coulombs decreases with the exponential equation (1);whereR^2 is equal to 0.9951 correlation coefficient:

$$y = 1623.7e^{-21.66x} \quad (1)$$

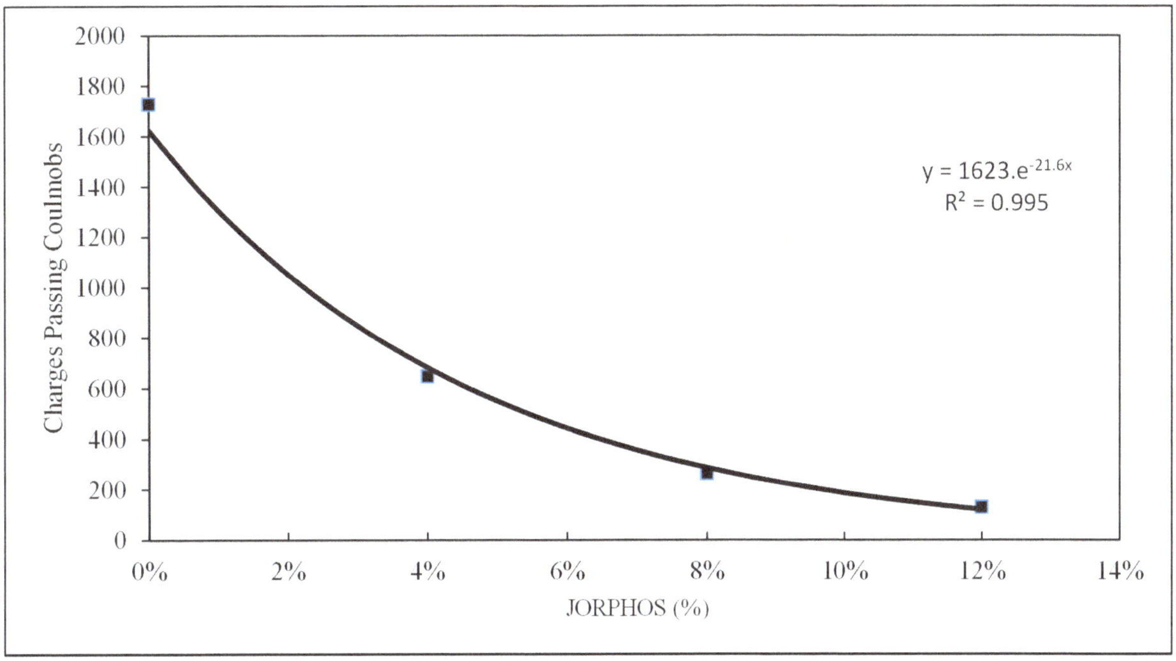

Figure 3. Average Test Results Chart of percentage of JORPHOS vs. Charges passed.

4. Discussion and Conclusion

From the Preceding results and according to the rating of the RCPT as shown in Table 4. Chloride Permeability of the 0% mix is low. Chloride Permeability of the 4% mix is very low. Chloride Permeability of the 8% mix is very low. Chloride Permeability of the 12% mix is very low. The 12% mix is the one with the lowest permeability to chloride. It must be taken into account that the RCPT does not measure the chloride ions alone but it is a measure of all the ions moving in the circuit and logged by the data logger.

Table 4. Rating of Chloride Permeability of Concrete According to the RCPT

Chloride Permeability	Charge Passing Coulombs	Typical Concrete Type
High	>4000	High W/C ratio (>0.6) Conventional PC Concrete
Moderate	2000 to 4000	High W/C ratio (0.4 to 0.5) Conventional PC Concrete
Low	1000 to 2000	Low W/C ratio (< 0.4) Conventional PC Concrete
Very Low	100 to 1000	Latex-Modified Concrete
		Internally Sealed Concrete
Negligible	< 100	Polymer-impregnated concrete.
		Polymer Concrete

This research determined the ratio of JORPHOS that best resists chloride penetration (0%, 4%, 8%, and 12%) to reduce the effects of steel corrosion in reinforced concrete structures. Figure 3 shows the charge (coulombs) passing the cross section of the core is inversely proportional to the ratio of the JORPHOS in the mix. Therefore, the following conclusion can be drawn: the percentage of JORPHOS is inversely proportional to chloride permeability and the 12% mix is the best mix in this experiment. It is highly recommended for future researches to use ratios of JORPHOS greater than 12% to estimate the best proportion to be used to avoid chloride penetration. The higher the JORPHOS ratio, the lower the chloride penetration into concrete. This behavior (i.e. the relationship) between the ratio of JORPHOS and the charge passing is a result of the following factors:

- The PO_4^{-3} is a strong negative ion, which leads to a repulsion with the $Cl;^-$
- The small grain size of the JORPHOS particles forms a filling material that prevents the ions from moving in the small pores or cracks of the concrete;
- The correlation coefficient is 99.15%.

References

Alexander, M. G., Santhanam, M., & Ballim, Y. (2010). Durability design and specification for concrete structures-the way forward. *International Journal of Advances in Engineering Sciences and Applied Mathematics, 2*(3), 95-105.

Brian, B. H., & Alan, C. I. (1987). Corrosion of Steel in Concrete Made With Slag Cement. *Materials Journal, 84*(6). http://dx.doi.org/10.14359/2470

Concrete, C., & Australia, A. (2009). *Chloride resistance of concrete.* Sydney: Cement Concrete & Aggregate Australia.

Dhir, R. K., Jones, M. R., & McCarthy, M. J. (online 12 February 1999). Binder content influences on chloride ingress in concrete. *Cement and Concrete Research, 26*(12), 5. http://dx.doi.org/10.1016/S0008-8846(96)00179-2

Gjørv, O. E., & Vennesland, Ø. (1979). Diffusion of chloride ions from seawater into concrete. *Cement and Concrete Research, 9*(2), 229-238. http://dx.doi.org/10.1016/0008-8846(79)90029-2

Gu, P., Beaudoin, J., Zhang, M.-H., & Malhotra, V. (2000). Performance of reinforcing steel in concrete containing silica fume and blast-furnace slag ponded with sodium-chloride solution. *ACI Materials Journal, 97*(3).

Guirguis, S., Cement, & Australia, C. A. O. (1989). Durable Concrete Structures: Cement and Concrete Association of Australia.

Jaegermann, C. (1990). Effect of water-cement ratio and curing on chloride penetration into concrete exposed to Mediterranean sea climate. *ACI Materials Journal, 87*(4).

Jensen, O. M., Hansen, P. F., Coats, A. M., & Glasser, F. P. (1999). Chloride ingress in cement paste and mortar. *Cement and Concrete Research, 29*(9), 1497-1504. http://dx.doi.org/10.1016/S0008-8846(99)00131-3

Mehta, P. K. (1988). Durability of Concrete Exposed to Marine Environment-A Fresh Look. *Special Publication, 109.* http://dx.doi.org/10.14359/1899

Sandberg, P., Tang, L., & Andersen, A. (1998). Recurrent studies of chloride ingress in uncracked marine concrete at various exposure times and elevations. *Cement and Concrete Research, 28*(10), 1489-1503. http://dx.doi.org/10.1016/S0008-8846(98)00124-0

Stark, D. (1995). *Long-Time Performance of Concrete in a Seawater Exposure.*

Tarek Uddin Mohammed, H. H., & Toru, Y. (2004). Concrete after 30 Years of Exposure—Part II: Chloride Ingress and Corrosion of Steel Bars. *Materials Journal, 101*(1). http://dx.doi.org/10.14359/12983

13

Structure and Properties of Cast Iron DESIGNATED for Working Layer of Rolls

K. N. Vdovin[1], A. N. Zavalishchin[1], D. A. Gorlenko[1] & N. A. Feoktistov[1]

[1] FSBEI HBE Nosov Magnitogorsk State Technical University, Magnitogorsk, Russia

Correspondence: K. N. Vdovin, FSBEI HBE Nosov Magnitogorsk State Technical University, Magnitogorsk, Russia. E-mail: kn.vdovin@gmail.com

Abstract

The formation of the structure and properties of cast irons designated for the working layer during solidification and subsequent tempering of Indefinite Chill Double Poured Centrifugal Casting Rolls (ICDP) has been studied.

Graphitization of cast irons designated for the working layer of rolls occurs under isothermal conditions at high temperatures, which vary over the cross section and resulted from casting the roll core. At subsequent slow cooling, secondary carbides precipitate and austenite partially transforms into martensite; the resulting end structure consists of martensite and austenite metallic base (primary austenite dendrites) with 11.4 per cent of retained austenite, 26 per cent of eutectic and secondary carbides, and 2.6 per cent of flake graphite.

The amount of retained austenite in the metallic base of the cast structure of cast iron designated for the working layer, heated to 430 °C, decreases from 11.4 per cent to 3.2 per cent due to the partial transformation into bainite. The martensite tetragonality decreases due to the carbide precipitation; the existing excess phases grow and the new ones are formed; the total volume fraction of the carbide phase increases to 29.8 per cent and graphite to 3.7 per cent.

Keywords: Roll, cast iron, solidification, tempering, microstructure, martensite, retained austenite, carbides, graphite

1. Introduction

The improvement of the rolled products quality and production efficiency is one of the most important tasks for the metallurgical sector under fierce competition on the rolled products market. Rolls, the key tool of rolling-mill machinery, are significant in the solution of these tasks with the expenses of 15 ... 20 per cent in the cost structure of rolled products, moreover, their durability has considerable impact on the metal product quality and steel mill productivity (Traino 2008). More severe operating conditions of modern hot-rolling mills cause a significant increase in the requirements for the roll barrel surface, its axial part and trunnions (Ray et al., 2000).

Centrifugally cast iron rolls meet these requirements best of all due to the absence of the transition zone from the core to the working layer, made of high-alloy cast iron, which is characterized by a homogeneous structure and has higher density, strength, and ductility, as well as keeps its hardness unchanged throughout the entire operating period that results in the constant level of roll wear resistance (Martini & Gostev, 1999; Ray, Prasad, Barhai, & Mukherjee, 2004; Sinha, Indimath, Mukhopadhyay, & Bhattacharyya, 2014; Martini, 1999). Rolls made of high chrome cast iron (Hi-Cr) are installed in the first three stands of the finishing group, and rolls made of indefinite (mottled) ICDP iron are installed in the last three stands (Gostev, 2008; Palit, Jugade, Jha, Das, & Mukhopadhyay, 2015). The working layer made of the Ni-Cr white iron that contains the martensite, eutectic and secondary carbides, retained austenite, and graphite can help to achieve the high quality of the work rolls for the last finishing stands of rolling mills (Asensio-Lozano & Álvarez-Antolín, 2008).

Carbides, metallic base, and graphite are the main structural constituents of wear-resistant mottled cast irons.

The carbide phase in cast iron is determinative for the abrasive wear processes and the wear resistance coefficient. The wear resistance coefficient depends on the amount of carbides and the increased size of carbide inclusions decreases wear resistance of cast irons (Kang, Suh, Oh, & Lee, 2014).

The degree of influence of carbide sizes depends on the conditions and nature of abrasive wear. This is associated with the stress distribution between the carbide and metallic base: large carbide inclusions, especially in a soft matrix, crack and crumble under stresses, generated by an abrasive particle; small carbide inclusions transfer some portion of the stresses to the metallic base and they do not crumble (Ray, Mukherjee, Sarkar, & Mishra, 1994; Lee, Kim, Ryu, & Shin, 1997; Nilsson & Olsson, 2013). The high wear resistance of iron with the most suitable carbide structure is feasible only when the metallic base itself has a high degree of microhardness, provides strong carbide fixation, and is not undergone deformation when loading in operation. The sizes of the base areas between carbides should be sufficiently small in order to minimize the selective abrasive wear of the base, carbide denudation, their washout, or breaking off. Martensite and retained austenite, unstable under these wear conditions, meet these requirements to the greatest extent but a ferrite, soft, pearlite or stable austenitic metallic base is less suitable.

The metallic base of cast iron designated for the working layer, as well as the carbide phase, influences on the forming roll resistance. At the same characteristics of the carbide constituent, wear resistance is linearly dependent on microhardness of the base and is independent of its phase composition. Thus, martensite is the best base metal in order to obtain maximum wear resistance (Al-Rubaie & Pohl, 2014; Jinzhu, Shizhuo, & Yongfa, 1993).

The influence of the graphite, contained in the structure of mottled cast irons, on wear resistance of rolls may be both positive and negative and depends on the shape and number of inclusions, as well as the working conditions.

A negative impact should be noted that graphite possesses poor resistance to extension and shearing, consequently, graphite inclusions are unable to carry heavy loads. Uneven distribution of graphite inclusions decreases all mechanical properties of cast iron and influences on its fatigue strength, but the growth of graphite inclusions and the decreased shape of graphite inclusions increase the micro defects and degree of wear.

The positive impact of graphite on the wear resistance of rolls is more prevalent due to the following advantages: the places of occurrence of spherical graphite are lubricant reservoirs; large graphite inclusions absorb solid abrasive particles; increased thermal conductivity of cast iron; improved kinetics of the oxide film formation on the surface of rolls; increased thermal wear resistance. The occurrence of thermal fatigue cracks and graphite precipitation promotes oxidation of the structural constituents of cast iron, and under the influence of the friction forces in the zone of deformation it causes the separation of cementite particles, but thanks to the graphite this process does not spread in depth. The thermal endurance of cast iron with evenly distributed graphite is 40 ... 100 per cent more than the thermal endurance of cast iron with graphite concentration in the form of nests and sockets, and finely crushed inclusions provide a minimum tendency to heat checking (Kuskov, 2006; Asensio-Lozano, Álvarez-Antolín, & Vander Voort, 2008).

The purpose of this work is to study the structure and properties of cast iron designated for the working layer of rolls in as - cast state and after tempering at 430° C.

2. Methods of Carrying out Research

Cast iron indefinite (mottled) rolls are produced by centrifugal casting method, and two electric furnaces (induction and electric) are used to smelt metal for their production at Magnitogorsk Rolls Plant JSC. A centrifugal machine is used to cast the surface work layer of the roll of alloyed cast iron, and the core of the roll is poured into a special caisson hole to form a core of grey cast iron, the upper and lower necks.

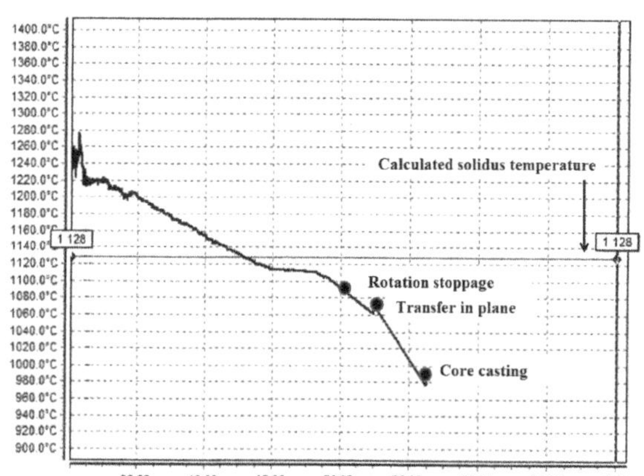

Figure 1. Changes in the temperature of cast iron designated for the working layer after casting into a chill before pouring the core, optical pyrometer data

The temperature and time required for the necessary operations are under control in the process of roll manufacturing. Platinum - rhodium / platinum thermocouples measure the temperature changing of melted irons, and a special optical pyrometer measures the temperature of the interior surface of iron, which is in the mould of a centrifugal machine. The data obtained using this pyrometer is formed a diagram of temperature changing of iron during crystallization of the working layer of the roll (Figure 1).

The chemical composition of the experimental cast iron, shown in Table 1, was analysed by spectral methods using the spectrometers: OBLF QSN 750 (GOST 18895-97) and OBLF GS 1000 (GOST 27611-88).

Table 1. Chemical composition of cast iron designated for the working layer

C	Si	Mn	S	P	Cr	Ni	V	Ti	Nb	Cu
3.00-3.20	0.7-1.0	0.75-0.95	0.015	0.045	1.5-1.85	4.0-4.6	0.15-0.2	0.01-0.02	0.45-0.70	0.11-0.12

The microstructure was examined using the optical microscope Meiji-2700 at 50- to 1000-fold magnification and a computer system of images analysis Thixomet Standard PRO. Marble's Reagent (50 ml HCl, 2 g $CuSO_4$, 50 ml C_2H_5OH, 50 ml H_2O) was applied for revealing the microstructure of the cast iron samples under investigation that allowed the boundaries of the various structural constituents to be clearly defined.

X-ray diffractometer DRON-3 was used for X-ray phase analysis of the total X-ray spectrum of cast iron at a voltage of 32 kV and a current of 10 mA with Mo – k_α radiation; on the potentiometer operable in the record mode, and automatic memory on the PC with the help of the program package DIFWIN1 (step scanning mode: step shooting of 0,05 ° (in the scale of 2Θ), the exposure time of 10 sec., angles within a range of 10 to 40°).

The amount of retained austenite was calculated from the formula:

$$K_A = \frac{1 - K_K}{1 + \frac{F_{пр}}{F_A} * \frac{\Psi_A}{\Psi_{пр}} * \frac{P_{(HKL)A}}{P_{(HKL)пр}} * \frac{V_{пр}}{V_A} * \frac{S_A^2}{S_{пр}^2} * \frac{\sin^2\Theta_A}{\sin^2\Theta_{пр}}}, \qquad (1)$$

where Ψ is an angular factor of the diffraction peak in polycrystalline samples;

$P_{(HKL)}$ is the factor for recurrence of planes with Miller indices (HKL);

$V_i = a^3$ is the volume of the unit cell in angstroms;

S^2_i is a structural factor: for the body-centered cubic lattice is 4, for the face-centered cubic lattice is 16;

$Sin^2\Theta_i$ is squared half the Bragg angle.

The interference lines {112} recorded for α-Fe (Θ=17°40') and the interference lines {113} recorded for γ-Fe (19°07') are planimetrically measured and limited areas F_A and $F_{пр}$ are determined with the use of the program package DIFWIN1.

The amount of martensite is determined from the equation:

$$K_A + K_{пр} + K_к = 1, \qquad (2)$$

where K_A, $K_{пр}$, $K_к$ are the volume fractions of retained austenite, the decomposition products of austenite and carbides, respectively.

The chemical composition of the phases was analysed by X-ray spectrum method using scanning electron microscopes (SEM) TESCAN VEGA II XMU (Czech Republic) equipped with an INCA energy-dispersive X-ray microanalysis system (EDS) ENERGY 450, with a detector of OXFORD (UK) and a software package of INCA (Ekaterinburg), as well as SEM of JEOL, the model JSM-6490LV in the laboratory of NMSTU (Magnitogorsk, Russia).

A dilatometric method was used to study the start and end temperatures of phase transitions in the sample of cast iron designated for the working layer, followed by a change in the specific volume. Investigations were carried out on a Gleeble 3500, the sample is heated by direct passing a current through the vacuum.

The method of computer simulation with the help of a licensed software LWMFlow 4.2r2 was used to analyse thermal processes occurring in the body of the solidifying casting.

3. Research Results and Discussion

The research study found that crystallization of the working layer of alloyed cast iron starts at 1234 ° C in a rotating chill under the effect of centrifugal force due to the heat removal through the mould walls.

The intensity of heat removal varies with time and is dependent on the thickness of the solidified metal, which in turn affects the size of the dendrites. Randomly oriented primary austenite dendrites are crystallized directly in a thin surface layer. Due to centrifugal force elongated columnar crystals, formed forth and oriented in a direction perpendicular to the mould cavity wall at ordinary crystallization, lose stability relatively to the direction of heat removal and become oriented perpendicular to this direction (Figure 2).

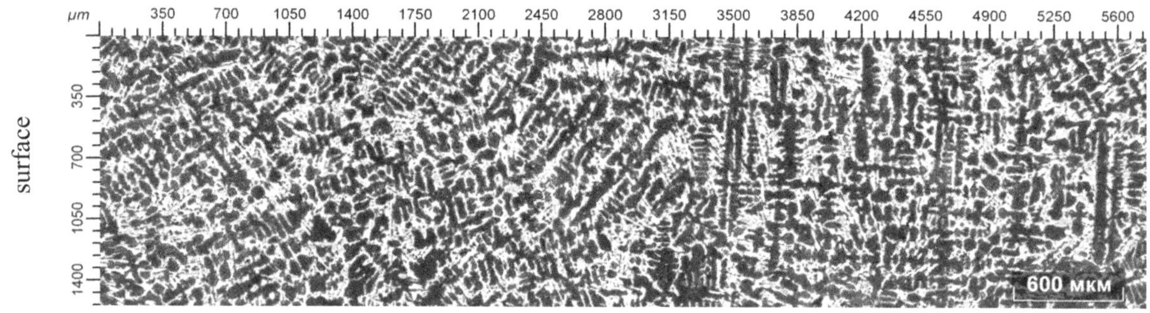

Figure 2. Panoramic image of the microstructure of cast iron designated for the working layer

After the separation of primary austenite dendrites from the liquid phase crystallization ends in the formation of eutectics (Figure 3) at a temperature of 1118 ° C, corresponding to the horizontal portion of the cooling curve of cast iron designated for the working layer (Figure 1). For crystallization under these conditions, it was possible to achieve 10 ° C of supercooling and 1128 ° C of the solidus temperature.

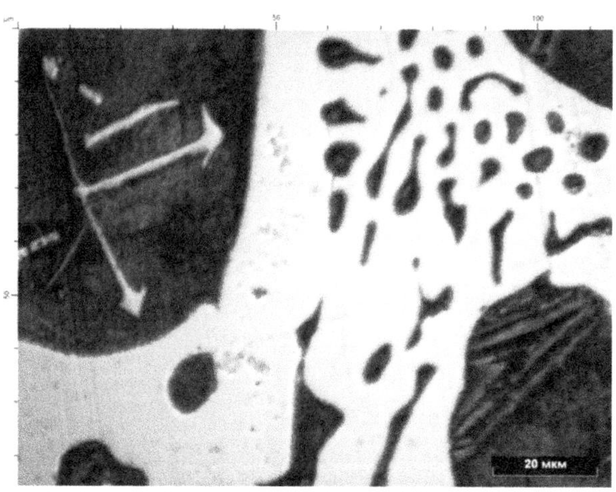

Figure 3. Eutectic in cast iron designated for the working layer

After the temperature of the crystallized working layer reduced to 980 ... 1000 ° C (Figure 1) a chill is transferred to the caisson by a crane and mounted in the moulding box of the lower roll neck. The mould of the upper roll neck is installed on the chill, as well as the funnel for pouring iron into the core at temperature of 1370 ... 1380 ° C; as a result, the working layer of alloyed cast iron is heated and partially melted. Since an optical pyrometer is unfit for a further determination of the change in the temperatures of the working layer, theoretical calculations of the temperature fields in the cross section of the roll in the casting mould at different times, obtained with the help of LVMFlow program, were used (Figure 4).

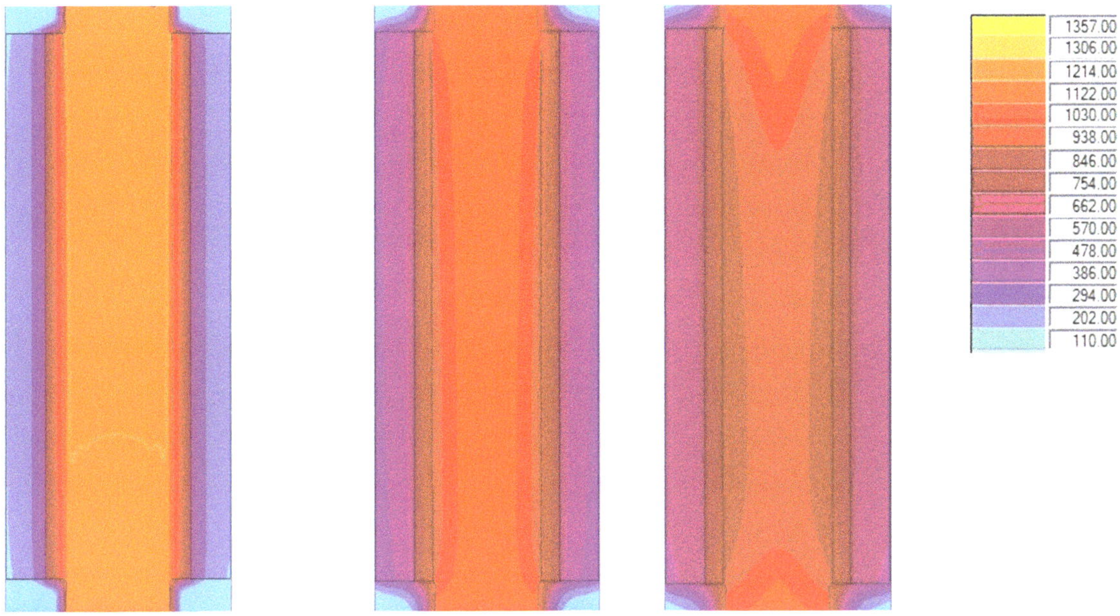

a b c d

Figure 4. Graphical representation of the temperature distribution over the cross section of the roll: a - in18 minutes; b - in 105 minutes; c - in 196 min; d - colour-grade matching to temperature, ° C

Temperatures in the fusion area of the working layer with the core, the centre and the surface of the working layer, defined by calculation, are shown in Figure 5. According to the calculations, the working layer of cast iron for a long time is under isothermal conditions at high temperatures, which vary from the layer thickness, resulted from the roll core pouring, that causes partial decomposition of carbides existing in the structure with the release of carbon in the form of flake graphite (Figures 6, 7).

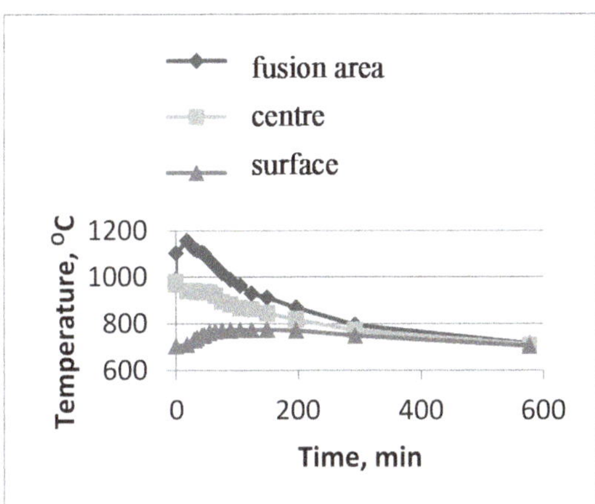

Figure 5. Changes in the temperature of the working layer zones after pouring the core

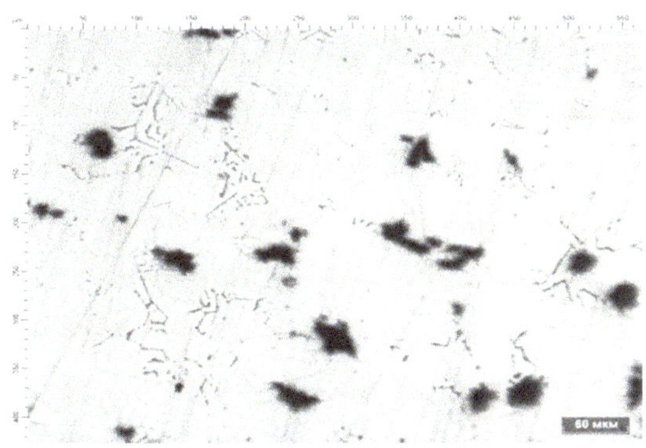

Figure 6. Microstructure of iron designated for the working layer after cooling with isothermal holding: large dark inclusions - graphite; thin dark inclusions - secondary carbides (not pickled)

The carbon solubility decreases during cooling after crystallization throughout the whole temperature range and the secondary carbides of various shapes precipitate from the metallic base, as well as the flake graphite precipitation is continuing.

The structure of iron designated for the working layer, cooled without being isothermally hold, in which there are no graphite inclusions proves out that the partial graphitization of iron cementite designated for the working layer is the result of isothermal holding. When cooling the excess phase precipitates only in the form of secondary carbides (Figure 8).

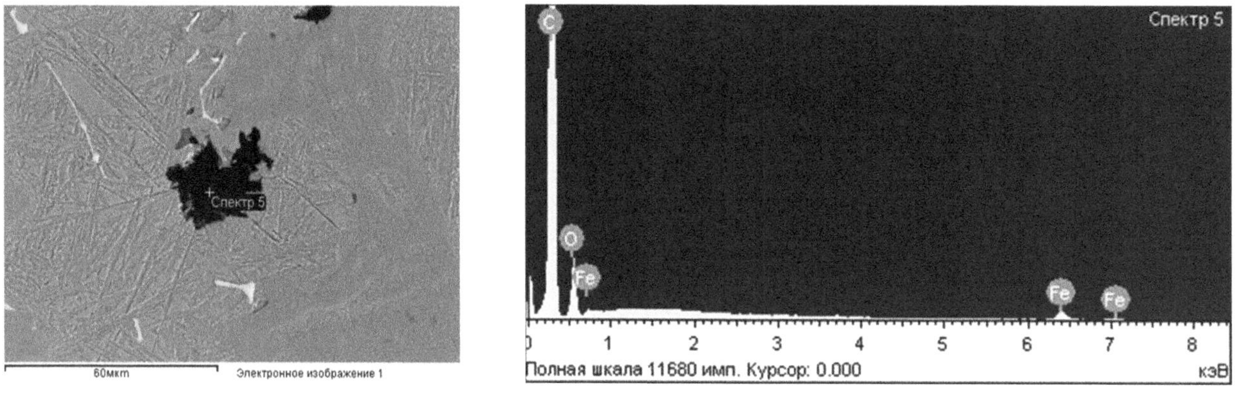

a b

Figure 7. The electron image (a) and spectrum of graphite inclusions (b)

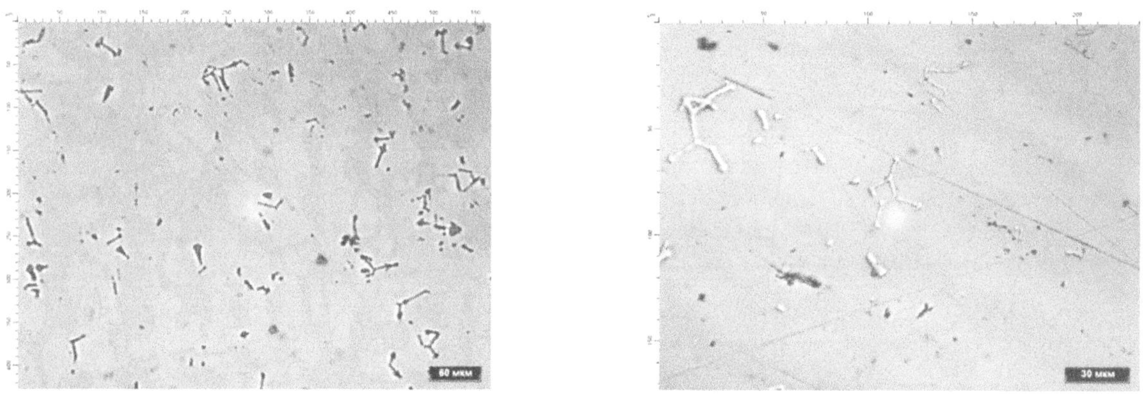

Figure 8. Microstructure of iron designated for the working layer after cooling without isothermal holding: thin inclusions - secondary carbides, (not pickled)

The formation of the structure of iron designated for the working layer in the process of roll manufacturing of grade LPHNd-71 ends with the partial transformation of austenite into martensite when a martensite start temperature is 180 ° C (Figure 9).

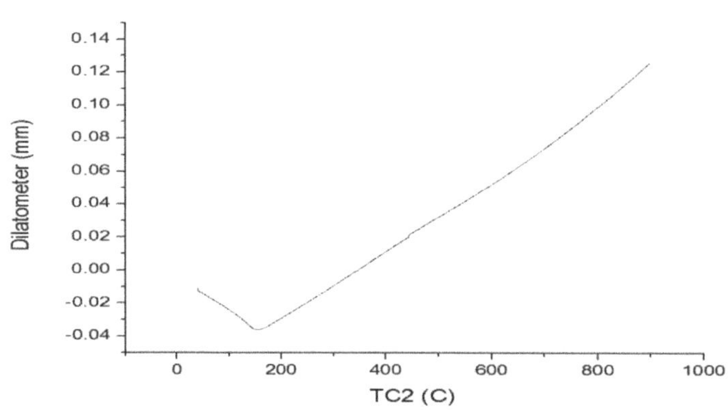

Figure 9. Dilatometric curve of cast iron designated for the working layer obtained in cooling at a rate of 5 ° C / min

Thus, it was found that the final structure of cast iron designated for the working layer consists of the martensite-austenite metallic base (primary austenite dendrites), eutectic carbides, and excess phases in the form of secondary carbides and graphite inclusions (Figure 10).

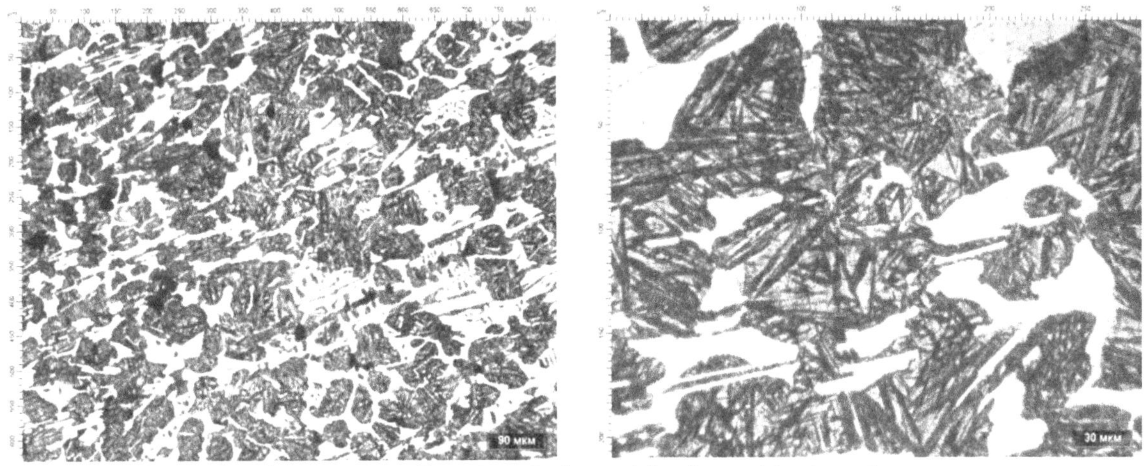

Figure 10. Microstructure of cast iron designated for the working layer in as-cast state

The amount of martensite and retained austenite was determined with the help of X-ray structural analysis; their content in the metallic base is equal to 60.0 per cent and 11.4 per cent, respectively.

The size of the martensite needles was defined in cast iron designated for the working layer, which ranges from a long size of grade 9, growing through the former austenite dendrite, to a short size of grade 7, growing in the former austenite dendrite at an angle to the existing needles.

The analysis of the diffractogram allowed one to determine that the martensite formation results in a tetragonal splitting of distinctive maximum of α-iron (2.01 Å) and its shift towards larger interplanar distances – 2.029 Å (Figure 12).

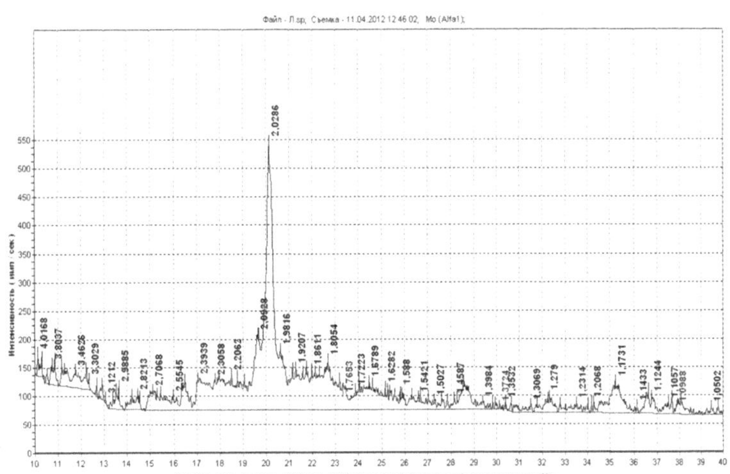

Figure 12. Diffractogram of iron designated for the working layer in as-cast state

Shifting is caused by the occurrence of double peak in the martensite lattice: the left one corresponds to the diffraction from the crystal plane (011), the right one corresponds to the diffraction from the plane (110); herewith the intensity at the maxima, forming this pair, is not the same as the repeatability multiplication factor for these planes is 8 and 4, respectively (Figure 13). Having determined the periods "a" and "c", we calculated the degree of the tetragonal martensite c / a = 1.069, a decrease in the value of which gives evidence of the martensite decay.

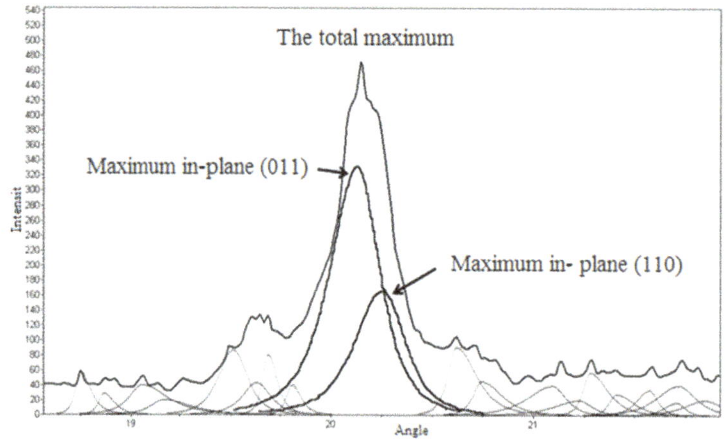

Figure 13. A fragment of diffractogram of cast iron designated for the working layer in as - cast state: distinctive maximum of α-iron with martensitic splitting

The volume fraction, the average area, and the density of the distribution of graphite and carbides were determined by metallographic analysis method. The hardness of iron designated for the working layer, and the average microhardness of the base and carbides were determined. Data obtained by metallographic analysis and mechanical test results of cast iron designated for the working layer in as - cast state are represented in Tables 2 and 3.

The working layer of the roll while in operation undergoes thermal loads related to heating during the contact with the rolled strip and subsequent water cooling, resulting in austenite disintegration in cast iron designated for the working layer, the amount of which in as-cast state is more than 11 per cent with the martensite formation. This process results in the occurrence of stresses, cracks on the roll surface and with its subsequent spalling. To reduce the amount of retained austenite and relieve stresses, which arise during the formation of its degradation products, rolls are subjected to tempering with slow heating to 430 ° C.

Table 2. Data obtained by metallographic analysis

Volume fraction of carbides, %	26
The average area of carbides, mkm^2	146
The density of distribution of carbides, 1 / mm^2	1783
Volume fraction of graphite inclusions, %	2,6
The average size of graphite inclusions, mkm^2	125
The density of distribution of graphite inclusions, 1 / mm^2	235

Table 3. The test results on the hardness and microhardness

Hardness, HRC	52,5
The average microhardness of the base, HV	497
The average microhardness of carbides, HV	942

This processing resulted in the cast iron structure, which consists of the metallic base, formed by the needle phases and retained austenite, eutectic and secondary carbides and graphite (Figure 14).

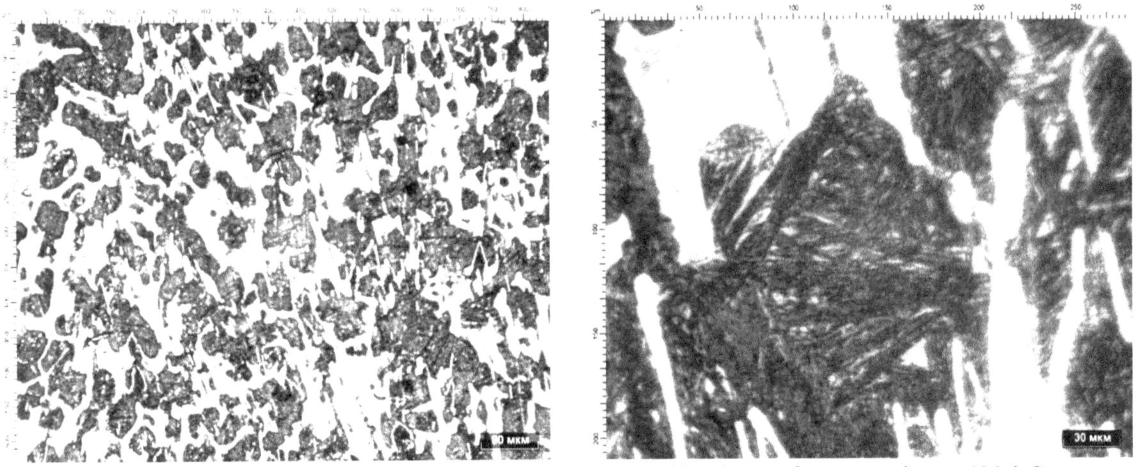

Figure 14. Microstructure of iron designated for the working layer after tempering at 430 ° C

Qualitatively, the structure of iron designated for the working layer after tempering at 430°C is not different from the structure of iron in the as-cast state. The difference can be observed for the quantitative relationships of the phases (Vdovin, Gorlenko, & Zavalishchin, 2013).

Tempering results in an increase in the size of excess phases, precipitated during cooling after solidification, and the formation of new additional secondary carbides and graphite inclusions in the martensitic-austenitic metallic base. At the same time, austenite, impoverished in alloying elements due to the excess phase release during heating, becomes less stable to the disintegration, and is partially transformed to bainite during cooling, which together with the tempered martensite is included in the composition of the acicular structure of the metallic base.

Using X-ray diffraction analysis we determined that the amount of retained austenite is reduced from 11.4 per cent in the as-cast state to 3.0 ... 3.2 per cent after tempering. Heating to 430°C decreases the shifting of α-iron specific peak to values of 2.015 Å, compared with a cast state (2,028 Å) (Figure 15).

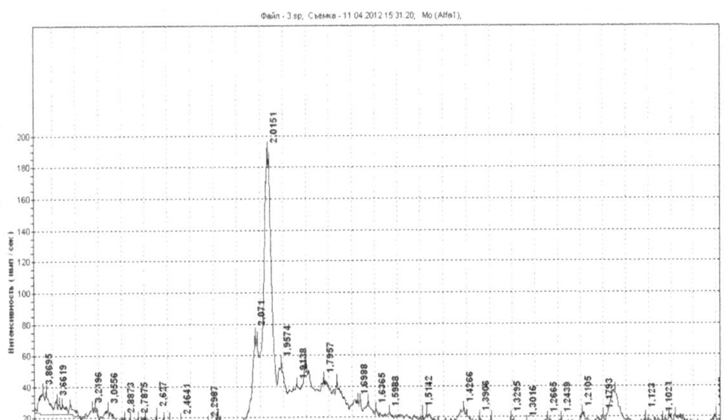

Figure 15. Diffractogram of cast iron designated for the working layer after tempering at 430 ° C

As a result of martensite desintegration the ratio of the axes of its tetragonal lattice relative to the as-cast state decreases, and the tetragonal degree c/a reduces from 1,069 to 1,013, correspondingly (Figure 16).

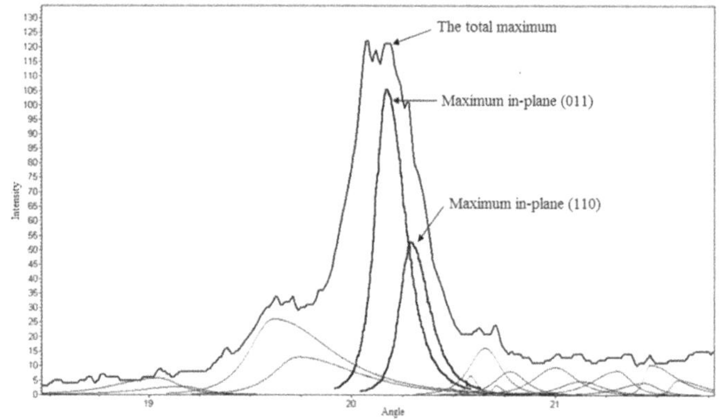

Figure 16. A fragment of diffractogram of iron designated for the working layer after tempering at 430 ° C:distinctive maximum of α-iron with martensitic splitting

In addition to reducing the amount of retained austenite, changes occur in the volume fraction, the number of impurities per unit area, and the average area of excess phases after heating cast iron designated for the working layer to 430 ° C (Vdovin, Gorlenko, & Zavalischin, 2013). The volume fraction of graphite increases more than 1 per cent, and becomes equal to 3.7 per cent (2.6 per cent in as-cast). The number of graphite impurities per unit area increases more than twice, and the average area of inclusions decreases due to the fact, that the precipitated graphite particles are less than "as-cast" (Figure 17).

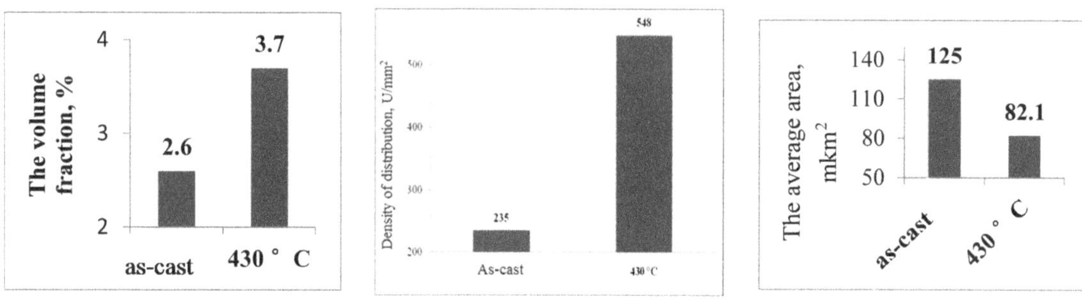

a b c

Figure 17. Characteristics of graphite inclusions: a - the volume fraction, %; b - the density of distribution, units / mm²; c - the average area, mkm²

The total volume fraction of eutectic and secondary carbides increases from 26 to 29.8 per cent, the number of inclusions per unit area increases almost twice from 1783 units / mm² to 3181 units / mm², herewith, the amount of eutectic carbides does not changed. The increase in the total volume fraction of the carbide phase occurs due to the additional extraction of dispersed secondary carbides, which also results in a decrease in values of the average area of the carbide phase (Figure 18).

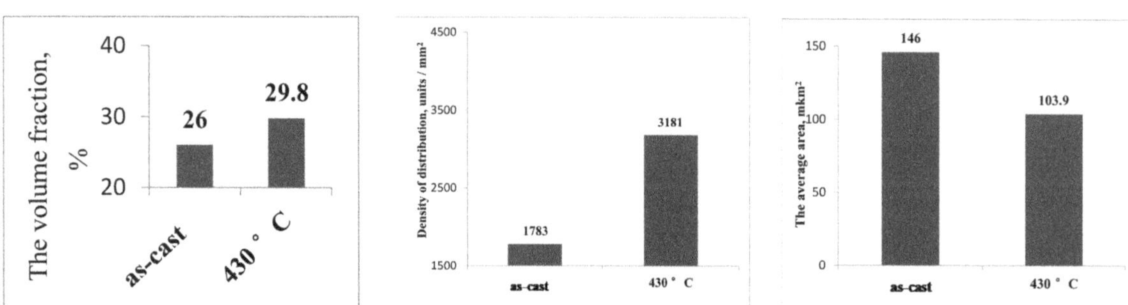

a b c

Figure 18. Characteristics of the carbide phases: a - the volume fraction, %; b - the density of distribution, units / mm²; c - the average area, mkm²

The average microhardness of the metallic base and carbides, as well as hardness in general of iron designated for the working layer after tempering at 430 ° C remain practically unchanged. Two processes of changes in hardness, which compensate each other, occur in the metallic base: decreasing the number of retained austenite increases microhardness, and the disintegration of martensite reduces microhardness (Figure 19 a). Microhardness of the carbide phase does not change for the newly precipitated excess carbides have the same microhardness as "as-cast" (Figure 19 b). The precipitation of solid secondary carbides and soft graphite and bainite does not result in the change of hardness in iron after tempering at a temperature 430°C as a whole (Figure 19).

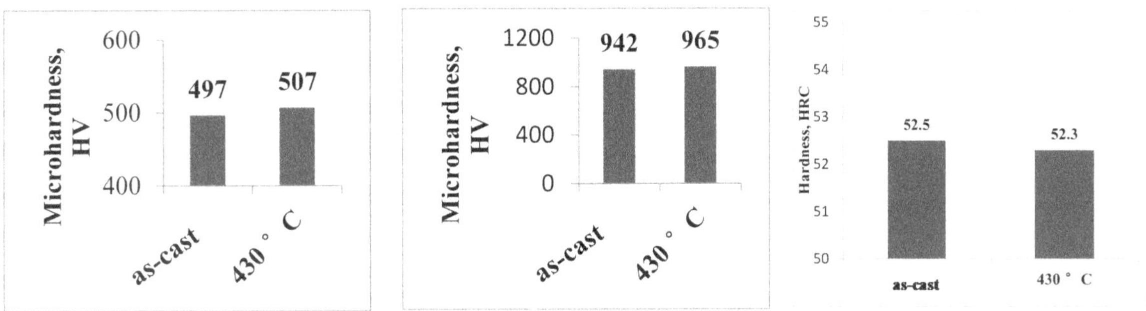

a b c

Figure 19. Mechanical properties after tempering at 430 ° C: a - the average microhardness of the metallic base, HV; b - the average microhardness of carbides, HV; c - hardness, HRC

4. Conclusions

The following conclusions have emerged from the present study.

The microstructure of cast iron designated for the working layer, which consists of 60 per cent of martensite, 11.4 per cent of retained austenite, 26 per cent of eutectic and secondary carbides, 2.6 per cent of graphite, is formed during crystallization and further cooling with further isothermal holding.

Partial graphitization of cast iron cementite designated for the working layer is a result of isothermal holding, in the case of its absence while cooling the excess phase precipitates only in the form of secondary carbides.

Cast iron tempering at a temperature of 430°C changes the quantitative phase relationship relative to the as-cast state, herewith the structure consists of 63.4 per cent of martensite, 3.1 per cent of retained austenite, 29.8 per cent of eutectic and secondary carbides, and 3.7 per cent of graphite.

References

Al-Rubaie, K. S., & Pohl, M. (2014). Heat treatment and two-body abrasion of Ni-Hard 4. *Wear*, *312*(1), 21-28.

Asensio-Lozano, J., & Álvarez-Antolín, J. F. (2008). Saturated fractional design of experiments: Toughness and graphite phase optimizing in Nihard cast irons. *Journal of Materials Engineering and Performance*, *17*(2), 216-223. DOI 10.1007/s11665-007-9138-8.

Asensio-Lozano, J., Álvarez-Antolín, J. F., & Vander Voort, G. F. (2008). Identification and quantification of active manufacturing factors for graphite formation in centrifugally cast Nihard cast irons. *journal of materials processing technology*, *206*(1), 202-215. DOI:10.1016/j.jmatprotec.2007.12.015.

Gostev, K. A. (2008). Heat treatment in the production of cast rolls. *Steel, 9*, 79-84.

Jinzhu, L., Shizhuo, L., & Yongfa, M. (1993). Wear resistance of Ni-hard 4 and high-chromium cast iron re-evaluated. *Wear*, *166*(1), 37-40.

Kang, M., Suh, Y., Oh, Y. J., & Lee, Y. K. (2014). The effects of vanadium on the microstructure and wear resistance of centrifugally cast Ni-hard rolls. *Journal of Alloys and Compounds*, *609*, 25-32.

Kuskov, Y. M. (2006). Influence of graphite inclusions on the durability of the cast and hardfaced hot rolling rolls. *Steel, 2*.

Lee, S., Kim, D. H., Ryu, J. H., & Shin, K. (1997). Correlation of microstructure and thermal fatigue property of three work rolls. *Metallurgical and materials transactions A*, *28*(12), 2595-2608. DOI 10.1007/s11661-997-0017-6.

Martini, F. (1999). Main manufacturing and service requirements for the backup rolls and work rolls of modern hot-strip mills. *Metallurgist, 43*(8), 365-370. DOI 10.1007/BF02463657

Martini, F., & Gostev, K. A. (1999). New generation of wear-resistant bimetallic rolls for sheet mills. *Metallurgist, 43*(5), 200-204.

Nilsson, M., & Olsson, M. (2013). Microstructural, mechanical and tribological characterisation of roll materials for the finishing stands of the hot strip mill for steel rolling. *Wear, 307*(1), 209-217. DOI: 10.1016/j.wear.2013.09.002.

Palit, P., Jugade, H. R., Jha, A. K., Das, S., & Mukhopadhyay, G. (2015). Failure analysis of work rolls of a thin hot strip mill. *Case Studies in Engineering Failure Analysis, 3*, 39-45. DOI:10.1016/j.csefa.2015.01.001.

Ray, A., Mukherjee, D., Sarkar, B., & Mishra, S. (1994). Influence of microstructure on the premature failure of a second-intermediate Sendzimir mill drive roll. *Journal of materials engineering and performance, 3*(5), 649-656. DOI 10.1007/BF02645263.

Ray, A., Prasad, M. S., Barhai, P. K., & Mukherjee, S. K. (2004). Metallurgical investigation of prematurely failed hot-strip mill work-rolls: some microstructural observations. *Journal of Failure Analysis and Prevention, 4*(3), 58-66. DOI 10.1007/s11668-996-0016-8.

Ray, A., Prasad, M. S., Dhua, S. K., Sen, S. K., & Jha, S. (2000). Microstructural features of prematurely failed hot-strip mill work rolls: Some studies in spalling propensity. *Journal of materials engineering and performance, 9*(4), 449-456. DOI 10.1361/105994900770345854.

Sinha, P., Indimath, S. S., Mukhopadhyay, G., & Bhattacharyya, S. (2014). Failure of a Work Roll of a Thin Strip Rolling Mill: A Case Study. *Procedia Engineering, 86*, 940-948. DOI:10.1016/j.proeng.2014.11.117

Traino, A. I. (2008). Rational roller operation and restoration. *Steel in Translation, 38*(10), 871-875.

Vdovin, K. N., Gorlenko, D. A., & Zavalischin, A. N. (2013). Structure changes of chromium-nickel indefinite cast irons in heating. *Vestnik of Nosov Magnitogorsk State Technical University, 5*(45), 9-11.

Vdovin, K. N., Gorlenko, D. A., & Zavalishchin, A. N. (2013). Influence of industrial tempering on the composition of complex cast iron. *Steel in Translation, 43*(5), 288-290.

14

Determining the Directionally-Dependent Effective Hydrogen Diffusivity of High Strength Steel Strand

Joseph R. Fernandez[1]

[1] University of South Florida, 4202 East Fowler Ave ENL 111, Tampa, Florida 33784, USA

Correspondence: Joseph R. Fernandez, University of South Florida, 4202 East Fowler Ave ENL 111, Tampa, Florida 33620, USA. E-mail: jfernandez@mail.usf.edu

Abstract

An important factor in the Hydrogen Embrittlement (HE) of cold-worked carbon steel strand is the effective hydrogen diffusivity of the strand material. The speed at which HE can occur is determined by the transport of the hydrogen to locations of stress concentrations. A Scanning Kelvin Probe (SKP) was used to measure the flattened surface of strand wires before and after being charged with hydrogen to determine a relationship between the delta work function measured (and indirectly the potential of the specimen) and the presence of hydrogen near the surface of that area of the specimen. Numerous scans were performed to determine a change in delta work function between the probe and the specimen. This change in delta work function was related to the presence of hydrogen at the specimen surface over time, giving a rate of hydrogen movement within a specific area of the specimen (both in the radial and drawn directions). The change in delta work function over time was then related to the diffusion of hydrogen over time in a given area resulting in a directional effective diffusivity of hydrogen through the strand which includes the effects of hydrogen trapping and method of hydrogen transport.

The results indicate that the effective or apparent hydrogen diffusivity in the drawn direction in pearlitic, cold worked, carbon steel strand is 5.24E-08 cm^2/sec. The effective hydrogen diffusivity in the radial direction was faster (3.56E-07 cm^2/sec) as compared to the wire drawn direction.

Keywords: carbon steel, hydrogen embrittlement, Scanning Kelvin Probe, work function.

1. Introduction and Background

Hydrogen Embrittlement (HE) can affect heavily cold worked, high strength (1.86 GPa) carbon steel (0.75% Carbon, 0.6% Manganese, and the balance Iron) strands that actively corrode (due to surface reactions with salt and/or water on the surface of the strand) while under stress. HE occurs when atomic hydrogen has entered the carbon steel bulk and built up to a critical concentration (which can be very small, hence easily reached) at points of stress concentration in the material (e.g. crack tips). HE can lead to a normally ductile material in tension failing in a violent brittle manner. On application of sufficiently high tensile stress, the hydrogen experiences preferential transport (often by diffusion or dislocation motion) to those regions of stress concentration, where cracks develop and propagate. One aspect of the HE of the strand due to hydrogen generation from surface reactions (Sagüés, Fernandez, Hutchinson, & Mullins, 2014) is the diffusion of hydrogen within the strand wires. While the diffusion transport of hydrogen through the steel is normally represented simply by Fick's second law of diffusion (Jones, 1996), the effects of hydrogen traps and the number of them (which vary in number and size between the radial and drawn directions) on the effective diffusivity (the rate of transport of hydrogen including delays due to hydrogen trapping and dislocation motion) requires Hydrogen (H) diffusivity measurement. Due to the cold working, which adds dislocation in the drawn direction as well as thinning the grains in the radial direction, the effective diffusivity should be faster in the radial direction as compared to the drawn direction. The validation of the different directional effective diffusivities of the strand as well as the faster diffusivity direction will give insight into the fracture morphology of the strand material under stress. A Scanning Kelvin Probe (SKP) has been used to determine effective diffusivity in high-strength steels (Schaller & Scully, 2013).

The adverse effect has been explained alternatively by the phenomena of Hydrogen Induced Decohesion (HID), Hydrogen Enhanced Localized Plasticity (HELP), and Adsorption Induced Dislocation Emission (AIDE) (Jones, 1996; Moro et al., 2010; Lynch, 2012). All of these phenomena have in common the movement of hydrogen to a location in front of the crack tip growth front is required for crack propagation. Therefore, the speed of crack growth as well as the path of the crack growth should be dominated by the direction of hydrogen transport that is the fastest, which will allow for faster hydrogen buildup at the location near the crack tip.

Hydrogen becomes naturally available at a wet steel surface when cathodic, hydrogen evolution reactions take place there as part of corrosion processes. Corrosion reactions can occur in humid air, distilled water, acid environments, neutral environments, or alkaline environments (Mietz, 2000). The cathodic reaction removes the electrons that are created by the anodic reaction (metal dissolution);

$$Fe \rightarrow Fe^{2+} + 2e^{-} \tag{1}$$

on the steel surface. The cathodic reaction is often oxygen reduction, expressed as;

$$\frac{1}{2}O_2 + H_2O + 2e^{-} \rightarrow 2OH^{-} \tag{2}$$

if the water solution is neutral or alkaline, or;

$$H^{+} + e^{-} \rightarrow \frac{1}{2}H_2 \tag{3}$$

(hydrogen ion reduction) if acidic. If oxygen is depleted and the solution is neutral or alkaline, the reduction reaction consists of direct water decomposition (Hartt, Kumria, & Kessler, 1993).

$$H_2O + 2e^{-} \rightarrow \frac{1}{2}H_2 + OH^{-} \tag{4}$$

Both hydrogen evolution reactions are written above as forming hydrogen molecules. However, those reactions are known to involve intermediate steps where atomic hydrogen is formed first as a surface adsorbate which then recombines as H_2 (Oriani, 1993). Most of the evolved hydrogen ends up as H_2 (Oriani, 1993) which are too large to enter into the interstitial sites of the crystal structure of the steel, and thus bubble off the surface into the electrolyte. Atomic hydrogen is small enough in size to become an interstitial atom in the steel (entering the tetrahedral sites of the lattice), so a fraction (e.g., 1/100) of the total hydrogen produced by the cathodic reactions actually enters the steel bulk (Oriani, 1993).

Although the fraction of H penetrating the steel is small, the amount of hydrogen required to initiate HE at a location of stress concentration is equally very small. This can be attributed to the smaller interstitial space of the tetrahedral sites and the inability for much expansion before the bonds are weakened (Oriani, 1993). A concentration of ~0.03 ppm of diffusible hydrogen is sufficient to initiate HE in post tensioning steel strand (Enos, Williams, Clemeña, & Scully, 1998). Detecting this hydrogen content before HE occurs using a non-destructive method that does not negatively alter the surface chemistry is important to prevent any future failures due to HE.

A Scanning Kelvin Probe (SKP) can determine the difference in work functions ($\Delta\varphi$) between two surfaces.

$$\Delta\varphi = \varphi_w - \varphi_p \tag{5}$$

The probe surface or reference surface (φ_p) is placed at a distance h away from the specimen surface or working surface (φ_w). The reference surface is 0.5 mm in diameter (d), and is placed approximately 0.05 mm away from the working surface. The small distance between the working and reference surfaces allows for the approximation of the system as a parallel-plate capacitor. As shown in earlier work (Sagüés & Walsh, 2011), a relationship between the distance h and the capacitance C can be related to the electrical charge Q and potential E.

The SKP reference surface is then oscillated at a frequency of 81 Hz which then causes the Q to change cyclically resulting in an alternating current I_A. Using a variable potential source (in series with the working/reference surface circuit) with potential E', the potential between the reference and the working surface is $E + E'$. When E' equals $-E$, the potential across the capacitor will be zero. The potential E can be determined by varying E' until the I_A equals zero. The difference in work function measured by the SKP is also the difference between the Volta (Sagüés & Walsh, 2011) potential Ψ_p of the reference surface and the Volta potential of the working surface Ψ_w. Both the metal reference point surface and the working specimen surface have a ψ average value over their surface areas.

$$\Delta\varphi = \Psi_p - \Psi_w = \varphi_w - \varphi_p \tag{6}$$

The effect that hydrogen has on the surface of the steel is an increase in potential, which leads to a decrease in the steel surface work function. This decrease in the steel's work function can be detected by the decrease in the delta work function between the probe and the steel. Hydrogen reaches the surface of the steel, reducing the ratio of Fe^{3+} to Fe^{2+} iron oxide, which leads to the change in the Fermi level position, and in turn decreases the work function of the surface of the steel (Evers, Senöz, & Rohwerder, 2013). This decrease in work function at the surface of the steel can also be related to the increase in Volta potential of the steel as related to the potential of the H/H^{+} electrode using the Nernst equation (Evers, Senöz, & Rohwerder, 2013). Also, a change in the pH of the surface of the specimen occurs due to atomic hydrogen at the surface oxidizing and increases the OH^{-}/O^{2-} ratio in the passive film weakening and eventually destroying it (Guo, Liang, Bai, Miao, & Volinsky, 2014). As the probe work function is constant, any change in the delta work function in the negative direction is due to a decrease in the surface work function, which can be caused by the presence of hydrogen at the surface.

2. Experimental Procedure

Testing of strand specimens for hydrogen effective diffusivity was performed by charging specimens with hydrogen using a hydrogen charging solution with 875 ml of DI water, 125 ml of Sulfuric Acid (H_2SO_4), and 0.5 grams of Antimony Trioxide (Sb_2O_3) [as a hydrogen recombination poison] while cathodic polarization to about 0.5 A/cm^2 occurred at room temperature 298.15 K.

The specimens were initially charged for as much as 24 hours, but shorter charging times while ensuring hydrogen charging sufficient to detect hydrogen concentration change over time was desired. It was discovered through numerous iterations, that sufficient hydrogen concentrations could be routinely achieved by charging the specimens for as little as 2 hours at 0.5 A/cm^2. This combination was chosen for the subsequent tests as it reduced the testing time and was sufficient to initiate hydrogen concentration gradients over time detectable by the SKP as desired.

A Princeton Applied Research SKP100E Scanning Kelvin Probe with a platinum tip was used. The SKP 100E scanned 256 points along the x-axis. These line scans were performed once every hour for 24 hours. Scans were performed along the center of the specimens, from the one end to the other. A topography scan is done prior to delta work function scans to obtain an accurate measure of the variation of the surface height, and is used to mitigate the change in surface height artifact on the readings.

Early work on scanning charged carbon steel showed little to no clear effect. Noting that most of the previous work using SKPs to detect hydrogen either plated the carbon steel with Palladium, or used stainless steel, a set of scanning experiments using stainless steels were performed. Two specimens of 316L high-strength stainless steel and two specimens of XM 29 high-strength stainless steel were charged with hydrogen using the same method as the charging of the U-bend specimens in the prior section. Figures 1 through 4 show a significant drop in the region that was charged with hydrogen. The location of hydrogen charging was at one end of the specimen, and they were scanned from the uncharged end (left end of Figures 1 through 4) to the charged end (right end of Figures 1 through 4) to ensure plenty of uncharged surface of the specimen scanned for comparison.

Variations in the scan pattern over the uncharged areas differ between 316L and XM29 are possibly due to the difference in material composition. Specifically, the XM29 uses manganese and nitrogen, rather than nickel as the austenitic stabilizer. Nitrogen is known to have an effect on the concentration of hydrogen in steel (Lynch, 2012). Even though the scan pattern in the uncharged region is different between the two stainless steel alloys, the dip in the charged region is very similar and is clear that some effect, due to the hydrogen present on the surface, is being detected by the SKP.

This data indicated that we could detect hydrogen on the surface, but also that a protective passive layer on the metal gave clearer results. Therefore the need to form a protective passive layer on the carbon steel was required to obtain clear scan data. Carbon steel forms a protective passive layer at a pH above 9 (Gellings, 2005). Therefore, to obtain better results, all of the carbon steel specimens scanned were first placed in a $Ca(OH)_2$ solution (~pH 12) for 30 minutes to allow the formation of a protective passive layer.

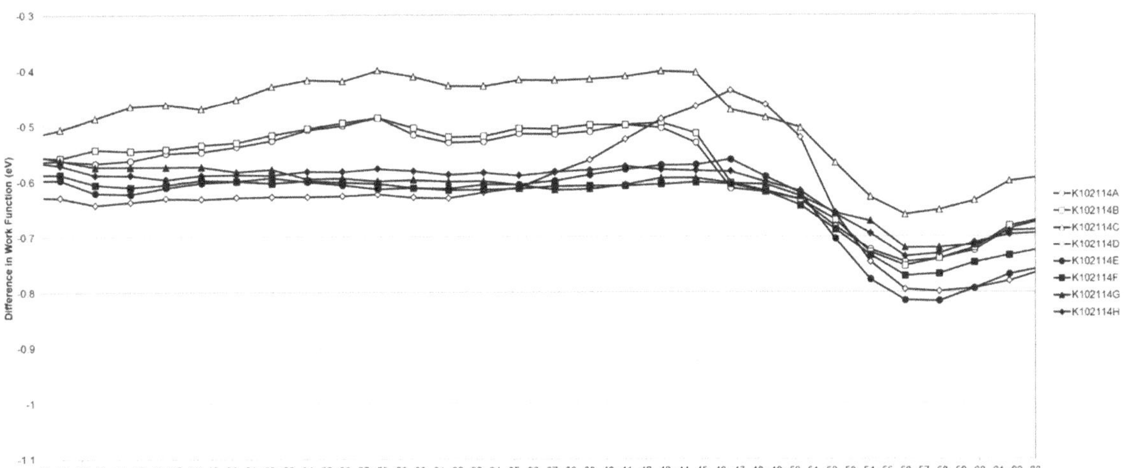

Figure 1. Scans of Charged 316L Specimen 1

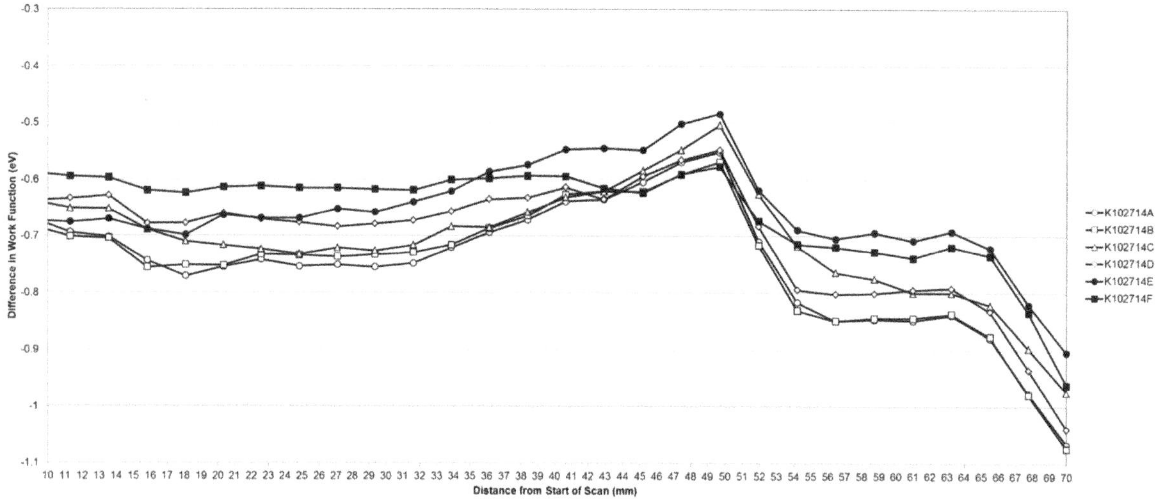

Figure 2. Scans of Charged 316L Specimen 2

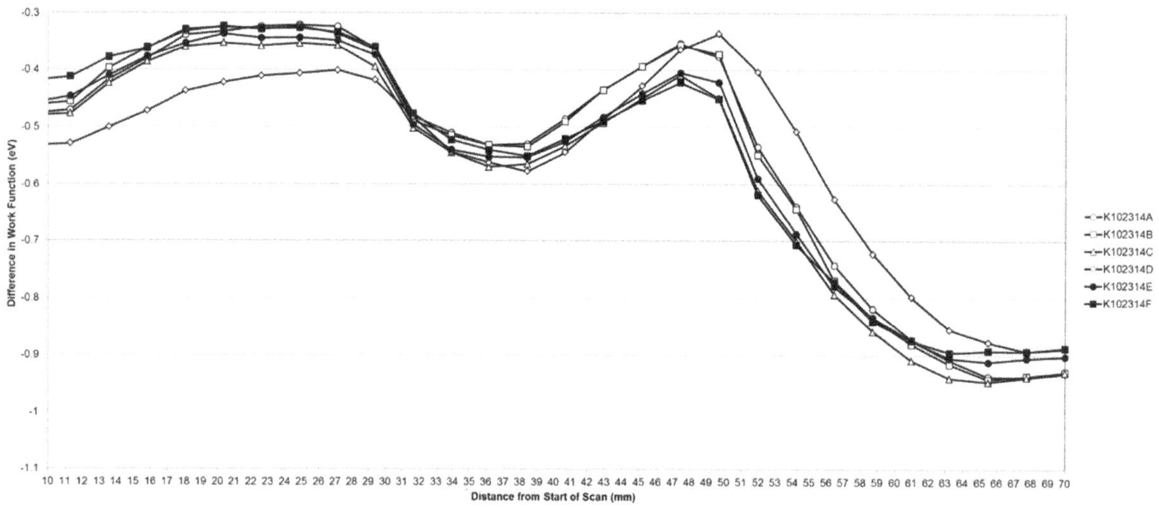

Figure 3. Scans of Charged XM 29 Specimen 1

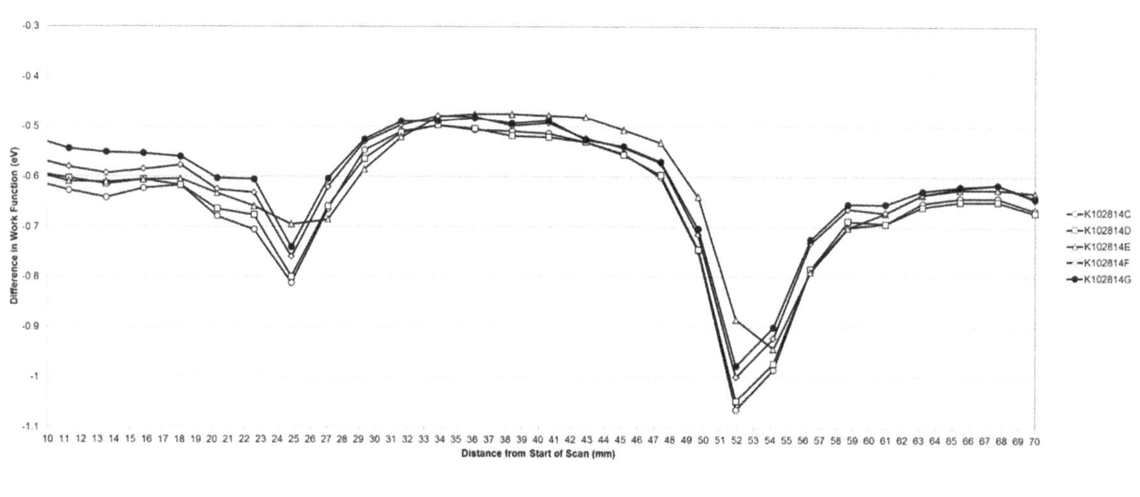

Figure 4. Scans of Charged XM 29 Specimen 2

Specimens were charged at a spot along the scan surface using the charging setup shown in Figure 5. A total of eight set of scans, were performed on different wire specimens of 1080 high-strength carbon steel. Four set of scans used flattened specimens along the drawn direction, so as to ensure a flat surface for scanning. Drawn direction scan data was truncated 20 mm from the start and end of the scans to focus on the data in the region charged.

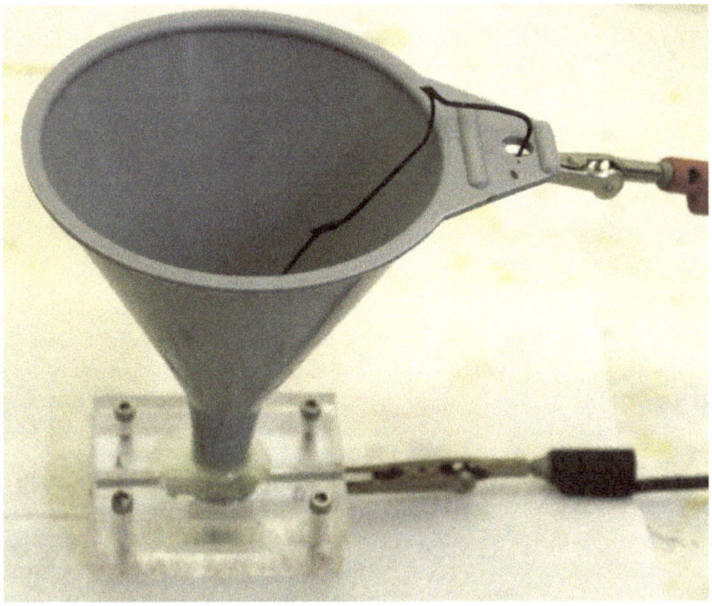

Figure 5. Setup of Specimen Spot Charging Device for Drawn Direction Scans

The specimens were charged at 0.3 A over an area of 0.6 cm^2 (0.5 A/cm^2) for 2 hours at in the hydrogen charging solution used previously. Then, the specimens were removed from that setup, degreased with acetone and polished at 600 grit in oil (to prevent water contact before scanning), placed in a solution of $Ca(OH)_2$ for 30 minutes as well as ultrasonic cleaned and mounted for scanning.

Four set of scans used specimens cut into small cylinders (0.635 cm (0.25 inches)) and the surfaces polished flat. Specimens were charged at the top of the cylinders using the same charging setup shown previously. These specimens were also charged for 2 hours at 0.5 A/cm^2 and then cleaned, placed in a solution of Ca $(OH)_2$ for 30 minutes and mounted. These four set of scans used flattened specimens along the radial direction, which were scanned along the radial direction as opposed to the drawn direction.

As the effective diffusivity is related to the concentration of hydrogen at the surface, for a specimen with low trap occupancy (this is true until steady state is achieved when trap occupancy is at its maximum), the delta work function is linearly related to the hydrogen concentration (Gangloff & Somerday, 2012a & 2012b). This allows the SKP data to be used to determine the effective diffusivity of the specimen by relating the hydrogen concentration at the surface to the delta work function at the surface. The ratio of the hydrogen concentration at a point x on the surface of the specimen at time t to the hydrogen concentration at a uncharged point x' on the surface of the specimen was directly related to the ratio of the work function of the point x at time t to the work function of the surface of the specimen away from the charged region (x') (Schaller & Scully, 2013). This relationship was also shown to be a function of the effective diffusivity at the point x at time t (Schaller & Scully, 2013).

The delta work function is shown to be the work function of the surface of the specimen at a certain location minus the work function initially measured, which is also the work function of the probe (as the work function of the probe is constant). Therefore the delta work function of the surface not charged with hydrogen, the location of this measurement was chosen to be the farthest point from the center of the charged region while being the closest point to the start of the scan, is the equivalent of the difference between the work function of the specimen at a point x at time t minus the initial work function of the specimen at point x ($\Delta\varphi_{ncp}$). The delta work function of the surface charged with hydrogen, the location of this measurement was chosen to be in the center of the charged region, is then the equivalent of the difference between the work function of the specimen at the surface charged minus the initial work function of the specimen at the surface ($\Delta\varphi_{cp}$). The equation that was used to determine the effective diffusivity is:

$$D_{eff} = \frac{\left(\frac{x}{2*erf^{-1}\left(1-\frac{\Delta\varphi ncp}{\Delta\varphi cp}\right)}\right)^2}{t} \tag{7}$$

5. Results and Discussion

The calculations of the effective diffusivity in the drawn and radial direction were performed based on the sets of SKP scan data. The effective diffusivity equation using the delta work function ratio of charged and uncharged location at a point x and time t was used with the data obtained from each scan in a set. The uncharged scan location used for the calculation was the farthest point from the center of the charged location area, closest to the start of the scan. The charged scan location used for the calculation was the center of the charged location area. Those effective diffusivities were averaged together to obtain that specimen's effective diffusivity in either the drawn or radial direction.

The diffusivities of the four sets of scan (all scans in a set were averaged together to get the set diffusivity), for each direction scanned (four sets for the drawn direction and four sets for the radial direction) were then averaged together to get the final effective diffusivity in the drawn (or radial) direction. As shown in Figures 6 through 13, the region charged has a significant dip initially, which then widens and shrinks. This is indicative of the hydrogen dissipating outward as related to the scan direction. Therefore, if the hydrogen dissipates quickly (in the scanned direction as hydrogen loss through the surface takes considerably more time due to desorption characteristics) then the effective diffusivity of hydrogen in that direction is fast (or high in values). If the hydrogen dissipates slowly, then the effective diffusivity of hydrogen in that direction is slow (or low in value).

The work function of the surface is dependent not only on the presence of hydrogen or lack of hydrogen at the surface, but other factors such as water vapor or other surface contaminants that can lead to fluctuation of the exact work function value at a point on the surface of the specimen. Therefore, each specimen delta work function scan will be different due to these fluctuation, as well as scans over time, as water in the air settles onto the surface of the specimen.

While the topography scan is used to compensate for the variations in the surface height, it is not possible to adjust for the changes in surface condition, especially over time. This is one of the reasons that the set of scans were averaged together, and then the four sets of scans were averaged together again to obtain the effective diffusivity for that direction. The delta work function values were also related to the time after charging ended so as to attempt to reduce these artifacts in the measurements.

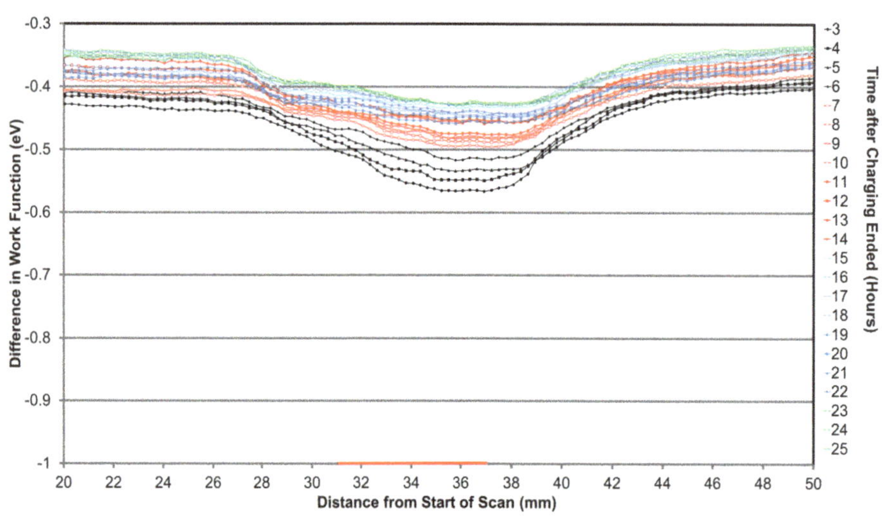

Figure 6. Scans of Charged Carbon Steel Specimen C10. Red Line at Bottom Indicates Region of Charging

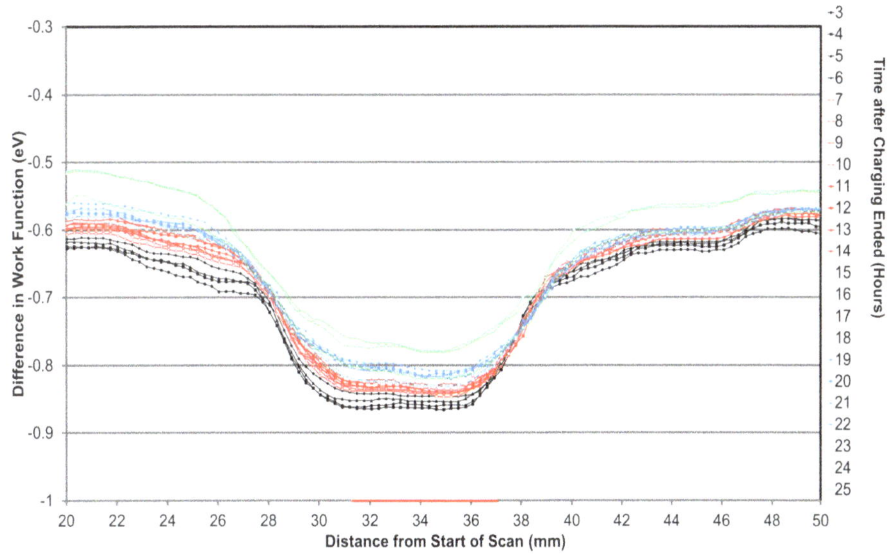

Figure 7. Scans of Charged Carbon Steel Specimen C11. Red Line at Bottom Indicates Region of Charging

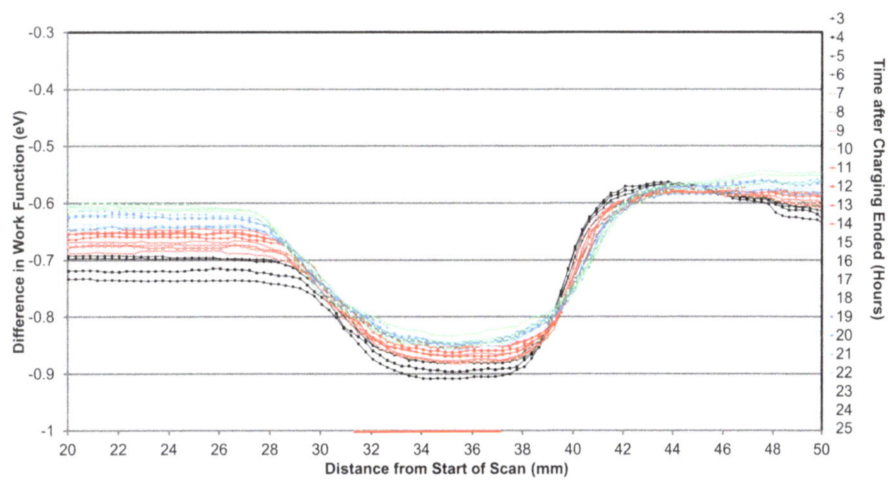

Figure 8. Scans of Charged Carbon Steel Specimen C12. Red Line at Bottom Indicates Region of Charging

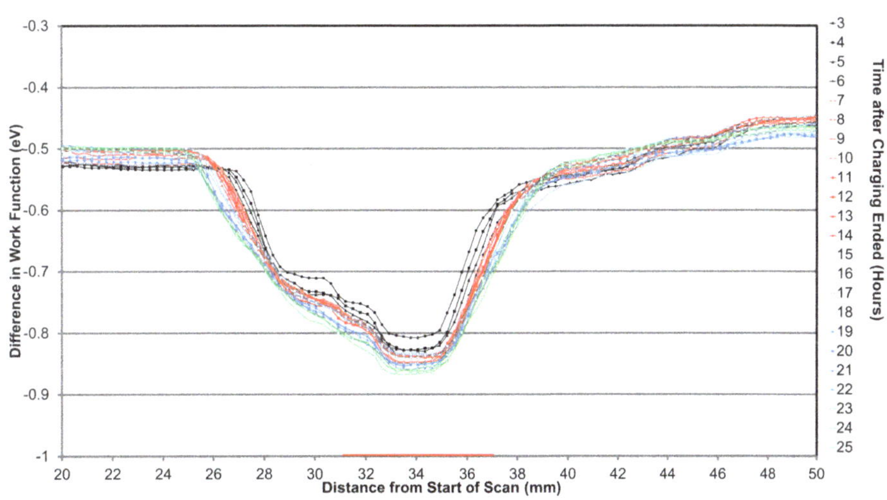

Figure 9. Scans of Charged Carbon Steel Specimen C13. Red Line at Bottom Indicates Region of Charging

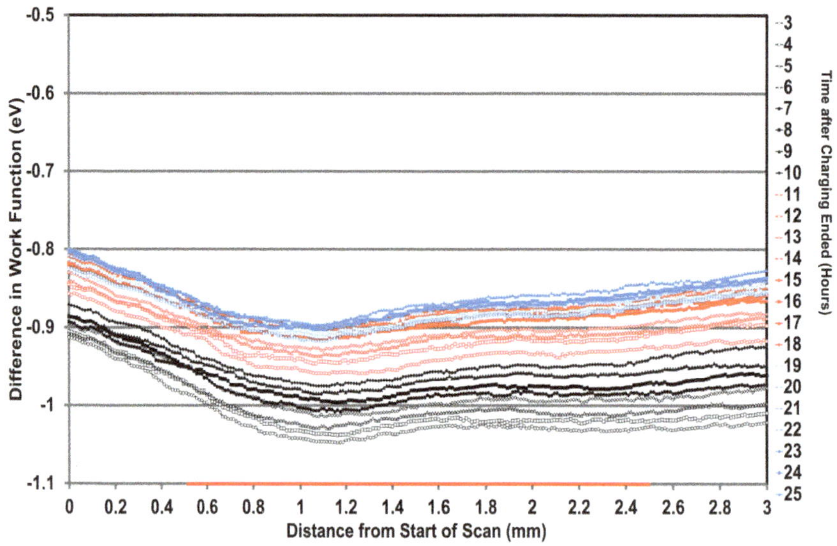

Figure 10. Scans of Charged Carbon Steel Specimen C14. Red Line at Bottom Indicates Region of Charging

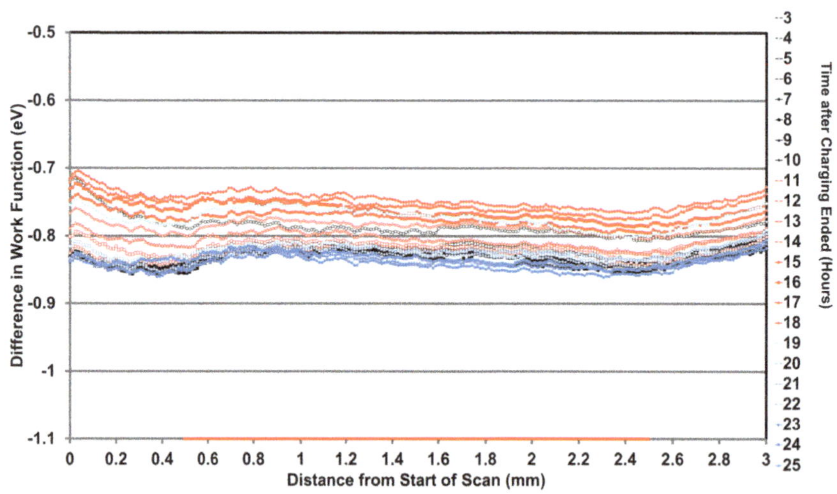

Figure 11. Scans of Charged Carbon Steel Specimen C15. Red Line at Bottom Indicates Region of Charging

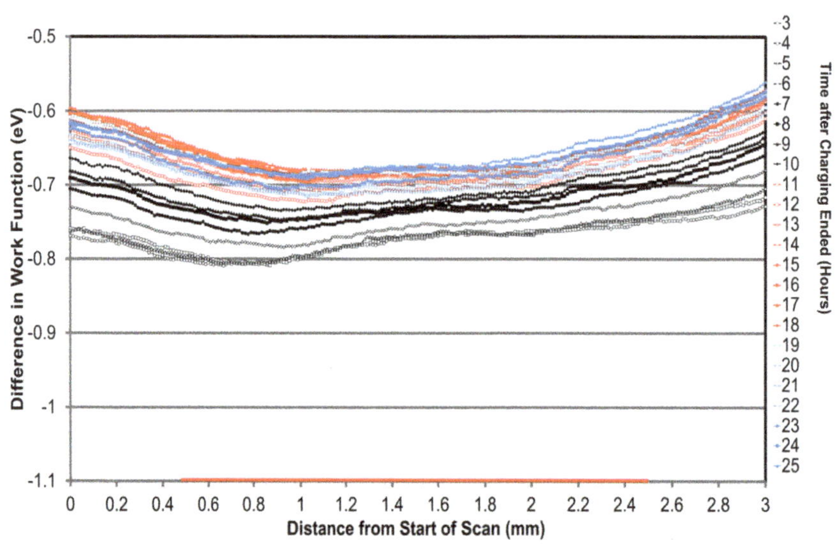

Figure 12. Scans of Charged Carbon Steel Specimen C16. Red Line at Bottom Indicates Region of Charging

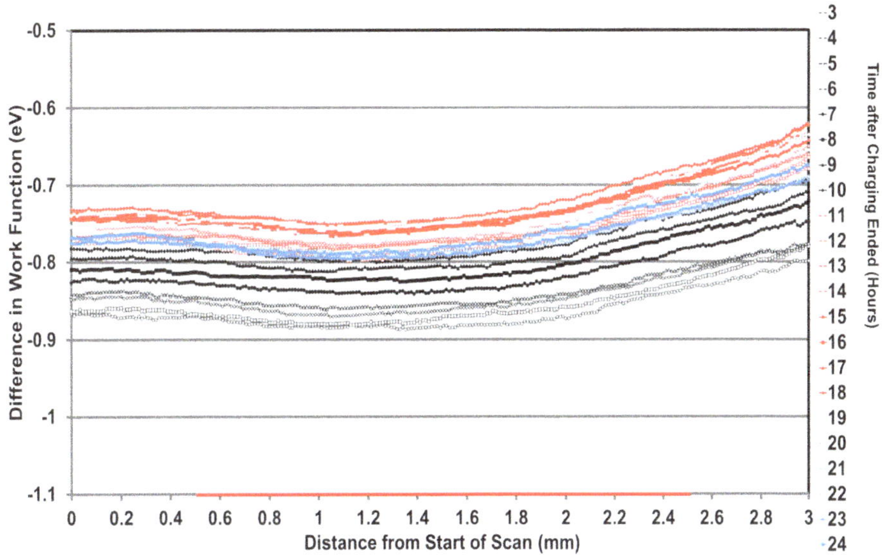

Figure 13. Scans of Charged Carbon Steel Specimen C17. Red Line at Bottom Indicates Region of Charging

Figures 6 through 9 are for the scans in the drawn direction and Figures 10 through 13 are for the scans in the radial direction. While the drop in the drawn direction scans appears to be larger than the radial direction scans (indicating more hydrogen at the surface where the specimen was charged), the dissipation of the hydrogen outward away from the charged region is much less than that of the radial direction scans. It is the speed of this dissipation (over time) in the scan direction that determines the effective diffusivity in that scanned direction.

Using the data from these eight scans summarized in Table 1, gives an effective diffusivity of 5.24E-08 cm^2/sec in the drawn direction and 3.56E-07 cm^2/sec in the radial direction. Comparing the final effective diffusivities in the drawn and radial direction shows that the transport of hydrogen in the radial direction is faster than in the drawn direction. This was expected as previous noted due to the radial direction having less hydrogen traps and thinner grains as compared to the drawn direction. This also shows that the effective diffusivity is directionally dependent in cases of cold worked high strength carbon steel.

Table 1. Summary of the Effective Diffusivities Calculated from the Data Obtained during the Eight Scans

Specimen Name	Effective Diffusivity (cm^2/sec)	Specimen Name	Effective Diffusivity (cm^2/sec)
CS 10	4.52E-08	CS 14	1.32E-07
CS 11	2.84E-08	CS 15	6.16E-07
CS 12	8.23E-08	CS 16	5.74E-07
CS 13	5.39E-08	CS 17	1.01E-07
Drawn Direction Average	5.24E-08	Radial Direction Average	3.56E-07

The confirmation of the effective diffusivity of the strand in the radial direction being faster seems to contradict the HE fracture pattern experienced in the strand tested. Test involving the charging of the strand while under stress show a fracture pattern that is not preferentially in the radial direction. Strand which was uncharged and pulled to failure showed indications of cap and cone breaking morphology as well as some radial cracking as shown in Figure 14.

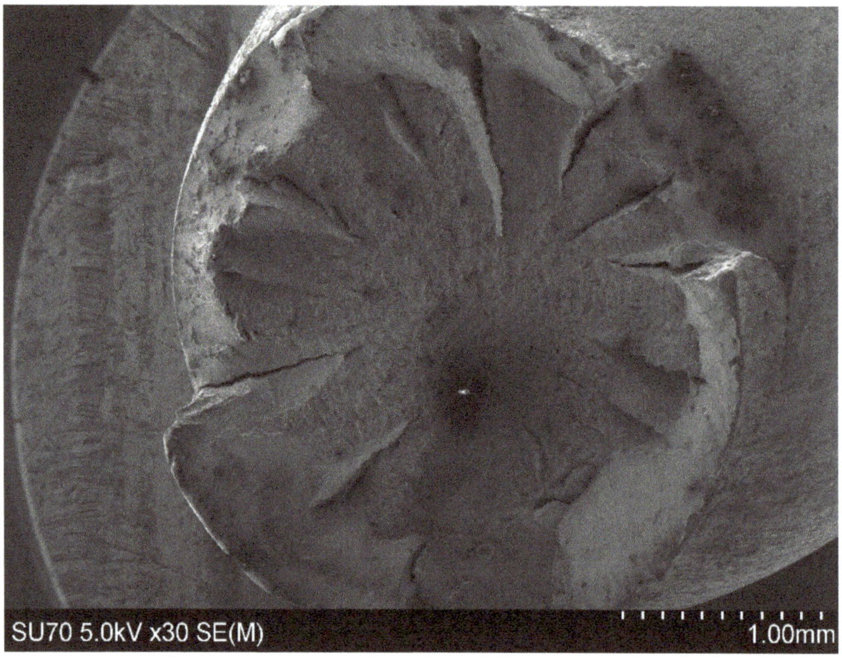

Figure 14. SEM of Uncharged Strand Failed after ASTM Pull Test

Strand that was charged then stressed showed a fracture morphology that was preferentially in the drawn directions, with some signs of shear cracking and longitudinal splitting. The cracking of the strand initially starts in the radial direction (as shown by the arrow in Figure 15), but quickly shift to mostly the drawn direction as shown in Figure 16.

Therefore, while the effective diffusivity would suggest crack propagation in the radial direction, the crack propagation is in the drawn direction for this cold worked strand exposed to hydrogen, while would be due to dislocation motion being the predominate mode of transport for the hydrogen. As hydrogen in the metal facilitates dislocation motion be HELP or AIDE, it appears that dislocation motion path dictates the path of crack propagation rather than the effective diffusivity.

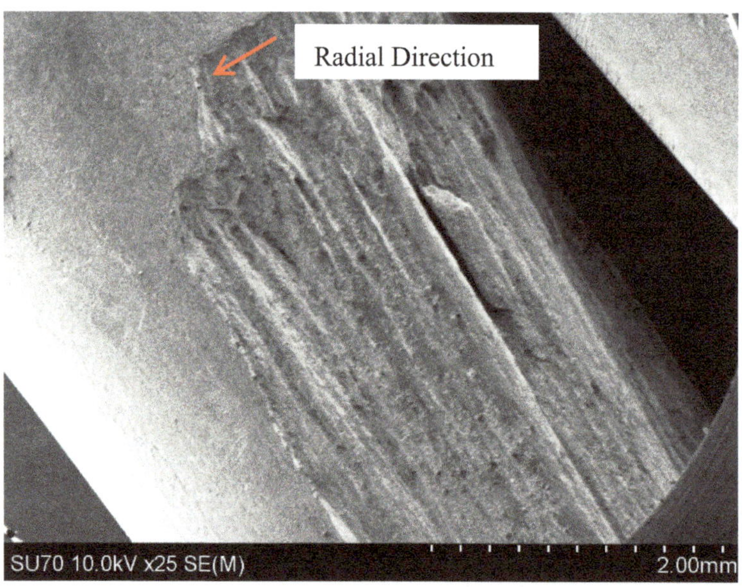

Figure 15. SEM of Charged Strand HE Failure Showing Crack Initiation

Figure 16. SEM of Charged Strand HE Failure showing mostly Drawn Direction Crack Progression

6. Conclusions

- Kelvin probes can be used to determine the presence of hydrogen at the surface of high strength carbon steel strand material in the drawn form in both the radial and drawing direction, as long as a passive film is formed due to contact with a high pH solution. Estimation of the effective or apparent diffusivity is possible with the use of a scanning kelvin probe on bare steel surface in ambient conditions.
- The effective diffusivity with trapping on high strength steel strand in the radial direction is 3.56E-07 cm^2/sec while is it 5.24E-08 cm^2/sec in the drawing direction. Therefore, hydrogen transport should be dominant in the radial direction. Dislocation motion may allow for hydrogen transport in another direction(s) to be faster than both the drawn and radial diffusional speeds. It appears that the dislocation in the drawn direction is dominant and dictates the path of crack propagation in this heavily cold worked pearlitic steel strand.

Acknowledgements

The author thanks the Florida Department of Transportation, as well as the Alfred P. Sloan Foundation for their material support of this study (*The opinions, findings, and conclusions expressed here are those of the author and not necessarily those of any supporting organizations*).

Reference

Enos, D. G., Williams Jr, A. J., Clemena, G. G., & Scully, J. R. (1998). Impressed-current cathodic protection of steel-reinforced concrete pilings: Protection criteria and the threshold for hydrogen embrittlement. *Corrosion, 54*(5), 389-402. http://dx.doi.org/10.5006/1.3284867

Evers, S., Senöz, C., & Rohwerder, M. (2013). Hydrogen detection in metals: a review and introduction of a Kelvin probe approach. *Science and Technology of Advanced Materials, 14*(1), 1-12. http://dx.doi.org/10.1088/1468-6996/14/1/014201

Gangloff, R. P., & Somerday, B. P., (2012a). *Gaseous Hydrogen Embrittlement of Materials in Energy Technologies Volume 1: The Problem, its Characterization and Effects on Particular Alloy Classes.* Woodhead Publishing.

Gangloff, R. P., & Somerday, B. P., (2012b). Gaseous Hydrogen Embrittlement of Materials in Energy Technologies Volume 2: Mechanisms, Modeling and Future Development. *Woodhead Publishing.*

Gellings, P. J. (2005). *Introduction to Corrosion Prevention and Control* (2nd ed.). University of Twente, Enschede, the Netherlands.

Guo, L. Q., Liang, D., Bai, Y., Miao, L. J., & Volinsky, A. A., (2014). Effects of Hydrogen and Chloride Ions on Automobile Interstitial-Free Steel Corrosion. *Corrosion Science, 70,* 1024-1030. http://dx.doi.org/10.5006/1126

Hartt, W. H., Kumria, C. C., & Kessler, R. J. (1993). Influence of Potential, Chlorides, pH, and Precharging Time on Embrittlement of Cathodically Polarized Prestressing Steel. *Corrosion Journal, 49*, 377-385. http://dx.doi.org/10.5006/1.3316064

Jones, D. A. (1996). Principles and Prevention of Corrosion (2nd ed.) New Jersey: Prentice Hall.

Lynch, S. (2012). Hydrogen Embrittlement Phenomena and Mechanisms. *Journal of Corrosion Review, 30*, 105-123. http://dx.doi.org/10.1533/9780857093769.1.90

Mietz, J. (2000). Investigations on Hydrogen-Induced Embrittlement of Quenched and Tempered Prestressing Steels. *Materials and Corrosion Journal, 51*, 80-90. http://dx.doi.org/10.1002/(sici)1521-4176(200002)51:2<80::aid-maco80>3.0.co;2-a

Moro, I., Briottet, L., Lemoine, P., Andrieu, E., Blanc, C., & Odemer, G. (2010). Hydrogen Embrittlement Susceptibility of a High Strength Steel X80. *Materials Science and Engineering A, 527*, 7252-7260. http://dx.doi.org/10.1016/j.msea.2010.07.027

Oriani, R. A. (1993). The Physical and Metallurgical Aspects of Hydrogen in Metals. *Fourth International Conference on Cold Fusion.*

Sagüés, A. A, Fernandez, J., Hutchinson, M., & Mullins, G. (2014). *Corrosion Characteristics of Unprotected Post-Tensioning Strands under Stress*. Florida Department of Transportation BDJ84 977-22.

Sagüés, A. A., & Walsh, M. T., (2011). Kelvin Probe Electrode for Contactless Potential Measurement on Concrete- Properties and Corrosion Profiling Application. *Corrosion Science, 56*, 26-35. http://dx.doi.org/10.1016/j.corsci.2011.11.007

Schaller, R.F., & Scully, J. R., (2013). Measurement of Effective Hydrogen Diffusivity using the Scanning Kelvin Probe. *Journal of Electrochemistry Communications, 40*, 42-44. http://dx.doi.org/10.1016/j.elecom.2013.12.025

15

Studies on Amorphous Alloy Dispersed Aluminium Matrix Composite Prepared by High Pressure Torsion

S. Hembrom[1], B. N. Roy[1], N. Khobragade[2] & D. Roy[2]

[1] Department of Metallurgical Engineering, BIT, Sindri, Dhanbad-828123, India

[2] Department of Materials and Metallurgical Engineering, NIFFT, Ranchi-834003, India

Correspondence: D. Roy, Department of Materials and Metallurgical Engineering, NIFFT, Ranchi-834003, India. E-mail: droy2k6@gmail.com

Abstract

Aluminium-based composite reinforced with Cu base amorphous alloy dispersed composite was prepared by means of high pressure torsion between a powder mix of aluminium and amorphous Cu base alloy. The X-ray diffraction pattern of powdered and consolidated composites shows the aluminium phase while the thermal stability of the amorphous alloy was studied with the aid of differential scanning calorimetry (DSC). The microstructural feature of the composite through scanning electron microscope reveals the well-distributed reinforcements in the host aluminium matrix. The hardness measurement on the as prepared composites shows significant increase in hardness with increase in reinforced amorphous alloy. Wear property of the synthesized composites were measured by using ball on plate wear tester which shows increase in wear resistance with increase in reinforced amorphous alloys.

Keywords: metal matrix composite, amorphous alloy, mechanical properties, electron microscopy, high pressure torsion

1. Introduction:

Aluminum (Al)-based metal matrix composites (MMCs) make up a distinct category of advanced engineering materials that provide unique advantages over conventional Al alloys. With the development of new forming methods and the use of low cost particulate material, the use of these composites is increasing in a wide variety of industries. Particulate reinforced composites consist of a uniform distribution of strengthening particles within a matrix. In general, these materials exhibit good wear and erosion resistance, as well as higher stiffness, hardness and strength at a lower density when compared to the unreinforced matrix material (Wielage, Zschunke, Henker, & Steinhauser, 1996). Recently metal matrix composites (MMCs) are increasingly used for critical structural and wear resistant applications because of their excellent strength to weight ratio (Clyne & Withers, 1995) and interesting physical properties. In the last few decades, MMCs, and in particular aluminium (Al)-based MMCs, are receiving increasing application as aerospace and automotive components because of high specific strength and stiffness (Harrington, 1994). Normally MMCs are reinforced with continuous fibers, discontinuous particles or whiskers. However, particle-reinforced MMCs posses some distinct advantages over fiber-reinforced composites in terms of low cost and isotropic mechanical property. In general, reinforcement particulates are prepared independently prior to composite fabrication and are incorporated into the metal matrix at a later stage. The composites so developed are called ex situ MMCs. Al composite have the advantage of low cost over most other metal matrix composites. In addition to the above they offer excellent abrasion resistance, high temperature operation, non-flammability, minimum attack by fuels & solvents, and their ability to be formed and treated on conventional equipments. The performance of MMCs can further be improved by refining the grain structure and the size of reinforcement particles to nano-metric range (Harrington, 1994) and (Yamasakiet al., 2003). Recent investigations find that incorporation of nano-particles into the aluminum matrix could enhance the hardness, the yield and ultimate tensile strength considerably, while the ductility retained (Mazahery, Abdizadeh, & Baharvandi, 2009).

It has been investigated earlier that amorphous materials can be synthesized by mechanical alloying of the elemental powder constituents of different Al-based binary and higher order alloys, e.g. Al-Ti (Lee, Sim, Heo, Cho, & Kwon, 2000), Al-Fe (Kleiner, Bertocco, Khalid, & Beffort, 2005), and Al-Ti-Si (Lv et al., 2001). However, in none of these alloys, amorphous microstructures could be obtained in bulk quantity at relatively slow or normal solidification rate. Cu–Zr based bulk metallic glasses (BMGs) attracted high attention because of their relatively low costs (Pauly, Das, Mattern, Kim, & Eckert, 2009; Louzguine-Luzgin et al., 2012; Zhang, Chen, Zheng, & Chen, 2013). Aluminum has been regarded as useful element to improve the plasticity of the Cu-Zr-based BMGs (Wu et al., 2011). The addition of minor Al quantities (up to 10 at.%) to the Cu–Zr glassy alloys may improve their thermal stability, mechanical properties and glass forming ability (GFA) (Cheung & Shek, 2007; Lee, Jo, & Lee, 2013). Research is therefore in progress to find out other possible Al-based alloy compositions, which can either retain a completely amorphous structure or an amorphous-nanocrystalline aggregate. Such developments are expected to pave the way for fabricating bulk amorphous materials in future. In the present work aluminium based composites synthesized by reinforcing Cu base amorphous alloy by high pressure torsion and the microstructure and mechanical properties have been investigated.

2. Experimental

2.1 Processing of Amorphous $Cu_{50}Zr_{45}Al_5$ Powder

Amorphous $Cu_{50}Zr_{45}Al_5$ powder was prepared from commercially available micron sized powders of $Cu_{50}Zr_{45}Al_5$ by dry mechanical milling (cryo milling, **-196 °C**) performed in a planetary ball mill. During milling, the weight ratio of the ball to the powder was maintained at 10:1. The balls and vials are made of tungsten carbide. The rotating speed of the vial was maintained at 300 rpm. Milling was stopped after 4 hr and powder samples were collected.

2.2 Processing of Composites

Aluminium powder of 99% purity with an average particle size ~10μm (Loba Chemie Pvt. Ltd., Mumbai) and amorphous $Cu_{50}Zr_{45}Al_5$ powder prepared by dry mechanical milling were mixed thoroughly. The different weight percentages of amorphous $Cu_{50}Zr_{45}A_{15}$ powder were mixed with aluminium for composite as shown in the Table:.1

Table 1. Different wt% of amorphous alloys in different samples

Sl. No.	Weight % of Aluminium	Weight % of amorphous $Cu_{50}Zr_{45}Al_5$ powder
1.	100	**0**
2.	90	10
3.	80	20
4.	70	30
5.	60	40

The powder **mixture**, so prepared, was subsequently compacted in cylindrical (10mm dia.) graphite dies under an applied pressure of 50 MPa in a hot press (Fritschkg Postfech-500428, Germany) at 200°C. The pressing load on the specimen was applied for 15 min after the desired temperature was attained. The sample was then allowed to cool to room temperature and the cylindrical samples were stripped from the graphite die for characterization. Hot pressing was followed by high pressure torsion (HPT) under an applied pressure of 1GPa was used for full densification and Al-matrix grain refinement.

2.3 Characterization

XRD studies on the $Cu_{50}Zr_{45}Al_5$ powder prior to and after ball milling were recorded by using Cu-K_α radiation. Hot pressed composite samples were also characterized by XRD technique using a Panalytical X'Pert Pro diffractrometer with Cu Kα radiation. Prior to the analyses, the XRD patterns were corrected for the effects of the $K_{\alpha2}$ radiation. The particle size of the ball milled $Cu_{50}Zr_{45}Al_5$ powders were evaluated from the XRD patterns. Microscopic studies to examine the morphology and distribution of particles in the synthesized composites were done by a JEOL 6480 LV scanning electron microscope (SEM). The thermal stability of the amorphous powder, were studied under argon gas atmosphere with the aid of a differential scanning calorimeter (DSC, Mettler, TA400). The flow rate of argon gas was maintained at 80cc/min throughout the experiment. The powder were heated from room temperature to 800°C at a predetermined heating rate of 10°C/min and then cooled to room temperature at a cooling rate of 10°C/min. The density of synthesized aluminium based composites was calculated by using Archimedes principle.

Average values of hardness on different phases of the composite samples were measured with the help of a standard Vickers hardness tester (Type 3212, Zwick, Ulm Germany) using a load of 25 gf applied for 15s. Sliding wear resistance of the samples was evaluated using a ball on plate wear testing instrument having a hardened ball (SAE 52100) indenter of 6mm diameter. DUCOM TR-208-M1 ball on plate wear tester was used for evaluating the wear resistance of Al-matrix dispersed with amorphous reinforcement samples. The arrangement of the machine is such that it ensures a precise measurement of the tangential frictional force. This test method involves a ball shaped upper specimen that slides against a rotating disk as a lower specimen under a prescribed set of conditions. The load is applied vertically downward with a motor driven carriage that uses the force/load sensor for feedback to maintain a constant load. The experimental process was carried out at 10 N loads for 15 minutes at 15 rpm at an average humidity of 50-55% and at an average temperature of 25°C with steel ball. Surface damage caused by wear testing was subsequently analyzed using a scanning electron microscopy to get an idea about the wear mechanism.

3. Results and Discussion

XRD pattern reveals the phases present and the sequence of their evolution in course of mechanical alloying of the $Cu_{50}Zr_{45}Al_5$ elemental powder blend shown in Figure 1. The diffraction pattern shows the peaks of the constituent elements present in the initial mixture either disappear due to alloying, or undergo broadening and reduction in their intensity with increase in the milling time. XRD pattern reveals that the $Cu_{50}Zr_{45}Al_5$ powders prior to milling consist of peaks of Al, Cu and Zr. The similar characteristic peaks of respective elements have been reported (Azimi, Shokuhfar, & Zolriasatein, 2014). After 4 h of milling, the powders appear to constitute amorphous solid solution. The XRD diffraction pattern of milled powder shows the amorphization of $Cu_{50}Zr_{45}Al_5$ powder. Guoqiang Xie et al. (2010) studied the similar composite and reported the amorphization of milled powder.

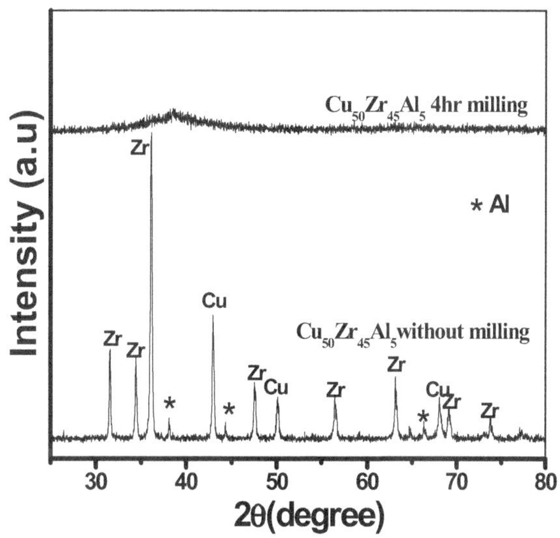

Figure 1. XRD pattern of milled and without milled $Cu_{50}Zr_{45}Al_5$ powders

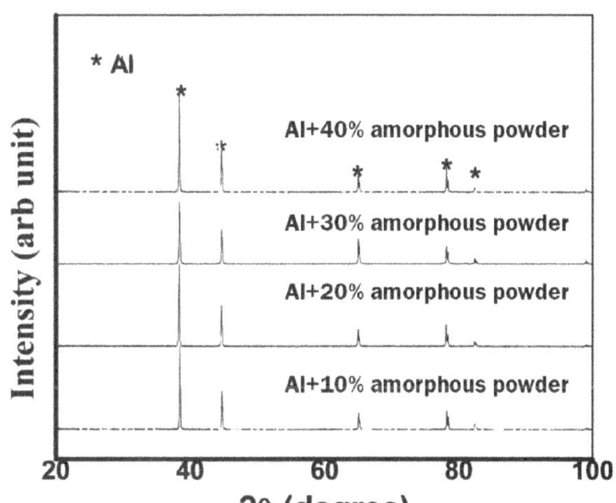

Figure 2. XRD **pattern** of Al+ different wt% of amorphous powder

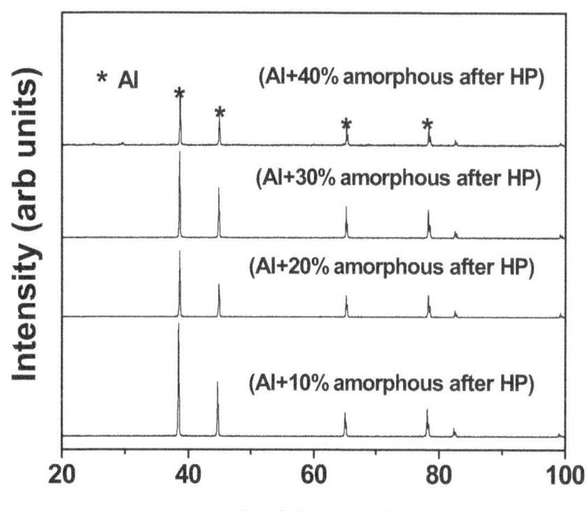

Figure 3. XRD **pattern** of the **sample** prepared by hot pressing

Figure 2 and Figure 3 show the XRD patterns of the sample in powdered form and consolidated form prepared by hot pressing respectively. Both XRD pattern revealing the presence of sharp peaks representing the phases of aluminium. The results of XRD analysis of the samples in powdered and consolidated form representing the same structure. Sharifi and Karimzadeh (2011) reported the similar diffraction peaks of aluminium in the synthesized composites.

Figure 4 shows the DSC plot obtained during heating and cooling of green compact from room temperature to 800°C and from 800°C to room temperature at a heating/cooling rate of 10°C/min under argon gas atmosphere. It is seen that, the peak obtained at 475°C, representing the amorphous- crystalline transformations. Kukuła-Kurzyniec et al. (2014) reported that the crystallization begins in the ribbon at the temperature of 437°C which is higher for the amorphous aluminium based composites. He, Bian, and Chen (2000) determined the DSC curve of bulk metallic glass $Zr_{52.5}Ni_{14.6}Al_{10}Cu_{17.9}Ti_5$ by which the Tg and Tx are reported as 631K and 710K respectively.

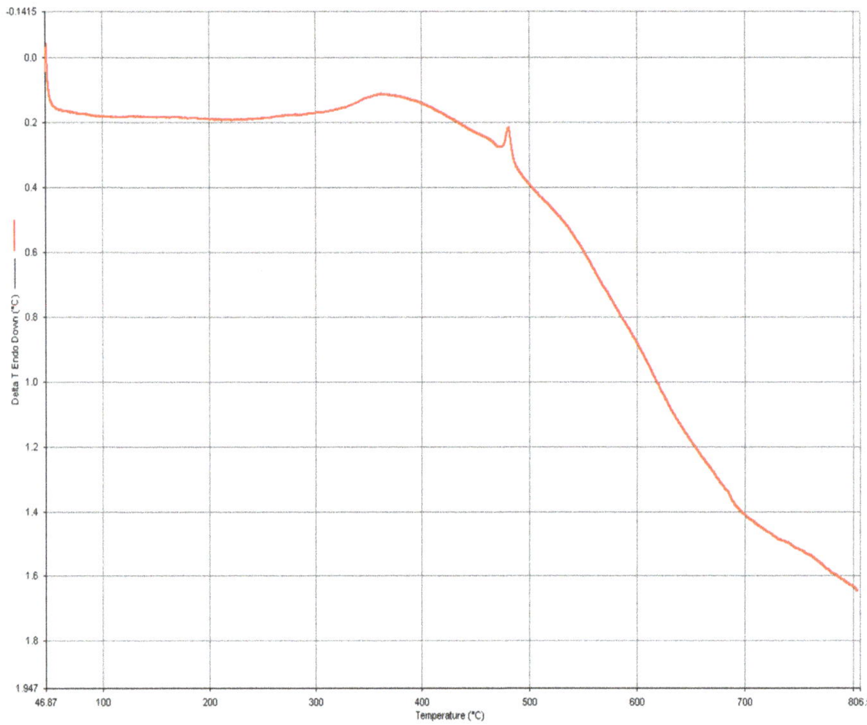

Figure 4. DSC curve for the $Cu_{50}Zr_{45}Al_5$ mechanically alloyed powders subjected to heating and cooling

SEM micrographs of the pure aluminium and composites with different weight% of amorphous alloy prepared by hot pressing at 400°C are illustrated in Figure 5(a), Figure (b), Figure (c), Figure (d), Figure (e). The microstructural feature of the composite reveals the well-distributed reinforcements (bright contrast) in the host aluminium matrix. Inference drawn from the XRD results (Figure 3) investigation suggests the absence of any constituent phase except aluminium in bright areas. Roy, Ghosh, Basumallick, and Basu (2007) reported the SEM micrographs of 10 vol% and 30 vol% reinforced Ti-aluminide content in aluminium matrix prepared by hot pressing at 700°C.

The hardness of the composites has been measured by Vickers hardness tester while densities were calculated by using Archimedes principle. Figure 6 shows the variation of density and hardness with the wt% of reinforcement of the composite samples prepared by hot pressing. On examination of the plots, it is obvious that the samples with 40% of amorphous alloy have demonstrated the highest density, which also coincides with the highest hardness. The lowest density is observed in the samples, containing 0 wt% of amorphous alloy (pure Al) due to absence of reinforcement. It has been found that the hardness of the samples increases remarkably with the increase in reinforcement content. Roy, Basu, and Mallick (2005) reveals the similar results which shows increase in hardness with increase in reinforced content in Ti-aluminide reinforced Al-based insitu metal matrix composites.

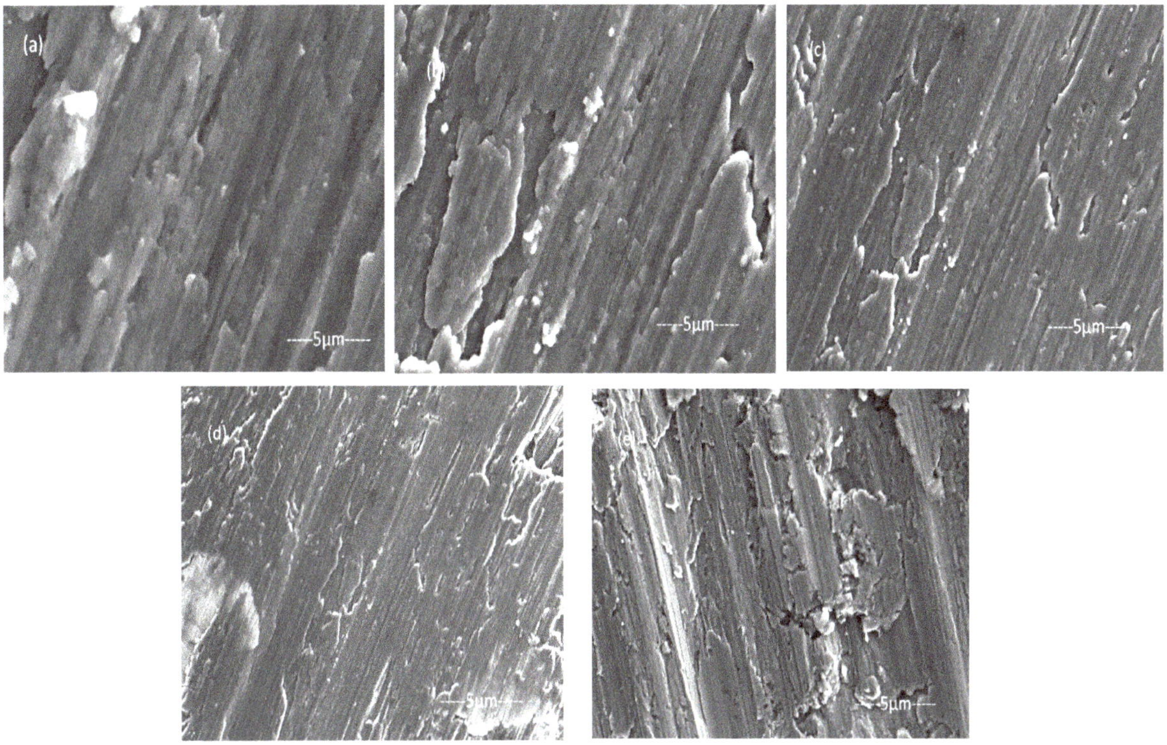

Figure 5. SEM micrographs of pure Al and composites prepared by hot pressing(a)pure Al, (b)10% , (c)20%, (d)30%, (e)40% amorphous alloy

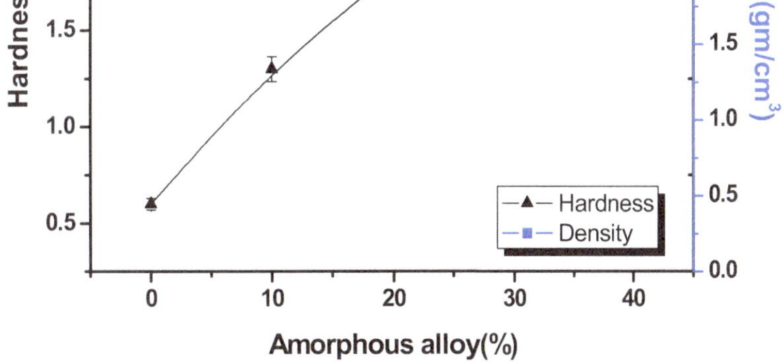

Figure 6. Density and hardness variation with respect to wt% of reinforcement of the composite samples prepared by hot pressing at 400°C

Wear properties of the designed composites were measured by using ball on plate wear test at an applied load of 10 N at 15 rpm sliding speed on a 6 mm diameter track for 15 minutes. Figure 7 shows the plot for wear depth Vs time for synthesized composites. As the microhardness results shows that hardness of the samples increases with increase in reinforced content. The similar trend was observed in wear test, 40wt % reinforced Al-matrix composite exhibit the high hardness and wear resistance. Roy, Basu, Mallick, Kumar, & Ghosh (2006) represents the similar results in Fe-aluminide reinforced Al matrix composites.

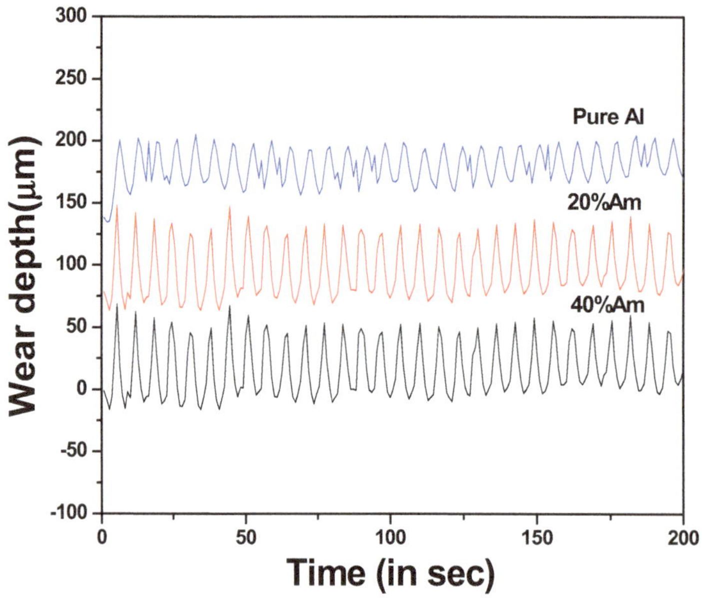

Figure 7. **Plot** of wear depth Vs time (in sec) for amorphous $Cu_{50}Zr_{45}Al_5$ powder content in Al-matrix

The topographical features of the worn surfaces are investigated using scanning electron microscopy in an attempt to study the wear mechanism. Figure 8 presents the overall topography of pure Al and Al-based composite reinforced with 20 and 40 wt% ex situ $Cu_{50}Zr_{45}Al_5$ powder formed, hot pressed at 400°C. For 20 wt% reinforced Al matrix, extensive plastic deformation was observed to cause more wear. The improvement in hardness due to harder amorphous $Cu_{50}Zr_{45}Al_5$ reinforcements causes increase in wear resistance of 40 wt% reinforced composite. For 40 wt% reinforcement, a transition in wear mechanism from predominantly abrasion and deformation to predominant abrasion and tribochemical wear takes place. The similar wear behavior of Al based composites were shown (Roy, Basu, & Mallick, 2005) in which Ti-aluminide is used as reinforcement in Al-matrix composite.

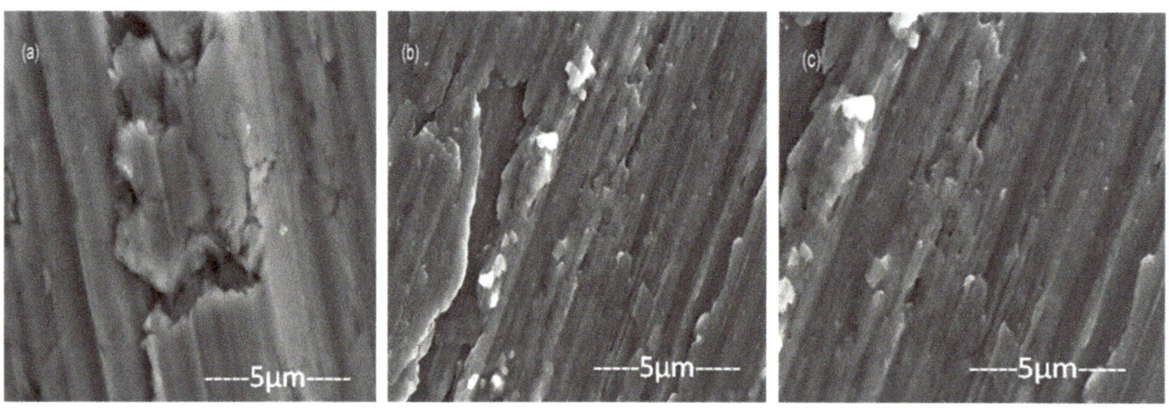

Figure 8. SEM **micrograph** of wear track of (a)20wt% amorphous alloy, (b)40 wt% amorphous alloy and (c) Pure Al alloy

4. Conclusion

Aluminium matrix composites reinforced with well dispersed amorphous powder prepared by hot pressing followed by high pressure torsion method. XRD pattern of both powdered and consolidated composites shows the phases of aluminium with same structure. DSC reveals that peak obtained at 475°C represents the amorphous to crystalline transformation. Microstructural architecture of amorphous alloy remains same before and after hot pressing followed by high pressure torsion. Microhardness of aluminium matrix composite reinforced with amorphous alloy shows increase in hardness with increase in reinforcement content and similarly increase in wear resistance upto 40vol% reinforcement in synthesized composites.

References

Azimi, A., Shokuhfar, A., & Zolriasatein, A. (2014). Nanostructured Al–Zn–Mg–Cu–Zr alloy prepared by mechanical alloying followed by hot pressing. *Materials Science and Engineering: A*, *595*, 124-130.

Cheung, T. L., & Shek, C. H. (2007). Thermal and mechanical properties of Cu–Zr–Al bulk metallic glasses. *Journal of alloys and compounds*, *434*, 71-74.

Clyne, T. W., & Withers, P. J. (1995). *An introduction to metal matrix composites*. Cambridge University Press.

Harrington, Jr. W. C. (1994). Metal matrix composite application. In S. Ochiai (Ed.), *Mechanical properties of metallic composites* (pp. 759-73). New York: Marcel-Dekker.

He, G., Bian, Z., & Chen, G. L. (2000). Structure evolution of bulk metallic glass Zr52. 5N14. 6Al10Cu17. 9Ti5 during annealing. *JOURNAL OF MATERIALS SCIENCE & TECHNOLOGY*, *16*(4), 357-361.

Kleiner, S., Bertocco, F., Khalid, F. A., & Beffort, O. (2005). Reactively Synthesized Nanostructured PM Aluminium Composite–Microstructure Stability and Elevated Temperature Hardness Response. *Advanced Engineering Materials*, *7*(5), 380-383.

Kukuła-Kurzyniec, A., Dutkiewicz, J., Góral, A., Coddet, C., Dembinski, L., & Perrière, L. (2014). Aluminium based composites strengthened with metallic amorphous phase or ceramic (Al 2 O 3) particles. *Materials & Design*, *59*, 246-251.

Lee, K. B., Sim, H. S., Heo, S. W., Cho, S. Y., & Kwon, H. (2000). Fabrication of Al alloys reinforced with AlN particles formed byin-situ reaction. *Metals and Materials*, *6*(1), 25-32.

Lee, K. S., Jo, Y. M., & Lee, Y. S. (2013). Crystallization and high-temperature deformation behavior of $Cu_{49}Zr_{45}Al_6$ bulk metallic glass within supercooled liquid region. *Journal of Non-Crystalline Solids*, *376*, 145-151.

Louzguine-Luzgin, D. V., Xie, G. Q., Gonzales, S., Wang, J. Q., Nakayama, K., Perepezko, J. H., & Inoue, A. (2012). Nano-crystallization behavior of Zr–Cu–Al bulk glass-forming alloy. *Journal of Non-Crystalline Solids*, *358*(2), 145-149.

Lv, L., Lai, M. O., Su, Y., Teo, H. L., & Feng, C. F. (2001). In situ TiB 2 reinforced Al alloy composites. *Scripta Materialia*, *45*(9), 1017-1023.

Mazahery, A., Abdizadeh, H., & Baharvandi, H. R. (2009). Development of high-performance A356/nano- Al_2O_3 composites. *Materials Science and Engineering: A*, *518*(1), 6164.

Pauly, S., Das, J., Mattern, N., Kim, D. H., & Eckert, J. (2009). Phase formation and thermal stability in Cu–Zr–Ti (Al) metallic glasses. *Intermetallics*, *17*(6), 453-462.

Qin, S., Chen, C., Zhang, G., Wang, W., & Wang, Z. (1999). The effect of particle shape on ductility of SiCp reinforced 6061 Al matrix composites. *Materials Science and Engineering: A*, *272*(2), 363-370.

Roy, D., Basu, B., & Mallick, A. B. (2005). Tribological properties of Ti-aluminide reinforced Al-based in situ metal matrix composite. *Intermetallics*, *13*(7), 733-740.

Roy, D., Basu, B., Mallick, A. B., Kumar, B. M., & Ghosh, S. (2006). Understanding the unlubricated friction and wear behavior of Fe-aluminide reinforced Al-based in-situ metal–matrix composite. *Composites Part A: Applied Science and Manufacturing*, *37*(9), 1464-1472.

Roy, D., Ghosh, S., Basumallick, A., & Basu, B. (2007). Preparation of Ti-aluminide reinforced in situ aluminium matrix composites by reactive hot pressing. *Journal of alloys and compounds*, *436*(1), 107-111.

Sharifi, E. M., & Karimzadeh, F. (2011). Wear behavior of aluminum matrix hybrid nanocomposites fabricated by powder metallurgy. *Wear*, *271*(7), 1072-1079.

Wielage, B., Zschunke, M., Henker, A., & Steinhauser, S. (1996). Manufacture and characterization of particle reinforced aluminum coatings. In C. C. Berndt (Ed.), *Thermal Spray: Practical Solutions for Engineering Problems* (pp. 333-337). Ohio: ASM International.

Wu, Y., Wang, H., Wu, H. H., Zhang, Z. Y., Hui, X. D., Chen, G. L., ... & Lu, Z. P. (2011). Formation of Cu–Zr–Al bulk metallic glass composites with improved tensile properties. *Acta Materialia*, *59*(8), 2928-2936.

Xie, G., Louzguine-Luzgin, D. V., Fukuhara, M., Kimura, H., & Inoue, A. (2010). Cu particulate dispersed $Cu_{50}Zr_{45}Al_5$ bulk metallic glassy composite with enhanced electrical conductivity. *Intermetallics*, *18*(10), 1973-1977.

Yamasaki, T., Zheng, Y. J., Ogino, Y., Terasawa, M., Mitamura, T., & Fukami, T. (2003). Formation of metal–TiN/TiC nanocomposite powders by mechanical alloying and their consolidation. *Materials Science and Engineering: A*, *350*(1), 168-172.

Zhang, L. K., Chen, Z. H., Zheng, Q., & Chen, D. (2013). Isochronal and isothermal phase transformation of $Cu_{45}Zr_{45}Ag_7Al_3$ bulk metallic glass. *Physica B: Condensed Matter*, *411*, 149-153.

16

Effect of Sintering Temperature on the Microstructure and Electrical Characteristics of Low Clamping Voltage Zinc Oxide- Based Ceramic Varistor

J. O. Akinnifesi[1] & O. O. Akinwunmi[1]

[1] Department of Physics and Engineering Physics, Obafemi Awolowo University, Ile-Ife, Nigeria

Correspondence: J. O. Akinnifesi, Department of Physics and Engineering Physics, Obafemi Awolowo University, Ile-Ife, Nigeria. E-mail: akinnifesi@yandex.com

Abstract

Pelletized samples of polycrystalline zinc oxide ceramics doped with 2mol% MnO_2 and 3mol% PbO were prepared by the conventional method of ceramic processing. The requisite composition was obtained by the direct mixing of constituent phases (DMCP) technique. Sintering was carried out at various temperatures ranging from 650°C to 850°C. The effects of sintering on the non-linear characteristics of the doped ceramic samples were investigated. The threshold or breakdown voltages were found to decrease as sintering temperature increased.

Microstructural investigation revealed improved homogeneity and increased grain sizes with increasing sintering temperature. The random distribution of secondary phases also featured prominently. These findings were observed to correlate with the evolution of electrical characteristics while the sample sintered at 850°C exhibited the best electrical response suitable for varistor behaviour.

Keywords: ZnO varistor, sintering temperature, densification, microstructure, electrical characteristics.

1. Introduction

A number of studies have successfully proved that n- type doped zinc oxide ceramics are effective material for varistors or voltage- dependent resistors (VDR), using transition metal oxides as the dopant. The role of the individual dopants in forming potential barriers to electron flow has been reported (Clark, 1999; Haskell et al., 1999; Duran et al., 2003). Varistors have applications in the protection of electrical power transmission and electronic circuit devices against transient voltage surges (Gupta, 1990; Leach, 2005; Annas et al., 2007). Their current-voltage characteristics exhibit very high resistance at low voltages and low resistance at high voltages. The threshold voltage of transition to increased conduction, also known as clamping or breakdown voltage occur in a region of non-linearity (Matsuoka, 1991; Pillar et al., 2003). There have been reports in the literature on the basic physics of doped varistors (Levison et al., 1975; Entage, 1977), in addition to which a brief review will be pertinent here.

The microstructure of a varistor is made of thin intergranular phase separating ZnO grains which cause localized defects on the interface. These defects give rise to a schottky barrier which limits the current flow at low temperatures. When higher voltage is applied across the boundary layer, minority carriers are generated, thereby increasing conductivity significantly (Mahan et al., 1993; Pillat et al., 2003). The electronic potentials that build up at the grain boundaries arise from the charges trapped at the interface states (Fernandez- Hevia et al., 2004; Leach, 2005). Consequently, the equivalent circuit of the varistor can be represented as a parallel combination of resistors and capacitors, while the bulk varistor consists of several series and parallel combinations of micro-varistors (Kourdi et al., 1992). In this way, they are equivalent to back-to-back zener diodes, with greater current and energy handling capabilities (Peitado et al., 2007).

The mixing of two or more components to form a multicomponents system separated by distinct interfaces produce superior material properties. The behaviour of such system depends on the composition and temperature of processing and when the free energy of the system is minimum, the system remains stable (Vijaya and Rangarajan, 2003). The microstructure of the polycrystalline material largely depends on the number, size, shape and rate of formation of phases present (Barrett et al., 1973). A previous study of a ternary varistor system sintered above 800°C showed that liquid phase reactions produced functional microstructure, promoting the densification of the ceramics and the growth of ZnO grains (Leite et al., 1996). It has also been reported that mechano-chemical activation (MCA) of microstructure resulted from using the direct mixing of constituent phases (DMCP) technique with high sintering temperature. This was found to significantly reduce particles' sizes and their compactness, giving rise to more homogeneous samples with higher density and finer distribution of secondary phases (Bernik et al., 2008). This finding confirmed an earlier report by kutty (2000).

In our previous investigations (Akinnifesi et al., 2014) we studied the electrical properties of Mn^{2+}- Pb^{2+}- doped ZnO varistors and found this to be considerably influenced by the microstructure. Onreabroy et al. (2006) has also reported that the non-linear coefficient depended primarily on the sintering temperature and significantly decreased at higher temperatures. Breakdown voltages had equally been found to decrease linearly as average size of ZnO grains increased (Hozer, 1994). Interestingly, variation in microstructure and electrical properties at different sintering temperatures has been reported (Xu et al., 2009; Nahm, 2006) while homogeneity and development of interfaces have been related to sintering (Leach et al., 2000; Bernik et al., 2001). Significantly, the evolution of particle size and distribution during processing is temperature-sensitive. Consequently, the breakdown voltages can be varied by changing the grain size under sintering condition (Lee et al., 2007).

The objective of this work is therefore to investigate the effect of varying sintering temperature on the microstructure and electrical characteristics of bulk specimen of Mn^{2+}- Pb^{2+}- doped zinc oxide ceramics. This is carried out by the conventional means of electron microscopy and X-ray diffraction techniques, as well as current-voltage characterization.

2. Experimental Procedure

2.1 Sample Preparation:

Reagent grade (Sigma Aldrich) ZnO powder of 99.9% purity and mean particle size $0.28\mu m$ were doped with requisite amounts of MnO_2 and PbO powders, by direct mixing of constituent phase procedure (DMCP). This method allows the preparation of bulk-type varistors with precisely defined components. Each phase is prepared separately and the final varistor is formed by sintering the powder mixture of constituent phases (Brankovic et al., 2007). Mixing was thoroughly done in a ceramic mortar for considerable homogeneity. The mixture was then entirely scooped into the mould and pelletized with Caver Hydraulic unit with a load of 7.49MPa for about 4 minutes to enhance effective compression. The resulting pellets were disk-shaped with about 2.00mm in thickness and 10.0mm in diameter. The thickness/diameter measurements were averaged, with no substantial deviation found. The composition was 95% ZnO of 95mol% ZnO with 2mol% MnO_2 and 3mol% PbO. Specimens of these pellets were sintered at different temperatures varying from 650°C to 850°C for about 6 hours in each case. Each specimen was inserted in quartz-tube before placing in the preheated oven, to ensure there was no reaction between the furnace interior and samples, thereby avoiding any contamination. It was then inserted at the preset temperature at a heating rate of 5°C/min and subsequently oven-cooled to about 280°C. It was finally removed outside the oven to cool to room temperature. As the cooling to room temperature was gradual, no disc bending resulted.

2.2 Density, ρ Determination

The density of the sintered pellets were determined firstly by precise geometrical measurements of their dimension using the micrometer screw gauge. This was followed by a precise weighing of the pellet with an electronic balance of sensitivity 10^{-4}g. The mass to volume ratio of each of the samples prepared at different sintering temperatures was then evaluated to obtain the densities

2.3 Measurements of Microstructure:

To observe the microstructure of the sintered doped-ZnO ceramic samples, the fractured surface was coated with gold for examination under a Scanning Electron Microscope. The micrographs were produced by the CARL-ZEISS machine, model EVOMA10. Average grain size was determined by the linear intercept method. The crystalline phases were identified by X-ray Diffractometry. For this, a Radicon MD 10 diffractometer making use of CuKα radiation at 1.5418A°, in the detection range between 16° and 72° was employed.

2.4 Measurement of Electric Field –Current Density (E-J) Characteristics

The DC current-voltage characteristics of the sintered samples were measured by coating silver paste onto the opposite faces of the samples. The silver electrodes of 5mm in radius were formed by heating at 500°C for about 10minutes, to establish ohmic contact. The current and voltages were obtained with a Keitheley electrometer, model 2636A and the values were expressed as current densities (in terms of the unit length of the varistor thickness respectively). Hence the breakdown voltages of the samples were obtained for different sintering temperatures.

3. Results and Discussion

3.1 Densification

The variation of the density of the varistor samples sintered at different temperatures is shown in Figure 1. An initial increase in densification with increasing temperature is revealed. This is followed by a drop in value at a temperature of 800°C. The initial increase may have arisen from the annihilation of probable voids in the microstructure, as well as from the formation of new phases and growth of ZnO grains (Han et al., 1999; Han et al., 2000). The drop is also indicative of the onset of significant loss in dopant materials at higher sintering temperature. This loss is manifested by the decrease in XRD peak intensity of PbO_2 from 0.75 at 800 °C to 0.35 at 850 °C and that of MnO_2 from 0.60 to 0.10 over the same range of temperature increase. This could serve as a guide to the choice of temperature of sintering. Densification at higher sintering temperatures can be improved by increasing the Mn doping level as this has been found to promote grain growth in the final stages of sintering (Han et al., 2002).

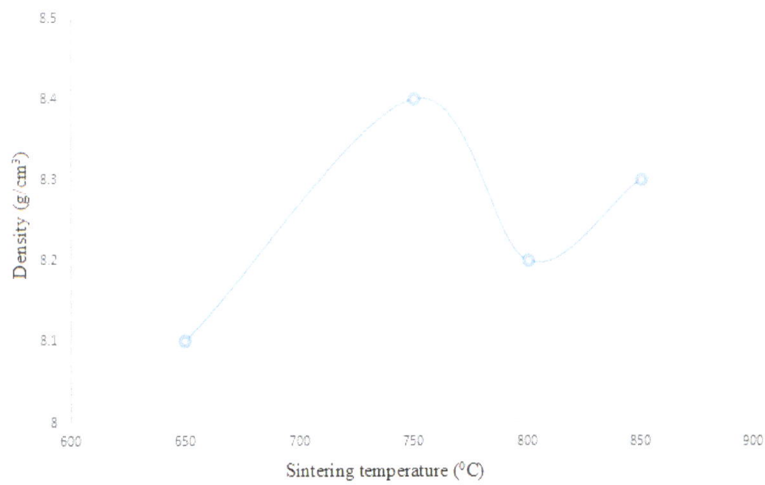

Figure 1. Variation 0f densification with sintering temperature

3.2 Microstructural Studies

Figure 2(a-d) show the scanning electron micrographs of the sintered doped ZnO ceramic samples at different temperatures of sintering. These basically consist of different phases with ZnO as the main phase. From the appearance of the phase distributions the microstructures seem identical, though with trends of improved matrix homogeneities with increasing sintering temperature. Better homogeneities have been reported to confer better varistor characteristics (Bernik et al., 2008). The variation of average grain size with sintering temperature is shown in Figure 3, in which the grain sizes increase with the sintering temperature. The emergence of the different phases with the different sintering temperatures is revealed by the x-ray diffraction spectra of the doped ceramic samples shown in Figure 4(a-d). More phases are observed to form with increasing sintering temperature, thus contributing to the creation of interface barriers to electrical conduction. The improvement in varistor behaviour would consequently be enhanced.

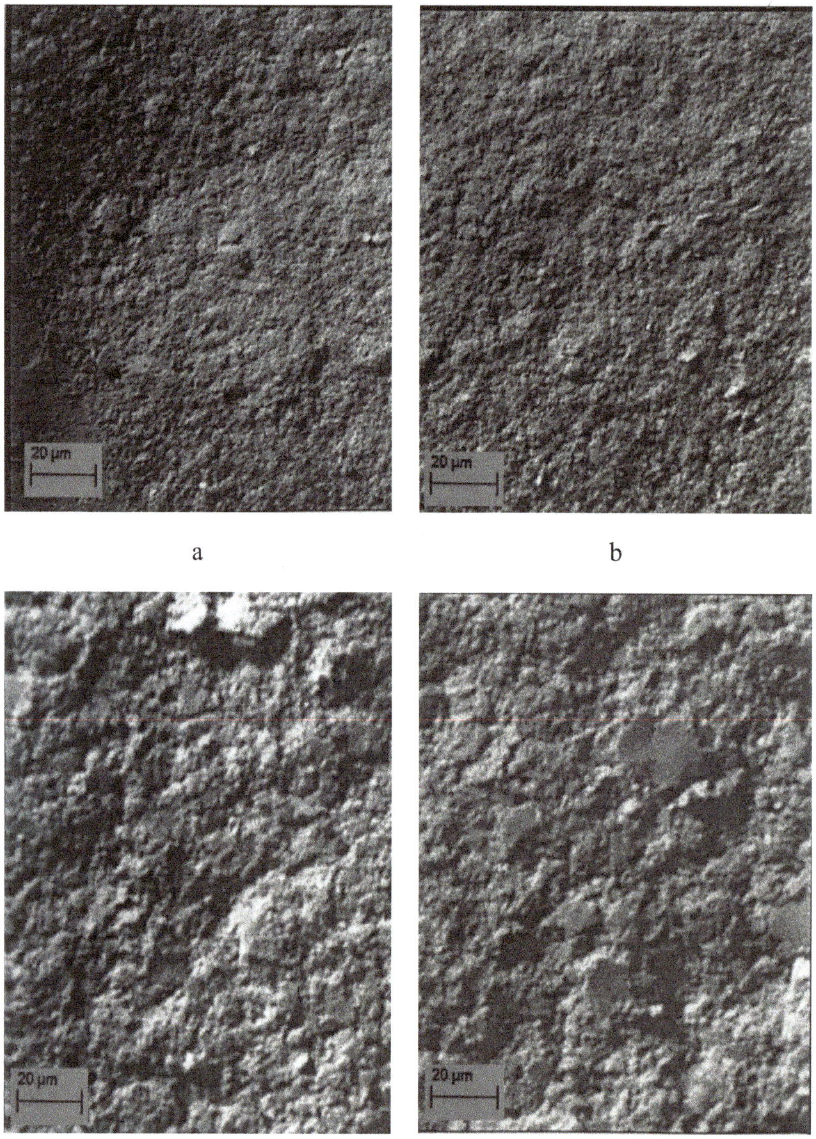

Figure 2. SEM Micrographs of samples sintered at (a) 650°C, (b) 750°C, (c) 800°C and (d) 850°C

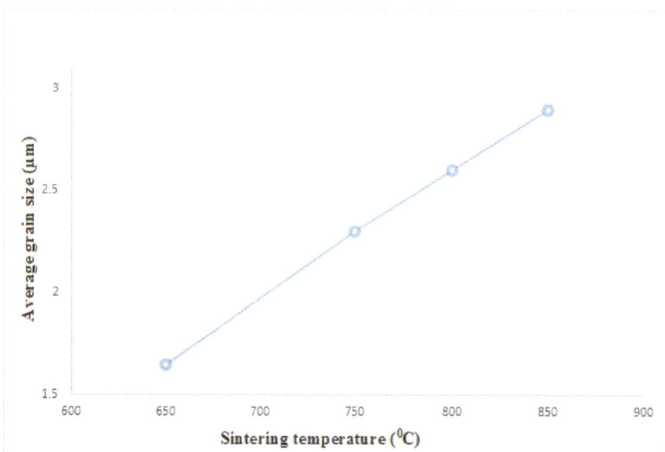

Figure 3. Variation of average grain size with sintering temperature

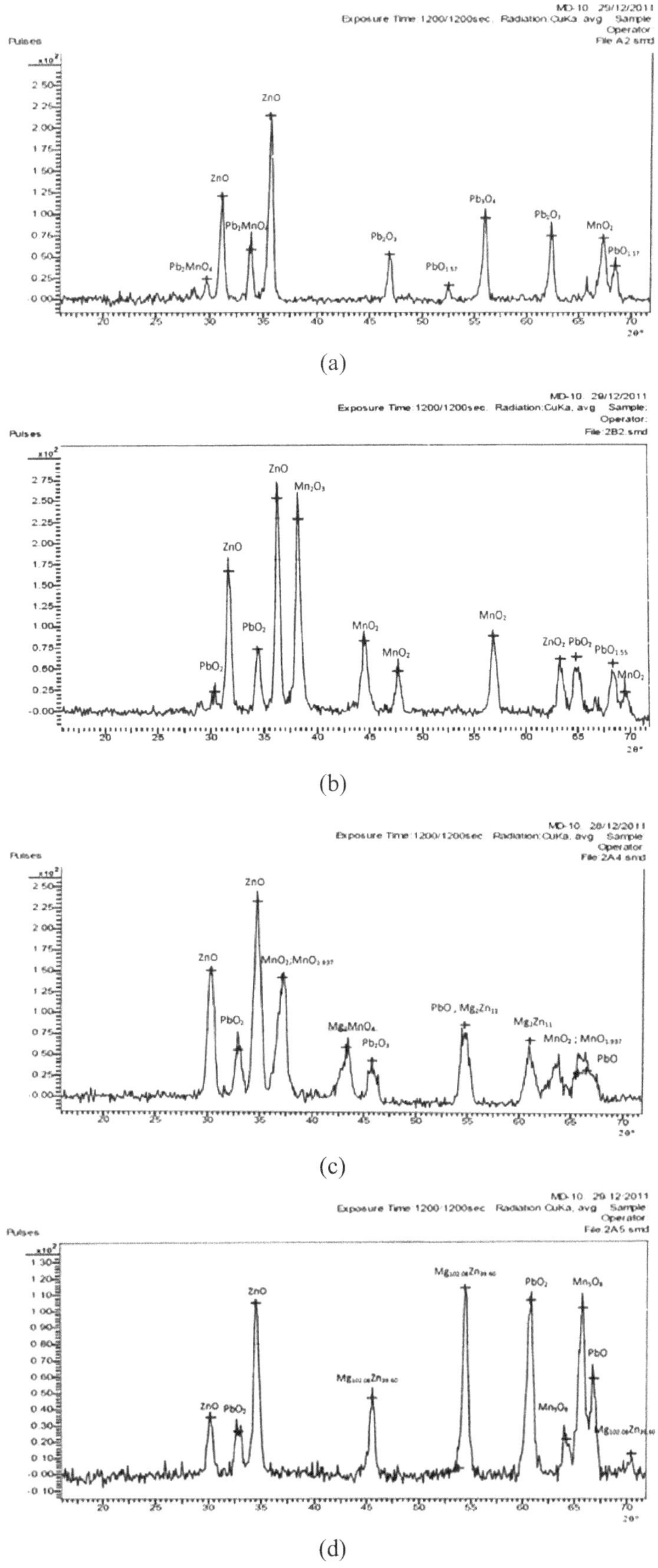

Figure 4. XRD Spectra of the sintered samples at (a) 650°C, (b) 750°C, (c) 800°C and (d) 850°C.

3.3 Electrical Characteristics

The properties of varistors are best highlighted by the electrical characteristics which result from the fabrication processes, particularly sintering. Figure 5 shows the current-voltage characteristics of the doped ZnO ceramic samples sintered at the various temperatures. These exhibit a sharp transition in the electric field from low current zones to non-linear region of higher currents. The associated threshold or breakdown voltages V_b, vary with the sintering temperatures as shown in Figure 6, where V_b decreases with increasing sintering temperatures. This variation is traceable, on one hand to the formation and consequent growth of secondary phases at the different temperatures. According to Zhang et al. (2005), the growth of secondary phases involves the dissolution of cations which makes the interphase junctions more electrically active. On the other hand, the increased average grain sizes would lead to increased conductivity and a consequent reduction in the breakdown voltage. This phenomenon was previously observed by Cheng et al. (2007). Hence, the formation of several polymorph phases and increase in grain sizes in the sample sintered at 850°C explains the occurrence of a minimum breakdown voltage at this temperature. However, it is evident from Figure 5 that the increase in breakdown voltage is accompanied by increased leakage currents. This is indicative of a deterioration tendency in the varistor microstructure. Nahm (2004), Leach et al. (2006) and Nahm (2007) have all reported a lowering of the barrier height and poorer non-linearity in the electrical response of excessively sintered varistors.

In the overall analysis, considering the knee of the non-linearity in electrical response and other important parameters, the best varistor behavior is exhibited by the sample sintered at 850 °C.

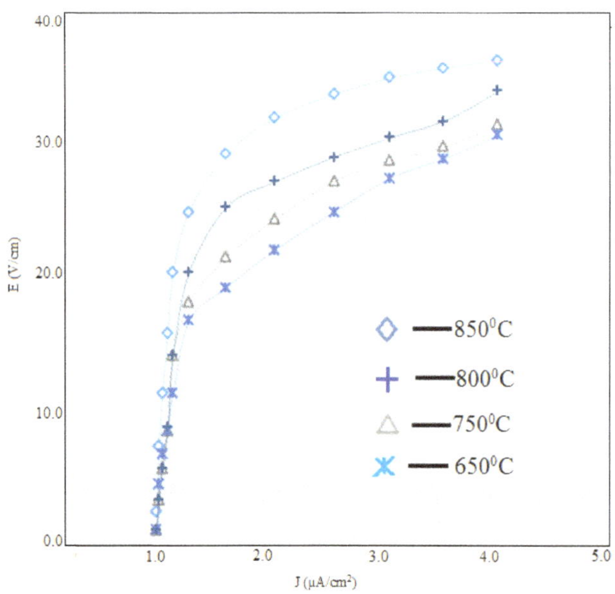

Figure 5. Current- voltage (J-E) characteristics of the sintered samples

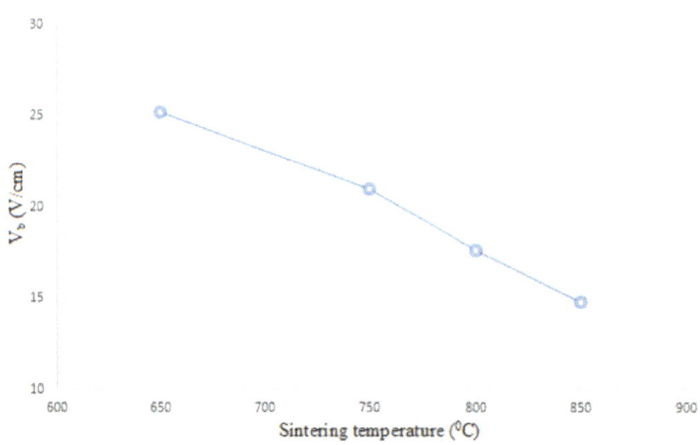

Figure 6. Variation of breakdown voltage, V_b with sintering temperature

4. Conclusion

The influence of sintering temperature on the microstructure and electrical response of MnO_2-PbO- doped ZnO varistor ceramics were investigated between 650°C and 850°C. The samples exhibited initial densification with increased sintering temperature.

This was followed by a reduction in densification at 800°C and a subsequent improvement at a sintering temperature of 850°C. SEM studies showed improved homogeneity with increasing temperature. The grain sizes also increased accordingly while XRD spectra revealed the formation of secondary polymorphic phases. The electrical characteristics exhibited sharp transitions from ohmic to non-ohmic response and the associated breakdown voltage decreased with increased sintering temperature.

The improvement in electrical response with increased sintering temperature was attributed to the improved microstructural homogeneities as indicated by more uniform phase distribution and narrower grain boundaries. The best characteristics found suitable for varistor behaviour was exhibited by the sample sintered at 850°C for 6 hours. However, the duration of sintering may have been crucial to the microstructural evolutions of the ZnO-based ceramic varistor samples. As the microstructural findings correlated with the electrical characteristics, it follows that breakdown voltages can be significantly controlled by means of the sintering process.

Acknowledgements

The authors would like to thank the XRD staff of Center for Energy Research and Development (CERD), Obafemi Awolowo University, Ile-Ife; Nigeria. We are also grateful to the SEM staff of Sheda Science and Technology Complex (SHESTCO), Abuja Nigeria; and to Dr Ilori of the Electrical and Electronics Engineering Department, Obafemi Awolowo University, Ile-Ife for their technical supports.

References

Akinnifesi, J. O., Erinfolani, F. B., & Akinwunmi, O. O. (2014). Influence of microstructure on the non- ohmic behavior of zinc oxide varistor ceramics prepared by direct mixing of constituent phases. *Ife Journal of Science, 16*, 91- 98.

Anas, S., Mangalaraja, R. V., Poothayal, M., Shukla, S. K., & Ananthakumar, S. (2007). Direct synthesis of varistor-grade doped nanocrystalline ZnO and its densification through a step-sintering technique. *Acta Materialia, 55*(17), 5792-5801. http://dx.doi.org/10.1016/j.actamat.2007.06.047

Barett, C., Nix, W., & Tetelman, A. (1973). *The principles of engineering materials*. New Jersey: Prentice Hall.

Bernik, S., Brankovic, G., Rustja, S., Zunic, M., & Pollogar, Z. (2008). Compositional aspects of ZnO-based varistor ceramics prepared by direct mixing of the constituent phases and high energy milling. *Ceramics international, 34*,1495-1502. http://dx.doi.org/10.1016/j.ceramint.2007.04.003

Bernik, S., Marcek, S., & Bui, A. (2001). Microstructural and electrical characteristics of V_2O_3- doped ZnO-Bi_2O_3- based varistor ceramics. *Journal of the European Ceramic Society, 21*, 1875-1878. http://dx.doi.org/10.1016/S0955-2219(01)00134-0

Brankovic, Z., Brankovic, G., Bernik, S., & Zunic, M (2007). ZnO varistors with reduced amount of additives prepared by direct mixing of constituent phases. *Journal of the European Ceramic Society, 27*, 1101-1104. http://dx.doi.org/10.1016/j.jeurceramsoc.2006.05.040

Cheng, P. F., Li, S. T., & Alim, M. A. (2007). Soft core behaviour in ZnO- Bi_2O_3 based varistors containing oxides of Ce and Gd. *Physica Status Solidi A- Applications and materials science, 204*, 887-899. http://dx.doi.org/10.1002/pssa.200622180

Clarke, D. R. (1999). Varistor Ceramic *Journal of the American Ceramic Society, 82*, 2106-2110. http://dx.doi.org/10.1111/j.1151-2916.1999.tb01793.x

Duran, P., Tarjaj, J., & Moure, C. (2003). Fully dense, fine-grained doped zinc oxide varistors with improved non-linear properties by thermal processing optimization. *Journal of the American Ceramic Society, 86*, 1326-1329. http://dx.doi.org/10.1111/j.1151-2916.2003.tb03470.x

Entage, R. (1977). Physics of zinc oxide varistors. *J. Appl. Phys., 48*, 4372-4384. http://dx.doi.org/10.1063/1.323391

Fernandez-Hevia, D., Peitado, M., De Frustos, J., Caballero, A., & Fernandez, J. (2004). Wide range dielectric spectroscopy of ZnO- based varistors as a function of sintering time. *Journal of the European Ceramic Society, 24*, 1205-1208. http://dx.doi.org/10.1016/S0955-2219(03)00411-4

Gupta, T. K. (1990). Application of zinc oxide varistors. *J. Am. Ceram. Soc., 7*, 1814-1840. http://dx.doi.org/10.1111/j.1151-2916.1990.tb05232.x

Han, J., Mantas, P. Q., & Senos, A. M. R. (2000). Grain growth in Mn- doped ZnO ceramics. *J. Eur. Ceram Soc., 20*, 2753-2758. http://dx.doi.org/10.1016/S0955-2219(00)00220-X

Han, J., Senos, A. M. R., & Mantas, P. Q. (1999). Non – isothermal sintering of Mn-doped ZnO ceramics. *J. Eur. Cera Soc., 19*, 1003-1006. http://dx.doi.org/10.1016/S0955-2219(98)00362-8

Han, J., Senos, A. M. R., & Mantas, P. Q. (2002) : Vaistor behavior of Mn-doped ZnO ceramics. *J. Eur. Ceram Soc., 20*, 1653-1660. http://dx.doi.org/10.1016/S0955-2219(01)00484-8

Haskell, B. A., Souri, S. J., & Helfand, M. A. (1999). Varistor behaviour at grain- boundaries in ZnO. *J. Am. Ceram. Soc., 82*, 2106-2110. http://dx.doi.org/10.1111/j.1151-2916.1999.tb02048.x

Hozer, L. (1994). Metal oxide varistors, semiconductor ceramics: grain boundary effects (pp. 44-109). Warszawa: Polish Scientific Publishers.

Kourdi, M., Bui, A., louibierre, A., & Khedim, A. (1992). Behaviour of metal oxide based varistors subjected to partial discharge in air. *J. Phys. D. Appl. Phys., 25*, 548. http://dx.doi.org/10.1109/ICSD.1992.224942

Kutty, T. (2000). Effect of sintering temperature on microstructure of ceramics. *Journal of Materials Chemistry and Physics, 63*, 218.

Leach C., Ling, Z., & Freer, R. (2000). The effect of sintering temperature variation on the development of electrically active interfaces in zinc oxide based varistors. *Journal of the European Ceramic Society, 20*, 2759. http://dx.doi.org/10.1016/S0955-2219(00)00223-5

Leach, C. (2005). Grain boundary structures in zinc oxide varistors. *Acta Materiala, 53*, 237-245. http://dx.doi.org/10.1016/j.actamat.2004.07.006

Lee, W. S, Chen, W. T., Lee, Y. C., Yang, T., Su, C. Y., & Hu, C. L. (2007). Influence of sintering on microstructures and electrical properties of ZnO- based multilayer varistors. *Ceramics International, 33*, 1001-1005. http://dx.doi.org/10.1016/j.ceramint.2006.02.017

Leite, E. R., Nobre, M. A. L., Longo, E., & vareta, J. A. (1996). Microstructural development of ZnO Varistors during reactive liquid phase sintering . *J. mater. Sci., 31*, 5391-5398 http://dx.doi.org/10.1007/BF01159308

Levinson, L. M., & Philipp, H. R. (1975). Physics of metal oxide varistors. *J. Appl. Phys., 46*, 1332-1341. http://dx.doi.org/10.1063/1.321701

Mahan, G., Levinson, M., & Philipp, H. (1993). Theory of conduction in ZnO varistors. *J.Appl. Phys., 54*, 2799-2812.

Matsuoka, M. (1991). Non-ohmic properties of zinc oxide ceramics. *Jpn. J. Appl. Phys., 10*, 736- 746. http://dx.doi.org/10.1143/JJAP.10.736

Nahm, C. (2006). Effects of sintering temperature on microstructure and electrical properties of Zn, Pr, Co, La oxide – based varistors. *Mater. Lett., 60*, 3394. http://dx.doi.org/10.1007/s10853-005-2185-3

Nahm, C. (2007). The effect of sintering temperature on electrical properties and accelerated aging behaviour of PCCL – doped ZnO varistors. *Materials Science and Engineering. B., 136*, 34–139. http://dx.doi.org/10.1016/j.mseb.2006.09.010

Nahm, C. W. (2004). Effect of sintering time on varistors properties of Dy_2O3 – doped ZnO – Pr_6O_{11} – based ceramics. *Materials letters, 58*, 3297-3300. http://dx.doi.org/10.1016/j.matlet.2004.06.020

Onreabroy, W., Sirikulrat, N., Brown, A. P., Hammond, C., & Milne, S. J. (2006). properties and intergranular phase analysis of a ZnO - CoO – Bi_2O_3 varistors. *Solid State Ionics, 177*, 411–420. http://dx.doi.org/10.1016/j.ssi.2005.10.032

Peitado, M., Fernandez, J., & Caballero, A. (2007). Varistor based in the ZnO –Bi_2O_3 system -microstructure control and properties. *Journal of the European Ceramic Society, 27*, 3867-3872. http://dx.doi.org/10.1016/j.jeurceramsoc.2007.02.046

Pillar, S., Kelly, J., Mccormack, D., O'brien, P., & Ramesh, R. (2003). The effect of processing conditions on varistors prepared from nanocrystalline ZnO. *Journal of Materials Chemistry, 13*, 2586-2590.

Vijaya, M., & Rangarajan, G. (2003). *Materials Science* (pp. 122-123). New Delhi: McGraw Hill Co.

Xu, D., Shi, L., Wu, Z., Zhang, Q., & Wu, X. (2009). Microstructure and electrical properties of ZnO – Bi_2O_3 – based varistor ceramic by different sintering process. *Journal of the European Ceramic Society, 29*, 1789-1794. http://dx.doi.org/10.1016/j.jeurceramsoc.2008.10.020

Zhang, J., Cao, S., Zhang, R., Yu, L., & Jing, C. (2005). Effect of fabrication conditions on the I-V properties for ZnO varistors with high concentration of additives by sol – gel technique. *Current Applied Physics, 5*, 381-386. http://dx.doi.org/10.1016/j.cap.2004.03.004

17

A Modified Stillinger-Weber Potential for TlBr, and Its Polymorphic Extension

X. W. Zhou[1], M. E. Foster[2], Reese Jones[1], P. Yang[3], H. Fan[4,5] & F. P. Doty[6]

[1] Mechanics of Materials Department, Sandia National Laboratories, Livermore, California 94550, USA

[2] Materials Chemistry Department, Sandia National Laboratories, Livermore, California 94550, USA

[3] Electronic, Optical, and Nano Department, Sandia National Laboratories, Albuquerque, NM 87106, USA

[4] Advanced Materials Laboratory, Sandia National Laboratories, Albuquerque, NM 87106, USA

[5] Center for Micro-Engineered Materials, Department of Chemical and Biological Engineering, University of New Mexico, Albuquerque, NM 87106, USA

[6] Radiation and Nuclear Detection Materials and Analysis Department, Sandia National Laboratories, Livermore, California 94550, USA

Correspondence: X. W. Zhou, Mechanics of Materials Department, Sandia National Laboratories, Livermore, California 94550, USA. E-mail: xzhou@sandia.gov

Abstract

TlBr is promising for γ- and x- radiation detection, but suffers from rapid performance degradation under the operating external electric fields. To enable molecular dynamics (MD) studies of this degradation, we have developed a Stillinger-Weber type of TlBr interatomic potential. During this process, we have also addressed two problems of wider interests. First, the conventional Stillinger-Weber potential format is only applicable for tetrahedral structures (e.g., diamond-cubic, zinc-blende, or wurtzite). Here we have modified the analytical functions of the Stillinger-Weber potential so that it can now be used for other crystal structures. Second, past modifications of interatomic potentials cannot always be applied by a broad community because any new analytical functions of the potential would require corresponding changes in the molecular dynamics codes. Here we have developed a polymorphic potential model that simultaneously incorporates Stillinger-Weber, Tersoff, embedded-atom method, and any variations (i.e., modified functions) of these potentials. We have implemented this polymorphic model in MD code LAMMPS, and demonstrated that our TlBr potential enables stable MD simulations under external electric fields.

Keywords: interatomic potential, molecular dynamics, TlBr crystal, structure evolution, radiation detection semiconductor

1. Introduction

Thallium bromide (TlBr) has emerged to be one of the most promising semiconductors for γ- and x- radiation detection in recent years, achieving resolution as high as 1% at 662 keV (Shorohov et al., 2009; Hitomi, Matsumoto, Muroi, Shoji, & Hiratate, 2002). To collect the charges created during the radiation events, the material must be subject to an external electric field. Unfortunately, the performance of TlBr degrades under external electric fields after operation times as short as a few hours to a few weeks (Hitomi, Kikuchi, Shoji, & Ishii, 2009). It is unclear if this degradation comes from the structure evolution due to the ionic conduction under the electric fields. As ionic conduction may be affected by defective structures such as the open channels of edge dislocations, molecular dynamics (MD) that allows extended defects to be included in simulated crystals becomes a useful method to study the ionic conduction induced structure evolution of TlBr. Such MD simulations are not yet possible due to the lack of an interatomic potential for the Tl-Br system that has a CsCl type of crystal structure.

Past MD simulations (Zheng-Johansson & McGreevy, 1996; Zheng-Johansson, Ebbsjö, & McGreevy, 1995; Huang, Wei, Chen, & Chen, 2011; Tung, Chang, Hsiung, Chiang, & Li, 2010; Yamamura, Kawasaki, & Sakai, 1999; Chung, & De Leeuw, 2004; Beckers, Van Der Bent, & De Leeuw, 2000; Lindan & Gillan, 1993) mainly focused on ionic conductivity, but not the structural evolution under external electric fields. As a result, these studies typically do not apply external electric fields. Instead, the system is simply annealed at a sufficiently high temperature to cause thermally activated diffusion of ions. The trajectories of ions are then used to calculate diffusion coefficients of ions. These diffusion coefficients are in turn used to calculate ionic conductivity through Nernst-Einstein relation. Our interest is to understand the structural evolution under external electric fields. Hence, we will apply an external electric field to accelerate the ionic motion, therefore leading to sufficient structural changes needed for the analysis within the short time scale of MD simulations. This, however, requires a robust interatomic potential that can at least maintain the crystal structure at a large external field.

While relatively complex many-body potential (Marrocchelli, Madden, Norberg, & Hull, 2011) can be applied, it is not uncommon to model ionic materials with simple force fields such as pair potentials (Zheng-Johansson, Ebbsjö, & McGreevy, 1995; Huang, Wei, Chen, & Chen, 2011; Tung, Chang, Hsiung, Chiang, & Li, 2010; Yamamura, Kawasaki, & Sakai, 1999), or pair potentials plus angular energy penalty interactions (Chung, & De Leeuw, 2004; Beckers, Van Der Bent, & De Leeuw, 2000). The angular energy penalty approach is very similar to the widely-used Stillinger-Weber (SW) potential (Stillinger & Weber, 1985), and has the advantage to stabilize relatively complex crystal structures that the pair potentials cannot. For example, SW potentials use a parabolic energy penalty term to penalize non-tetrahedral bond angles. As a result, SW potentials have been successfully applied to tetrahedral structures such as diamond-cubic, zinc-blende, and wurtzite[Note 1]. Depending on parameterization, SW potentials can also be used for fcc elements (Zhou et al., 2013; Zhou, Jones, Kimmer, Duda, & Hopkins, 2013; Zhou, Jones, Duda, & Hopkins, 2013). However, the conventional SW potentials have two limitations (Zhou et al., 2013; Zhou, Jones, Kimmer, Duda, & Hopkins, 2013; Zhou, Jones, Duda, & Hopkins, 2013): (1) they significantly overestimate the elastic constants of closely packed (e.g., fcc) elements; and (2) they usually cannot be used for other (non-tetrahedral) structures such as sc elements, and NaCl and CsCl compounds. Note that although most elements do not exhibit the lowest energy for the sc crystal, the sc crystal can have a near-lowest energy that cannot be captured by SW potentials. Potentials capable of prescribing a low energy for the sc structure, therefore, improve upon SW potentials in terms of the general energy trends when a variety of configurations are considered (Zhou, Foster, Van Swol, Martin, & Wong, 2014).

Many literature potentials are constructed using particular analytical functions. These potentials can be easily improved if alternative functions are used. For example, Rockett (Wang & Rockett, 1991) used alternative functions in the Tersoff (Tersoff, 1989; Albe, Nordlund, Nord, & Kuronen, 2002) potential format to better treat short- and long- range interactions in covalent systems, and we used alternative functions in the Tersoff potential to improve its prediction on thermal conductivity (Zhou & Jones, 2011). Although alternative functions do not change the potential format and should, therefore, be easily applied, such an approach has not been widely used because it does require new molecular dynamics codes to be developed for each modified function.

With the recognition that the problems described above can limit the atomistic level studies of emerging important materials, the objective of the present work is fourfold: (1) modify the SW potential so that it can better describe elastic constants of elements and be applicable to a wide range of crystal structures including sc, NaCl, and CsCl; (2) develop a polymorphic potential model that incorporates simultaneously SW potential, modified SW potential, Tersoff potential (Tersoff, 1989; Albe, Nordlund, Nord, & Kuronen, 2002), modified Tersoff potential (Wang & Rockett, 1991), and embedded-atom method (EAM) potential (Daw & Baskes, 1984); (3) implement this polymorphic potential model in the public MD code LAMMPS (Plimpton, 1995) so that future development of any alternative functions for these types of potentials no longer requires modification of the molecular dynamics codes; and (4) parameterize the modified SW potential for TlBr system and demonstrate the utility of the resulting potential under external electric fields.

2. Modified SW Potentials

In SW potential (Stillinger & Weber, 1985), the total energy of a system of N atoms is expressed as

$$E = \frac{1}{2}\sum_{i=1}^{N}\sum_{j=i_1}^{i_N}\left[\varphi_{R,IJ}\left(r_{ij}\right) - \varphi_{A,IJ}\left(r_{ij}\right) + u_{IJ}\left(r_{ij}\right)\sum_{\substack{k=i_1\\ k\neq j}}^{i_N} u_{IK}\left(r_{ik}\right)\cdot g_{JIK}\left(\cos\theta_{jik}\right)^2\right] \tag{1}$$

Comment: In Eqs. (1), (2), (3), (9), (15), φ should all be changed to ϕ

where $i_1, i_2, \ldots, i_N$ is a list of neighbors of atom i, θ_{jik} is the bond angle formed by atoms j and k at the site of atom i, $\phi_{R,IJ}(r_{ij})$ and $\phi_{A,IJ}(r_{ij})$ are, respectively, pairwise repulsive and attractive functions, $u_{IJ}(r_{ij})$ is another pair function for the three-body term, $g_{JIK}(\cos\theta_{jik})$ is an angular energy penalty function, and subscripts i,j,k and I,J,K indicate, respectively, the atoms and the species of the atoms (note that three bodies JIK and KIJ are equivalent). The original SW potentials significantly overestimate the elastic constants of elements because the $\phi_{R,IJ}(r)$ and $\phi_{A,IJ}(r)$

functions used in these potentials do not allow independent adjustment of bond energy and its second derivative (Zhou et al., 2013). Here we propose to use modified Morse's functions capable of independent change of bond energy and its second derivative (Zhou et al., 2013) to represent $\phi_{R,IJ}(r)$ and $\phi_{A,IJ}(r)$:

$$\varphi_{R,IJ}(r) = \frac{E_{b,IJ} \cdot \beta_{IJ}}{\beta_{IJ} - \alpha_{IJ}} \cdot \exp\left(-\alpha_{IJ} \cdot \frac{r - r_{0,IJ}}{r_{0,IJ}}\right) \cdot f_{c,IJ}(r), \tag{2}$$

$$\varphi_{A,IJ}(r) = \frac{E_{b,IJ} \cdot \alpha_{IJ}}{\beta_{IJ} - \alpha_{IJ}} \cdot \exp\left(-\beta_{IJ} \cdot \frac{r - r_{0,IJ}}{r_{0,IJ}}\right) \cdot f_{c,IJ}(r), \tag{3}$$

where $E_{b,IJ}$, $r_{0,IJ}$, α_{IJ}, and β_{IJ} are four pair dependent parameters, and $f_{c,IJ}(r_{ij})$ is a cutoff function. Note that the parameters introduced here have physical meanings: $E_{b,IJ}$ and $r_{0,IJ}$ correspond respectively to the equilibrium bond energy and bond length, and $\alpha_{IJ} >> \beta_{IJ}$ control the curvature of the bond energy at the equilibrium bond length. The cutoff function $f_{c,IJ}(r_{ij})$ is expressed as:

$$f_{c,IJ}(r) = \begin{cases} \dfrac{\exp\left(-\varsigma_{IJ} \cdot r^{\upsilon_{IJ}}\right) - \exp\left(-\varsigma_{IJ} \cdot r_{c,IJ}^{\upsilon_{IJ}}\right)}{\exp\left(-\varsigma_{IJ} \cdot r_{s,IJ}^{\upsilon_{IJ}}\right) - \exp\left(-\varsigma_{IJ} \cdot r_{c,IJ}^{\upsilon_{IJ}}\right)}, & r \le r_{c,IJ} \\ 0, & r > r_{c,IJ} \end{cases} \tag{4}$$

where $r_{s,IJ}$ and $r_{c,IJ}$ ($r_{s,IJ} << r_{c,IJ}$) are two independent pair parameters, and ζ_{IJ} and υ_{IJ} are two dependent pair parameters $\upsilon_{IJ} = \dfrac{\ln\left[\ln(0.9)/\ln(0.00000001)\right]}{\ln\left(r_{s,IJ}/r_{c,IJ}\right)}$ and $\varsigma_{IJ} = -\dfrac{\ln(0.9)}{\left(r_{s,IJ}\right)^{\upsilon_{IJ}}}$. Note that the cutoff function approximately equals one at $r \le r_{s,IJ}$ and equals zero at $r = r_{c,IJ}$ (i.e., $r_{c,IJ}$ is the cutoff distance). Hence, multiplying any potential function with this cutoff function does not affect significantly the potential function at small distances but allows the potential function to be smoothly cut off at $r_{c,IJ}$. Such a cutoff method is superior to the spline approach used by the Tersoff potential as the latter does not have continuous second and higher order derivatives. A general exponential decay function is used to represent the $u_{IJ}(r)$ function:

$$u_{IJ}(r) = \exp\left(-\gamma_{IJ} \cdot \frac{r_{ij} - r_{0,IJ}}{r_{0,IJ}}\right) \cdot f_{c,IJ}(r_{ij}) \tag{5}$$

where γ_{IJ} is a pair parameter.

For the angular function, SW potentials use a parabolic energy penalty to the non-tetrahedral angle $g_{JIK}(\cos\theta_{jik}) = (\cos\theta_{jik} - \cos\theta_{0,JIK})^2$ where the parameter $\cos\theta_{0,JIK}$ is fixed at the tetrahedral bond angle $\cos\theta_{0,JIK} = -1/3$. This imposes two constraints: not only the favorable bond angle is fixed at the tetrahedral angle, but also the favorable bond angle does not depend on species I, J, and K. The $g_{JIK}(\cos\theta_{jik})$ function can be made more general by treating $\cos\theta_{0,JIK}$ as fitting parameters that depend on I, J, and K. Even so, the resulting function is still not fully flexible, for instance, it does not have a scaling factor, and its second derivative with respect to $\cos\theta_{jik}$ is fixed at 2. In addition, the function is symmetric at $\cos\theta_{jik} = \cos\theta_{0,JIK}$, and the energy penalty does not saturate (i.e., the absolute slope increases when the angle deviates from $\cos\theta_{0,JIK}$). These cause difficulties for capturing the angular function derived from quantum mechanical theories (Pettifor et al., 2004). Here we consider a new angular function $g_{JIK}(\cos\theta_{jik}) = \lambda_{JIK} \cdot \{1 - \exp[-\xi_{0,JIK} \cdot (\cos\theta_{jik} - \cos\theta_{0,JIK})^2]\}$, where λ_{JIK} is three-body dependent scaling factor and $\xi_{0,JIK}$ is another three-body dependent parameter. Note that for $\lambda_{JIK} = 1$, $\xi_{0,JIK} = 1$ and $\cos\theta_{0,JIK} = -1/3$, the modified energy penalty is equivalent to the parabolic function near $\cos\theta_{jik} = -1/3$ because the first term of the Taylor series of the modified function expanded at $\cos\theta_{jik} = \cos\theta_{0,JIK}$ is in fact the parabolic function. However, replacing the leading term with a full series does allow the energy penalty curvature to be adjusted through an added parameter $\xi_{0,JIK}$ and the value of function to be saturated (while retaining the "penalty" effect, i.e., the function minimizes at $\cos\theta_{jik} = \cos\theta_{0,JIK}$ and monotonically increases when $\cos\theta_{jik}$ deviates from $\cos\theta_{0,JIK}$). An even further flexible function will be to penalize the energy when the bond angle deviates from three independent values, which also results in asymmetric minimums. Based on this consideration, we propose a general angular function as:

$$g_{JIK}\left(\cos\theta_{jik}\right) = \lambda_{JIK} \cdot \frac{1-\exp\left[-\xi_{0,JIK}\cdot\left(\cos\theta_{jik}-\cos\theta_{0,JIK}\right)^2\right]}{1-\exp\left[-\xi_{0,JIK}\cdot\left(1-\cos\theta_{0,JIK}\right)^2\right]} \cdot \frac{1-\kappa_{1,JIK}\cdot\exp\left[-\xi_{1,JIK}\cdot\left(\cos\theta_{jik}-\cos\theta_{1,JIK}\right)^2\right]}{1-\kappa_{1,JIK}\cdot\exp\left[-\xi_{1,JIK}\cdot\left(1-\cos\theta_{1,JIK}\right)^2\right]} \cdot \frac{1-\kappa_{2,JIK}\cdot\exp\left[-\xi_{2,JIK}\cdot\left(\cos\theta_{jik}-\cos\theta_{2,JIK}\right)^2\right]}{1-\kappa_{2,JIK}\cdot\exp\left[-\xi_{2,JIK}\cdot\left(1-\cos\theta_{2,JIK}\right)^2\right]} \quad (6)$$

where λ_{JIK}, $\xi_{0,JIK}$, $\xi_{1,JIK}$, $\xi_{2,JIK}$, $\cos\theta_{0,JIK}$, $\cos\theta_{1,JIK}$, $\cos\theta_{2,JIK}$, $\kappa_{1,JIK}$, and $\kappa_{2,JIK}$ are all three-body dependent parameters. Note that when the parameters are given, the denominator in Equation (6) is essentially a normalization constant so that $g_{JIK}(\cos\theta_{jik}=1) = \lambda_{JIK}$. It can be seen that when $\kappa_{1,JIK} = \kappa_{2,JIK} = 0$, Equation (6) penalizes the energy when the bond angle deviates from a single value $\cos\theta_{jik} = \cos\theta_{0,JIK}$ as in the conventional SW potential. Otherwise Equation (6) can penalize the energy when the bond angle deviates from three values $\cos\theta_{jik} = \cos\theta_{0,JIK}$, $\cos\theta_{jik} = \cos\theta_{1,JIK}$, $\cos\theta_{jik} = \cos\theta_{2,JIK}$. Equations (1) – (6) fully define our modified SW (MSW) potential.

3. Polymorphic Potential Model

Any improved interatomic potentials will not be applied unless molecular dynamics codes are available to run them. To provide potential developers with a great flexibility for modifying the interatomic potentials without worrying about MD codes, we have constructed a polymorphic potential model. In this model, the energy of the system is expressed as

$$E = \frac{1}{2}\sum_{i=1}^{N}\sum_{j=1}^{N}\left[\left(1-\delta_{ij}\right)\cdot U_{IJ}\left(r_{ij}\right)-\left(1-\eta_{ij}\right)\cdot F_{IJ}\left(X_{ij}\right)\cdot V_{IJ}\left(r_{ij}\right)\right] \quad (7)$$

where δ_{ij} is Kronecker delta (i.e., $\delta_{ij} = 1$ when $i = j$ and $\delta_{ij} = 0$ when $i \neq j$), η_{ij} is an indicator of the potential type that can be set to either $\eta_{ij} = \delta_{ij}$ or $\eta_{ij} = 1- \delta_{ij}$, $U_{IJ}(r_{ij})$ and $V_{IJ}(r_{ij})$ are two pair functions, and $F_{IJ}(X_{ij})$ is a function of a local variable X_{ij} that will be discussed below. It can be seen that when $\eta_{ij} = \delta_{ij}$, the summation in equation (7) excludes the self-interaction term $i = j$ and is therefore over all pairs of different atoms. When $\eta_{ij} = 1 - \delta_{ij}$, the second term becomes $E = \frac{1}{2}\sum_{i=1}^{N} F_{II}\left(X_{ii}\right)\cdot V_{II}\left(r_{ii}\right)$, which can be used to incorporate the embedding energy of the embedded-atom method as will be clear later in this section. The variable X_{ij} essentially accounts for the environment surrounding the ij bond, and is defined as

$$X_{ij} = \sum_{\substack{k=i_1\\ k\neq i,j}}^{i_N} W_{IK}\left(r_{ik}\right)\cdot G_{JIK}\left(\theta_{jik}\right)\cdot P_{IK}\left(\Delta r_{jik}\right)\cdot \quad (8)$$

where $P_{IJ}(\Delta r_{jik})$ is a function of weighted difference between atomic spacing r_{ij} and r_{ik}, which is written as $\Delta r_{jik} = r_{ij} - \xi_{IJ}\cdot r_{ik}$ with the weighting factor ξ_{IJ} being either 0 or 1 to include or exclude r_{ik}, $W_{IK}(r_{ik})$ is another pair function, and $G_{JIK}(\theta_{jik})$ is a three-body function of bond angle θ_{jik}. It can be seen that this polymorphic interatomic potential model is fully defined when the indicators η_{ij} and ξ_{IJ}, and the six functions $U_{IJ}(r)$, $V_{IJ}(r)$, $P_{IJ}(\Delta r)$, $W_{IJ}(r)$, $F_{IJ}(X)$, and $G_{JIK}(\theta)$ (for all the species I, J, K = 1, 2, …) are given. Note that these six functions can all be supplied as one-dimensional tables and can therefore be implemented in MD codes using cubic spline interpolation and/or extrapolation. As a result, users can easily perform simulations using different potentials by tabulating these functions (in a MD read-in table file) accordingly. For instance, the polymorphic potential reduces to our MSW potential if we tabulate the functions according to:

$$\begin{cases} \eta_{ij} = \delta_{ij}, \xi_{IJ} = 0 \\ U_{IJ}(r) = \varphi_{R,IJ}(r) - \varphi_{A,IJ}(r) \\ V_{IJ}(r) = u_{IJ}(r) \\ F_{IJ}(X) = -X \\ P_{IJ}(\Delta r) = 1 \\ W_{IJ}(r) = u_{IJ}(r) \\ G_{JIK}(\theta) = g_{JIK}(\theta) \end{cases} \quad (9)$$

where $\phi_{R,IJ}(r)$, $\phi_{A,IJ}(r)$, $u_{IJ}(r)$, and $g_{JIK}(\cos\theta)$ are defined by Equations (2), (3), (5) and (6). The polymorphic potential reduces to a conventional SW (Stillinger & Weber, 1985) potential if we tabulate the functions according to:

$$\begin{cases} \eta_{ij} = \delta_{ij}, \xi_{IJ} = 0 \\ U_{IJ}(r) = A_{IJ} \cdot \varepsilon_{IJ} \cdot \left(\frac{\sigma_{IJ}}{r}\right)^q \cdot \left[B_{IJ}\left(\frac{\sigma_{IJ}}{r}\right)^{p-q} - 1\right] \cdot \exp\left(\frac{\sigma_{IJ}}{r - a_{IJ} \cdot \sigma_{IJ}}\right) \\ V_{IJ}(r) = \sqrt{\lambda_{IJ} \cdot \varepsilon_{IJ}} \cdot \exp\left(\frac{\gamma_{IJ} \cdot \sigma_{IJ}}{r - a_{IJ} \cdot \sigma_{IJ}}\right) \\ F_{IJ}(X) = -X \\ P_{IJ}(\Delta r) = 1 \\ W_{IJ}(r) = \sqrt{\lambda_{IJ} \cdot \varepsilon_{IJ}} \cdot \exp\left(\frac{\gamma_{IJ} \cdot \sigma_{IJ}}{r - a_{IJ} \cdot \sigma_{IJ}}\right) \\ G_{JIK}(\theta) = \left(\cos\theta + \frac{1}{3}\right)^2 \end{cases} \tag{10}$$

where A_{IJ}, B_{IJ}, ε_{IJ}, σ_{IJ}, λ_{IJ}, γ_{IJ}, a_{IJ}, p, and q are the normal parameters for the SW potential as described above. The polymorphic model represents Tersoff types of potential (Tersoff, 1989; Albe, Nordlund, Nord, & Kuronen, 2002) if we set

$$\begin{cases} \eta_{ij} = \delta_{ij}, \xi_{IJ} = 1 \\ U_{IJ}(r) = \frac{D_{e,IJ}}{S_{IJ} - 1} \exp\left[-\beta_{IJ}\sqrt{2S_{IJ}}\left(r - r_{e,IJ}\right)\right] \cdot f_{c,IJ}(r) \\ V_{IJ}(r) = \frac{S_{IJ} \cdot D_{e,IJ}}{S_{IJ} - 1} \exp\left[-\beta_{IJ}\sqrt{\frac{2}{S_{IJ}}}\left(r - r_{e,IJ}\right)\right] \cdot f_{c,IJ}(r) \\ F_{IJ}(X) = (1 + X)^{-\frac{1}{2}} \\ P_{IK}(\Delta r) = \exp\left[2\mu_{IK} \cdot \Delta r\right] \\ W_{IK}(r) = f_{c,IK}(r) \\ G_{JIK}(\theta) = \gamma_{IK}\left[1 + \frac{c_{IK}^2}{d_{IK}^2} - \frac{c_{IK}^2}{d_{IK}^2 + \left(h_{IK} + \cos\theta\right)^2}\right] \end{cases} \tag{11}$$

where $f_{c,IJ}(r_{ij})$ is a cutoff function defined as

$$f_{c,IJ}(r) = \begin{cases} 1, & r \le r_{s,IJ} \\ \frac{1}{2} + \frac{1}{2}\cos\left[\frac{\pi(r - r_{s,IJ})}{r_{c,IJ} - r_{s,IJ}}\right], & r_{s,IJ} < r < r_{c,IJ} \\ 0, & r \ge r_{c,IJ} \end{cases} \tag{12}$$

and $D_{e,IJ}$, S_{IJ}, $r_{e,IJ}$, β_{IJ}, μ_{IJ}, γ_{IJ}, c_{IJ}, d_{IJ}, h_{IJ}, $r_{s,IJ}$ and $r_{c,IJ}$ are all pairwise parameters. The polymorphic potential can also represent the Rockett-Tersoff potential (Wang & Rockett, 1991) if we set

$$\begin{cases} \eta_{ij} = \delta_{ij}, \xi_{IJ} = 1 \\ U_{IJ}(r) = \begin{cases} A_{IJ} \cdot \exp\left(-\lambda_{1,IJ} \cdot r\right) \cdot f_{c,IJ}(r), & r \le r_{s,1,IJ} \\ A_{IJ} \cdot \exp\left(-\lambda_{1,IJ} \cdot r\right) \cdot f_{c,IJ}(r) \cdot f_{c,1,IJ}(r), & r_{s,1,IJ} < r < r_{c,1,IJ} \\ 0, & r \ge r_{c,1,IJ} \end{cases} \\ V_{IJ}(r) = \begin{cases} B_{IJ} \cdot \exp\left(-\lambda_{2,IJ} \cdot r\right) \cdot f_{c,IJ}(r), & r \le r_{s,1,IJ} \\ B_{IJ} \cdot \exp\left(-\lambda_{2,IJ} \cdot r\right) \cdot f_{c,IJ}(r) + A_{IJ} \cdot \exp\left(-\lambda_{1,IJ} \cdot r\right) \cdot f_{c,IJ}(r) \cdot \left[1 - f_{c,1,IJ}(r)\right], & r_{s,1,IJ} < r < r_{c,1,IJ} \\ B_{IJ} \cdot \exp\left(-\lambda_{2,IJ} \cdot r\right) \cdot f_{c,IJ}(r) + A_{IJ} \cdot \exp\left(-\lambda_{1,IJ} \cdot r\right) \cdot f_{c,IJ}(r), & r \ge r_{c,1,IJ} \end{cases} \\ F_{IJ}(X) = \left[1 + \left(\beta_{IJ} \cdot X\right)^{n_{IJ}}\right]^{-\frac{1}{2n_{IJ}}} \\ P_{IK}(\Delta r) = \exp\left[\lambda_{3,IK} \cdot \Delta r^3\right] \\ W_{IK}(r) = f_{c,IK}(r) \\ G_{JIK}(\theta) = 1 + \frac{c_{IK}^2}{d_{IK}^2} - \frac{c_{IK}^2}{d_{IK}^2 + \left(h_{IK} + \cos\theta\right)^2} \end{cases} \tag{13}$$

where $f_{c,1,IJ}(r)$ is a cutoff function similar to equation (12) but operates at a different cutoff range:

$$f_{c,1,IJ}(r) = \begin{cases} 1, & r \le r_{s,1,IJ} \\ \frac{1}{2} + \frac{1}{2}\cos\left[\frac{\pi(r - r_{s,1,IJ})}{r_{c,1,IJ} - r_{s,1,IJ}}\right], & r_{s,1,IJ} < r < r_{c,1,IJ} \\ 0, & r \ge r_{c,1,IJ} \end{cases} \tag{14}$$

and A_{IJ}, B_{IJ}, $\lambda_{1,IJ}$, $\lambda_{2,IJ}$, $\lambda_{3,IJ}$, β_{IJ}, n_{IJ}, c_{IJ}, d_{IJ}, h_{IJ}, $r_{s,1,IJ}$, and $r_{c,1,IJ}$ are all pairwise parameters. To use the polymorphic model for the embedded-atom method potential (Daw & Baskes, 1984), we can simply set:

$$\begin{cases} \eta_{ij} = 1 - \delta_{ij}, \xi_{IJ} = 0 \\ U_{IJ}(r) = \varphi_{IJ}(r) \\ V_{IJ}(r) = 1 \\ F_{II}(X) = -2F_I(X) \\ P_{IK}(\Delta r) = 1 \\ W_{IK}(r) = f_K(r) \\ G_{JIK}(\theta) = 1 \end{cases} \tag{15}$$

where $\phi_{IJ}(r)$ is a pair function, $f_J(r)$ is an atomic electron density function, $F_I(X)$ is the embedding energy function, and X is used to represent electron density ($X = \rho$). We have implemented this polymorphic potential model in the public parallel MD code, LAMMPS (Plimpton, 1995). We plan to release the code in LAMMPS package after the present paper is published. Interested readers can contact us to get the code prior to the formal release of the code.

4. Parameterization

To enable structure evolution of ionic materials to be studied within the short time scales of MD simulations, we attempt to accelerate ionic migration by external electric fields. Such electric fields are approximated by applying opposite external forces to cations and anions. To ensure that the interatomic potential is robust enough to maintain the equilibrium crystal structure of TlBr when Tl+ and Br^- ions are subject to large external forces, we require that the potential not only best captures the key experimental properties of the observed Tl, Br, and TlBr phases (e.g., Tl-hcp, TlBr-CsCl, etc.), but also predicts the crystalline growth of the ground state structures during MD simulations of growth (e.g., vapor deposition). Note that a potential is said to be incapable of crystalline growth simulations only when it predicts amorphous growth at all combinations of temperatures and growth rates possible with the MD simulations. At low temperatures and high growth rates where earlier adatoms are buried by later adatoms before the surface reaches a low energy configuration, amorphous growth is the correct prediction. The growth simulation tests are important because they sample a variety of configurations (at the growth surface) not considered a prior. If any of the random nuclei formed on the growth surface has a lower energy than the growth crystal, the simulations is likely to always give an amorphous growth regardless of temperature and growth rate. Hence, crystalline growth provides strong validation that the growth crystal has the lowest energy compared to any other configurations. When the equilibrium TlBr crystal has the lowest energy, large external forces can be applied to Tl+ and Br^- ions without causing phase transformation. Hence, the growth simulation capability is essential for our applications.

We proceed by parameterizing first the Tl and Br potentials, and then the TlBr potential at the known Tl and Br parameters. The observed room-temperature equilibrium phases are hcp for Tl (Donnay & Ondik, 1973), diatomic (Br_2) liquid for Br, and CsCl for TlBr (Donnay & Ondik, 1973). Note that solid Br has an orthorhombic crystal structure; however, Br_2 liquid is the stable room-temperature phase. Like many other potentials, our MSW model is not intended to capture the Br_2 molecules (it is possible to capture the Br_2 molecules, but this does not necessarily result in a better potential). On the other hand, our density function theory (DFT) calculations, employing the optB86b-vdW functional (see Appendix for computational details), indicated that the sc Br phase has a lower cohesive energy than dc, bcc, fcc, and hcp phases. Hence, we target sc as the lowest energy lattice phase for Br (at 0 K) while at the same time ensure that the stable Br phase at room temperature is liquid. Note that if a correct negative heat of formation of the TlBr compound is captured, elemental phases do not form in MD simulations under stoichiometric conditions. Because they do not form, particular elemental structures are not important for studying stoichiometric compounds.

Lattice constants, cohesive energies, and elastic constants for the model lowest energy lattices (Tl-hcp, Br-sc, and TlBr-CsCl) are fitted under the constraints that the energies of all the other phases (e.g., dc, bcc, fcc, NaCl, wz, etc.) are higher than the targeted lowest energy phases. For the experimentally observed structures such as Tl-hcp and TlBr-CsCl, the experimental lattice constants (Donnay & Ondik, 1973), cohesive energies (Barin, 1993), and elastic constants (Simmons, 1965; Morse & Lawson, 1967) are directly used as the target values for the fitting. For the phases that are not observed (e.g., Br-sc), the available experimental properties of other phases and DFT results are used to guide the selection of the target values. In particular, DFT results may be quite different from the experimental values. For example, the hcp Tl cohesive energy obtained from experiments and DFT calculations is -1.85 and -2.40 eV/atom respectively. As a result, we scale the DFT results so that for the observed structures, the scaled DFT results match the experiments.

The software package Mathematica (Wolfram, 2004) is used to perform the parameterization. To promote global optimization, four different numerical optimization routines, namely a conjugate gradient method (Hestenes & Stiefel, 1952), the downhill simplex method of Nelder and Mead (Olsson & Nelson, 1975), a genetic algorithm (Storn & Price, 1997), and a biased random walk (simulated annealing) (Kirkpatrick, Gelatt, & Vecchi, 1983), are all used to determine the parameters that minimize the weighted mean-square deviation between the target and predicted properties. The goal to capture the crystalline growth is more challenging, requiring a highly iterative parameterization process. After each fitting iteration, the four sets of parameters from the four optimization routines are tested for vapor deposition simulations. If the potential does not pass these tests, the entire process is repeated with an appropriate adjustment of parameters bounds, target structures and target properties. The iterations continue until one of the four optimization routines results in a satisfactory set of potential parameters. The MSW potential thus determined are listed in Tables I and II for two-body, and three-body parameters respectively.

Table 1. Two-body parameters of MSW potential (length in Å and energy in eV)

IJ	$r_{0,IJ}$	$r_{s,IJ}$	$r_{c,IJ}$	$E_{b,IJ}$	$\alpha_{,IJ}$	$\beta_{,IJ}$	$\gamma_{,IJ}$
TlTl	3.02409	3.97219	4.26900	-0.514663	5.60012	3.39397	1.94224
BrBr	2.83451	3.60000	4.05800	-0.336014	7.00000	4.37500	3.00000
TlBr	2.98521	3.90480	4.73600	-0.747107	6.20000	2.00000	2.20000

Table 2. Three-body parameters of MSW potential (length in Å and energy in eV)

IJK	TlTlTl	BrBrBr	TlTlBr (=BrTlTl)	TlBrBr (=BrBrTl)	TlBrTl	BrTlBr
λ_{JIK}	0.05000	0.13144	0.11804	0.05733	2.38107	2.38107
$\cos\theta_{0,JIK}$	0.07257	-0.50000	-0.53735	-0.54059	-0.29333	-0.29333
$\cos\theta_{1,JIK}$	-1.00000	-1.00000	-1.00000	-1.00000	-1.00000	-1.00000
$\cos\theta_{2,JIK}$	0.60000	0.60000	0.53947	0.56717	0.29333	0.29333
$\xi_{0,JIK}$	3.50000	0.62340	1.44291	2.79355	0.25182	0.25182
$\xi_{1,JIK}$	0.18625	2.60000	0.70848	1.42550	0.24026	0.24026
$\xi_{2,JIK}$	0.00000	0.00000	1.72178	2.04092	0.46849	0.46849
$\kappa_{1,JIK}$	0.20957	0.96000	1.00000	0.75390	1.00000	1.00000
$\kappa_{2,JIK}$	0.00000	0.00000	0.00000	0.27602	0.94796	0.94796

5. Evaluation of the Potential

5.1 Lattice Constants and Cohesive Energies of Various Phases

Our objective is to study the equilibrium TlBr crystal under the stoichiometric condition where no other phases occur, so that the critical properties are for the TlBr-CsCl crystal. With this in mind, calculations based on the polymorphic potential model are performed to evaluate our TlBr potential. First, lattice constants and cohesive energies of various Tl, Br, and TlBr lattices are calculated using energy minimization simulations. The results are summarized in Table III along with the available experimental data and our DFT values.

Table 3. Lattice constants a (for dimer, a refers to the dimer bond length) and c (Å), and cohesive energy E_c (eV/atom), obtained from the MSW potential, DFT calculations, and experiments for selected material structures

material	structure	MSW			DFT (HSE06)			Exp.		
		a	c	E_c	a	c	E_c	a (Donnay & Ondik, 1973)	c (Donnay & Ondik, 1973)	E_c (Barin, 1993)
Tl	dc	7.180	-----	-0.999	7.012	----	-1.526	-----	-----	-----
	sc	3.012	-----	-1.800	3.214	----	-1.827	-----	-----	-----
	bcc	3.854	-----	-1.805	3.971	----	-1.828	-----	-----	-----
	fcc	4.825	-----	-1.848	4.940	----	-1.927	-----	-----	-----
	hcp	3.408	5.582	-1.850	3.558	5.674	-1.874	3.450	5.520	-1.850
Br	di (gas)	-----	-----	-----	2.272	----	-1.132	-----	-----	-0.986
	dc	6.591	-----	-0.731	7.078	----	-0.405	-----	-----	-----
	fcc	4.878	-----	-0.748	4.803	----	-0.281	-----	-----	-----
	bcc	3.847	-----	-0.780	3.822	----	-0.361	-----	-----	-----
	hcp	3.328	5.743	-0.787	2.919	----	-0.840	-----	-----	-----
	sc	3.004	-----	-0.827	3.174	----	-0.459	-----	-----	-1.134*
TlBr	wz	4.877	7.964	-1.657	4.994	8.371	-2.473	-----	-----	-----
	zb	6.897	-----	-1.657	7.494	----	-2.450	-----	-----	-----
	NaCl	6.197	-----	-2.239	6.762	----	-2.529	-----	-----	-----
	CsCl	3.985	-----	-2.389	4.032	----	-2.487	3.985	-----	-2.389

*: Br-sc is the lowest energy lattice in models. No experimental data is available for the metastable Br-sc phase. Instead, we list the experimental cohesive energy of the lowest energy phase (Br_2-liquid) as a reference.

Comment: In Tables 3 and A-I, use horizontal lines to separate Tl, Br, and TlBr data. In the present version, it is not clear which structures are for Tl, Br, and TlBr.

Table III indicates that of the lattice structures explored, our MSW potential captures the Tl-hcp and TlBr-CsCl lattice crystals as the lowest energy phases, in agreement with experiments that Tl-hcp and TlBr-CsCl phases are observed at room temperature. Table III also indicates that our MSW potential captures the lowest energy for the Br-sc crystal, which is really the target of the MSW model. We cannot include the Br_2 liquid observed in experiments in Table III. Instead, we will show below through growth simulations that our potential gives Br liquid as the most stable phase at room temperature. Here we can see that in addition to capturing appropriate lowest energy phases, our MSW model also reproduces the experimental cohesive energies of Tl-hcp and TlBr-CsCl. Our energy of -0.827 eV/atom for Br-sc is not unreasonable compared with the experimental cohesive energy of -1.134 eV/atom for a more stable Br_2-liquid, especially considering that this difference between MSW and experiments is not much more significant than that between DFT and experiments, not to mention that the elemental Br phase does not occur for our intended problem. Note that MSW model also accurately captures the experimental lattice constant of TlBr-CsCl. The lattice constants of Br-sc and Tl-hcp as determined by our MSW is not far off from the experimental or DFT values either (in case of Tl-hcp, for example, while the lattice constant a is slightly under-estimated, the lattice constant c is slightly over-estimated resulting in a good description of atomic volume).

Capturing non-tetrahedral crystals (such as hcp, sc, CsCl) as the lowest energy phases proves that the MSW potential is more flexible than the conventional SW potentials. SW potentials are not designed for capturing property trends of a variety of metastable structures. The purpose of listing some selected metastable phases in Table III is to show that the equilibrium phases have lower energies than these metastable phases. Interestingly, however, our MSW potential reproduces the DFT order of Tl structures with increasing (more negative) cohesive energies as in dc $\rightarrow$ sc $\rightarrow$ bcc $\rightarrow$ fcc. Furthermore, our MSW model correctly captures the lowest energy for Tl-hcp whereas the DFT does not. This observation suggests that MSW potential can possess good transferability to a variety of other phases. Usually such transferability is only possible with more sophisticated potentials. Since our MSW is designed for studying the TlBr bulk, the transferability to many metastable phases is not relevant and so we do not exploit this further for the Br and TlBr phases which would necessarily sacrifice the properties of the equilibrium phases.

Finally, we point out that Table III only explores limited number of phases and therefore does not prove that the predicted lowest energy phases indeed have the lowest energies as compared to any other configurations. As will be discussed below, we will prove this rigorously through vapor deposition simulations.

5.2 Elastic Constants and Melting Temperature of the Observed Tl and TlBr Phases

Elastic constants and melting temperature of the experimentally observed Tl-hcp and TlBr-CsCl phases are calculated, and the results are summarized in Table IV along with the experimental values. Here the melting temperature calculations follow the same method described previously (Zhou et al., 2013; Zhou, Foster, Van Swol, Martin, & Wong, 2014; Ward, Zhou, Wong, Doty, & Zimmerman, 2012; Ward, Zhou, Wong, Doty, & Zimmerman, 2012; Ward, Zhou, Wong, & Doty, 2013). It can be seen that our MSW model reproduces the experimental elastic constants for TlBr-CsCl. The model elastic constants for Tl-hcp are generally lower than the experimental values. Note that we could have fit exactly the experimental elastic constants for Tl-hcp. However, this would result in a melting temperature that is significantly above the experimental value. On the other hand, the conventional SW potentials tend to significantly overestimate elastic constants for closely packed elements and hence the ability to prescribe low elastic constants is significant for the MSW model.

Table 4. Elastic constants C_{11}, C_{12}, C_{13}, C_{33}, C_{44}, bulk modulus B (eV/Å^3), and melting temperature T_m (K) of the observed Tl and TlBr crystalline phases, obtained from the MSW potential and experiments

structure	method	C_{11}	C_{12}	C_{13}	C_{33}	C_{44}	B	T_m
Tl	MSW	0.240	0.140	0.137	0.242	0.053	0.133	691
(hcp)	Exp.	0.277	0.235	0.187	0.376	0.055	0.239	577
	(Simmons, 1965; Afinogenov & Larionova, 1992)							
TlBr	MSW	0.275	0.104	0.104	0.275	0.067	0.161	1442
(CsCl)	Exp.	0.275	0.104	0.104	0.275	0.067	0.161	753
	(Morse & Lawson, 1967; Kim et al., 2011)							

We get the melting temperature of Tl-hcp reasonably close to the experimental value. Despite a prolonged effort, our current parameterization of the MSW potential still significantly overestimates the melting temperature of TlBr-CsCl. We can better capture the melting temperature by reducing the elastic constants. Considering that our objective is to study the behavior of the TlBr-CsCl crystal at ambient temperature, we chose to capture exactly the elastic constants at the cost of overestimating the melting temperature. We feel that this problem cannot be resolved by parameterization alone, and further modifications of the MSW format are needed in future efforts in order to capture both elastic constants and melting temperature.

D. Point Defects

Native point defects in TlBr-CsCl, including Tl vacancy V_{Tl}, Br vacancy V_{Br}, Tl at Br antisite Tl_{Br}, Br at Tl antisite Br_{Tl}, Tl interstitial between Tl sites $Tl_{i,1}$, Tl interstitial between Br sites $Tl_{i,2}$, Br interstitial between Tl sites $Br_{i,1}$, and Br interstitial between Br sites $Br_{i,2}$, are all studied. In particular, intrinsic defect energies $E_D^{'}$ are calculated as (Zhang & Northrup, 1991; Northrup & Zhang, 1993)

$$E_D^{'} = E - \left(N_{Tl} + N_{Br}\right) \cdot E_{TlBr} - 0.5 \cdot \left(N_{Tl} - N_{Br}\right) \cdot \left(E_{Tl} - E_{Br}\right) \quad (16)$$

where E, N_{Tl}, and N_{Br} are total energy, number of Tl atoms, and number of Br atoms of the system containing the defect, and E_{TlBr}, E_{Tl}, and E_{Br} are cohesive energies (per atom unit) for the lowest energy phases of TlBr, Tl, and Br respectively. The results obtained from MSW and DFT calculations are summarized in Table V. It can be seen that MSW potential predicts a lower energy for Tl vacancy than for Br vacancy, a low energy for Br at Tl antisite than for Tl at Br antisite, and a lower energy for $Br_{i,2}$ interstitial than for both Tl interstitials. All these are in good agreement with the DFT calculations. Our MSW potential indicates that the $Br_{i,1}$ interstitial has a lower energy than both Tl interstitials, which differs from the DFT result that the $Br_{i,1}$ interstitial energy is only lower than the $Tl_{i,2}$ interstitial energy. Again here we only compare the trends, and do not make conclusions on the absolute values due to the lack of experimental data.

Table 5. Intrinsic defect energy $E_D^{'}$ (eV) obtained from different models for TlBr-CsCl. The DFT results were determined using the HSE06 functional based on the optB86b-vdW optimized geometries (see Appendix for details)

method	V_{Tl}	V_{Br}	Tl_{Br}	Br_{Tl}	$Tl_{i,1}$	$Tl_{i,2}$	$Br_{i,1}$	$Br_{i,2}$
MSW	1.04	1.29	2.75	1.05	2.93	3.03	2.00	2.00
DFT	1.24	1.81	3.59	3.02	1.85	2.82	2.69	1.74

6. Molecular Dynamics Vapor Deposition Verifications

As mentioned above, only when a potential captures the crystalline growth during MD vapor deposition simulations will it capture the lowest energy phase and be robust enough to allow applications of high external electric fields. Here we perform vapor deposition simulations to validate that our TlBr MSW potential captures the crystalline growth of the lowest energy Tl-hcp and TlBr-CsCl crystals. In addition, we will demonstrate that our potential gives the crystalline growth of Br-sc at a low temperature (this is the lowest energy "model" phase as opposed to the orthorhombic experimental lowest energy phase) but gives a liquid Br structure as the stable phase at room temperature.

6.1 Tl-hcp Growth

For Tl-hcp growth, an initial substrate of an hcp crystal containing 1008 Tl atoms with 28 $(2\bar{1}\bar{1}0)$ layers in the x direction, 9 (0002) layers in the y direction, and 8 $(0\bar{1}10)$ layers in the z direction is used. Here layers refer to crystallographic planes so that one (0001) layer is equivalent n (000n) layers etc. The substrate temperature is set at T = 300 K by assigning velocities to atoms according to the Boltzmann distribution. During simulations, the bottom (-y) 2 (0002) layers are held fixed to prevent crystal shift upon adatom impact on the top surface. The next 3 (0002) layers are isothermally controlled at the substrate temperature. This leaves the top 4 layers free where the motion of atoms is solely determined by Newton's law. Injection of Tl adatoms from random locations far above the surface simulates the growth. All adatoms have an initial far-field incident kinetic energy $E_i = 0.05$ eV and an incident angle $\theta = 0^\circ$ (i.e., the moving direction is perpendicular to the surface). The adatom injection frequency is chosen to give a deposition rate of R = 2.5 nm/ns. To approximately maintain a constant thickness of the free surface region, the isothermal region expands upward during simulations. Since surface roughness might develop, the isothermal region expands at about 80% of the surface growth rate to ensure that the upper boundary of the isothermal region never exceeds the surface valley locations. Fig. 1 shows the resulting configuration obtained after 0.66 ns deposition, where the original substrate is shaded in purple. It can be seen that the MSW potential correctly captures the crystalline growth of the Tl-hcp phase. This strongly validates that Tl-hcp has the lowest energy at room temperature as compared to any other configurations.

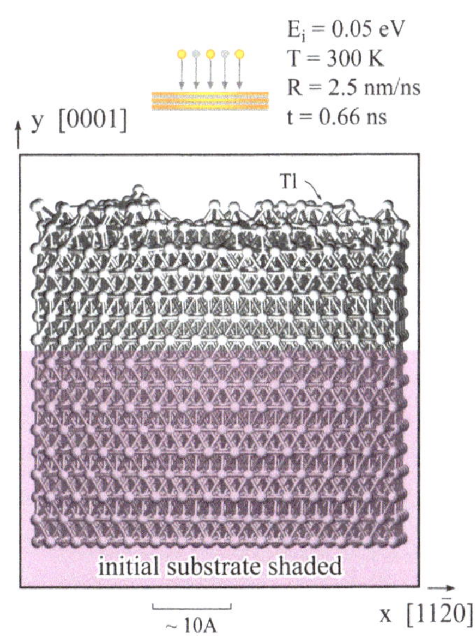

Figure 1. Vapor deposited hcp Tl film obtained from MD simulations

6.2 Br-sc Growth

MD simulations are also performed to grow Br on an Br-sc substrate. A sc crystal containing 1008 Br atoms with 14 (100) layers in the x direction, 9 (010) layers in the y direction, and 8 (001) layers in the z direction is used as the initial substrate. During simulations, the bottom 2 (010) layers are held fixed. The next 4 (010) layers are controlled at the desired growth temperature. Using the same approach as described above, the growth simulation is performed at two substrate temperatures T = 150 K and T = 300 K, an incident energy E_i = 0.05 eV, an incident angle $\theta = 0^o$, and a deposition rate R = 2.5 nm/ns. The resulting configurations obtained after 1.20 ns deposition are shown in Figs. 2(a) and 2(b) respectively for the 150 K and 300 K temperatures. It can be seen from Fig. 2(a) that the sc- crystalline growth is achieved with our potential, strongly validating that the Br-sc crystal has the lowest (free) energy at 150 K as compared to any other configurations. Fig. 2(b), on the other hand, shows that amorphous growth is achieved at 300 K. This strongly validates that Br exhibits a liquid type phase at room temperature.

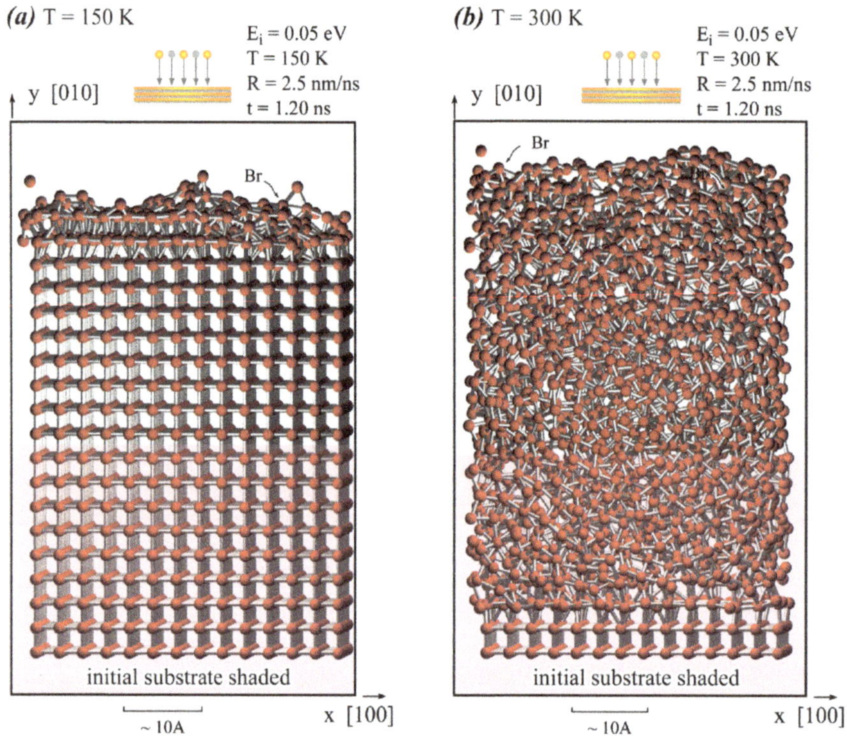

Figure 2. MD simulations of Br growth on a Br-sc substrate at a temperature of (a) 150 K and (b) 300 K. Our potential prescribes a liquid Br as the most stable phase at room temperature

6.3 TlBr-CsCl Growth

For TlBr-CsCl growth, an initial TlBr substrate of a CsCl type of crystal containing 300 Tl atoms and 300 Br atoms with 20 (200) layers in the x direction, 10 (020) layers in the y direction, and 12 (002) layers in the z direction is used. Initially, Br terminates the top y surface. During the simulations, the bottom 3 (020) layers are fixed. The next 4 (020) layers are used to control the growth temperature. Following the same approach as described above, the growth simulations are performed at a substrate temperature T = 700 K, an incident energy E_i = 0.05 eV, an incident angle $\theta = 0^o$, a deposition rate R = 2.75 nm/ns, and a stoichiometric vapor flux ratio Tl:Br = 1:1. Fig. 3 shows the system configuration obtained at 0.78 ns deposition time. It is seen again that our MSW potential correctly captures the crystalline growth of the equilibrium (CsCl) phase of TlBr. In particular, the randomly injected Tl and Br atoms are reconstructed correctly to their corresponding sublattices. Because the potential captures the crystallization from a rather stochastic vapor phase, it enables robust simulations of structural evolution of TlBr-CsCl under conditions where the presence of both dislocations and external electric fields may induce configuration disorders if the potential has any deficiencies in capturing the lowest energy phase.

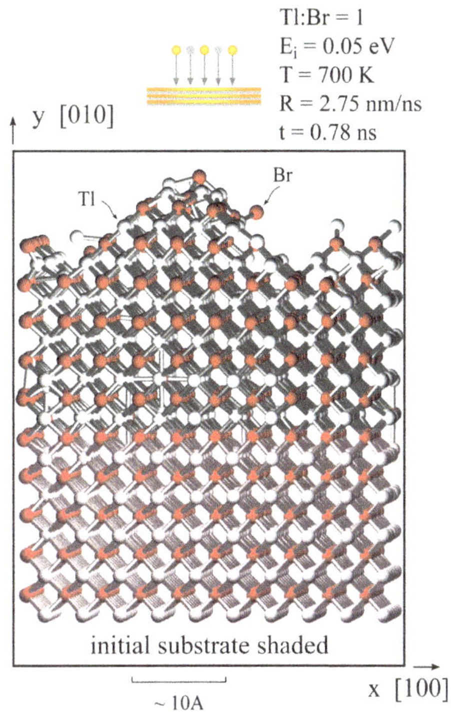

Figure 3. Vapor deposited CsCl phase of a TlBr film obtained from MD simulations

7. Molecular Dynamics Structural Evolution Under Fields

Our objective is to allow direct MD simulations of structural evolution of TlBr-CsCl crystal under an external electric field. The external electric field can be simulated by applying opposite biased forces to Tl and Br atoms. Note that in our model, we do not directly address charges. This is a reasonable approximation because charges only give two forces: the Coulomb forces between atoms, and biased forces under external fields. The Coulomb forces between atoms are digested into the interatomic potential in our model. This allows us to use biased forces to independently simulate the external electric field.

The simulations may become challenging at large external electric fields because the biased forces may induce phase transformation when the potential does not capture the lowest energy for the equilibrium phase. Here we demonstrate two cases to demonstrate that our potential allows simulations to be performed when atoms are subject to large forces of ± 0.4 eV/Å. These forces correspond to a high electric field of 4×10^6 V/mm assuming Tl and Br atoms adopt full charges of ± 1 e. Again note that atoms are not strictly point charges and atoms in perfect bulk may not be subject to big forces from the external field. Nonetheless, the model becomes robust if we pass this test.

A TlBr-CsCl crystal containing 16128 Tl atoms and 16128 Br atoms with 84 (110) layers in the x direction, 24 $(\bar{1}10)$ layers in the y direction, and 32 (002) layers in the z direction is used. To prevent system from shifting, the bottom region of about 10 Å wide is fixed. To remove the boundary effects, periodic boundary conditions are used in all three coordinate directions. An MD simulation is then performed at 1200 K (a homologous temperature 0.832 T_m) with a biased force of ± 0.4 eV/Å using the NVT ensemble (i.e., number of atoms, volume, and temperature are all constant).

In the first case, we assume that the system contains a pair of Tl and Br vacancies by removing a Tl and a Br atom far away from the fixed region. Comparison of atomic configurations between a time span of 0.06 ns is shown in Fig. 4. Fig. 4 verifies that our potential allows for stable TlBr-CsCl MD simulations to be performed with a high biased force. Interestingly, we found that Tl and Br vacancies are not very mobile even at the high biased force and temperature. In fact, during the 0.06 ns span, Tl vacancy jumped by one lattice spacing whereas the Br vacancy did not jump.

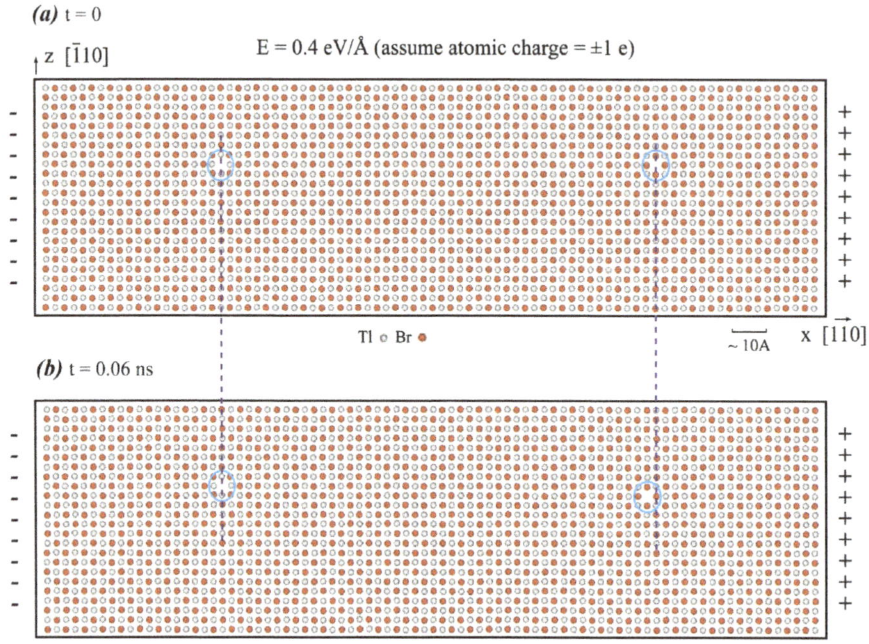

Figure 4. Structure evolution of a TlBr-CsCl crystal containing vacancies at a temperature of 0.832 T_m and a biased force of ±0.4 eV/Å: (a) Starting time of observation; and (b) 0.06 ns later. The TlBr-CsCl crystal remains intact at the large electric field. During the 0.06 ns span, Tl vacancy jumped by one lattice spacing whereas the Br vacancy did not jump

In the second case, we assume that the system contains a pair of Tl and Br interstitials by inserting a Tl atom and a Br atom at locations far away from the fixed region. Comparison of atomic configurations between a time span of 0.12 ns is shown in Fig. 5. Fig. 5 again verifies that our potential allows stable TlBr-CsCl MD simulations to be performed with a high biased force. Unlike the vacancy case, we found that Tl and Br interstitials are very mobile, and both interstitials moved a significant distance during the 0.12 ns time span.

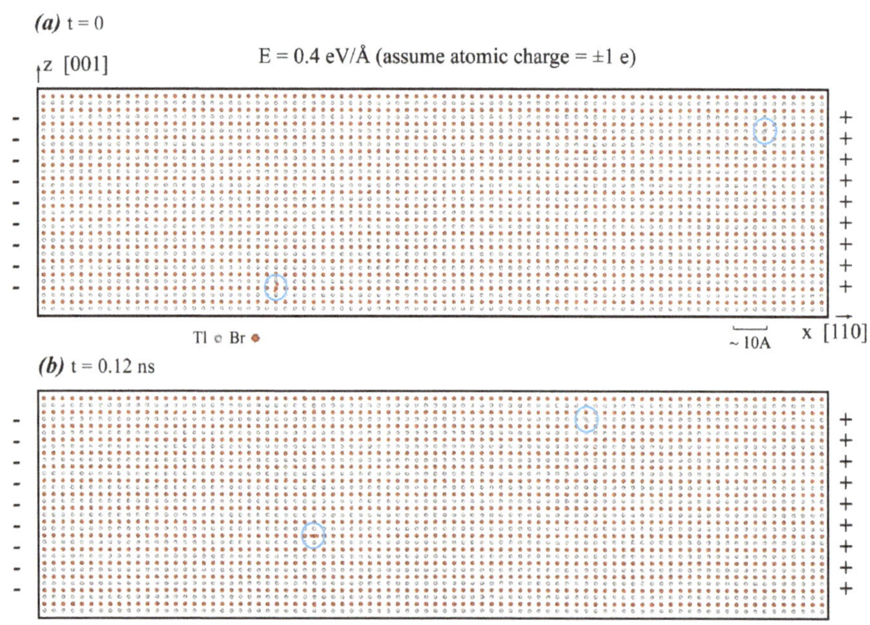

Figure 5. Structure evolution of a TlBr-CsCl crystal at a temperature of 0.832 T_m and a biased force of ±0.4 eV/Å: (a) Starting time of observation; and (b) 0.12 ns later. The TlBr-CsCl crystal remains intact at the large electric field. During the 0.12 ns span, Tl and Br interstitials migrated significant distances

The cases shown in Figs. 4 and 5 are designed to validate the robustness of our potential, but not to launch a thorough study of structure deterioration under external fields. We are currently using our model to study how defects (dislocations, vacancies, interstitials) evolve under external electric fields, and how diffusion of various species contribute to mass transport and changes of stoichiometry of the material. These studies are beyond the scope of this paper and will therefore be published separately.

8. Conclusions

Herein we have developed a new modified Stillinger-Weber potential. Unlike the conventional SW potential that significantly overestimates the elastic constants of closely packed elements and are limited mainly to tetrahedral structures, our modified potential can capture very low elastic constants for closely packed elements and can be used for many non-tetrahedral crystal structures. We have parameterized the modified SW potential for TlBr. Through rigorous vapor deposition simulation tests, we have demonstrated that our potential captures the experimental properties of the observed Tl and TlBr phases, and predicts the Br liquid to be the most stable phase at room temperature. Moreover, we have demonstrated that our potential is robust enough for challenging simulations of structure evolution of TlBr crystals under very high external electric fields. Test simulations indicate that interstitials migrate much faster than vacancies. We have also developed a polymorphic potential model and implemented it in the public MD code, LAMMPS. This essentially enables future material research to be performed at a higher fidelity level because improved potentials no longer require modification of the MD codes and therefore can immediately be utilized by a broader materials and physics community.

Acknowledgements

Sandia National Laboratories is a multiprogram laboratory managed and operated by Sandia Corporation, a wholly owned subsidiary of Lockheed Martin Corporation, for the US Department of Energy's National Nuclear Security Administration under contract DE-AC04-94AL85000. This work was performed under a Laboratory Directed Research and Development (LDRD) project.

Appendix

Density functional theory (DFT) calculations were performed within the spin-polarization formalism using the optB86b-vdW functional (Klimeš, Bowler, & Michaelides, 2011), and the hybrid HSE06 (Heyd, Scuseria, & Ernzerhof, 2003) functional with and without the inclusion of the empirical dispersion correction developed by Grimme (Grimme, 2006) (the one with the dispersion correction is notated as HSE06+D). The calculations were carried out using projector-augmented-wave (PAW) pseudopotentials and a plane-wave energy cutoff of 500 eV. The atom positions and primitive unit cells of the bulk materials were optimized using the three methods (optB86b-vdW, HSE06, and HSE06+D) until all forces were less than 0.01 eV/Å. The Brillouin zone was sampled using a 10×10×10 and a 5×5×5 gamma-centered Monkhorst-Pack grid respectively. The optimized lattice parameters and cohesive energies are reported in Table A-I. For the point defect calculations (see Table A-II), 5×5×5 supercells were created from the optB86b-vdW optimized TlBr-CsCl primitive unit cell and the systems were relaxed (atoms only) until all forces were less than 0.01 eV/Å. For these calculations, the Monkhorst-Pack grid was reduced to 2×2×2. The point defect energies ($E_D^{'}$) are reported in Table A-II. The point defect energies were also determined at the HSE06 level of theory using the optB86b-vdW optimized geometries and a 1×1×1 k-point grid. Since the point defect structures were based on the optB86b-vdW optimized geometries, the cohesive energies of the lowest energy bulk materials, as needed for the determination of $E_D^{'}$ (see Equation 16), were also determined based on the optB86b-vdW optimized geometries. This was done to avoid the inclusion of lattice deformation energy in the point defect energy. All calculations were performed using VASP 5.3.5 (Kresse, & Hafner, 1993; Kresse & Hafner, 1994; Kresse & Furthmüller, 1996; Kresse & Furthmüller, 1996).

The calculations herein were carried out using different methodologies for a couple reasons. First, we wanted to consider methods (optB86b-vdW and HSE06+D) capable of modeling dispersion interactions (van der Waals forces) since bromine molecules (Br_2) are held together by such interactions. Although, calculations were not explicitly performed on such a system, we found that the inclusion of van der Waals forces had significant effects on the cohesive energies of the bulk Br structures (see Table A-I). Second, Klimeš et al. (Klimeš, Bowler, & Michaelides, 2011) showed that the optB86b-vdw functional outperforms several traditional generalized-gradient approximation (GGA) functionals (e.g. PBE) for the prediction of lattice constants, bulk moduli, and atomization energies for a range of solids and we wanted to identify a GGA based functional capable of accurately predicting the lattice constants of TlBr-CsCl. This was desirable because performing hybrid DFT optimizations on large supercells, required for studying point-defects, is extremely computationally demanding. However, hybrid functionals are arguably more accurate than traditional GGA functionals accredited to the inclusion of exact Fock exchange. The results obtained using the optB86b-vdW functional are in relative good agreement with those obtained using the HSE06+D method. Moreover, the lattice constants obtained for TlBr-CsCl and Tl-hcp agree very well with experiment (~1%), and even slightly better than those obtained using the HSE06 and HSE06+D methods. Note, the PBE functional was also considered but the predicted lattice constants were in greater disagreement with experiment. Regarding the prediction of the cohesive energies, the HSE06 functional performs the best. This is attributed to a decrease in self-interaction error as a result of using a functional that incorporates exact Fock exchange. We have decided to included only the HSE06 values in the main text for internal consistency and simplicity; however, we have included the results obtained using the different methods for completeness and for interested readers.

Table A-I. Lattice constants a (for dimer, a refers to the dimer bond length) and c (Å), and cohesive energy E_c (eV/atom) determined using optB88b-vdw, HSE06, and HSE06+D functionals

material	structure	optB86b-vdw			HSE06			HSE06+D		
		a	c	E_c	a	c	E_c	a	c	E_c
Tl	dc	6.965	----	-1.922	7.012	----	-1.526	6.681	----	-1.899
	sc	3.185	----	-2.238	3.214	----	-1.827	3.082	----	-2.371
	bcc	3.868	----	-2.368	3.971	----	-1.828	3.973	----	-2.294
	fcc	4.873	----	-2.340	4.940	----	-1.927	4.868	----	-2.349
	hcp	3.500	5.581	-2.399	3.558	5.674	-1.874	3.521	5.614	-2.289
Br	di (gas)	2.309	----	-1.265	2.272	----	-1.132	2.280	----	-1.132
	dc	6.898	----	-0.961	7.078	----	-0.405	7.031	----	-0.489
	fcc	4.644	----	-1.123	4.803	----	-0.281	4.852	----	-0.496
	bcc	3.703	----	-1.160	3.822	----	-0.361	3.820	----	-0.580
	hcp	2.945	7.049	-1.210	2.919	----	-0.840	2.902	----	-0.947
	sc	3.006	----	-1.218	3.174	----	-0.459	3.150	----	-0.612
TlBr	wz	4.682	8.724	-2.767	4.994	8.371	-2.473	4.897	8.209	-2.686
	zb	7.371	----	-2.662	7.494	----	-2.450	7.371	----	-2.598
	NaCl	6.622	----	-2.866	6.762	----	-2.529	6.616	----	-2.794
	CsCl	3.959	----	-2.880	4.032	----	-2.487	4.066	----	-2.822

Table A-II. Intrinsic defect energies $E_D^{'}$ (eV) for TlBr-CsCl determined using optB88b-vdw, HSE06, and HSE06+D functionals

method	V_{Tl}	V_{Br}	Tl_{Br}	Br_{Tl}	$Tl_{i,1}$	$Tl_{i,2}$	$Br_{i,1}$	$Br_{i,2}$
optB86b-vdw	0.66	1.56	3.47	1.72	2.16	2.55	2.32	1.59
HSE06	1.24	1.81	3.59	3.02	1.85	2.82	2.69	1.74
HSE06+D	1.51	2.24	3.79	2.66	1.70	2.66	2.35	1.70

Note

Note 1. For convenience, we will use abbreviation to represent structures in the following: di: dimer; dc: diamond-cubic; sc: simple-cubic; bcc: body-centered-cubic; fcc: face-centered-cubic; hcp: hexagonal-closely-packed; zb: zinc-blende; wz: wurtzite; NaCl: B1; and CsCl: B2.

References

Afinogenov, Y., & Larionova, T. (1992). Triple intersystem of 3TlBr + Al = $AlBr_3$ + 3Tl displacement. *Zhurnal Neorganicheskoi Khimii, 37*(11), 2606-2608.

Albe, K., Nordlund, K., Nord, J., & Kuronen, A. (2002). Modeling of compound semiconductors: Analytical bond-order potential for Ga, As, and GaAs. *Physical Review B, 66*(3), 035205. http://dx.doi.org/10.1103/PhysRevB.66.035205

Barin, I. (1993). *Thermochemical Data of Pure Substances*. Wiley-VCH.

Beckers, J. V. L., Van Der Bent, K. J., & De Leeuw, S. W. (2000). Ionic conduction in Na+-β-alumina studied by molecular dynamics simulation. *Solid State Ionics, 133*(3), 217-231. http://dx.doi.org/10.1016/S0167-2738(00)00685-8

Chung, R. W. J. M. H. F., & De Leeuw, S. W. (2004). Ionic conduction in LiI–α, γ-alumina: molecular dynamics study. *Solid state ionics, 175*(1), 851-855. http://doi:10.1016/j.ssi.2004.09.047

Daw, M. S., & Baskes, M. I. (1984). Embedded-atom method: Derivation and application to impurities, surfaces, and other defects in metals. *Physical Review B, 29*(12), 6443-6453. http://dx.doi.org/10.1103/PhysRevB.29.6443

Donnay, J. D. H., & Ondik, H. M. (1973). *Crystal data, determinative tables* (3rd ed., Vol. 2, inorganic compounds). U. S. Department of Commerce, National Bureau of Standards, and Joint Committee on Power Diffraction Standards, U. S. A.

Grimme, S. (2006). Semiempirical GGA - type density functional constructed with a long - range dispersion correction. *Journal of computational chemistry, 27*(15), 1787-1799. http://dx.doi.org/10.1002/jcc.20495

Hestenes, M. R., & Stiefel, E. (1952). Methods of conjugate gradients for solving linear systems. *J. Research Nat. Bur. Standards, 49*(6), 409-436. http://dx.doi.org/10.6028/jres.049.044

Heyd, J., Scuseria, G. E., & Ernzerhof, M. (2003). Hybrid functionals based on a screened Coulomb potential. *The Journal of Chemical Physics, 118*(18), 8207-8215. http://dx.doi.org/10.1063/1.1564060

Hitomi, K., Kikuchi, Y., Shoji, T., & Ishii, K. (2009). Polarization phenomena in TlBr detectors. *Nuclear Science, IEEE Transactions on, 56*(4), 1859-1862. http://dx.doi.org/10.1109/TNS.2009.2013349

Hitomi, K., Matsumoto, M., Muroi, O., Shoji, T., & Hiratate, Y. (2002). Thallium bromide optical and radiation detectors for X-ray and gamma-ray spectroscopy. *Nuclear Science, IEEE Transactions on, 49*(5), 2526-2529. http://dx.doi.org/10.1109/TNS.2002.803851

Huang, C. W., Wei, W. C. J., Chen, C. S., & Chen, J. C. (2011). Molecular dynamics simulation on ionic conduction process of oxygen in $Ce_{1-x}M_x O_{2-x/2}$. *Journal of the European Ceramic Society, 31*(16), 3159-3169. http://dx.doi.org/10.1016/j.jeurceramsoc.2011.05.029

Kim, H., Churilov, A., Ciampi, G., Cirignano, L., Higgins, W., Kim, S., … Shah, K. (2011). Continued development of thallium bromide and related compounds for gamma-ray spectrometers. *Nuclear Instruments and Methods in Physics Research Section A: Accelerators, Spectrometers, Detectors and Associated Equipment, 629*(1), 192-196. http://dx.doi.org/10.1016/j.nima.2010.10.097

Kirkpatrick, Gelatt, C. D., S., & Vecchi, M. P. (1983). Optimization by simmulated annealing. *Science, 220*(4598), 671-680. http://dx.doi.org/10.1126/science.220.4598.671

Klimeš, J., Bowler, D. R., & Michaelides, A. (2011). Van der Waals density functionals applied to solids. *Physical Review B, 83*(19), 195131. http://dx.doi.org/10.1103/PhysRevB.83.195131

Kresse, G., & Furthmüller, J. (1996). Efficiency of ab-initio total energy calculations for metals and semiconductors using a plane-wave basis set. *Computational Materials Science, 6*(1), 15-50. http://dx.doi.org/10.1016/0927-0256(96)00008-0

Kresse, G., & Furthmüller, J. (1996). Efficient iterative schemes for ab initio total-energy calculations using a plane-wave basis set. *Physical Review B, 54*(16), 11169-11186. http://dx.doi.org/10.1103/PhysRevB.54.11169

Kresse, G., & Hafner, J. (1993). Ab initio molecular dynamics for liquid metals. *Physical Review B, 47*(1), 558-561. http://dx.doi.org/10.1103/PhysRevB.47.558

Kresse, G., & Hafner, J. (1994). Ab initio molecular-dynamics simulation of the liquid-metal–amorphous-semiconductor transition in germanium. *Physical Review B, 49*(20), 14251-14269. http://dx.doi.org/10.1103/PhysRevB.49.14251

Lindan, P. J. D., & Gillan, M. J. (1993). Shell-model molecular dynamics simulation of superionic conduction in CaF_2. *Journal of Physics: Condensed Matter, 5*(8), 1019-1030. http://dx.doi.org/10.1088/0953-8984/5/8/005

Marrocchelli, D., Madden, P. A., Norberg, S. T., & Hull, S. (2011). Structural disorder in doped zirconias, Part II: Vacancy ordering effects and the conductivity maximum. *Chemistry of Materials, 23*(6), 1365-1373. http://dx.doi.org/10.1021/cm102809t

Morse, G. E., & Lawson, A. W. (1967). The temperature and pressure dependence of the elastic constants of thallium bromide. *Journal of Physics and Chemistry of Solids, 28*(6), 939-950. http://dx.doi.org/10.1016/0022-3697(67)90209-0

Northrup, J. E., & Zhang, S. B. (1993). Dopant and defect energetics: Si in GaAs. *Physical Review B, 47*(11), 6791-6794. http://dx.doi.org/10.1103/PhysRevB.47.6791

Olsson, D. M., & Nelson, L. S. (1975). The Nelder-Mead simplex procedure for function minimization. *Technometrics, 17*(1), 45-51. http://dx.doi.org/10.1080/00401706.1975.10489269

Pettifor, D. G., Finnis, M. W., Nguyen-Manh, D., Murdick, D. A., Zhou, X. W., & Wadley, H. N. G. (2004). Analytic bond-order potentials for multicomponent systems. *Materials Science and Engineering: A, 365*(1), 2-13. http://dx.doi.org/10.1016/j.msea.2003.09.001

Plimpton, S. (1995). Fast parallel algorithms for short-range molecular dynamics. *Journal of computational physics, 117*(1), 1-19. http://dx.doi.org/10.1006/jcph.1995.1039

Shorohov, M., Kouznetsov, M., Lisitskiy, I., Ivanov, V., Gostilo, V., & Owens, A. (2009). Recent results in TlBr detector crystals performance. *Nuclear Science, IEEE Transactions on, 56*(4), 1855-1858. http://dx.doi.org/10.1109/TNS.2009.2022015

Simmons, G. (1965). *Single crystal elastic constants and calculated aggregate properties*. Southern Methodist University Press.

Stillinger, F. H., & Weber, T. A. (1985). Computer simulation of local order in condensed phases of silicon. *Physical review B, 31*(8), 5262-5271. http://dx.doi.org/10.1103/PhysRevB.31.5262

Storn, R., & Price, K. (1997). Differential evolution–a simple and efficient heuristic for global optimization over continuous spaces. *Journal of global optimization, 11*(4), 341-359. http://dx.doi.org/10.1023/A:1008202821328

Tersoff, J. (1989). Modeling solid-state chemistry: Interatomic potentials for multicomponent systems. *Physical Review B, 39*(8), 5566-5568. http://dx.doi.org/10.1103/PhysRevB.39.5566

Tung, K. L., Chang, K. S., Hsiung, C. C., Chiang, Y. C., & Li, Y. L. (2010). Molecular dynamics simulation of the complex dopant effect on the super-ionic conduction and microstructure of zirconia-based solid electrolytes. *Separation and Purification Technology, 73*(1), 13-19. http://dx.doi.org/10.1016/j.seppur.2009.07.026

Wang, J., & Rockett, A. (1991). Simulating diffusion on Si (001) 2×1 surfaces using a modified interatomic potential. *Physical Review B, 43*(15), 12571-12579. http://dx.doi.org/10.1103/PhysRevB.43.12571

Ward, D. K., Zhou, X. W., Wong, B. M., Doty, F. P., & Zimmerman, J. A. (2012). Analytical bond-order potential for the Cd-Zn-Te ternary system. *Physical Review B, 86*(24), 245203. http://dx.doi.org/10.1103/PhysRevB.86.245203

Ward, D. K., Zhou, X. W., Wong, B. M., Doty, F. P., & Zimmerman, J. A. (2012). Analytical bond-order potential for the cadmium telluride binary system. *Physical Review B, 85*(11), 115206. http://dx.doi.org/10.1103/PhysRevB.85.115206

Ward, D. K., Zhou, X., Wong, B. M., & Doty, F. P. (2013). A refined parameterization of the analytical Cd–Zn–Te bond-order potential. *Journal of molecular modeling, 19*(12), 5469-5477. http://dx.doi.org/10.1007/s00894-013-2004-8

Wolfram, S. (2004). The Mathematica Book (5th ed.). Wolfram Research, Inc., Champaign, IL, 2004.

Yamamura, Y., Kawasaki, S., & Sakai, H. (1999). Molecular dynamics analysis of ionic conduction mechanism in yttria-stabilized zirconia. *Solid State Ionics, 126*(1), 181-189. http://dx.doi.org/10.1016/S0167-2738(99)00227-1

Zhang, S. B., & Northrup, J. E. (1991). Chemical potential dependence of defect formation energies in GaAs: Application to Ga self-diffusion. *Physical review letters, 67*(17), 2339-2342. http://dx.doi.org/10.1103/PhysRevLett.67.2339

Zheng-Johansson, J. X. M., & McGreevy, R. L. (1996). A molecular dynamics study of ionic conduction in CuI. II. Local ionic motion and conduction mechanisms. *Solid State Ionics, 83*(1), 35-48. http://dx.doi.org/10.1016/0167-2738(95)00218-9

Zheng-Johansson, J. X. M., Ebbsjö, I., & McGreevy, R. L. (1995). A molecular dynamics study of ionic conduction in CuI. I. Derivation of the interionic potential from dynamic properties. *Solid State Ionics, 82*(3), 115-122. http://dx.doi.org/10.1016/0167-2738(95)00205-0

Zhou, X. W., & Jones, R. E. (2011). Effects of cutoff functions of Tersoff potentials on molecular dynamics simulations of thermal transport. *Modelling and Simulation in Materials Science and Engineering, 19*(2), 025004. http://dx.doi.org/10.1088/0965-0393/19/2/025004

Zhou, X. W., Foster, M. E., Van Swol, F. B., Martin, J. E., & Wong, B. M. (2014). Analytical bond-order potential for the Cd–Te–Se ternary system. *The Journal of Physical Chemistry C, 118*(35), 20661-20679. http://dx.doi.org/10.1021/jp505915u

Zhou, X. W., Jones, R. E., Duda, J. C., & Hopkins, P. E. (2013). Molecular dynamics studies of material property effects on thermal boundary conductance. *Physical Chemistry Chemical Physics, 15*(26), 11078-11087. http://dx.doi.org/10.1039/c3cp51131f

Zhou, X. W., Jones, R. E., Kimmer, C. J., Duda, J. C., & Hopkins, P. E. (2013). Relationship of thermal boundary conductance to structure from an analytical model plus molecular dynamics simulations. *Physical Review B, 87*(9), 094303. http://dx.doi.org/10.1103/PhysRevB.87.094303

Zhou, X. W., Ward, D. K., Martin, J. E., van Swol, F. B., Cruz-Campa, J. L., & Zubia, D. (2013). Stillinger-Weber potential for the II-VI elements Zn-Cd-Hg-S-Se-Te. *Physical Review B, 88*(8), 085309. http://dx.doi.org/10.1103/PhysRevB.88.085309

18

Chitosan Nanocomposite Mesoporous Membranes: Mechanical and Barrier Properties as a Function of Temperature

Irene S. Fahim[1,2], Wael Mamdouh[2,3] & Hanadi G. Salem[1,2]

[1] Department of Mechanical Engineering, School of Sciences and Engineering (SSE), The American University in Cairo (AUC), AUC Avenue, P.O. Box 74, New Cairo 11835, Egypt.

[2] Yousef Jameel Science and Technology Research Center (YJSTRC), School of Sciences and Engineering (SSE), The American University in Cairo (AUC), AUC Avenue, P.O. Box 74, New Cairo 11835, Egypt.

[3] Department of Chemistry, School of Sciences and Engineering (SSE), The American University in Cairo (AUC), AUC Avenue, P.O. Box 74, New Cairo 11835, Egypt.

Correspondence: Irene Fahim, Department of Mechanical Engineering, School of Sciences and Engineering (SSE), The American University in Cairo (AUC), AUC Avenue, P.O. Box 74, New Cairo 11835, Egypt. E-mail: rico4000@aucegypt.edu

Abstract

This work investigates the influence of the type and wt.% of nanofillers on the tensile strength behaviour and barrier properties of fabricated chitosan (CS) mesoporous membranes and their nanocomposites using graphene (G) and fullerene (F) nanofillers. Non cross-linked chitosan (NCLCS) as well as cross-linked chitosan (CLCS) solutions with sodium tripolyphosphate (TPP) were both mixed with G and F nanofillers with different wt.%. NCLCS membranes displayed yield tensile strength of 24MPa while the CLCS membranes displayed a much lower yield tensile strength of 2.87MPa. The addition of G and F nanofiller enhanced the yield tensile strength of the CS membranes up to 45MPa. However, the increase in % elongation for CLCS membranes was 75% higher than that for NCLCS ones. Furthermore, the results revealed that there was a significant effect of the operating temperature on the membrane pore size, which decreased the tensile strength and the barrier of the produced membranes. The enhancement of the tensile properties of polymer nanocomposites membranes (PNC) membranes is crucial to avoid film fracture or delamination. Moreover, the addition of nanofillers improve the barrier properties of PNC membranes, thus the membranes are used in packaging and current filtration techniques. In this work, experimental as well as statistical analysis of the fabricated CS membranes and their yield tensile strength and % elongation data are presented. This study presents the effect of the filler type, the filler content and the cross-linking of CS membranes on tensile strength behaviour both experimentally as well as theoretically.

Keywords: pore size, nanocomposites, nanoparticles, membranes

1. Introduction

Membrane technology is one of the most promising solutions to air pollution. Membranes are fabricated either from natural or synthetic polymers (Mackenzie, 1999). Although synthetic membranes are widely used as valuable scientific and technical tools, they are not well defined in terms of their structure and function (Olness, 1995). They are usually less selective and less energy efficient. The preparation of natural polymer membranes and their utilization on a large industrial scale is a recent development that gained substantial importance for sustainable development and environmental preservation. In addition, natural polymers are becoming attractive for separation of molecular mixtures due to their porous nature (Strathmann, 2006). Most importantly is that the separation is performed by physical means without chemically altering the constituents of the mixture. Natural polymer membranes attracted significant interest in a broad range of applications such as water treatment, separation membranes, food packaging, tissue engineering, and drug delivery (Mackenzie, 1999). The mechanical properties of PNC membranes (% elongation and tensile strength) determine their application (Strathmann, 2006). The objective of this work is to study the tensile properties of natural polymer membranes used for packaging and separation, where membranes are subjected to mechanical stresses arising from the particles that need to be filtered. PNC membranes reach their optimum performance characteristics with enhanced tensile strength. This increases the membrane's durability and resistance to wear in service (Strathmann, 2006). However, it was reported that natural polymer membranes had their functional limitations that restricted their widespread application in packaging and separation. The main limitations are their brittleness, thermal instability, low tensile strength, high vapor and oxygen permeability, and poor mechanical stiffness (Kim et al, 1995). Weak mechanical strength of natural polymers was due to their porous structure. To overcome the mechanical strength deterioration, several physical and chemical treatments were performed to natural polymers such as chitosan, to enhance the mechanical properties due to their importance in different applications. CS is one of the most promising polymers that have been widely used in the preparation of CS membranes for various uses: for example for pervaporation, ultra filtration, reverse osmosis, gas separation, and purification processes or drug delivery (Kim et al., 1995).

CS contains abundant amino and hydroxyl groups, which enable nanoparticles formulation via both physical and chemical cross-linking. Covalent cross-linking is usually achieved by treatment of CS solutions by glutaraldehyde, which reacts with the amino groups on the CS chains (Leo et al., 2005). Ionic cross-linking of chitosan, on the other hand, is a typical non-covalent interaction, which can be realized by the association with negatively charged multivalent ions such as TPP. Physical cross-linking is more promising method since cross-linking is a reversible process and could largely avoid potential toxicity of the reagents (Kawashima et al., 1985). Diverse efforts have been made to obtain CS nanoparticles via TPP cross-linking following the pioneering work of Calvo (Calvo et al., 1997). A number of studies on the cross-linking reaction have been reported and showed that it is mainly influenced by the size and type of the cross-linker agent and the functional groups of CS. The smaller the molecular size of the cross-linker, the faster the cross-linking reaction since its diffusion is easier (Murphy, 1994; Chang et al., 2000). Different studies showed that CS cross-linked with TPP has high chemical stability together with membrane flexibility compared to other cross-linking agents (Muzzarelli, 1986; Basan & Gümüsdereliolu, 2002; Hamdi et al., 2001; Knaul et al., 1999; Kumar, 2000).

The incorporation of inorganic nanofillers dispersed within natural polymeric matrices at a nano-scale level produces polymer nanocomposite (PNCs) membranes with enhanced mechanical properties (Koros, 2002). The two-dimensional single-layer G, the basic building block for naturally occurring graphite, has currently attracted tremendous attention especially because of its unique structure and extraordinary electronic and mechanical properties. The nanocomposite materials combine the advantages of the matrix ranging from flexibility, processability of polymers, and the selectivity, together with mechanical strength and thermal stability of the inorganic fillers (Morooka & Kusakabe, 1999; Tsapatis & Gavalas, 1999). The sheet-like nanoparticles (such as graphene) offer the weakest increase in viscosity that causes clustering, and accordingly have the smallest fraction of bridging chains, and provide the greatest reinforcement (Vaia & Giannelis, 2001; Starr et al., 2004; Hooper & Schweizer, 2005). On the other hand, fullerene interactions are highly favorable, as they promote nanoparticles dispersion due to chain bridging between the nanoparticles and better absorption and transportation of penetrants (Hooper & Schweizer, 2006; Gersappe, 2002; Patel et al., 2004; Shekhawat et al., 2003). This results in favorable selectivity and permeability of the matrix (Zimmerman et al., 1997).

Another significant property to evaluate the performance of the membranes is their ability to control the rate of permeation of different species. The two models used, which are also mentioned in literature to describe the mechanism of permeation, are illustrated in Figure 1a. One model is the pore flow model in which permeants are transported by pressure driven convective flow through tiny pores. Separation occurs because one of the permeants is filtered from some of the pores in the membrane through which other permeants move. The second model shown in Figure 1b. is the solution diffusion model in which permeants dissolve in the membrane material and then diffuse through the membrane down a concentration gradient. In case of gas diffusion, the gas flux is driven by the concentration gradient of absorbed molecules within the polymer matrix. The permeants are separated because of the differences in the solubility of the materials in the membrane and the differences in the rates at which materials diffuse through the membrane (Baker, 2000). In 1940, the solution diffusion model was used to explain the transport of gases through polymeric membranes. Diffusion, the basis of the solution diffusion model, is the process by which matter is transported from one part of the system to another by a concentration gradient. If a concentration gradient is formed in the medium, the transport of matter will occur from high concentration to low concentration region. This concept was first recognized by Ficks (Baker, 2000).

The model utilized in this work to evaluate the barrier properties is the solution diffusion model. It was used to determine the barrier property in terms of permeability, which is dependent on the diffusion coefficient of gas and the solubility coefficient of that gas in the polymer matrix, and compare it with experimental results. Finite element modeling was used in the current work to predict the diffusion behaviour of a porous material through the entire thickness of the membrane. The diffusion model was formed in ABAQUS, a finite element solver commonly used for mass diffusion modeling. It creates simulations to predict the change in concentration of gases passing through the membrane thickness. The governing equations for this steady state mass diffusion model are extensions of

Fick's equations: they allow for non-uniform solubility of the diffusing substance in the membrane and for mass diffusion driven by gradients of temperature and pressure. The basic solution variable is normalized concentration. It was compared with experimental results and the simulations were refined accordingly (Kumar, 2011). The simulations took into account the change of the diffusion coefficient and the solubility with the increase in temperature, the type of filler, and the wt.% of nanofiller. To the best of our knowledge, this is a unique study that combines theoretical mass diffusion results and experimental tensile strength results.

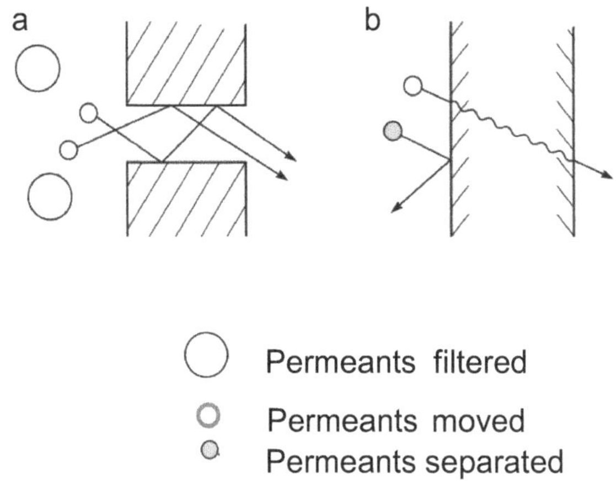

Figure 1a. Pore flow model and 1b. Solution diffusion model (Baker, 2000)

2. Materials & Experimental Procedures

2.1 Materials

Shrimp Chitosan, ≥ 93% (w/w), was purchased from Primex as shown in Figure 2. Chemicals used for dissolving chitosan were purchased from Sigma Aldrich. These included NaOH; HCl; acetic acid (Ac-OH, 99% purity); sodium tripolyphosphate (TPP), used to synthesize chitosan nanoparticles; phenolphthalein (phph, 99% purity), used as an indicator for NaOH in testing liquid permeability of thin films; and methanol (99.9% purity), used for cleaning glassware.

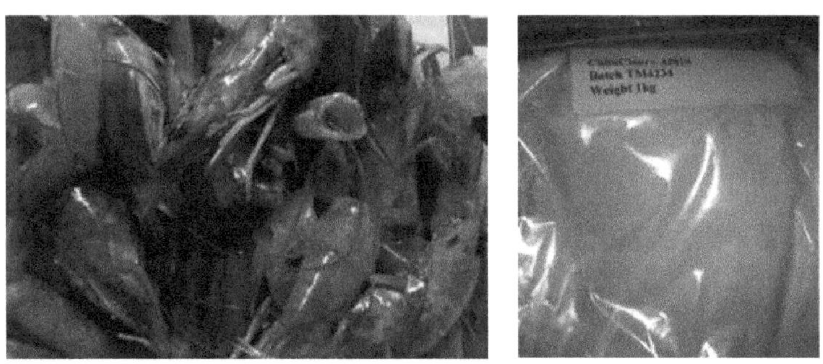

Figure 2. Shrimp Shells toused in Chitosan Production

To prepare the CLCS membranes, 0.2gm of CS was dissolved in 2% acetic acid at room temperature with continuous stirring. 0.033gm TPP were dissolved in 11mL distilled H_2O and added drop-wise onto the CS solution during the homogenization at 10 000rpm for 30min using a Polytron homogenizer PT 10-35GT.

G-nanofiller, shown in Figure 3a. (Sky Spring Nanomaterials, Inc. USA), and F-nanofiller, shown in Figure 3b. (Carbon 60, 99.5+%, SES Research, USA), were used to produce the PNC thin films. Graphene is a one-atom-thick planar sheet of carbon atoms, densely packed together into a honeycomb shaped crystal lattice. It is the basic structural element of several carbon allotropes including graphite, carbon nanotubes and fullerenes. The difference in morphology (Flaky versus spherical) and nano cluster size between F and G had a significant effect on the tensile and barrier properties of the membranes.

Figure 3. SEM Image for a. G-nanofiller b. F-nanofiller

2.2 Film Processing

CLCS and NCLCS solutions were mixed with two nanofillers (F and G) with 0.1, 0.5, 1 wt. % by solvent mixing. Mixing occurred with constant stirring for 60 minutes with a VWR® Standard Analog Shaker to form a clear homogeneous solution. The solutions were used to produce non-cross-linked CS nanocomposites (NCLCS/G and NCLCS/F), as well as the cross-linked CS nanocomposite (CLCS/G and CLCS/F) thin films. The NCLCS and CLCS filtrates were poured into two separate flattened containers and left to dry at room temperature to form CS thin films with a 2mm thickness. The dry, thin films obtained from the NCLCS and CLCS solutions are shown in Figures 4a and 4b. The same procedure was used for the CLCS/G, CLCS/F, NCLCS/G, and NCLCS/F. The as-received F and G-nanofiller diameters were 10µm and 15µm respectively. Upon sonication and forming a solution, the F and G were transformed to clusters of size 2nm and 10 nm respectively (Perry et al., 2013)

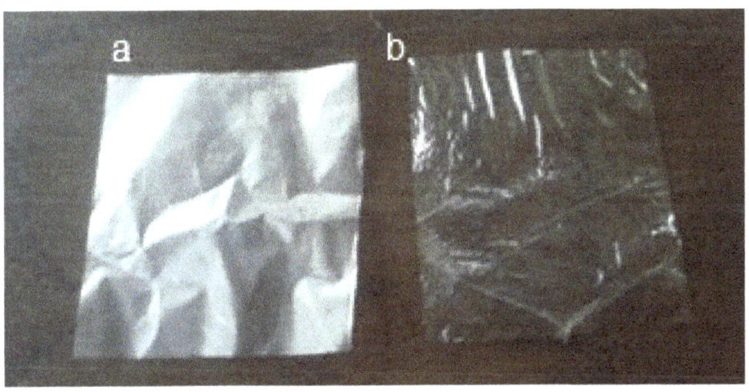

Figure 4. Macrographs for the Fabricated a. NCLCS b. CLCS Membranes, Respectively

2.3 Tensile Properties Characterization

The yield tensile strength and % elongation (ductility) of the previously prepared CS membranes and nanocomposites was evaluated through an axial tensile test. The test was conducted using an Instron universal testing machine with 100KN capacity. To insure the axial uniform load distribution on the membrane thin sheets, a fixture was designed as a transition between the flat grips of the machine and the thin films. The % elongation for all CS membranes at three different temperatures (23, 30, and 60 °C) was also measured during tensile testing. The guidelines for the dimensions of the membrane samples were cut in accordance with ASTM Standard Method D 882-91 (ASTM, D882-91). The initial grip separation was set at 30mm. The tensile tests were conducted at a preselected strain of 0.4mm/mm for the NCLCS and 1.8 for the CLCS membranes. The testing was performed at the preselected strain values to insure that testing occurred within the uniform plastic deformation. A strain rate of 10mm/min was employed for all tested samples. The tensile behaviour of the fabricated membranes was tested as a function of temperature at 23, 30 and 60 °C to investigate the influence of stretching of the membranes as a function of the operating temperatures on the pore size and shape.

2.4 Finite Element Analysis

The Finite Element Analysis (FEA) is a numerical method for solving problems of engineering and mathematical physics. Abaqus/Standard provides a modeling of steady-state diffusion of the fabricated membranes using governing equations that are an extension of Fick's equations. The input data for ABAQUS software included the geometry of the membrane, which was constructed with a specific thickness, after which the material properties were specified. These included diffusivity and the solubility of oxygen gas through the fabricated membranes (Duncan et al., 2005). The input data varied according to the material it would diffuse in, whether pure polymer or polymer reinforced nanocomposite. The first step was constructing the membrane followed by assigning each material to the solid homogenous section. This was followed by specifying the steady state mass diffusion and then specifying a time increment and a time period for the step. Abaqus/Standard then proceeded through the step accordingly. The boundary conditions were applied at the inlet to the nodes in the mass diffusion element to prescribe values of normalized concentration as shown in Figure 5. This was followed by applying a surface concentration flux, as concentration fluxes were the only loads that could be applied in a mass diffusion analysis step. Finally, meshing was optimized to assure convergence of the solution when choosing the DCC3D8 element (the 8-node convection/diffusion brick). There were no applicable element controls for this type of element (Duncan et al., 2005). The normalized concentration output shown in Figure 5. was measured through the thickness of the one-layer membrane (2mm). The normalized concentration of the diffusing molecules passes through a diffusion pathway from high concentration at the inlet to lower concentration at the outlet. The output was calculated at three different temperatures (23, 30, 60 °C).

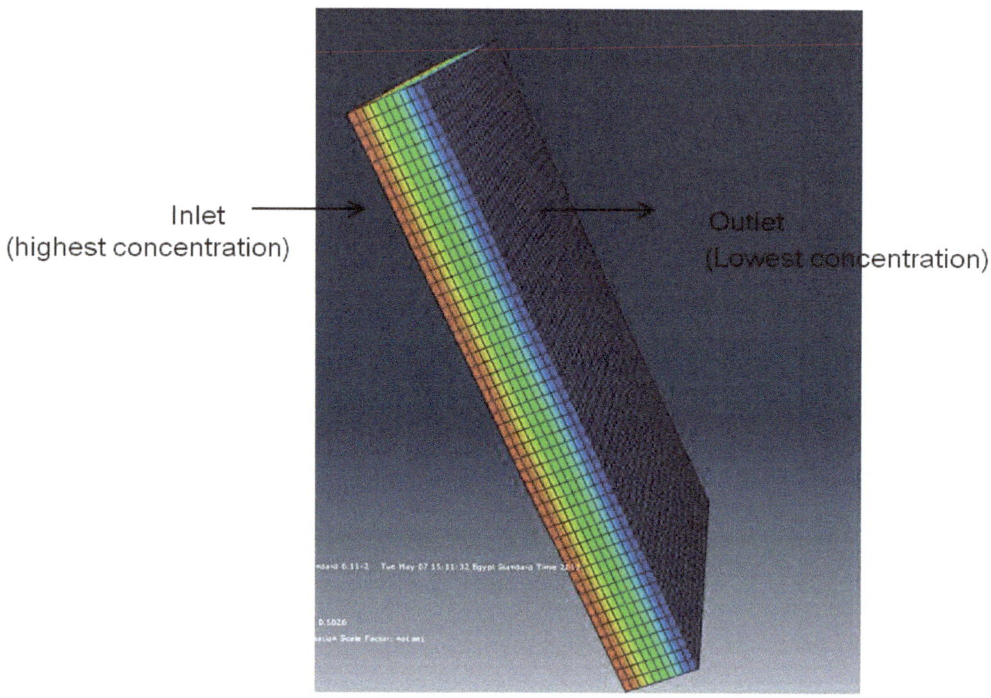

Figure 5. Inlet and Outlet Normalized Concentration through the Membrane Thickness

2.5 Statistical Analysis

For better interpretation of data and to derive the representative conclusions about the influence of the various parameters on the membrane properties, a statistical analysis was performed using Design Expert software (Myers & Montgomery, 2002). The parameters included in the statistical analysis are illustrated in Table 1. The output of Design Expert 9.0.4.1.included three responses: tensile strength, diffusion time, and pore size. The analysis of variance showed the parameters that have significant effect. The significant factors have values with low p-value (Probability>F) where p was less than 0.0001 (Myers & Montgomery, 2002).

Table 1.The Factors and Their Combinations Used for the Statistical Analysis

A	Temperature (23, 30, 60 °C)
B	Polymer (NCLCS-CLCS)
C	Type of filler (Graphene, Fullerene)
D	Wt. % of filler (0.1, 0.5, 1 %)
AB	Interaction between temperature and type of polymer
AC	Interaction between temperature and type of filler
AD	Interaction between temperature and wt. % of filler
BC	Interaction between type of polymer and type of filler
BD	Interaction between type of polymer and wt.% of filler
CD	Interaction between type of filler and wt. % of filler

Response surface methodology (RSM) was useful in this work, as it modeled three different responses: tensile strength, pore size and diffusion time. The eventual objective of RSM was to determine the optimum operating conditions for the membrane fabrication. Fitting and analyzing response surfaces was greatly facilitated by the proper choice of factorial design as it provides distribution of data points, allows model adequacy, provides precise estimates of model coefficients, provides a good profile of the prediction variance throughout the experimental region, provides reasonable robustness to insure simplicity of calculation of model parameters (Montgomery & Runger, 2003).

3. Results

3.1 Pore Size

Our group has recently published the work related to the effect of nanofiller materials (G and F) on the pore sizes of CS membranes. The fabricated CS membranes were found to be mesoporous; NCLCS membranes showed the formation of pore size of 10nm, while CLCS membranes displayed 200% coarser pore sizes and 70% higher permeability. NCLCS and CLCS membranes reinforced with G and F nanofillers showed an improvement of the barrier properties compared to plain NCLCS and CLCS membranes. This was done through filling up of the pores upon the addition of the nanofillers with different wt.%. There was a decrease in the pore size in NCLCS membranes by 70% (with 0.1wt.% G) and 80% (with 1wt.% G), respectively. This was mainly due to the barrier effect of G-nanofiller morphology and size. There was also a decrease in the pore size in NCLCS membranes by 50% (with 0.1wt.% F) and 60% (with 1wt.% F), respectively. Thus the nanofiller wt.% played a crucial role in controlling the pore size of CS membranes. The cluster effect of G-nanofiller on the CS pores that led to their accumulation inside the pores of the CS membranes. This decreased the pore size of the NCLCS membranes, whereas F- nanofiller was dispersed inside the pores to an extent that allows higher permeability compared with G-nanofiller. A thorough investigation of the produced CS nanocomposite membranes by using experimental techniques such as scanning electron microscopy (SEM), average pore size measurement, liquid and gas permeability testing were reported (Fahim et al., 2015). The SEM representative images showed the dispersion of nanofillers within the polymer matrix and the nanofillers effects on the barrier properties of the membranes. The reported results helps in explaining the tensile properties in this work.

CLCS and NCLCS nanocomposite membranes with F and G-nanofillers were fabricated, and their tensile properties were investigated. The influence of the membrane chemical structure on the tensile strength and % elongation was studied. Physical cross-linking of CS by TPP and the addition of F and G-nanofillers were found to be crucial factors affecting the tensile strength and % elongation of the fabricated CS membranes. In the following subsections, the influence of the membrane crosslinking condition with and without nanofillers on the tensile behaviour of the films was displayed.

3.2 Tensile Properties

3.2.1 Tensile Behaviour of NCLCS and CLCS Membranes

The tensile behaviour versus strain of the plain NCLCS and CLCS membranes is shown in Figure 6. It clearly revealed higher tensile strength of 24MPa for the NCLCS compared to 2.8MPa for the CLCS membranes. CLCS membranes had coarser pore sizes compared to NCLCS. The formation of ionic cross-links between amino groups of CS and TPP groups decreased the surface area of the polymer leading to an increase in the surface area of the membranes' pores. Surface area was measured (using a porosimeter), it is defined as area / mass (m^2/g) between the absolute surface area of a solid and its mass (sample weight). The surface area includes all parts of accessible inner surfaces (mainly pore wall surfaces). The surface area of the NCLCS membrane was 0.373 m^2/g while the one for CLCS membrane was 0.1574 m^2/g. The measurements revealed the decrease in the surface area of CS upon crosslinking (Fahim et al., 2015). The decrease in tensile strength could be attributed to the coarse pore size formed in the CLCS, which promoted pore size growth and coalescence resulting in premature failure (Muzzarelli, 1986). However, in NCLCS membranes, the pore size was smaller. Therefore, the stability of the pores against growth and coalescence due to increased bonding interaction between the CS chains could have resulted in the enhanced tensile strength. (Sannan et al., 2003).

The correlation between elongation and cross-linking of polymers was not as straight-forward as was the relation between cross-linking and tensile strength. There are two important observations when using crosslinking agents that should be taken into account. Some research carried out in which crosslinking increased the tensile strength of polymers. While other research reports revealed that addition of cross linking agent decreased the tensile strength depending on several factors. % elongation either increased or decreased depending on the chemical backbone structure of the polymer, type and density of crosslinking agent. For example, crosslinking of chitosan with glutaraldyde enhanced the tensile strength by 25 % (Chan et al., 2013). On the other hand, it has been reported that crosslinking increased the elasticity of LDPE/wax blends. When the amount of crosslinker increased, a decrease in the crystallinity of the polymer occurred leading to a decrease in tensile strength (Krupa et al., 2001). Another example of chitosan films that were greatly enforced by the introduction of crosslinking agent, achieving an increased elongation of about 80%, was mentioned by Kiuchi et al. (Kiuchi et al., 2008). In this work, chitosan was chemically cross-linked with TPP at (0.3% the weight of chitosan). Increasing this degree of crosslinking (which is not included in this chart) lead to a decrease in elongation. This conclusion agreed with the findings of Daniels et al., where increasing the concentration of the crosslinking agent increased the stiffness of the membrane (Daniels, 1989). % elongation increased from 0.5 to 2 % upon cross-linking as shown in Figure 6. Crosslinking increased elongation since the structure of the overall material changed from individual chains linked only with van der Waals forces to covalent bonds, which are few but strong. The polymer acts as a single molecule as soon as the covalent bond occurs. A further explanation of the relationship between crosslinking and increasing % elongation can be attributed to the more flexible molecular structure resulting from ionic cross-linking (Wang et al., 1998).

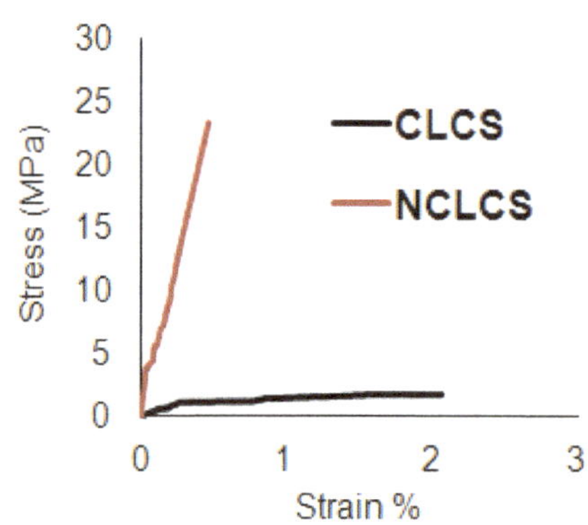

Figure 6. Stress-strain Curve for NCLCS and CLCS Membranes

3.2.2 Effect of F and G-nanofiller Content on the Tensile Properties of NCLCS Membranes

Figure 7. reflected the effect of addition of nanofillers (F or G) to CS matrices membranes. On using F as a nanofiller in NCLCS, the tensile strength increased from 24MPa for the plain matrices up to 45MPa with increasing F-content up to 1 wt.%. with an average of 46% as illustrated in Figures 7a. The increase in tensile strength upon addition of F-nanofiller was most probably due to the rigidity of the F nanofiller and the strong interaction between the CS polymeric chains within. The clusters of F-nanofiller were dispersed within the CS polymeric chain (Shlykov et al., 2010). They reported that F improved the tensile strength of polymers to 30-40 % (Sung, 2001).

On the other hand, addition of G-nanofiller, the tensile strength of NCLCS/0.1 wt.% G displayed was about 30MPa, which showed an increase in the tensile strength compared to NCLCS (24MPa). The strength was further increased by adding 1wt. % of G up to 40MPa as shown in Figure 7b. Accordingly, increasing G-content up to 1 wt.% to the

NCLCS matrices resulted in 40% increase in tensile strength. It was reported that G improved the tensile strength of polymers to 29%. The enhancement was easily explainable with the large aspect ratio and high interfacial contact area of G-nanoparticles with the polymers (Chaharmahali et al., 2014).

The increase in tensile strength with the addition of different wt.% of G suggested that G-nanofiller was mechanically dispersed into the NCLCS during the wet mixing process forming a carbon network in the polymer structure. The structural arrangement of G-nanoparticles within the NCLCS offers higher interfacial adhesion then CS without G nanofillers. one of the possible binding scenarios could be related to the layered structure of G-nanofiller that facilitate sliding of polymer chains next to each other. However, the higher tensile strength upon addition of F-nanofiller was attributed to the small clusters of F-nanofiller that were better dispersed within the CS polymeric chains.

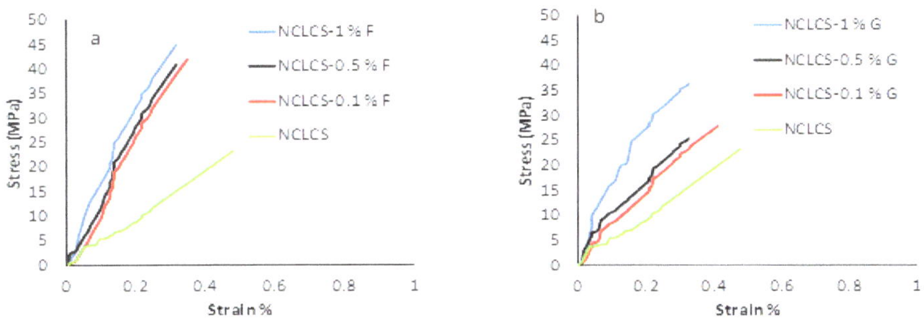

Figure 7. Stress-strain Curve for NCLCS Matrices as a Function of Increasing the Wt.% of a. F and b. G.

The reduction in the ductility of the composite with increase in the F and G-nanofillers was due to the increase in the deformation of a rigid interfacial interaction between F and the polymer matrix, and G and the polymer matrix. The rigidity of bonding between F, G and the NCLCS matrix led to a weak % elongation, as illustrated in Figure 7b. However, upon comparing the % elongation in NCLCS/F membranes and NCLCS/G membranes, one would conclude that the same behaviour occurred and the nanocomposite was tending toward brittle behavior (Siracusa, 2008).

3.2.3 Effect of F and G-nanofiller Content on the Tensile Properties of CLCS Membranes

The addition of nanofillers (F or G) to CLCS matrices membranes enhanced their tensile strength. The tensile strength of CLCS/0.1 wt.% F displayed about 10MPa. The tensile strength was further increased by adding 1wt. % of F up to 20MPa, as shown in Figure 8a. The tensile strength of CLCS/0.1 wt.% G displayed 9MPa, which showed an increase in the tensile strength compared to CLCS (2.8MPa). The tensile strength was further increased by adding 1wt. % of G up to 17MPa, as shown in Figure 8b. Accordingly, increasing G-content up to 1 wt.% to the CLCS matrices resulted in 80 % increase in tensile strength, while causing a decrease of 20 % in elongation due to the rigidity of the nanocomposite (Siracusa, 2008). However, the higher tensile strength upon addition of F-nanofiller was attributed to the small clusters of F-nanofiller that were better dispersed within the CS polymeric chains.

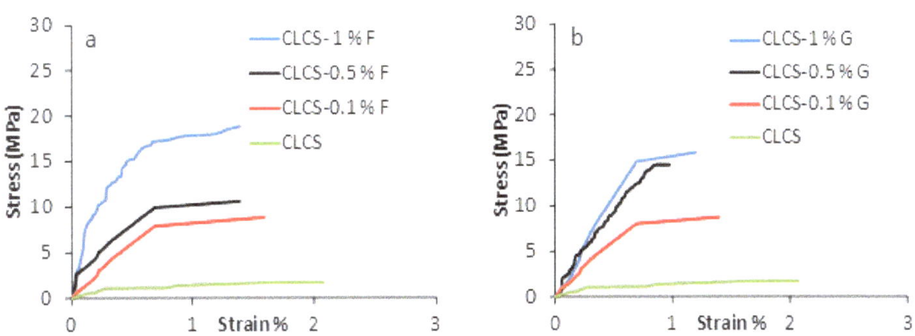

Figure 8. Stress-strain Curve for CLCS with Different Wt.% of a. F and b. G

3. 2.4 Effect of Increasing Temperature on the Tensile Properties of CS Membranes

The results referred to in the previous sections were recorded at room temperature. However, in order to study the influence of temperature on the tensile strength and % elongation of the CS mesoporous membranes as a function of the cross-linking and the nanofillers content, the tensile testing of the fabricated membranes were also carried out at temperatures of 30 and 60 °C. The tensile strength and pore size for all CS membranes were measured at a preselected strain of (Strain = 0.4% for NCLCS - 1.8% for CLCS membranes) and listed in Table 2. It is worth mentioning that these tensile strength results were recorded prior to membrane failure.

It was clear from the listed results in Table 2. that the NCLCS tensile strength was higher than that of the CLCS, while the average pore size for the NCLCS was lower than that of the CLCS. Increasing the temperature resulted in the decrease of tensile strength of NCLCS membranes by 8% and 20 % and an increase in the pore size by 30% and 90% at 30 and 60 °C, respectively as shown in Table 2. This could be due to the increase in the movement of CS chains leading to enlargement of the pore sizes in the different CS membranes (Alexeev, 2010). The effect of increasing temperature on polymer nanocomposites was also reported by Nam et al., high temperatures decrease the polymer nanocomposites stiffness and it becomes more ductile leading to a decrease in tensile strength (Nam et al., 2001). (Appendix B include SEM images showing the pore size of NCLCS and CLCS membranes).

Upon the addition of 0.1wt.% F to NCLCS membranes there was a decrease in tensile strength of 30 % and 20 % and an increase in pore by size 70% and 100% at 30 and 60 °C, respectively. Upon the addition of 1wt.% F to NCLCS membranes there was a decrease in tensile strength of 45% and 53% and an increase in pore by size 62% and 91% at 30 and 60 °C, respectively as shown in Table 2. The same behavior of decreasing tensile strength with an increase in pore size was observed in CLCS membranes upon the addition of F-nanofiller. However, the decrease of tensile strength is larger than that of NCLCS, as explained in section 3.2.1.

On the other hand, upon the addition of 0.1wt.% G to NCLCS membranes there was a decrease in tensile strength of 10% and 30% and an increase in pore by size 22% and 66% at 30 and 60°C, respectively. Upon the addition of 1wt.% G to NCLCS membranes there was a decrease in tensile strength of 5% and 28% and an increase in pore by size 50% and 50% at 30 and 60 °C respectively, as shown in Table 2. The same behavior was observed in CLCS membranes upon the addition of G-nanofiller. Although the tensile strength decreased as a result of pore enlargement, the tensile strength of NCLCS/1 wt.% G at 60°C (29 MPa) was still considered a reliable membrane, suitable for packaging applications even at ambient temperatures up to 60° C (Siracusa et al., 2008).

Table 2.Tensile Strength for CS Membranes at 23, 30, 60 °C

Temperature	**23°C**		**30°C**		**60°C**	
Membranes	**T.S. MPa**	**P.S. nm**	**T.S. MPa**	**P.S. nm**	**T.S. MPa**	**P.S. nm**
NCLCS	24.0	10.6	22.8	13.4	19.4	15.4
NCLCS-0.1%F	40.0	2.6	27.5	4.4	19.9	5.2
NCLCS-0.5%F	43.0	2.5	30.5	4.1	20.2	4.9
NCLCS-1%F	45.0	2.4	36.0	3.9	21.1	4.6
NCLCS-0.1%G	30.8	1.8	27.2	2.2	24.2	3.0
NCLCS-0.5%G	35.3	1.1	32.5	1.6	26.2	2.5
NCLCS-1%G	40.1	0.0	38.8	1.0	29.8	2.0
CLCS	2.8	19.3	1.2	29.5	0.8	33.3
CLCS-0.1%F	9.0	18.1	8.6	26.4	7.9	29.4
CLCS-0.5%F	18.0	17.2	11.5	18.2	11.5	26.4
CLCS-1%F	20.0	16.0	17.3	4.5	15.2	25.1
CLCS-0.1%G	10.9	15.3	8.6	17.0	5.2	20.0
CLCS-0.5%G	14.2	9.2	10.1	11.5	7.4	15.2
CLCS-1%G	17.0	4.0	13.2	5.0	10.2	7.0

Note. T.S=Tensile Strength, P.S.=Pore size.

3.2.5 Effect of Increasing Temperature on Barrier Properties of CS Membranes Using FEA

The increase in temperature had a direct effect on the mass diffusion. High diffusion rates occurred at low % of nanofillers as a function of increasing temperature, as shown in Appendix A. Tables A1. and A2. show the effect of increasing temperature on the normalized concentration of both NCLCS and CLCS and their nanocomposite membranes along the membrane thickness. It is clear that the increase in temperature from 23 ºC to 30 ºC and 60 ºC revealed higher (normalized concentration) diffusion due to increase in the pore size. The normalized concentration at 0.2mm in NCLCS was 767 (C/S) and increased to 777 (C/S) at 30 ºC. The same behavior was displayed upon raising the temperature to 60 ºC. A normalized concentration of 797 (C/S) was measured as shown in Table A1.

The normalized concentration, shown in Table A2. for CLCS membranes at 0.2mm distance, was higher (780 (C/S)) than that of NCLCS membranes (767 (C/S)), which agreed with the exhibited increase of pore size upon cross-linking. The similar effect of increasing temperature was illustrated in Table A2. for CLCS membranes at 0.2mm distance; the increase of normalized concentration from 780 (C/S) to 800 (C/S) at 30 ºC was followed by an increase to 840 (C/S) at 60 ºC. This established the correlation between pore size increase and the increase in temperature (Morehouse et al. 2006), where the increase in temperature increased the normalized concentration of the membranes due to increase in pore size with stretching the membranes.

3.3 Statistical Analysis

In order to find the optimum and most desirable membranes for filtration and packaging applications, statistical analysis was performed. The input for the analysis was the experimental results of the collective properties (porosity and tensile strength) of CLCS and NCLCS with different wt.% of G and F at 23, 30, and 60°C. The experimental results used are mentioned in sections 4.3.2, 4.3.3 and 4.3.8.

RSM was useful in applying the statistical analysis as it modeled three different responses: i) tensile strength, ii) pore size and iii) diffusion time. The responses were influenced by several parameters including temperature (23, 30, and 60°C), polymer (CLCS and NCLCS), nanofiller type (F and G) and wt.% of the nanofiller (0.1, 0.5, and 1 wt.%). It is worth mentioning that the input data for tensile strength were prior to membrane failure. The ultimate objective of RSM was to determine the optimum operating conditions for the membrane fabrication (Bettinger & Chinnici, 1991).

3.3.1 Analysis of Variance

The main aim for this experiment was investigating the effect of F and G-nanofillers and temperature on the tensile strength, porosity, and diffusion time of CS nanocomposite membranes. There were 2 replicates to each data point to ensure that the data was correct. The 48 runs were entered in random order. This randomized test sequence was necessary to prevent the effects of unknown nuisance variables and avoid contaminating the results. To be more objective in this work, an analysis was performed to test the differences between the mean of each factor at every specified response. The appropriate procedure for testing the equality of several means was the analysis of variance (ANOVA) (Montgomery & Runger, 2003). The ANOVA was suitable for analysis of these types of experiments. The name analysis of variance was derived from the partitioning of total variability into its component parts. It states that the total variability in the given data as measured by the total corrected sum of squares can be partitioned into a sum of squares of the differences between the treatment average and the grand average plus a sum of squares of the differences of observations within treatments from the treatment average. A further step in this method was calculating the mean square of the treatments and the error. F is the ratio of the mean square of the treatments to the mean square error. In this work, the ANOVA showed that the mean square of treatments is larger than the error mean square and the p value can be computed (if it is less than 0.0001 than the factor is significant) (Bettinger & Chinnici, 1991). The factors and their interactions highlighted in Table 3 are the most significant.

The Design Expert output chart shown in Figure 9. highlighted the significant factors within each response. The significant factors for tensile strength response included temperature, crosslinking of polymer, and the wt.% of filler. The higher the temperature, the lower the tensile strength in NCLCS membranes as shown in Figure 9a. CLCS has lower tensile strength than NCLCS, as illustrated in Figure 9b. The increase in the wt.% of the nanofiller led to an increase in the tensile strength as shown in Figure 9b. Figure 9c shows the wt % of filler with temperature intersecting with the lines representing tensile strength. The analysis of variance outcomes corresponded to the experimental results in section 3.2.3 where CLCS had a lower tensile strength than NCLCS. In section 3.2.2 and 3.2.3 the addition of nanofiller increased the tensile strength of both NCLCS and CLCS membranes. The results in section 3.2.4 established the same effect of temperature on lowering tensile strength of the membranes as the software output.

Table 3. ANOVA for the Three Responses

	F	P value	F	P value	F	P value
A-Temp	35.62	< 0.0001	46.26	< 0.0001	45.38	< 0.0001
A^2	6.7	0.015	17.69	0.0002	26.34	< 0.0001
B-Polymer	118.42	< 0.0001	337.43	< 0.0001	47.03	< 0.0001
C-Filler	0.1	0.75	57.37	< 0.0001	34.88	< 0.0001
D-% of filler	23.35	< 0.0001	40.92	< 0.0001	24.72	< 0.0001
AB	1.49	0.23	12.12	0.002	0.12	0.735
AC	3.53	0.07	3.21	0.083	3.56	0.069
AD	2.79	0.06	2.57	0.074	1.69	0.196
BC	1.29	0.26	10.54	0.003	1.93	0.175
BD	0.85	0.48	6.58	0.002	2.64	0.069
CD	0.15	0.93	12.61	< 0.0001	7.65	0.0007

Note. F= Ratio of Mean Square, P=Probability.

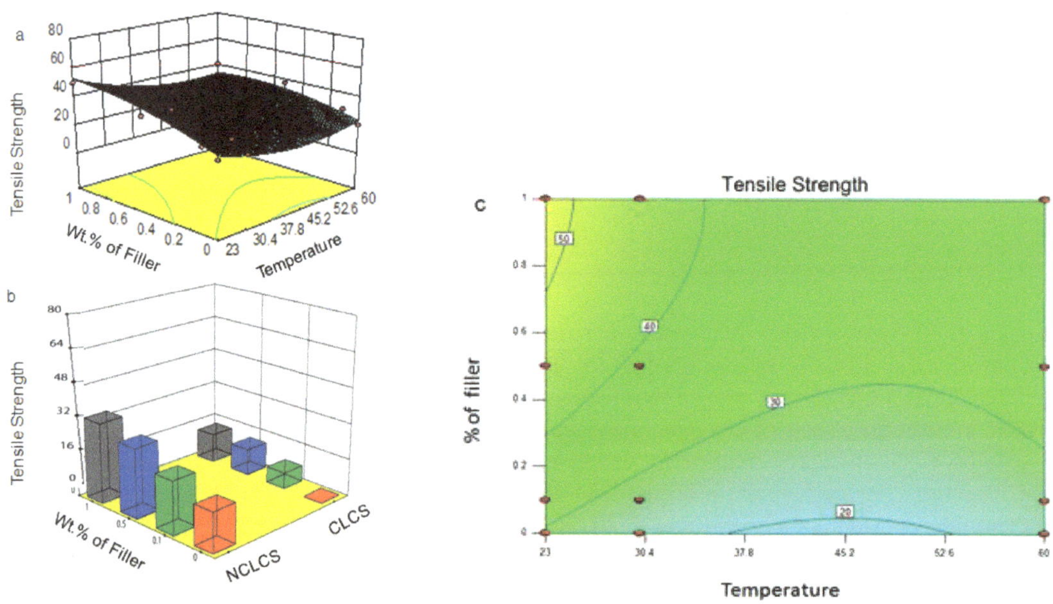

Figure 9. Output Charts for NCLCS Membranes Showing Effect of Temperature, Cross-linking of the Polymer, and the Wt.% of the Nanofiller on the Tensile Strength Response

For the pore size response, there are several factors which were significant, including temperature, cross-linking of polymer, nanofiller type, the wt.% of filler, and the interaction between the nanofiller type and the wt.% of the filler. As the temperature increased, pore size increased in NCLCS membranes, as shown in Figure 10a. The effect of increase of wt.% of filler is quite clear in Figure 10b. Figure 10c shows the wt % of filler with temperature intersecting with the lines representing pore size. As wt.% of filler increased, the pore size decreased, thus increasing the barrier effect of the fabricated membranes. The experimental results were similar in sections 3.2.4. where the pore size decreased in Table 2. due to effect of nanofillers.

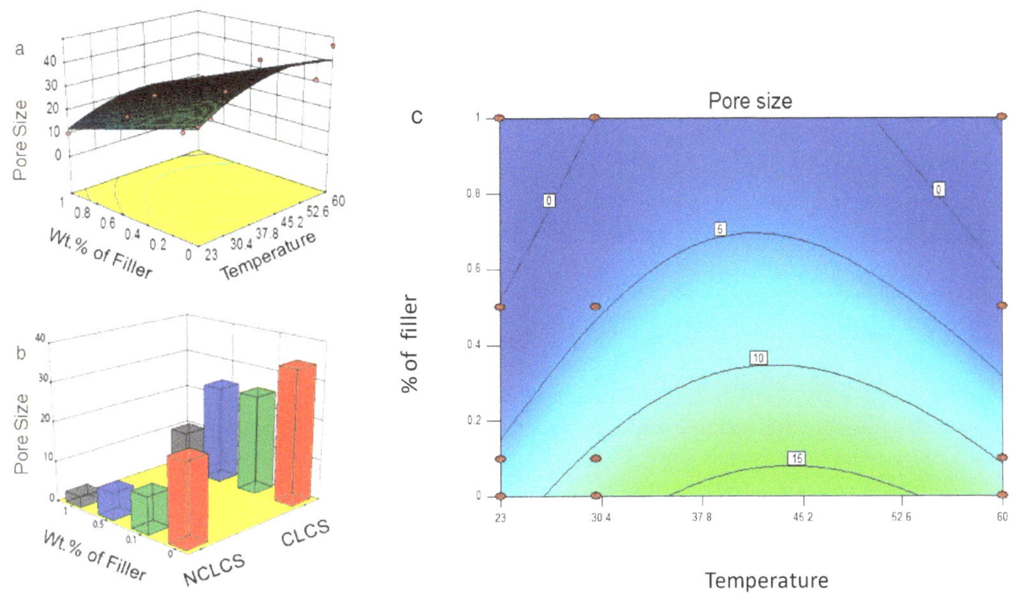

Figure 10. Output Charts for NCLCS Membranes Showing Effect of Temperature and the Wt.% of the Nanofiller on the Pore Size Response

The significant factors for the third response, the diffusion time, were the temperature, cross-linking of polymer, nanofiller type, and the wt.% of filler. The increase in temperature decreased the diffusion time due to enlargement of the pore size in NCLCS membranes because of the heating effect as revealed in Figure 11a. The increase in the wt.% of the nanofiller showed a prolonged diffusion time due to blocking of the pores as illustrated in Figure 11b. Figure 11c shows the wt % of filler with temperature intersecting with the lines representing diffusion time.

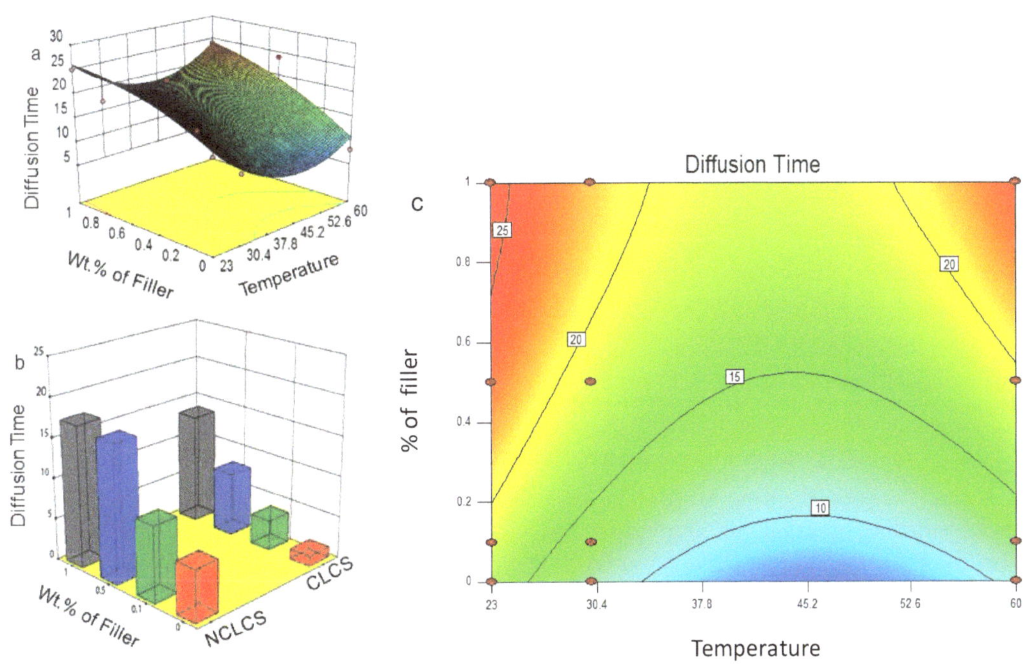

Figure 11. Output Charts for NCLCS membranes Showing Effect of Temperature and Wt.% of Filler on Diffusion Time Response

4. Conclusions

G and F addition resulted in the enhancement of the tensile properties of CS membranes by increasing the yield tensile strength of NCLCS membranes by 46 % for F and 40 % for G. This was due to the structural arrangement of nanoparticles within the CS membranes offers higher interfacial adhesion within the polymer matrix. This could be related to the G-nanofiller flaky structure versus the interaction of F nanofillers with the surrounding atoms. The crosslinking of CS membranes increased the % elongation for CLCS by more than 75 %. depending on the time and density of the crosslinking agent. Moreover the addition of nanofillers improved the membranes' gas barrier properties by decreasing the pore size in the NCLCS, CLCS conditions. The CS membranes' pore size decreased due to the filling up of pores with G and F nanoparticles The preferred temperature range from 23 to 30° C offered the most suitable environment in which tensile strength and pore size were adequate to packaging applications. Upon increasing the temperature to 60 ° C, the tensile strength decreased due to pore size increase which had a negative effect on the performance of the membranes. However, NCLCS/1 wt.% G retains its properties (permeability and tensile strength). The optimum membrane pore size, tensile strength, and operating temperature were calculated by RSM. This method highlighted some of the filtration applications of the fabricated PNC membranes. The optimum membranes were selected according to the size of the permeant that needs to be filtered out. NCLCS/1wt. % G could be used to filter gases. CLCS/1 wt. % F could be used for smoke filtration.

Acknowledgments

The authors acknowledge the financial support received from the American University in Cairo (AUC) through Faculty Support Research Grant and also the Arab Academy for Science and Technology (ASRT), Egypt, through Ph.D. support grant.

References

Alexeev, V., Keberg, E., & Evmenenko, G. (2010). Development and characterization of chitosan film. *International Journal of Engineering Research and Applications, 1*(2), 292-299. Retrieved from http://www.ijera.com/papers/vol%201%20issue%202/012292299AL.pdf.

Baker, R. (2004). *Membrane technology and applications* (2nd ed.). Chichester: J. Wiley. http://dx.doi.org/10.1002/0470020393.

Basan, H., Gümüşderelíoğlu, M., & Orbey, T. (2002). Diclofenac sodium releasing pH-sensitive monolithic devices. *International Journal of Pharmaceutics, 245*, 191-198. http://dx.doi.org/10.1016/s0378-5173(02)00350-2

Bettinger, D., & Chinnici, J. (1991). Utiliization of response surface modeling to evaluate the interaction between aflatoxin B1 and caffeine on egg adult viability in drosophila meganogaster. *Virginia Journal of Science, 42*. Retrieved from http://www.vacadsci.org/vjsArchives/V42/42-3/42-311.pdf

Calvo, P., Remunan-Lopez, C., Vila-Jato, J., & Alonso, M. (1997). Novel hydrophilic chitosan-polyethylene oxide nanoparticles as protein carriers. *Journal of Applied Polymer Science, 63*, 125-132. http://dx.doi.org/10.1002/(sici)1097-4628(19970103)63:1<125::aid-app13>3.0.co;2-4

Chaharmahali, M., Hamzeh, Y., Ebrahimi, G., Ashori, A., & Ghasemi, I. (2014). Effects of nano-graphene on the physico-mechanical properties of bagasse/polypropylene composites. *Polymer Bulletin, 71*, 337-349. http://dx.doi.org/10.1007/s00289-013-1064-3

Chan, M. Y., Husseinsyah, S., & Sam S. T. (2013). Chitosan/Corn Cob Biocomposite Films by Cross-linking with Glutaraldehyde. *Biosources, 8*, 2. http://dx.doi.org/10.15376/biores.8.2.2910-2923

Chang, K., Ming-Chih T., & Cheng, F. (2000). Kinetics and products of the degradation of chitosan by hydrogen peroxide. *Journal of Agriculture Food Chemistry, 49*, 4845-4851. http://dx.doi.org/10.1021/jf001469g

Choi, M., Kwon, K., Park, C., Kim, S., & Hahn, S. (1999). Estimate of the thermal diffusivity of films with a sandwich structure by using pulsed transient analysis and AC calorimetry. *Journal of the Korean Physical Society, 34*. Retrieved from http://www.springer.com/physics/journal/40042

D882-91 Standard test method for tensile testing of thin plastic sheeting (film). (n.d.). Retrieved from http://www.astm.org/Standards/D882.htm

Daniels, C. (1989). *Polymers: Structure and properties,* Lancaster: CRC Press. http://dx.doi.org/10.1002/pi.4980220314

Duncan, B., Urquhart, J. & Roberts, S. (2005). Review of measurement and modeling of permeation and diffusion in polymers. NPL Report. Retrieved from http://resource.npl.co.uk/materials/polyproc/iag/october2005/depc_mpr_012.pdf

Fahim, I., Marei, N., Salem H., G., & Mamdouh, W. (2015). Effect of graphene and fullerene nanofillers on controlling the pore size and physicochemical properties of chitosan nanocomposite mesoporous membranes. *Journal of Nanomaterials, 10*. http://dx.doi.org/10.1155/2015/979561

Gersappe, D. (2002). Molecular mechanisms of failure in polymer nanocomposites. *Physical Review Letters, 89*, 058301. http://dx.doi.org/10.1103/physrevlett.89.058301

Hamdi, G., Ponchel, G., & Duchêne, D. (2001). Formulation of epichlorohydrin cross-linked starch microspheres. *Journal of Microencapsulation, 18*, 373-383. http://dx.doi.org/10.1080/02652040010019505

Hooper, J. & Schweizer, K. (2005). Contact aggregation, bridging and steric stabilization in dense polymer- matrix mixtures. *Macromolecules, 38*, 8858-8869. http://dx.doi.org/10.1021/ma051318k

Hooper, J., & Schweizer, K. (2006). Theory of phase separation in polymer nanocomposites. *Macromolecules, 39*, 5133-5142. http://dx.doi.org/10.1021/ma060577m

Kawashima, Y., Handa, T., Kasai, A., Takenaka, H., Lin, S., & Ando, Y. (1985). Novel method for the preparation of controlled-release theophylline granules coated with a polyelectrolyte complex of sodium polyphosphate–chitosan. *Journal of Pharmaceutical Sciences, 74*, 264-268. http://dx.doi.org/10.1002/jps.2600740308

Kiuchi, H., Kai W., & Inoue, Y. (2007). Preparation and characterization of poly(ethylene glycol) crosslinked chitosan films, *Journal of Applied Polymer Science, 107*, 3823-3830. http://dx.doi.org/10.1002/app.27546

Knaul, J. Z., Hudson, S. M., & Creber, K. A. M. (1999). Mechanical properties of chitosan fibers. *Journal of Applied Polymer Science, 72*, 1721-1732. http://dx.doi.org/10.1002/(sici)1097-4628(19990624)72:13<1721::aid-app8>3.0.co;2-v

Koros, W. (2002). Gas separation membranes: Needs for combined materials science and processing approaches. *Macromolecular Symposia, 188*, 13. http://dx.doi.org/10.1002/1521-3900(200211)188:1<13::aid-masy13>3.0.co;2-w

Krupa, I., & Luyt, S. (2001). Mechanical properties of uncrosslinked and crosslinked linear low-density polyethylene/wax blends, *Journal of Applied Polymer Science, 81*, 973-980. http://dx.doi.org/10.1002/app.1519

Lopez-Leo, T., Carvalho, E., Seijo, B., Ortega-Vinuesa, J., & Bastos-Gonzalez, D. (2005). Physicochemical characterization of chitosan nanoparticles: Electrokinetic and stability behaviour. *J. Colloid Interface Science, 283*, 344. http://dx.doi.org/10.1016/j.jcis.2004.08.186

Mackenzie, K. (1999). Parasites as pollution indicators in marine ecosystems: A proposed early warning system, *Marine Pollution Bulletin, 38*, 959. http://dx.doi.org/10.1016/s0025-326x(99)00100-9

Montgomery, D., & Runger, G. (2003). *Applied statistics and probability for engineers*. New York: John Wiley & Sons. http://dx.doi.org/10.1080/03043799408928333

Morehouse, J., Lloyd, D., Freeman, B., Lawler, D., Liechti, K., & Becker, E. (2006). Modeling the stretching of microporous membranes. *Journal of Membrane Science, 283*, 430-439. http://dx.doi.org/10.1016/j.memsci.2006.07.024

Morooka, S., & Kusakabe, K. (1999). Microporous inorganic membranes for gas separation. *Materials Research Bulletin, 24*, 25. Retrieved from http://journals.cambridge.org/action/displayAbstract?fromPage=online&aid=9028564&fileId=S088376 9400051873

Muzzarelli, R. (1986). *Chitin in nature and technology*. New York: Plenum Press. http://dx.doi.org/10.1007/978-1-4613-2167-5

Myers, R. H., & Montgomery, D. C. (1995). *Response surface methodology: Process and product optimization using designed experiments*. New York: Wiley. http://dx.doi.org/10.2307/1270613

Nam, P. H. Maiti, P., Okamoto, M., Kotaka, T., Hasegawa, N., & Usuki, A. (2001). A hierarchical structure and properties of intercalated polypropylene/clay nanocomposites. *Polymer, 42*, 9633. http://dx.doi.org/10.1016/s0 032-3861(01)00512-2

Olness, A. (1995). Water quality: Prevention, identification and management of diffuse pollution. *Journal of Environmental Quality, 24*, 383. http://dx.doi.org/10.2134/jeq1995.00472425002400020024x

Patel, N., Zielinski, J., Samseth, J., & Spontak, R. (2004). Effects of pressure and nanoparticle functionality on CO2-selective nanocomposites derived from crosslinked poly(ethylene glycol). *Molecular Chemistry and Physics, 205*, 2409-2419. http://dx.doi.org/10.1002/macp.200400356

Ravi Kumar, M. (2000). A review of chitin and chitosan applications. *Reactive and Functional Polymers, 46*, 1-27. http://dx.doi.org/10.1016/s1381-5148(00)00038-9

Ross-Murphy, S. (1994). Rheological characterization of polymer gels and networks. *Polymer Gels and Networks, 2*, 229-237. http://dx.doi.org/10.1016/0966-7822(94)90007-8

Sannan, T., Kurita, K., & Iwakura, Y. (1975), Studies on chitin, 1. Solubility change by alkaline treatment and film casting. *Makromolecular Chemistry, 176*, 1191–1195. http://dx.doi.org/10.1002/macp.1975.021760426

Shekhawat, D., Luebke, D., & Pennline, H. (2003). A review of carbon dioxide selective membranes. *U.S. Department of Energy*, 9-11. http://dx.doi.org/10.2172/819990

Siracusa, V., Rocculi, P., Romani, S., & Rosa, M. (2008). Biodegradable polymers for food packaging: A review. *Trends in Food Science & Technology, 19*, 634-643. http://dx.doi.org/10.1016/j.tifs.2008.07.003

Strathmann, H. (2011). *Introduction to membrane science and technology*. Weinheim, Germany: Wiley-VCH Verlag & Co. Retrieved from http://searchworks.stanford.edu/view/9371882

Sung, H., Liang, I., Chen, C., Huang, R., & Liang, H. (2001). Stability of a biological tissue fixed with a naturally occurring crosslinking agent (genipin). *Journal of Biomedical Materials Research, 55*, 538-546. http://dx.doi.org/10.1002/1097-4636(20010615)55:4<538::aid-jbm1047>3.3.co;2-u

Tsapatis, M., & Gavalas, G. (1999). Synthesis of inorganic membranes. *Journal of Materials Research Science Bulletin*, 24, 30-35. Retrieved from http://serials.unibo.it/cgi-ser/start/en/spogli/df-s.tcl?prog_art=4937890&language=ENGLISH&view=articoli

Vaia, R., & Giannelis, E. (2001). Structure and dynamics of polymer layered silicate nanocomposites. *Chemistry of Materials, 8*, 1728. http://dx.doi.org/10.1021/cm960127g

Wang, W., Roberts, G., Domard, A., Roberts, G., & Varum, K. (1998). *Advances in chitin science*. Lyon: Jacques Andre Publisher Press. http://dx.doi.org/10.1016/j.carbpol.2010.11.049

Zimmerman, C., Singh, A., & Koros, W. (1997). Tailoring mixed matrix composite membranes for gas separations. *Journal of Membrane Science, 137*, 145. http://dx.doi.org/10.1016/s0376-7388(97)00194

Zuev, V., Kostromin, S., & Shlykov, A. (2010). Mechanics of polymer nanocomposites modified with fulleroid nanofillers. *Polymer Science Series A, 52*(5), 532-536. http://dx.doi.org/10.1134/s0965545x10050081

Appendix A

Table A1. Normalized Concentration for NCLCS Membranes at 23, 30, 60 °C

23°C	Normalized Concentration(C/S)						
D. (mm)	NCLCS	NCLCS-0.1% F	NCLCS-0.5%F	NCLCS-1%F	NCLCS-0.1% G	NCLCS-0.5% G	NCLCS-1% G
0	850	850	850	850	850	850	850
0.2	767	766	764	763	765	762	754
0.4	684	682	678	676	680	674	658
0.6	601	598	592	589	595	586	562
0.8	518	514	506	502	510	498	466
1	435	430	420	415	425	410	370
1.2	352	346	334	328	340	322	274
1.4	269	262	248	241	255	234	178
1.6	186	178	162	154	170	146	82

1.8	103	94	76	67	85	58	0
30°C	Normalized Concentration (C/S)						
D.(mm)	NCLCS	NCLCS-0.1 % F	NCLCS-0.5 %F	NCLCS-1 %F	NCLCS-0.1% G	NCLCS-0.5 % G	NCLCS-1 % G
0	850	850	850	850	850	850	850
0.2	777	776	774	773	776	774	773
0.4	694	692	688	686	692	688	686
0.6	611	608	602	599	608	602	599
0.8	528	524	516	512	524	516	512
1	445	440	430	425	440	430	425
1.2	362	356	344	338	356	344	338
1.4	279	272	258	251	272	258	251
1.6	196	188	172	164	188	172	164
1.8	113	104	86	77	104	86	77
60°C	Normalized Concentration(C/S)						
D. (mm)	NCLCS	NCLCS-0.1 % F	NCLCS-0.5 %F	NCLCS-1 %F	NCLCS-0.1% G	NCLCS-0.5 % G	NCLCS-1 % G
0	850	850	850	850	850	850	850
0.2	797	796	794	793	796	794	793
0.4	714	712	708	706	712	708	706
0.6	631	628	622	619	628	622	619
0.8	548	544	536	532	544	536	532
1	465	460	450	445	460	450	445
1.2	382	376	364	358	376	364	358
1.4	299	292	278	271	292	278	271
1.6	216	208	192	184	208	192	184
1.8	133	124	106	97	124	106	97

Note. D=Distance, NCLCS=Non Cross-linked Chitosan, F=Fullerene, G=Graphene.

Table A2. Normalized Concentration for CLCS Membranes at 23, 30, 60 °C

23°C	Normalized Concentration (C/S)						
D. (mm)	CLCS	CLCS-0.1%F	CLCS-0.5 %F	CLCS-1 %F	CLCS-0.1%G	CLCS-0.5%G	CLCS-1 %G
0	850	850	850	850	850	850	850
0.2	780	778	788	776	775	756	750
0.4	710	706	726	702	700	662	650
0.6	640	634	664	628	625	568	550
0.8	570	562	602	554	550	474	450
1	500	490	540	480	475	380	350
1.2	430	418	478	406	400	286	250
1.4	360	346	416	332	325	192	150
1.6	290	274	354	258	250	98	50

1.8	220	202	292	184	175	4	0
30°C	Normalized Concentration (C/S)						
D.(mm)	CLCS	CLCS-0. 1%F	CLCS-0. 5 %F	CLCS-1 %F	CLCS-0. 1%G	CLCS-0. 5%G	CLCS-1 %G
0	850	850	850	850	850	850	850
0.2	800	798	808	796	795	776	770
0.4	730	726	746	722	720	682	670
0.6	660	654	684	648	645	588	570
0.8	590	582	622	574	570	494	470
1	520	510	560	500	495	400	370
1.2	450	438	498	426	420	306	270
1.4	380	366	436	352	345	212	170
1.6	310	294	374	278	270	118	70
1.8	240	222	312	204	195	24	20
60°C	Normalized Concentration (C/S)						
D.(mm)	CLCS	CLCS-0. 1%F	CLCS-0. 5 %F	CLCS-1 %F	CLCS-0. 1%G	CLCS-0. 5%G	CLCS-1 %G
0	850	850	850	850	850	850	850
0.2	840	838	848	836	835	816	810
0.4	770	766	786	762	760	722	710
0.6	700	694	724	688	685	628	610
0.8	630	622	662	614	610	534	510
1	560	550	600	540	535	440	410
1.2	490	478	538	466	460	346	310
1.4	420	406	476	392	385	252	210
1.6	350	334	414	318	310	158	110
1.8	280	262	352	244	235	64	60

Note. D=Distance, CLCS=Cross-linked Chitosan, F=Fullerene, G=Graphene

Appendix B

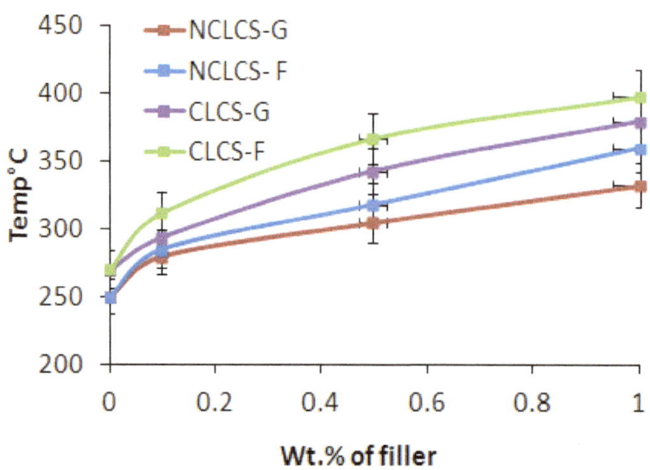

Figure 1. Glass transition temp versus wt.% of filler for CS nanocomposite membranes as a function of F and G

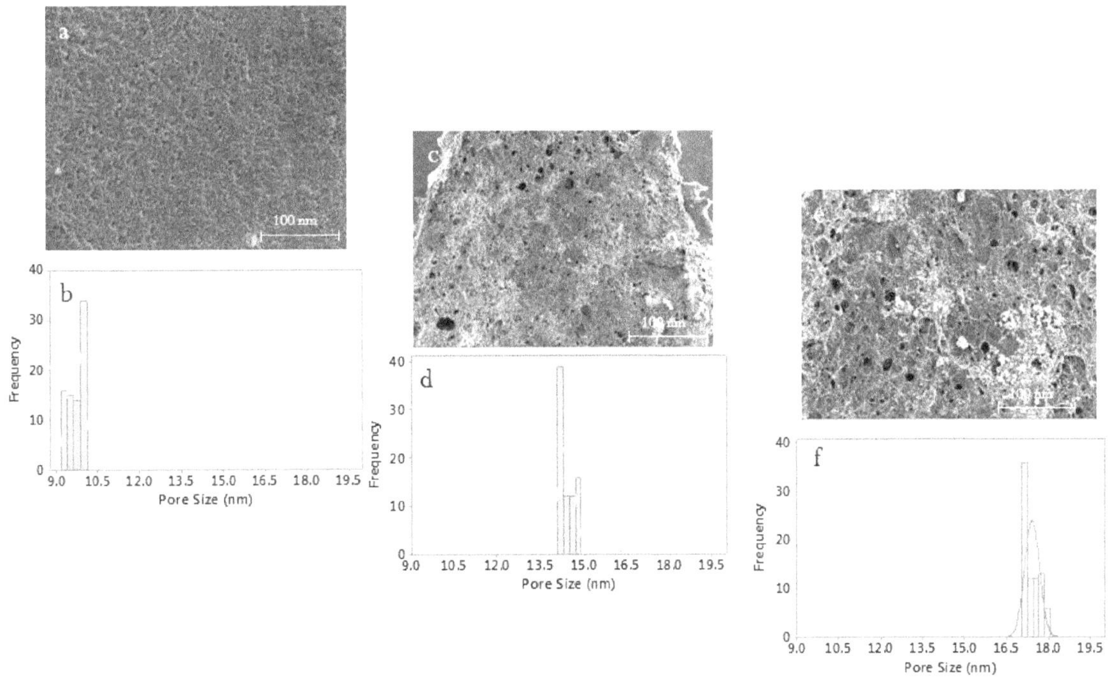

Figure 2. SEM images, pore size distribution of (a, b) NCLCS at 23°C (c, d) NCLCS at 30°C (e, f) NCLCS at 60 °C

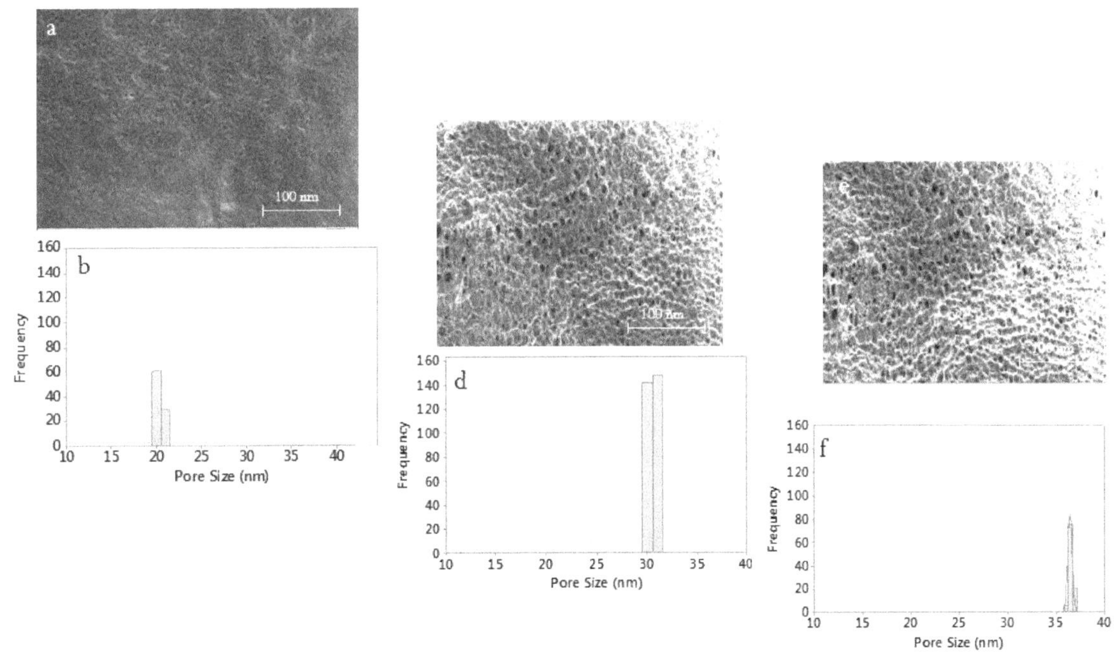

Figure 3. SEM images, pore size distribution of (a, b) CLCS at 23°C (c, d) CLCS at 30°F (e, f) CLCS at 60 °C

Cobalt(II) and Zinc(II) Complexes of Hexamethylenetetramine as Single Source Precursors for their Metal Oxide Nanoparticles

Che Dieudonne Tabong[1], Agwara Moise Ondoh[1], Divine Mbom Yufanyi[2] & Josepha Foba[3]

[1] Department of Inorganic Chemistry, Faculty of Science, University of Yaounde I, Yaounde, Cameroon

[2] Department of Chemistry, Faculty of Science, The University of Bamenda, Bamenda, Cameroon

[3] Department of Chemistry, Faculty of Science, University of Buea, Cameroon

Correspondence: Agwara Moise Ondoh, Department of Inorganic Chemistry, Faculty of Science, University of Yaounde I, Yaounde, Cameroon. E-mail: agwara29@yahoo.com

Abstract

Co_3O_4 and ZnO nanoparticles with different morphologies were synthesised by the thermal decomposition of single source precursors obtained from readily available and eco-friendly starting materials (hexamethylenetetramine and metal nitrates). The precursors, which were characterized by elemental analysis, Fourier transform infrared spectroscopy (FTIR), X-ray diffraction and thermal gravimetric analysis, were calcined at 500°C for 2 h, and the oxide samples obtained were characterized by FTIR, X-ray diffraction (XRD), field emission scanning electron microscopy (FESEM), transmission electron microscopy (TEM), selected area electron diffraction (SAED), and nitrogen physisorption. XRD showed that the oxides obtained were crystalline and free of extraneous impurity phases. The morphology of the nanoparticles obtained changed from cube-like (Co_3O_4) to hexagonal-prism like morphology (ZnO), while the particle size increased from 19.6 to 64.5 nm, respectively. The method used simple and cheap precursors, which should make it suitable for large-scale synthesis.

Keywords: Cobalt oxide, Zinc oxide, Nanoparticle, Thermal decomposition, Hexamethylenetetramine, XRD

1. Introduction

Metal oxide nanoparticles constitute a class of functional materials on which much attention has been focused recently due to their morphology and size-dependent physical and chemical properties (Devan, Patil, Lin, & Ma, 2012; Niederberger, 2007).

The size reduction of these metal oxides, which leads to novel properties that are different from those of the individual atoms as well as their bulk counterparts, makes them potential materials for application in the fields of catalysis, electronics, storage devices, gas sensors and magnetic resonance imaging (de Rivas, López-Fonseca, Jiménez-González, & Gutiérrez-Ortiz, 2012; Devan et al., 2012; Gomez & Tigli, 2013; Kahn et al., 2009).

Among these metal oxide nanoparticles, nanoscale cobalt and zinc oxides have received much scientific attention because they have interesting size-dependent physico-chemical properties for material as well as industrial applications. Cobalt oxide, especially the cubic Co_3O_4 phase with a spinel structure (Fd3m), is a p-type semiconductor which has been employed in several technological applications in the field of heterogeneous catalysis (Che, Liu, Fu, & Jiang, 2013; de Rivas et al., 2012), supercapacitors (Vijayakumar, Ponnalagi, Nagamuthu, & Muralidharan, 2013; H. Xu et al., 2014), anode materials in lithium ion rechargeable batteries (Huang et al., 2014; Rui et al., 2013), and magnetic materials (Moro, Yu Tang, Tuna, & Lester, 2013; Sharma & Jeevanandam, 2013).

ZnO is a thermally and chemically stable semiconducting material with a wide band gap of 3.37 eV and a large exciton binding energy of 60 meV at room temperature. It has found application in the domains of gas sensors, catalysis and optical devices (Faisal, Khan, Rahman, Jamal, & Abdullah, 2012; Gomez & Tigli, 2013; Guo & Peng, 2015; Hirai & Asada, 2005; Karami, Eskandari, Khodabakhshi, Hoseini, & Hashemian, 2013; Pál et al., 2012; Xiong, Ma, Wang, & Xia, 2011). ZnO is also biocompatible, nontoxic, chemically stable, and electrochemically active; properties which have enabled its application in biosensing (Gomez & Tigli, 2013).

Various methods have been employed for the morphology-controlled synthesis of Co_3O_4 and ZnO nanoparticles, which include microwave-assisted synthesis (Faisal et al., 2012; Vijayakumar et al., 2013), Chemical precipitation (Greene, Yuhas, Law, Zitoun, & Yang, 2006; Romo et al., 2011; Rui et al., 2013), hydrothermal (Dakhlaoui, Jendoubi, Smiri, Kanaev, & Jouini, 2009; Lester et al., 2012; Moro et al., 2013; Pál et al., 2012; Rui et al., 2013), sonochemical (Banerjee, Chakrabarti, Maitra, & Dutta, 2012), solvothermal synthesis (Y. Xu, Wang, Sun, Zhang, & Gao, 2010), pulsed layer deposition (Warang et al., 2012), and biosynthesis (Fu & Fu, 2015). These varied methods have led to nanoparticles with diverse morphologies like nanocubes (Lester et al., 2012), nanorods (Fu & Fu, 2015; Romo et al., 2011), nanospheres (Dakhlaoui et al., 2009; Darezereshki, Alizadeh, Bakhtiari, Schaffie, & Ranjbar, 2011; Hirai & Asada, 2005), nanowires (Greene et al., 2006), nanoflowers (Pál et al., 2012), and nanotubes (Li, Dou, & Bao, 2012).

These conventional methods, though successful, require specially designed and expensive equipment, extra purification steps, rigorous reaction conditions and/or relatively high temperatures. Some of the solution-based synthetic approaches employ surfactants and solvents which are toxic and not easily degraded in the environment. The synthesis of these nanomaterials with controlled size and shape is still a major challenge and large scale synthesis of phase pure Co_3O_4 and ZnO nanoparticles at relatively low temperatures, using readily available, environmentally benign and cost-effective precursors is a synthetic challenge. Thus, the preparation of cobalt oxide and zinc oxide nanoparticles by thermal decomposition of cobalt and zinc complexes becomes increasingly important mainly due to its cost-effectiveness, careful choice of precursors and calcination conditions, high yields, easy control of synthesis conditions, particle size, crystal structure, and purity. For example, the synthesis of metal oxide nanoparticles by the thermal decomposition of organometallic compounds or metal complexes has been reported (Ahmad, Ganguly, Ahmed, Ganguli, & Alhartomy, 2011; Huang et al., 2014; Kahn et al., 2009; Palacios-Hernández et al., 2012; Sharma & Jeevanandam, 2013). Porous Co_3O_4 with different morphologies has been prepared by the thermal decomposition of cobalt carbonates (300 - 500°C for 2 – 3 h) and oxalate precursors (400 - 500°C for 6 h) (Che et al., 2013; de Rivas et al., 2012; Nassar & Ahmed, 2011). Rivas *et al.* obtained Co_3O_4 by the direct calcination of cobalt nitrate and cobalt oxalate nanorods at 500°C (de Rivas et al., 2012). Co_3O_4 nanoparticles were obtained by solid-state thermal decomposition of a Schiff base complex [CoL_3] (L = N-salicylidin-2-chloroethylimine) at 450°C under air atmosphere for 3 h (Khalaji, Nikookar, Fejfarova, & Dusek, 2014). Hongwei and co-workers prepared Co_3O_4 microdiscs by a thermal decomposition of cobalt oxalate at 400°C, obtained by the solvothermal reaction of cobalt nitrate and ammonium oxalate in anhydrous alcohol (Che et al., 2013). Co_3O_4 nanoparticles were prepared via the decomposition of the pentammine(carbonato)cobalt(III) nitrate precursor complex $[Co(NH_3)_5CO_3]NO_3 \cdot 0.5H_2O$ at low temperature (175°C) (Farhadi & Safabakhsh, 2012). Palacios-Hernandez *et al.* reported the synthesis of Cu and Co metal oxide nanoparticles by the thermal decomposition of the corresponding tartrate complexes ($[M(C_4H_2O_6)_n]_m$, M = Cu(II), Co(II)) at 500°C for 5 h (Palacios-Hernández et al., 2012). Aghababazadeh *et al.* prepared ZnO nanoparticles by the thermal decomposition (170-380°C) of $ZnCO_3$ obtained by mechanochemical processing (Aghababazadeh, Mazinani, Mirhabibi, & Tamizifar, 2006). Hirai and co-workers obtained ZnO nanoparticles by the calcination of $Zn(OH)_2$ nanoparticles, which were prepared in a polyethylene glycol mono-4-nonylphenyl ether (NP-5)/cyclohexane reverse micellar system and incorporated into polyurea via an *in situ* polymerization of hexamethylene diisocyanate (Hirai & Asada, 2005). Rajesh *et al.* prepared ZnO nanoparticles by the mechanochemical synthesis of $ZnC_2O_4 \cdot 2H_2O$ nanoparticles followed by its thermal decomposition at 400°C (Rajesh, Vara Lakshmi, & Sunandana, 2012). Spherical ZnO nanoparticles have also been prepared by the thermal decomposition of $Zn_4(SO_4)(OH)_6 \cdot 0.5\ H_2O$ in air for 1 h at 825°C (Darezereshki et al., 2011).

The high costs, use of non-ecofriendly surfactants and solvents as well as the evolution of very toxic gaseous products during calcination, limits the use of some of these precursors. In continuation of our interest in the synthesis of phase pure metal oxide nanoparticles, we report herein a simple, environmentally benign and cost-effective method to synthesize Co_3O_4 and ZnO nanoparticles by the thermal decomposition of their respective metal-hexamethylenetetramine (M-HMTA) complexes. Hexamethylenetetramine (HMTA) is a cheap, ecofriendly and readily available heterocyclic organic compound with a cage-like structure. It is highly soluble in water and polar organic solvents. Some metal complexes of HMTA had previously been used as precursors for metal dispersions (Afanasiev et al., 2008), metal carbides and nitrides (Chouzier, Afanasiev, Vrinat, Cseri, & Roy-Auberger, 2006; Chouzier, Vrinat, Cseri, Roy-Auberger, & Afanasiev, 2011), and oxide nanoparticles (Yufanyi, Tendo, Ondoh, & Mbadcam, 2014).

2. Method

2.1 Materials

$Co(NO_3)_2.6H_2O$, $Zn(NO_3)_2.6H_2O$, hexamethylenetetramine and ethanol were obtained from Sigma Aldrich. All the chemicals were of analytical grade and were used as obtained without further purification.

2.2 Synthesis of Metal(II) - HMTA Complex

The precursors were synthesized according to a procedure previously reported for a polymeric $Co(HMTA)_2(NO_3)_2(H_2O)_2$ complex, with slight modifications (Ndifon et al., 2009).

HMTA (4 mmol; 0.5608 g) was dissolved in 20 mL of ethanol/water mixture (6:1 v/v) by sonicating for 20 minutes, at room temperature. The solution was magnetically stirred while stoichiometric amounts of the metal nitrate (2 mmol) in 10 mL of ethanol was added drop wise to the HMTA solution and the mixture was stirred for two hours. The pink precipitate (Co-HMTA) and the white precipitate (Zn-HMTA) that were obtained were filtered, washed several times with 99% ethanol and dried in a desiccator over silica gel. Pink crystals (Co-HMTA) and colourless crystals (Zn-HMTA) suitable for single crystal XRD were obtained from the filtrates within two weeks.

2.3 Synthesis of Cobalt and Zinc Oxide Nanoparticles

The dry M-HMTA complex (0.5 g) was ground and placed in a ceramic crucible that had previously been washed, dried and weighed. The crucible was placed into a furnace and heated at a ramp rate of 5°C per minute until the desired temperature was reached (500°C) and calcined in the presence of air for two hours. The sample was allowed to cool slowly to room temperature. The obtained black powder (case of cobalt) and the white product (case of zinc) were weighed and could easily be re-dispersed in water and ethanol. The samples are referred to as Co_3O_4-500 and ZnO-500.

2.4 Characterisation

Elemental analysis (C, H, N) of the precursor was carried out on a Flash 2000 Thermo Scientific analyser while the FT-IR spectra were recorded from 4000 to 400 cm^{-1} on a PerkinElmer Spectrum Two universal attenuated total reflectance Fourier transform infrared (UATR-FT-IR) spectrometer.

Single crystal X-ray data were collected using a graphite monochromatic Mo-Kα radiation (λ=1.54178 Å) from a Bruker APEX diffractometer. The structures were solved by direct methods and refined by full-matrix least squares on F^2 (Sheldrick, 1997a, 1997b). The non-hydrogen atoms were refined anisotropically. H atoms were included in calculated positions with C-H lengths of 0.95(CH), assigned isotopic thermal parameters and allowed to ride on their parent carbon atoms. All calculations were carried out using the SHELXTL package (Bruker, 2001).

Thermogravimetric measurements were obtained using a Pyris 6 PerkinElmer TGA 4000 thermal analyser. The TGA analyses were conducted between 30 and 900°C under nitrogen atmosphere at a flow rate of 20 mL/min and a temperature ramp of 10°C/min. The XRD diffractogram were recorded on a Bruker D8 Advance X-ray diffractometer using a Cu Kα radiation source (λ = 0.15406 nm, 40 kV and 40 mA). Scans were taken over the 2θ range from 10° to 100° in steps of 0.01° at room temperature in open quartz sample holders. The phases were identified with the help of the Bruker DIFFRACplus evaluation software in combination with the ICDD powder diffraction data base (International Centre for Diffraction Data). FESEM images and energy dispersive X-ray (EDX) spectra were obtained on a JEOL JSM-7600F field-emission scanning electron microscope (FESEM), with a lens diameter of 800 mm^2 (Oxford Instruments Ltd, UK) coupled to a silicon drift energy-dispersive X-ray detector. Transmission electron microscopy (TEM) was performed on a JEOL JEM-2100F microscope using a maximum acceleration voltage of 200 kV from the field emission gun. The particle size distributions were determined from the TEM images using the ImageJ software. N_2-physisorption experiments for the determination of the total surface area and the average pore diameter were conducted on a Micromeritics ASAP 2020 instrument. Prior to the measurement, the samples were degassed at 200°C for 6 h.

3. Results and Discussion

By careful choice of the precursor, coupled with a rational control of temperature and other conditions, nano-sized cobalt and zinc oxides were obtained by thermal decomposition of their corresponding hexamethylenetetramine (HMTA) complexes. These complexes were prepared in a single step, in high yield at room temperature.

The elemental analytical data of the precursors (Table 1) correspond closely to the empirical formulae $CoC_{12}H_{28}N_{10}O_8$ and $ZnC_{12}H_{44}N_{10}O_{16}$ which matches the structural formulae $[Co(HMTA)_2(H_2O)_2(NO_3)_2]$ and $[Zn(H_2O)_6](HMTA)_2(NO_3)_2 \cdot 4H_2O$, respectively, for cobalt and zinc.

Table 1. Elemental analytical data of the complexes

Complex	Yield (%)	Colour	Elemental Analysis: % Found (% Calc.)		
			% C	% H	% N
$[Co(HMTA)_2(H_2O)_2(NO_3)_2]$	85	Pink crystals	28.73 (28.86)	5.82 (5.65)	27.56 (28.05)
$[Zn(H_2O)_6](HMTA)_2(NO_3)_2 \cdot 4H_2O$	78	Colourless crystals	22.21 (22.18)	6.74 (6.82)	21.23 (21.55)

3.1 X-ray Crystal Structure

The crystal structures of the precursors Co-HMTA and Zn-HMTA were found to be the same as those of the complexes $[Co(HMTA)_2(H_2O)_2(NO_3)_2]$ (Ndifon et al., 2009) and $[Zn(H_2O)_6](HMTA)_2(NO_3)_2 \cdot 4H_2O$ (Singh et al., 2008) respectively, found in literature.

3.2 FTIR Analyses of the Precursor Complexes

The relevant infrared bands of HMTA and the precursor complexes are listed in Table 2, while the FTIR spectra of the precursor complexes and the corresponding oxides are shown in Figure 1. The broad bands at 3400 - 3501 cm^{-1} in the spectra of M-HMTA are attributed to ν(OH) of coordinated water , while that at 3383 cm^{-1} is assigned to ν(OH) of lattice water (Hee Ng, Guan Teoh, Moris, & Yang Yap, 2004). The band at 1235 cm^{-1}, assigned to the $\nu(CH_2)$ rocking vibration of the free HMTA ligand is observed at 1238 cm^{-1} in the Zn-HMTA complex, while it is split into 1240 and 1227 cm^{-1} in the Co-HMTA precursor suggesting that HMTA is coordinated to the cobalt ion (Ndifon et al., 2009). Strong prominent peaks at 812 and 1000 cm^{-1}due to the C-N stretching vibration of HMTA are shifted to 818-819 cm^{-1} and 1002 cm^{-1}, respectively in the Co-HMTA and Zn-HMTA precursor complexes (Jensen, 2002). The weak band observed at 1780 cm^{-1}shows the coordination of a monodentate nitrate ion, $Co\text{-}NO_3$ (Ndifon et al., 2009). The coordination of water molecules is also indicated by the IR bands in the region 400–600 cm^{-1}, assigned to $M\text{-}H_2O$ (Ndifon et al., 2009).

Table 2. Relevant FTIR bands of HMTA and the M-HMTA precursor complexes

HMTA	$[Co(HMTA)_2(NO_3)_2(H_2O)_2]$	$[Zn(H_2O)_6](HMTA)_2(NO_3)_2 \cdot 4H_2O$	Band Assignments
-	3501	3480	ν(OH) (coordinated water)
-	-	3383	ν(OH) (lattice water)
2955	-	2970	$\nu(CH_2)$
-	1780	-	$Co\text{-}NO_3$
-	1675	1674	HOH bend (lattice water)
1457	1475	1463	$\nu(CH_2)$ scissor (HMTA)
1370	1349	1343	$\nu(CH_2)$ wag (HMTA)
1236	1240	1238	$\nu(CH_2)$ rock (HMTA)
	1227		
1000	1002	1002	ν(CN) stretch (HMTA)
811	819	818	ν(CN) stretch (HMTA)
670	682	687	N-C-N bend (HMTA)
-	504	505	M-O stretch

3.3 Thermogravimetric Analysis

The TG curves of the precursor complexes are presented in Figure 2 and summarised in Table 3. The curves indicate that both complexes decompose in more than one step to give the metal oxides. The thermal decomposition of these precursors is accompanied by the evolution of a mixture of gases.

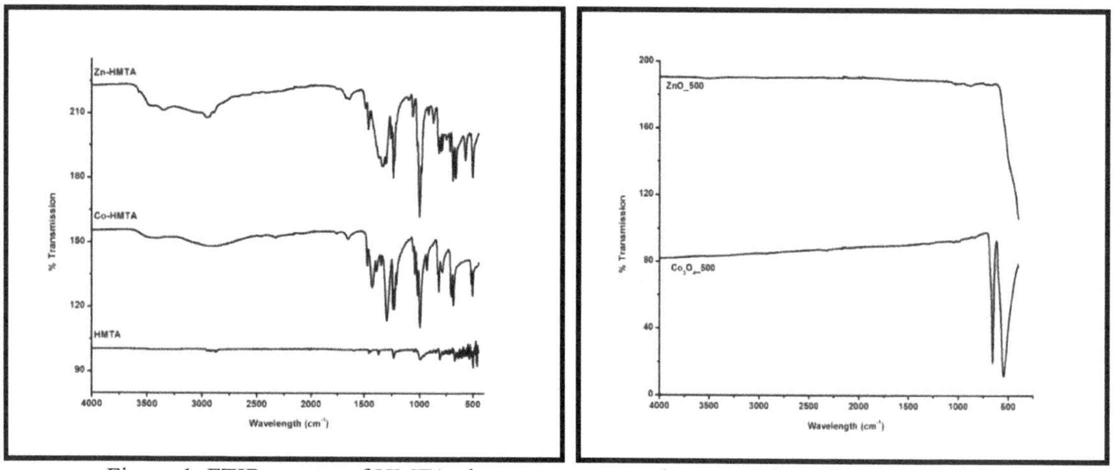

Figure 1. FTIR spectra of HMTA, the precursor complexes and the oxide nanoparticles

Table 3. Thermal decomposition data for the precursor complexes

Complex	Step	Temperature range (°C)	% Mass Loss	
			Obs.	Calc.
$[Co(HMTA)_2(NO_3)_2(H_2O)_2]$	I	135 - 198	8.5	10.02
	II	200 - 315	53.6	56
	III	320 - 540	13.6	12.4
$[Zn(H_2O)_6](HMTA)_2(NO_3)_2{\cdot}4H_2O$	I	40 - 100	14.8	13.8
	II	105 - 153	4.71	5.5
	III	160 – 420	42.4	43.1
	IV	420 - 740	23.4	27.4

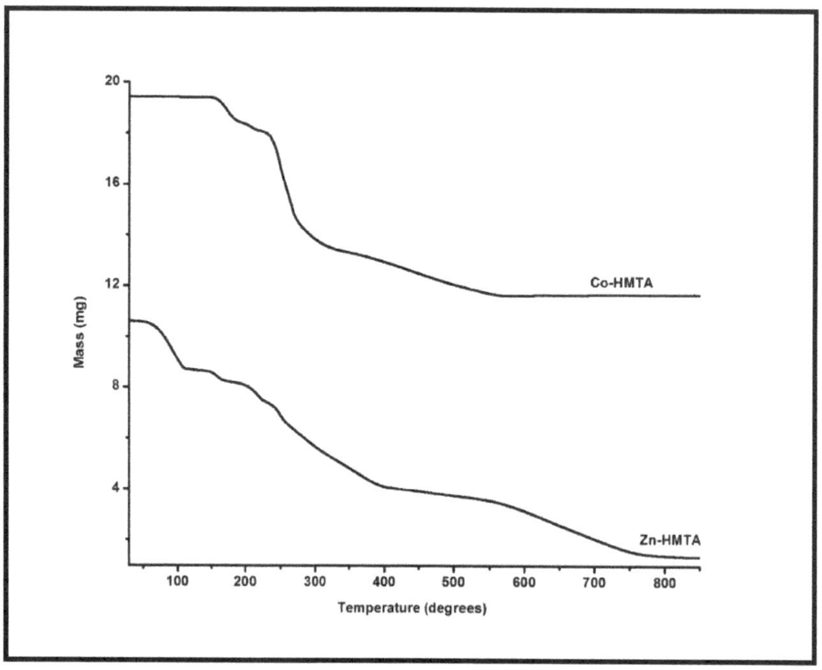

Figure 2. TG curves of the precursor complexes

For the Co-HMTA, the mass loss of 8.5% in the temperature region from 135 - 198°C is most probably due to the loss of two coordinated water molecules (10.02 %). The major mass loss of 53.6% takes place in the range 200 - 315°C, and is attributed to the complete decomposition of HMTA (56%) in the form of various gases (Afanasiev et al., 2008). In the range 320 – 540°C we have another distinctive decomposition step with a mass loss of 13.6% which can be assigned to the decomposition of the nitrate. A stable mass is reached at 540°C. The calcination temperature was chosen as 500°C from the derivative TG plot.

Considering the decomposition of the Zn-HMTA complex, the mass loss of 14.8% in the temperature region from 50 – 102°C is attributed to the loss of five water molecules (13.8%). The next decomposition step in the range 105 – 155°C with mass loss of 4.71% is probably due to the loss of two coordinated water molecules (5.5%). In the temperature range 155 – 380°C which, from the derivative TG plots, consists of several overlapping decomposition steps, there is a major mass loss of 42.4% which can be assigned to the decomposition of two HMTA molecules (43.1%) (Afanasiev et al., 2008; Hee Ng et al., 2004). The decomposition step in the range 400 – 740°C involves a mass loss of 23.4% which can be assigned to the decomposition of the two nitrates and two water molecules (calc. 27.4%). A residual mass of 14.6% is observed at 740°C which is probably due to zinc metal (10.1%).

The last decomposition step for Zn-HMTA is probably accompanied by the reduction of ZnO to zinc by carbon from 570˚C. From the Ellingham diagram, carbon will reduce zinc oxide above 950˚C for CO/CO_2 ratio of 1. In an oxygen deficient atmosphere, the ratio will probably decrease pushing the reduction temperature to the observed lower value.

The calcination temperature was chosen as 500°C from the derivative TG plot which indicates that 600°C is the optimum temperature. The decomposition pattern of the Zn-HMTA complex is slightly different from that reported in the literature, where the complex decomposes in two major steps in the temperature ranges 65 – 100°C and 600 – 645°C (Singh et al., 2008).

3.4 Structural Characterisation of the Oxide Nanoparticles

The FTIR spectra of the oxides are shown in Figure 1. The spectra show bands in the 500 - 650 cm^{-1} region which are attributed to the stretching vibration of the M-O bond. The bands at 547 and 645 cm^{-1} for cobalt oxide, suggest the formation of spinel Co_3O_4 that is $(Co^{2+})(Co^{3+})_2(O^{2-})_4$ (Makhlouf, Abu-Zied, & Mansoure, 2013; Nguimezong et al., 2014). These bands have been attributed to the (Co^{2+})-O-(Co^{3+}) and to (Co^{3+})-O-(Co^{3+}) vibrations, respectively (Makhlouf et al., 2013; Nguimezong et al., 2014).

The cobalt oxide and zinc oxide nanoparticles were obtained by thermal decomposition of their respective M-HMTA precursors at 500 °C. The XRD patterns of Co_3O_4 and ZnO (Figure 3) indicate that pure and crystalline metal oxide phases are formed in both cases. The samples have well defined diffraction patterns. The peaks for Co_3O_4 correspond to the (111), (220), (311), (222), (400), (511) and (440) crystals planes and match the spinel structure of cubic Co_3O_4 JCPDS card no 43-1003 (space group: Fd3m) (Lester et al., 2012). The peaks for ZnO correspond to the (100), (002), (101), (102), (110), (103), (200), (112), and (201) crystal planes and match the wurtzite structure of hexagonal ZnO JCPDS card no. 36-1451 (Zhang, Ram, Stefanakos, & Goswami, 2012). The average crystallite sizes determined by the Debye-Scherrer equation (Klug & (Eds.), 1974) from the peak-width at half-height of the (311) peak for Co_3O_4 and the (101) peaks for ZnO, were found to be 23 and 32.5 nm for Co_3O_4-500 and ZnO-500, respectively.

The morphologies and other structural features of the precursor complexes, Co_3O_4-500 and ZnO-500 samples were determined by FESEM, TEM and SAED. The FESEM images (Figure 4) show that the precursor complexes Co-HMTA (Figure 4a) and Zn-HMTA (Figure 4b) have spike-like and bar-like morphologies, respectively, while Co_3O_4-500 (Figure 4c) is highly porous and foam-like. The ZnO-500 particles (Figure 4d) exhibit a rod-like shape. The formation of highly porous foam can be attributed to the large amounts of gases evolved when HMTA and nitrates decompose (Afanasiev et al., 2008).

The corresponding TEM images (Figure 5) confirm cube-like (Co_3O_4) and hexagonal-prism like (ZnO) morphologies, with average particle sizes of 19.6 and 64.5 nm, respectively. The average particle diameter was determined after a log normal fitting of the data obtained from the TEM images. The SAED images (Figure 5) show patterns corresponding to the polycrystalline nature of the Co_3O_4 and ZnO formed.

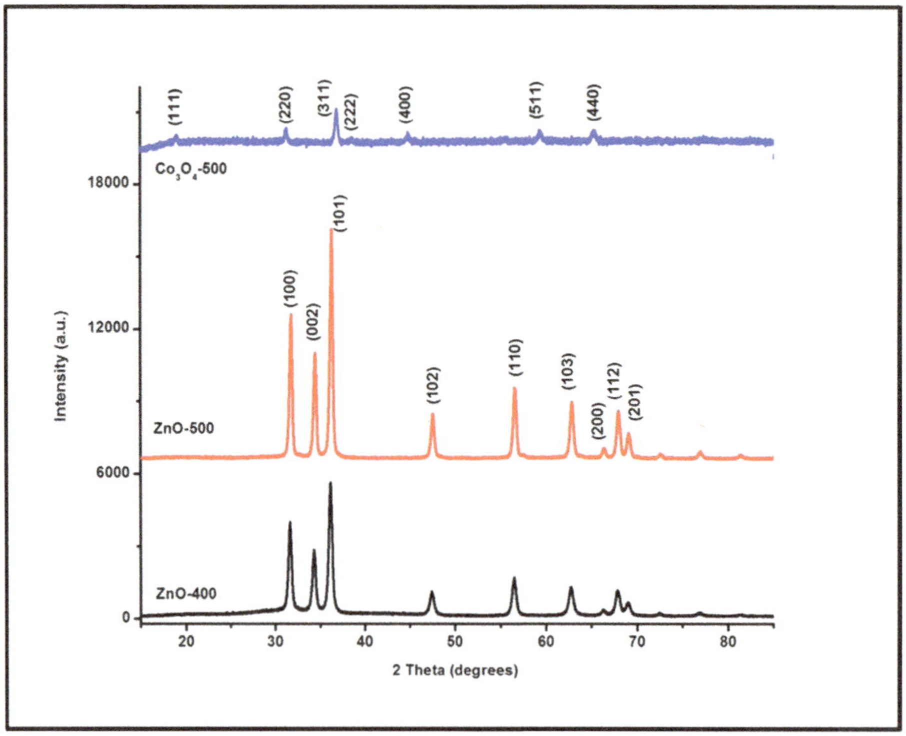

Figure 3. XRD patterns of Co_3O_4-500, ZnO-400 and ZnO-500 obtained by thermal decomposition

Figure 4. FESEM images of (a) Co-HMTA precursor; (b) Zn-HMTA precursor; (c) Co_3O_4-500 and (d) ZnO-500

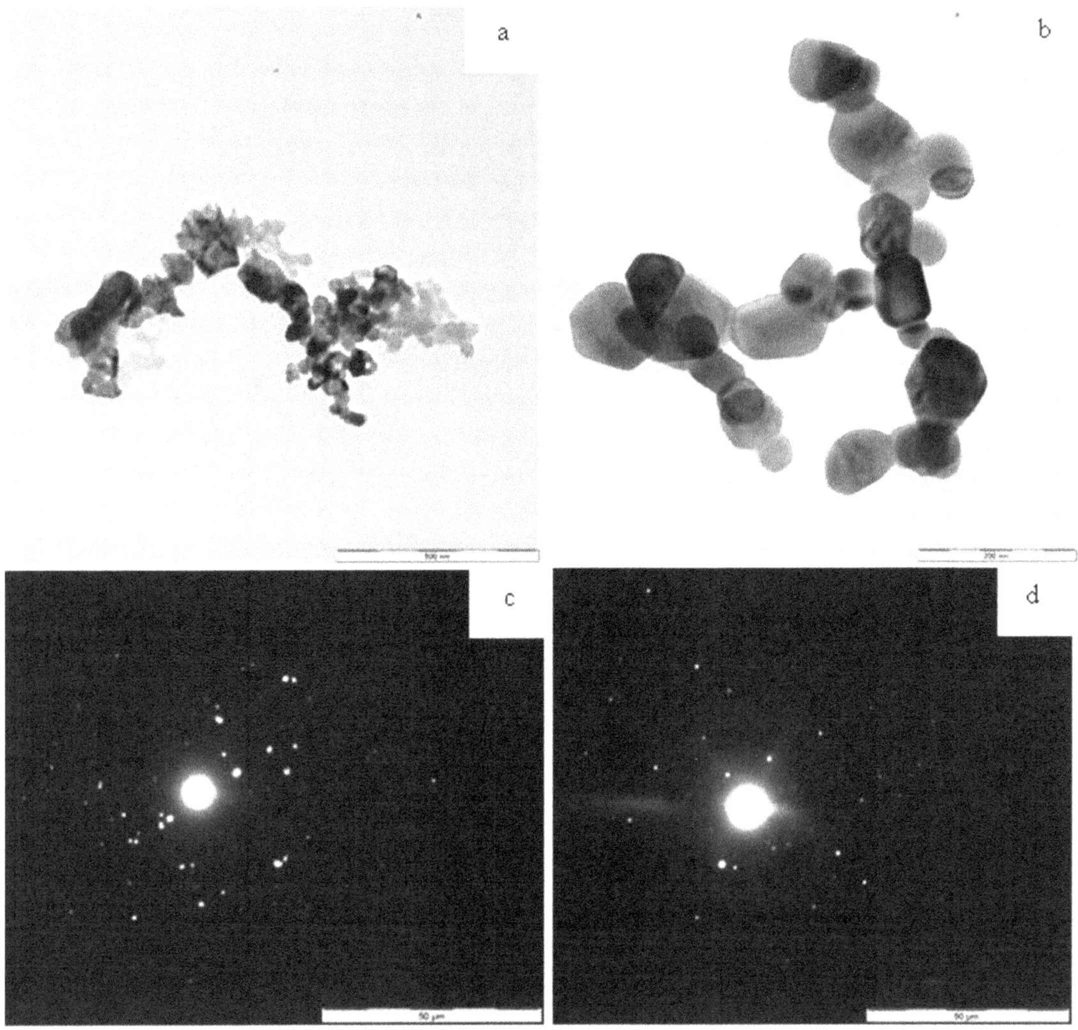

Figure 5. HRTEM images of (a) Co3O4 and (b) ZnO; SAED images of (c) Co3O4 and (d) ZnO

The particle size distributions and average particle sizes for Co_3O_4-500 and ZnO-500 nanoparticles are shown in Figure 6. From this histogram it is observed that the particle sizes for Co_3O_4-500 and ZnO-500 vary from 15 – 25 nm and 45 – 85 nm, respectively. The average particle sizes from TEM are consistent with values obtained by XRD.

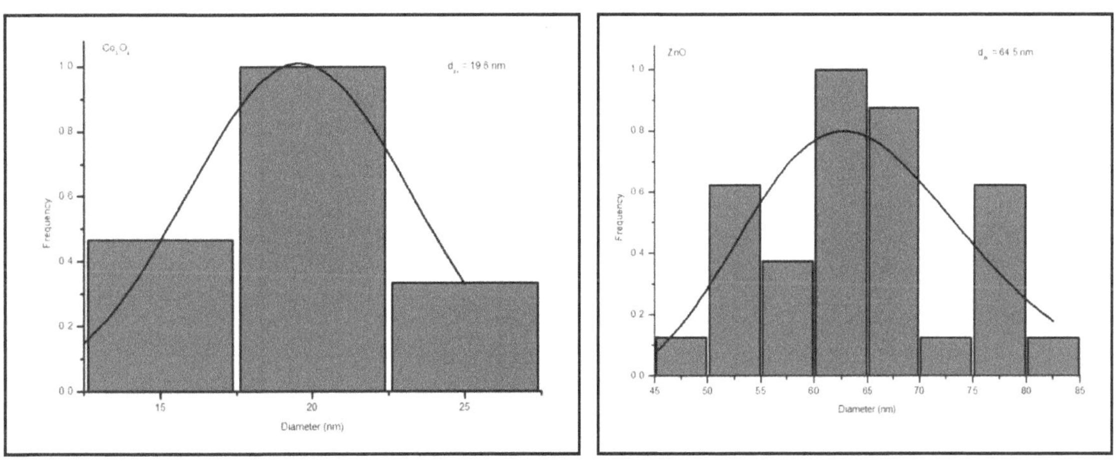

Figure 6. Particle size distribution for Co_3O_4-500 and ZnO-500 nanoparticles

For comparison, the precursors, decomposition time and temperatures used to obtain Co_3O_4-500 and ZnO-500 nanoparticles by thermal decomposition, with various morphologies and sizes, found in the literature are listed in Table 4. The results show that cube-like (Co_3O_4-500) and hexagonal-like (ZnO-500) nanoparticles with average particle sizes of 19.6 and 64.5 nm, respectively obtained from the M-HMTA precursors of this study compares favorably with those obtained from other starting materials. It can also be observed from Table 4 that our starting materials are the simplest, most readily available, and very cost-effective.

Table 4. Particle sizes (TEM) and morphologies of Co_3O_4-500 and ZnO-500 nanoparticles prepared by the thermal decomposition of various precursors at different calcination temperatures

Precursor	Calcination		Particle Size (TEM) (nm)	Morphology	Ref.
	Time (h)	Temp. (°C)			
$Co(NO_3)_2 \cdot 6H_2O + (NH_4)_2CO_3$	2 - 3	300 -500	12 - 14	Pseudo-hexagonal	(de Rivas et al., 2012)
$Co(acetate)_2$ + oxalic acid	2 -3	500	12 - 14	Pseudo-hexagonal	(de Rivas et al., 2012)
$CoCl_2$ or $Co(NO_3)_2$ or $Co(acetate)_2$ +urea + CTAB	6	400 - 500	30 -39	Rounded and squared	(Nassar & Ahmed, 2011)
$Co(NO_3)_2 + (NH_4)_2OX$	6	400	25	microdiscs	(Che et al., 2013)
$[Co(NH_3)CO_3]NO_3 \cdot 0.5H_2O$	1	175 - 300	10	Quasi-spherical	(Farhadi & Safabakhsh, 2012)
$Zn_4(SO_4)(OH)_6 \cdot 0.5\ H_2O$	1	825	92	Spherical	(Darezereshki et al., 2011)
$[Zn(H_2O)_6](HMTA)_2(NO_3)_2 \bullet 4H_2O$	2	500	64.5	Hexagonal-like	This work

3.5 Surface Area and Pore Size Analyses

The surface areas of Co_3O_4-500 and ZnO-500 were determined by N_2 physisorption. All isotherms (Figure 7) were typical of mesoporous solids (type IV), for which both the surface area and the pore size distribution can be determined (Greg & Sing, 1982). The hysteresis loops (according to the IUPAC classification) were of type H1, which are often obtained for agglomerates or compacts of spheroidal particles of fairly uniform size and array (Greg & Sing, 1982). The BET surface areas of the samples (according to Brunauer, Emmett and Teller) are 5.1 and 10.4 m^2/g, respectively. The corresponding Barrett-Joyner-Halenda (BJH) desorption pore sizes are 25 and 12 nm. Table 5 summarizes the results.

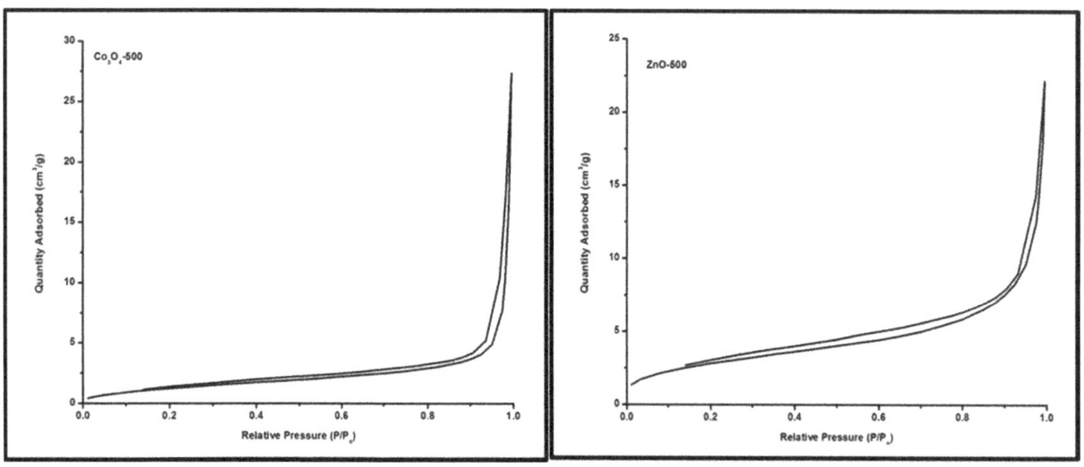

Figure 7. N_2 adsorption/desorption isotherms for the Co_3O_4-500 and ZnO-500 samples

Table 5. Surface area, average pore diameter (APD) and pore volume of the samples

Sample	BET Surface Area (m^2/g)	BJH Desorption APD (nm)	Pore Volume (cm^3/g)
Co_3O_4-500	5.1	25	0.043
ZnO-500	10.4	12	0.035

4. Conclusion

Pure and crystalline Co_3O_4 and ZnO nanoparticles, with defined morphologies, have been obtained from their corresponding metal-HMTA complexes by thermal decomposition. While Co_3O_4 is foam-like, ZnO nanoparticles have a rod-like shape according to FESEM. The corresponding TEM images confirm cube-like (Co_3O_4) and hexagonal-prism like (ZnO) morphologies, with average particle sizes of 19.6 and 64.5 nm, respectively. All oxide particles were mesoporous with BET surface areas of 5.1 m^2/g (Co_3O_4) and 10.4 m^2/g (ZnO). This technique that makes use of simple and readily available precursors for metal oxide nanoparticle synthesis could be extended to the synthesis of other metal oxide nanoparticles.

Acknowledgements

The authors acknowledge assistance by the Electron Microscopy and Tribocorrosion Facility of the Tshwane University of Technology (SEM analyses). The authors thank Dr. Khamlich Saleh (University of Pretoria) for assistance with the TEM images

References

Afanasiev, P., Chouzier, S., Czeri, T., Pilet, G., Pichon, C., Roy, M., & Vrinat, M. (2008). Nickel and Cobalt Hexamethylentetramine Complexes $(NO_3)_2Me(H_2O)_6(HMTA)_2 \cdot 4H_2O$ (Me = Co^{2+}, Ni^{2+}): New Molecular Precursors for the Preparation of Metal Dispersions. *Inorganic Chemistry, 47*(7), 2303-2311. http://dx.doi.org/10.1021/ic7013013

Aghababazadeh, R., Mazinani, B., Mirhabibi, A., & Tamizifar, M. (2006). ZnO Nanoparticles Synthesised by mechanochemical processing. *Journal of Physics: Conference Series, 26*(1), 312.

Ahmad, T., Ganguly, A., Ahmed, J., Ganguli, A. K., & Alhartomy, O. A. A. (2011). Nanorods of transition metal oxalates: A versatile route to the oxide nanoparticles. *Arabian Journal of Chemistry, 4*(2), 125-134. http://dx.doi.org/10.1016/j.arabjc.2010.06.041

Banerjee, P., Chakrabarti, S., Maitra, S., & Dutta, B. K. (2012). Zinc oxide nano-particles: Sonochemical synthesis, characterization and application for photo-remediation of heavy metal. *Ultrasonics Sonochemistry, 19*(1), 85-93. http://dx.doi.org/10.1016/j.ultsonch.2011.05.007

Bruker. (2001). SMART (Version 5.625), SADABS (Version 2.03a) and SHELXTL (Version 6.12). Madison, Wisconsin, USA: Bruker AXS Inc.

Che, H., Liu, A., Fu, Q., & Jiang, R. (2013). Facile synthesis of porous cobalt oxide microdiscs and their catalytic property in CO oxidation. *Materials Letters, 93*(0), 240-242. http://dx.doi.org/10.1016/j.matlet.2012.11.108

Chouzier, S., Afanasiev, P., Vrinat, M., Cseri, T., & Roy-Auberger, M. (2006). One-step synthesis of dispersed bimetallic carbides and nitrides from transition metals hexamethylenetetramine complexes. *Journal of Solid State Chemistry, 179*(11), 3314-3323. http://dx.doi.org/10.1016/j.jssc.2006.06.026

Chouzier, S., Vrinat, M., Cseri, T., Roy-Auberger, M., & Afanasiev, P. (2011). HDS and HDN activity of (Ni,Co)Mo binary and ternary nitrides prepared by decomposition of hexamethylenetetramine complexes. *Applied Catalysis A: General, 400*(1–2), 82-90. http://dx.doi.org/10.1016/j.apcata.2011.04.023

Dakhlaoui, A., Jendoubi, M., Smiri, L. S., Kanaev, A., & Jouini, N. (2009). Synthesis, characterization and optical properties of ZnO nanoparticles with controlled size and morphology. *Journal of Crystal Growth, 311*(16), 3989-3996. http://dx.doi.org/10.1016/j.jcrysgro.2009.06.028

Darezereshki, E., Alizadeh, M., Bakhtiari, F., Schaffie, M., & Ranjbar, M. (2011). A novel thermal decomposition method for the synthesis of ZnO nanoparticles from low concentration $ZnSO_4$ solutions. *Applied Clay Science, 54*(1), 107-111. http://dx.doi.org/10.1016/j.clay.2011.07.023

de Rivas, B., López-Fonseca, R., Jiménez-González, C., & Gutiérrez-Ortiz, J. I. (2012). Highly active behaviour of nanocrystalline Co_3O_4 from oxalate nanorods in the oxidation of chlorinated short chain alkanes. *Chemical Engineering Journal, 184*(0), 184-192. http://dx.doi.org/10.1016/j.cej.2012.01.027

Devan, R. S., Patil, R. A., Lin, J.-H., & Ma, Y.-R. (2012). One-Dimensional Metal-Oxide Nanostructures: Recent Developments in Synthesis, Characterization, and Applications. *Advanced Functional Materials, 22*(16), 3326-3370. http://dx.doi.org/10.1002/adfm.201201008

Faisal, M., Khan, S. B., Rahman, M. M., Jamal, A., & Abdullah, M. M. (2012). Fabrication of ZnO nanoparticles based sensitive methanol sensor and efficient photocatalyst. *Applied Surface Science, 258*(19), 7515-7522. http://dx.doi.org/10.1016/j.apsusc.2012.04.075

Farhadi, S., & Safabakhsh, J. (2012). Solid-state thermal decomposition of the $[Co(NH_3)_5CO_3]NO_3 \cdot 0.5H_2O$ complex: A simple, rapid and low-temperature synthetic route to Co_3O_4 nanoparticles. *Journal of Alloys and Compounds, 515*(0), 180-185. http://dx.doi.org/10.1016/j.jallcom.2011.11.135

Fu, L., & Fu, Z. (2015). *Plectranthus amboinicus* leaf extract-assisted biosynthesis of ZnO nanoparticles and their photocatalytic activity. *Ceramics International, 41*(2, Part A), 2492-2496. http://dx.doi.org/10.1016/j.ceramint.2014.10.069

Gomez, J. L., & Tigli, O. (2013). Zinc oxide nanostructures: from growth to application. *Journal of Materials Science, 48*(2), 612-624. http://dx.doi.org/10.1007/s10853-012-6938-5

Greene, L. E., Yuhas, B. D., Law, M., Zitoun, D., & Yang, P. (2006). Solution-Grown Zinc Oxide Nanowires. *Inorganic Chemistry, 45*(19), 7535-7543. http://dx.doi.org/10.1021/ic0601900

Greg, S. J., & Sing, K. S. W. (1982). *Adsorption, Surface Area and Porosity* (2nd ed.): Academic Press.

Guo, J., & Peng, C. (2015). Synthesis of ZnO nanoparticles with a novel combustion method and their C_2H_5OH gas sensing properties. *Ceramics International, 41*(2, Part A), 2180-2186. http://dx.doi.org/10.1016/j.ceramint.2014.10.017

Hee Ng, C., Guan Teoh, S., Moris, N., & Yang Yap, S. (2004). Structural, infrared spectral and thermogravimetric analysis of a hydrogen-bonded assembly of cobalt(II) and nickel(II) mixed complex cations with hexamethylenetetraamine and aqua ligands: $\{[M(hmt)_2(H_2O)_4][M(H_2O)_6]\}(SO_4)_2 \cdot 6H_2O$. *Journal of Coordination Chemistry, 57*(12), 1037-1046. http://dx.doi.org/10.1080/00958970412331281791

Hirai, T., & Asada, Y. (2005). Preparation of ZnO nanoparticles in a reverse micellar system and their photoluminescence properties. *Journal of Colloid and Interface Science, 284*(1), 184-189. http://dx.doi.org/10.1016/j.jcis.2004.09.069

Huang, Y., Chen, C., An, C., Xu, C., Xu, Y., Wang, Y., . . . Yuan, H. (2014). Synthesis of Cobalt based Complexes and conversion to Co_3O_4 nanoparticles as a high performance anode for lithium ion battery. *Electrochimica Acta, 145*(0), 34-39. http://dx.doi.org/10.1016/j.electacta.2014.08.085

Jensen, J. O. (2002). Vibrational frequencies and structural determinations of hexamethylenetetraamine. *Spectrochimica Acta Part A: Molecular and Biomolecular Spectroscopy, 58*(7), 1347-1364. http://dx.doi.org/10.1016/S1386-1425(01)00585-6

Kahn, M. L., Glaria, A., Pages, C., Monge, M., Saint Macary, L., Maisonnat, A., & Chaudret, B. (2009). Organometallic chemistry: an alternative approach towards metal oxide nanoparticles. *Journal of Materials Chemistry, 19*(24), 4044-4060. http://dx.doi.org/10.1039/b818935h

Karami, B., Eskandari, K., Khodabakhshi, S., Hoseini, S. J., & Hashemian, F. (2013). Green synthesis of three substituted methane derivatives by employing ZnO nanoparticles as a powerful and recyclable catalyst. *RSC Advances, 3*(45), 23335-23342. http://dx.doi.org/10.1039/c3ra42993h

Khalaji, A. D., Nikookar, M., Fejfarova, K., & Dusek, M. (2014). Synthesis of new cobalt(III) Schiff base complex: A new precursor for preparation Co_3O_4 nanoparticles via solid-state thermal decomposition. *Journal of Molecular Structure, 1071*(0), 6-10. http://dx.doi.org/10.1016/j.molstruc.2014.04.043

Klug, H. P., & (Eds.), L. E. A. (1974). *X-ray Diffraction Procedures for Polycrystalline and Amorphous Materials*. New York: Wiley.

Lester, E., Aksomaityte, G., Li, J., Gomez, S., Gonzalez-Gonzalez, J., & Poliakoff, M. (2012). Controlled continuous hydrothermal synthesis of cobalt oxide (Co_3O_4) nanoparticles. *Progress in Crystal Growth and Characterization of Materials, 58*(1), 3-13. http://dx.doi.org/10.1016/j.pcrysgrow.2011.10.008

Li, X., Dou, W., & Bao, N. (2012). Hydrothermal synthesis of tubular ZnO materials. *Materials Letters, 68*(0), 140-142. http://dx.doi.org/10.1016/j.matlet.2011.10.036

Makhlouf, M. T., Abu-Zied, B. M., & Mansoure, T. H. (2013). Direct Fabrication of Cobalt Oxide Nanoparticles Employing Sucrose as a Combustion Fuel. *Journal of Nanoparticles, 2013, Article ID 384350*, 7 pages. http://dx.doi.org/10.1155/2013/384350

Moro, F., Yu Tang, S. V., Tuna, F., & Lester, E. (2013). Magnetic properties of cobalt oxide nanoparticles synthesised by a continuous hydrothermal method. *Journal of Magnetism and Magnetic Materials, 348*, 1-7. http://dx.doi.org/10.1016/j.jmmm.2013.07.064

Nassar, M. Y., & Ahmed, I. S. (2011). Hydrothermal synthesis of cobalt carbonates using different counter ions: An efficient precursor to nano-sized cobalt oxide (Co_3O_4). *Polyhedron, 30*(15), 2431-2437. http://dx.doi.org/10.1016/j.poly.2011.05.039

Ndifon, P. T., Agwara, M. O., Paboudam, A. G., Yufanyi, D. M., Ngoune, J., Galindo, A., Alvarez, E., Mohamadou, A. (2009). Synthesis, characterisation and crystal structure of a cobalt(II)-hexamethylenetetramine coordination polymer. *Transition Metal Chemistry, 34*(7), 745-750. http://dx.doi.org/10.1007/s11243-009-9257-1

Nguimezong, M. B. N., Foba-Tendo, J., Yufanyi, D. M., Etape, E. P., Eko, J. N., & Ngolui, L. J. (2014). Averrhoa carambola: A Renewable Source of Oxalic Acid for the Facile and Green Synthesis of Divalent Metal (Fe,

Co, Ni, Zn, and Cu) Oxalates and Oxide Nanoparticles. *Journal of Applied Chemistry, 2014, Article ID 767695*, 9 pages. http://dx.doi.org/10.1155/2014/767695

Niederberger, M. (2007). Nonaqueous Sol–Gel Routes to Metal Oxide Nanoparticles. *Accounts of Chemical Research, 40*(9), 793-800. http://dx.doi.org/10.1021/ar600035e

Pál, E., Hornok, V., Kun, R., Chernyshev, V., Seemann, T., Dékány, I., & Busse, M. (2012). Growth of raspberry-, prism- and flower-like ZnO particles using template-free low-temperature hydrothermal method and their application as humidity sensors. *Journal of Nanoparticle Research C7 - 1002, 14*(8), 1-14. http://dx.doi.org/10.1007/s11051-012-1002-6

Palacios-Hernández, T., Hirata-Flores, G. A., Contreras-López, O. E., Mendoza-Sánchez, M. E., Valeriano-Arreola, I., González-Vergara, E., & Méndez-Rojas, M. A. (2012). Synthesis of Cu and Co metal oxide nanoparticles from thermal decomposition of tartrate complexes. *Inorganica Chimica Acta, 392*(0), 277-282. http://dx.doi.org/10.1016/j.ica.2012.03.039

Rajesh, D., Vara Lakshmi, B., & Sunandana, C. S. (2012). Two-step synthesis and characterization of ZnO nanoparticles. *Physica B: Condensed Matter, 407*(23), 4537-4539. http://dx.doi.org/10.1016/j.physb.2012.07.050

Romo, L. E., Saade, H., Puente, B., Lopez, M. L., Betancourt, R., & Lopez, R. G. (2011). Precipitation of Zinc Oxide Nanoparticles in Bicontinuous Microemulsions. *Journal of Nanomaterials, 2011, Article ID 145963*, 9 pages. http://dx.doi.org/10.1155/2011/145963

Rui, X., Tan, H., Sim, D., Liu, W., Xu, C., Hng, H. H., ... Yan, Q. (2013). Template-free synthesis of urchin-like Co_3O_4 hollow spheres with good lithium storage properties. *Journal of Power Sources, 222*(0), 97-102. http://dx.doi.org/10.1016/j.jpowsour.2012.08.094

Sharma, G., & Jeevanandam, P. (2013). Synthesis of MgO supported Co_3O_4 nanoparticles by a novel thermal decomposition approach and studies on their magnetic properties. *Microporous and Mesoporous Materials, 165*(0), 55-62. http://dx.doi.org/10.1016/j.micromeso.2012.07.034

Sheldrick, G. M. (1997a). SHELXL97, Program for crystal structure refinement. University of Göttingen, Germany.

Sheldrick, G. M. (1997b). SHELXS97, Program for Crystal Structure solution. University of Göttingen, Germany.

Singh, G., Baranwal, B. P., Kapoor, I. P. S., Kumar, D., Singh, C. P., & Fröhlich, R. (2008). Some transition metal nitrate complexes with hexamethylenetetramine: Part LV. Preparation, X-ray crystallography and thermal decomposition. *Journal of Thermal Analysis and Calorimetry, 91*(3), 971-977. http://dx.doi.org/10.1007/s10973-007-8615-5

Vijayakumar, S., Ponnalagi, A. K., Nagamuthu, S., & Muralidharan, G. (2013). Microwave assisted synthesis of Co_3O_4 nanoparticles for high-performance supercapacitors. *Electrochimica Acta, 106*(0), 500-505. http://dx.doi.org/10.1016/j.electacta.2013.05.121

Warang, T., Patel, N., Santini, A., Bazzanella, N., Kale, A., & Miotello, A. (2012). Pulsed laser deposition of Co_3O_4 nanoparticles assembled coating: Role of substrate temperature to tailor disordered to crystalline phase and related photocatalytic activity in degradation of methylene blue. *Applied Catalysis A: General, 423–424*(0), 21-27. http://dx.doi.org/10.1016/j.apcata.2012.02.037

Xiong, H.-M., Ma, R.-Z., Wang, S.-F., & Xia, Y.-Y. (2011). Photoluminescent ZnO nanoparticles synthesized at the interface between air and triethylene glycol. *Journal of Materials Chemistry, 21*(9), 3178-3182. http://dx.doi.org/10.1039/c0jm02577a

Xu, H., Gao, L., Zhang, Q., Li, J., Diwu, J., Chou, X., ... Xue, C. (2014, Article ID 723057). Preparation Method of Co_3O_4 Nanoparticles Using Degreasing Cotton and Their Electrochemical Performances in Supercapacitors. *Journal of Nanomaterials, 2014*, 9 pages. http://dx.doi.org/10.1155/2014/723057

Xu, Y., Wang, C., Sun, Y., Zhang, G., & Gao, D. (2010). Fabrication and characterization of nearly monodisperse Co_3O_4 nanospheres. *Materials Letters, 64*(11), 1275-1278. http://dx.doi.org/10.1016/j.matlet.2010.03.007

Yufanyi, D. M., Tendo, J. F., Ondoh, A. M., & Mbadcam, J. K. (2014). CdO Nanoparticles by Thermal Decomposition of a Cadmium-Hexamethylenetetramine Complex. *Journal of Materials Science Research, 3*(3), 1-11. http://dx.doi.org/10.5539/jmsr.v3n3p1

Zhang, Y., Ram, M. K., Stefanakos, E. K., & Goswami, D. Y. (2012). Synthesis, Characterization, and Applications of ZnO Nanowires. *Journal of Nanomaterials, 2012, Article ID 624520*, 22 pages. http://dx.doi.org/10.1155/2012/624520

20

Development of Effective Textile-Reinforced Concrete Noise Barrier

Henrik L. Funke[1], Sandra Gelbrich[1] & Lothar Kroll[1]

[1] Institute of Lightweight Structures, Technische Universität Chemnitz, Chemnitz, Germany

Correspondence: Henrik Funke, Institute of Lightweight Structures, Chemnitz University of Technology, Chemnitz, Germany, E-mail: henrik.funke@mb.tu-chemnitz.de

Abstract

Thin-walled, high-strength concrete elements exhibiting low system weight and great slenderness can be created with a large degree of lightweight structure using the textile-reinforced, load-bearing concrete (TRC) slab and a shell with a very high level of sound absorption. This was developed with the objective of lowering system weight, and then implemented operationally in construction.

Arising from the specifications placed on the load-bearing concrete slab, the following took place: an adapted fine-grain concrete matrix was assembled, a carbon warp-knit fabric was modified and integrated into the fine concrete matrix, a formwork system at prototype scale was designed enabling noise barriers to be produced with an application-oriented approach and examined in practically investigations within the context of the project. This meant that a substantial lowering of the load-bearing concrete slab's system weight was possible, which led to a decrease in transport and assembly costs.

Keywords: textile-reinforced concrete, noise barrier, high-performance concrete, durability

1. Introduction

Noise is a consequence of increasing mobility and has a negative effect on both human beings and the environment. The continual rise in traffic density, which grew by more than 70 percent on motor ways between 1980 and 2012 (Lohre et al., 2012), has led to greater noise emissions. Today, this makes sound protection measures more important than ever. It does not matter whether these noise barriers are for major roads, motorways or high-speed sections. For such systems, there are very complex and contain restrictions with regards to load-bearing capacity, serviceability and functionality. Alongside acoustic and mechanical specifications including durability, economic and ecological aspects additionally come to the fore. The objective is savings on resources not just based on optimizing weight and minimizing manufacturing costs but also leading to decreasing expenditure for transport and assembly.

In Germany, about 50 percent of noise barriers are erected in concrete. They are mostly composed of a load-bearing layer made from steel reinforced concrete and a noise-absorbing facing concrete shell set up on one or both sides and made from lightweight concrete. The usual material thickness with the load-bearing concrete layer is about 12 to 15 cm due to the concrete coverage needed in terms of passive protection against corrosion. Utilising textile reinforced concrete in this case offers great potential for reducing weight and constructing in a resource-efficient and sustainable manner. Textile reinforced concrete is an innovative high-performance composite material consisting of a fine concrete matrix and textile reinforcement (Curbach & Scheerer, 2011; Funke et al., 2014; Brameshuber, 2006). The development of this new type of composite material has made a construction method available which utilizes reinforcement that is generally not susceptible to corrosion due to environmental influences (e.g. air humidity and chloride) (Curbach et al., 2008; Curbach & Jesse, 2010). The concrete coverages needed in steel reinforced concrete construction can thus be substantially reduced by employing textile reinforced concrete (Curbach & Scheerer, 2012; Funke et al., 2013). Just a few millimeters minimum concrete cover is necessary to safeguard the transfer of bonding forces between the fine concrete matrix and the textile reinforcement (Mobasher et al., 2014; Curbach et al., 2013). This means that thin-walled, free-formed surfaces can be created with a great degree of lightweight structure that feature low system weight and great slenderness.

This paper reports the development of a noise barrier, which consists of a load-bearing concrete made of textile reinforced concrete and a noise-absorbent made of a lightweight concrete. An important part of this work is durability and so the lifespan of the noise barrier.

2. Materials and Methods

2.1 Components for Load-Bearing and Noise-Absorbent Concretes

Table 1 illustrates the qualitative and quantitative composition of the load-bearing concrete and the noise-absorbent concrete. The load-bearing concrete contains, besides type CEM I 52.5 R Portland cement according to DIN EN 206, pozzolanic fly ash and silica fume as binding agents (Table 1), which have a positive action on both fresh and hardened concrete characteristics in respect of sedimentation slability, durability and mechanical strength. Two quartz sands and dolomite powder are utilized as aggregate or filler. The alkali resistant (AR) glass fibres, which lower the propensity to shrinkage cracking and can enhance first crack tension in the concrete, are 12 mm long and have a length mass of 45 tex. The high-performance superplasticiser employed has a 30 percent by weight proportion of polycarboxylate ether (PCE). The noise-absorbent concrete has a monodisperse (single particle) aggregate made of round expanded clay in a fraction of 0 to 2 mm.

Table 1. Composition of the load-bearing and the absorber concrete

Component	Load-bearing in kg/m³	Absorber in kg/m³
White cement 52.5 R	495	250
Amorphous alumosilicate	150	-
Quartz sand 0/2	310	-
Quartz sand 0/1	900	-
Expanded clay	-	970
Dolomite powder ($x_{5ß}$= 70 μm)	190	-
Water	210	150
AR-glass fibres (12 mm, integral)	14	-
superplasticizers	15	-
Water-binder-ratio	0.38	0.60

2.2 Textile for Textile Reinforced Concrete

Two layers of a two-dimensional bidirectional warp-knit fabric of carbon were used for the reinforcement of TRC (Figure 1). The warp and weft yarn had a length weight of 3300 g/km (≙ 3300 tex) and a tensile strength of 1576 MPa. The mesh size of the 15 mass percent impregnated warp-knit fabric (measured by thermogravimetry) was 10.8×18.0 mm². The warp and weft yarn consisted of 50000 carbon filaments with a diameter of 7 μm. The degree of reinforcement in concrete was approximately 1.1% by volume.

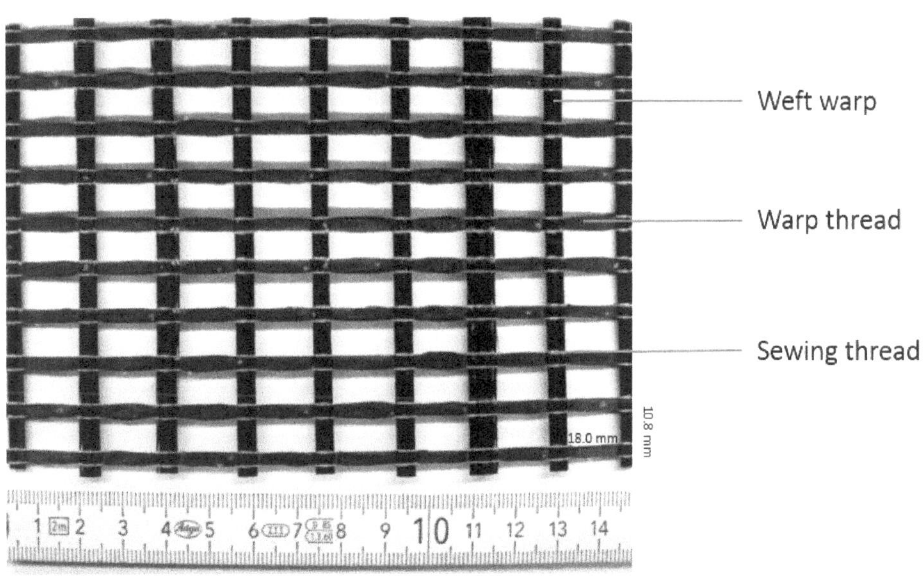

Figure 1. Schematic of the bidirectional warp-knit fabric of carbon fibres

2.3 Test Specimens

Several plates (50×50 cm²) were prepared for the test specimens. In a first step, the absorber concrete was mixed. Thereafter, the fresh absorber concrete was poured into the mold. After this, the fine grained concrete was mixed with the intensive mixer R05T by Erich. The mixing parameters are shown in Table 2.

Table 2. Mixing parameters for fine grained concrete

	component	mixing technology	mixing power in %	mixing time in s
1st	binders + aggregates	concurrent	15	60
2nd	75% of water	sequence	35	90
3rd	super plasticizer	sequence	35	60
4th	residual water	sequence	40	30
5th	AR-glass fibres	sequence	40	30

Subsequently, the fresh fine grained concrete was applied on the absorber concrete (Figure 2a). Finally, the test plates were demolded after 24h (Figure 2b).

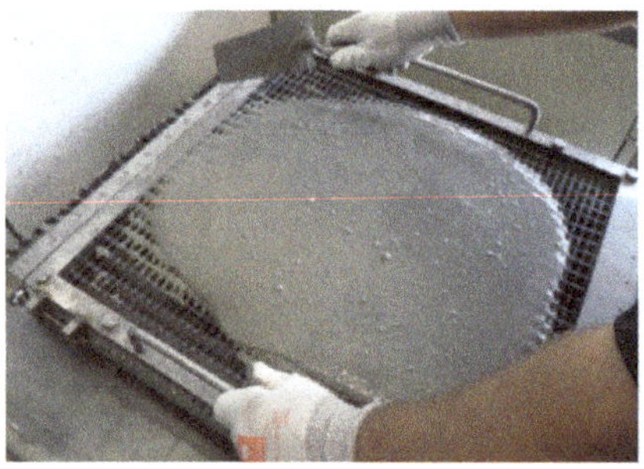

(a) Application of fine grained concrete

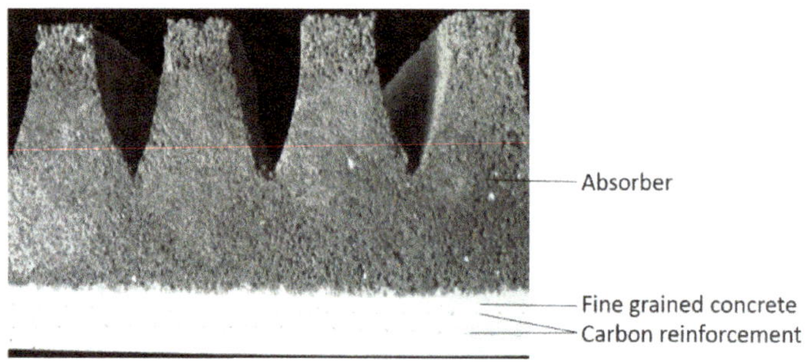

(b) Demolded test plate

Figure 2. Application of fine grained concrete (a) and demolded test plate (b)

2.3 Test Specimens Test Set-Up for TRC and Absorber

The samples for the tests to be performed on the hardened concrete were stored dry, according to DIN EN 12390-2. The 3-point bending tensile strength (Figure 3a) was determined by means of the Toni Technik ToniNorm with samples which measured 225 x 50 x 15 mm³ (length x width x height), based on DIN EN 12390-5. The span width set was 200 mm and the load speed 100 N/s constant. The compressive strength was determined by means of the Toni Technik ToniNorm (load frame 3000 kN) following DIN EN 12390-3, with cubes having an edge length of 150 mm (Figure 3b). The pre-load was 18 kN.

(a) 3-point bending tensile strength

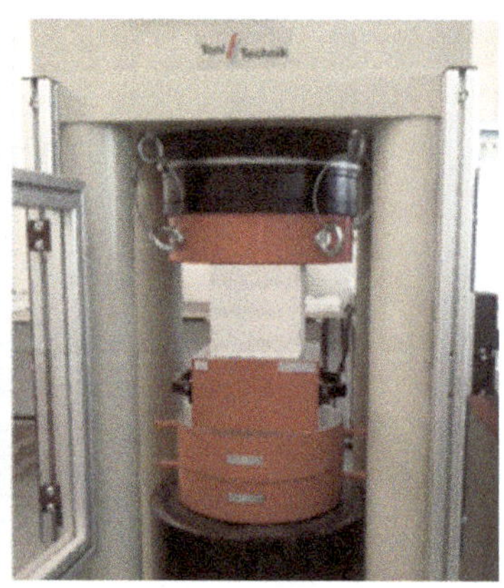

(b) compressive strength

Figure 3. Determination of 3-point bending tensile strength (a) and compressive strength (b)

To validate the durability of the fine grained concrete, the capillary suction of de-icing solution and freeze thaw test (CDF-Test) was measured by the Schleibinger Freeze-Thaw-Tester with standard agent solution according to the recommendations of RILEM TC 117-FDC (Figure 4).

Figure 4. Determination of the capillary suction of de-icing solution and freeze thaw test

3. Results

3.1 Properties of Fresh and Hardened TRC

Table 3 shows the fresh and hardened concrete characteristics after 28 days. The fresh fine grained concrete possessed great flowability (flow spread 640 mm) and corresponded to flow class F6. Using an air content tester, an air content of 2.7 volume percent and a gross geometric density of 2.31g/cm3 were determined in the fresh concrete. A shrinkage channel was utilized to determine the total shrinkage deformation at 0.68 mm/m. The reason for this high total shrinkage was the high binding agent content and the great chemical and autogenic strength of the hardened concrete at 28 days (gross density: 2.31 g/cm3) was 83.7 MPa; it was 25 MPa after 24 hours. The 3-point bending tensile strength of the unreinforced concrete was 11.73 MPa after 28 days.

Table 3. Properties of fresh and hardened fine grained concrete

characteristic	fresh concrete	hardened concrete
geometric bulk density	2.37 g/cm³	2.31 g/cm³
flow spread	640 mm	-
air content	2.7 Vol.-%	-
linear shrinkage	0.91 mm/m	
compressive strength	-	83.7 MPa
3-point bending tensile strength	-	11.73 MPa

In addition, the fine concrete exhibited great durability, which was verified through a successful CDF test with an average weathering of 912 g/m² and a relative dynamic modulus of elasticity of 100 percent after 28 freeze-thaw cycles (Table 4).

Table 4. Examinations of the durability of fine grained concrete

test method	*test value*
CDF test	m_{28} = 113 g/m² $R_{u,28}$ = 100%
water penetration depth	11 mm

Building in 2 layers of carbon meshes in with the flow of forces enabled the 3 point flexural strength to be enhanced to 24.51 MPa (Figure 5). In this case, the first crack tension was 11.21 MPa (Figure 5). By utilizing short fibres with a critical fibre volume content (fibres have a strengthening effect on the matrix) and prestressing the textile reinforcement, an increase in tension without slippage was observed in the concrete after macro-crack formation (see blue shading in Figure 5), i.e. the bending tensile stress was transferred without loss of force from the concrete into the textile reinforcement. Integrating short fibres in the concrete generated a fine crack pattern with a positive action on durability.

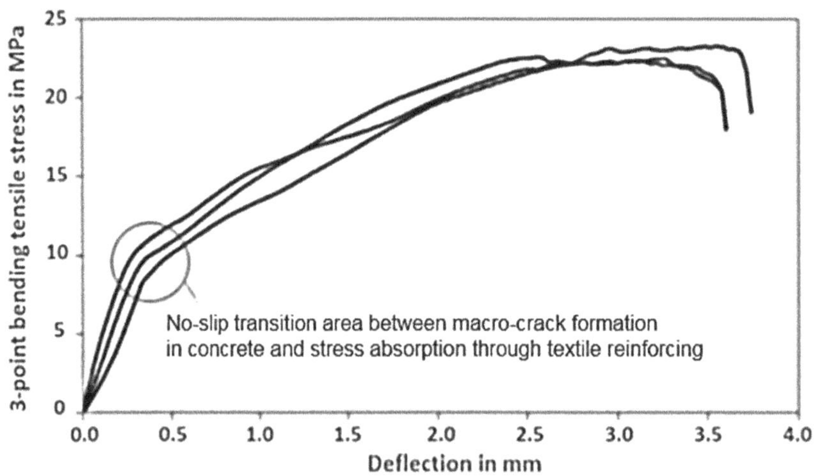

Figure 5. Stress-deflection diagram

3.2 Noise-Absorbent Concrete

The noise-absorbent concrete had a gross density of 1.31 g/cm3 and a compressive strength of 16.0 MPa (Table 5). A dynamic modulus of elasticity of 17 GPa was determined from the speed of sound using an ultrasonic measurement device.

Table 5. Properties of fresh and noise-absorbent concrete

characteristic	fresh concrete	hardened concrete
geometric bulk density	1.53 g/cm³	1.43 g/cm³
flow spread	650 mm	-
air content	11.5 Vol.-%	10.4
compressive strength	-	16.9 MPa
3-point bending tensile strength	-	2.0 MPa

The degree of sound absorption was ascertained by means of the reverberation time procedure according to DIN EN ISO 354:2003 with structured test specimens (Figure 6b). The results of this test showed that the degree of sound absorption is more than 0.7 at a frequency of 160 Hz (Figure 6a). The first local maximum sound absorption ($\alpha = 1$) was found at a frequency of 350 Hz. The decrease in sound absorption between 600 and 1000 Hz after 800

Hz means that great sound particle velocity prevails in this area. These types of curve progressions occur comparatively often with mono-modular and porous aggregate materials. They are indicative of insufficient flow resistance adaptation. The result is that the noise absorber acts like a resonator. Based on these measurements, the noise absorbent shell met the conditions for sound group 4 (highly absorbent).

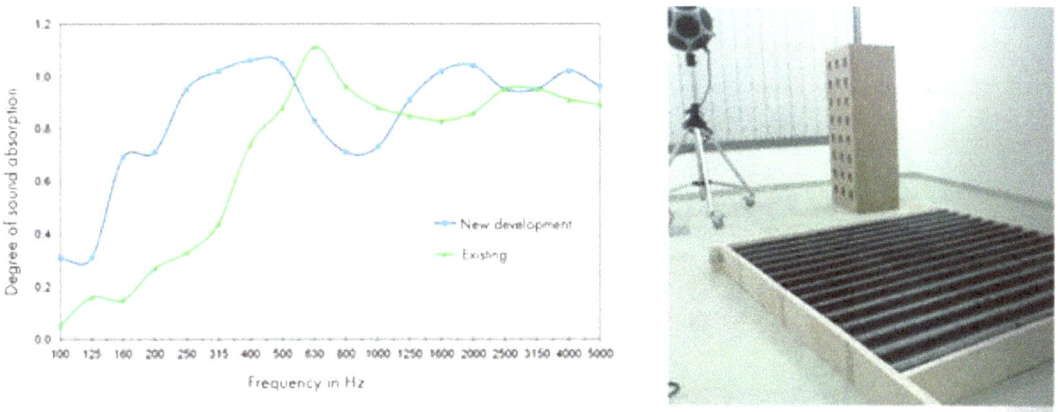

(a) Degree of sound absorption (b) measurement in echo chamber

Figure 6. Degree of sound absorption according to DIN EN ISO 354:2003

4. Conclusions

At the completion of this project, a noise barrier was able to be implemented using a textile-reinforced lightweight construction method under construction conditions. In the context of this present research and development project, a scientific basis was created for developing the materials: it was technically implemented in production, tested in numerous test series, adopted and practically investigated in the form of a prototype.

Integrating short fibres in the concrete generated a fine crack pattern with a positive action on durability, amongst other things. Taking the first crack computations into account, proof was furnished in mechanical investigations that the thickness of the textile-reinforced, load-bearing concrete slab can be reduced from 12 cm (steel reinforced concrete) to 5 cm (carbon reinforced concrete). Proof was also given of good post-cracking strength, low creep and shrinkage deformation, low susceptibility to cracking and excellent resistance to frost/de-icing salt (exposure class XF4).

This new type of textile-reinforced noise barrier meets requirements as regards resource efficiency, cost reduction, sustainability, minimization of the system slab weight and a decrease in transport and assembly expenses in practice under construction conditions.

The marginal conditions for manufacturing such noise barriers on an industrial scale in large quantities with high demands on equality are currently being created in preparation for its mass production and market launch. Beyond this, a key priority is obtaining general building authority approval or product approval.

Acknowledgements

This work was supported by the German Federation of Industrial Research Associations (AiF). The authors would like to acknowledge with gratitude the foundation's financial support. The publication costs of this article were funded by the German Research Foundation/DFG (Geschäftszeichen INST 270/219-1) and the Chemnitz University of Technology in the funding programme Open Access Publishing.

References

Brameshuber, W. (2006). Textile Reinforced Concrete. *RILEM Report 36. State-of-the-Art Report of RILEM Technical Committee, TC 201-TRC, 2006.*

Curbach, M., & Jesse, F. (2010). *Verstärken mit Textilbeton*. In K. Bergmeister, F. Fingerloos, & J.-D. Wörner (Eds.), *Betonkalender 2010*. Teil 1, Abschn. VII. Berlin: Ernst & Sohn.

Curbach, M., & Scheerer, S. (2011). Concrete light – Possibilities and Visions. In V. Šrůma (Ed.), *Proceedings of the fib Symposium Prague 2011: Concrete Engineering for Excellence and Efficiency* (pp. 29-44). 8-10. June 2011, DVD-ROM.

Curbach, M., Michler, H., Weiland, S., & Jesse, D. (2008). *Textilbewehrter Beton – Innovativ! Leicht! Formbar!* Z. *BetonWerk International, 11*(5), 62-72.

Curbach, M., Scheerer, S. (2012). *Wie die Baustoffe von heute das Bauen von morgen beeinflussen.* In *Baustoffe und Betonbau - Lehren, Forschen, Prüfen, Anwenden* (pp. 25-36).

Curbach, M., Scheerer, S., Hampel, T., Schmidt, N., & Wellner, S. (2013, May). Structural Reassessment of Existing Road Bridges in Germany. In *IABSE Symposium Report* (Vol. 99, No. 12, pp. 1282-1289). International Association for Bridge and Structural Engineering. http://dx.doi.org/10.2749/222137813806501948

Funke, H., Gelbrich, S., & Ehrlich, A. (2013). A New Hybrid Material of Textile Reinforced Concrete and Glass Fibre Reinforced Plastic. *Journal of Materials Science Research, 2*(3), 96-102. http://dx.doi.orrg/10.5539/jmsr.v2n3p96

Funke, H., Gelbrich, S., & Ehrlich, A. (2013). Development of a new hybrid material of textile reinforced concrete and glass fibre reinforced plastic. *Procedia Materials Science*, *2*, 103-110. http://dx.doi.org/10.1016/j.mspro.2013.02.013

Lohre, D., Bemecker, T., & Stock, W. (2012). ZF-Zukunftsstudie 2012. In *EuroTransport Media Verlags- und Veranstaltungs-GmbH*. Stuttgart, S. 17.

Mobasher, B., Dey, V., Cohen, Z., & Peled, A. (2014). Correlation of constitutive response of hybrid textile reinforced concrete from tensile and flexural tests. *Cement and Concrete Composites*, *53*, 148-161. http://dx.doi.org/10.1016/j. cemconcomp.2014.06.004

21

Comparative Assessment of Wiper and Conventional Carbide Inserts on Surface Roughness in the Turning of High Strength Steel

Adel Taha Mohamed Abbas[1]

[1] Department of Mechanical Engineering, College of Engineering, King Saud University, Saudi Arabia

Correspondence: Adel Taha Mohamed Abbas, Department of Mechanical Engineering, College of Engineering, King Saud University, Saudi Arabia. E-mail: atabbas1954@yahoo.com

Abstract

Wiper inserts have the characteristics of eliminating many grinding operations and achieving an excellent surface. They also improve component quality and roundness compared with grinding. This paper exposes a performance comparative analysis, involving the criteria of surface roughness (R_a, R_t and R_z) during high strength steel's turning operation, between the conventional and wiper inserts. The main parameters considered in this study are; the speed of cutting (CS), the feed rate (FR), and the cutting depth (DOC). The test plan was carried-out through (128) test specimens divided into two equal sample groups (A and B), each consists of (64) test specimens. Specimens of group (A) and (B) are tested using wiper and conventional inserts respectively. To apply the required testing conditions, a CNC Turning machine equipped with Sinumeric 840-D, was used. The surface roughness was evaluated using a surface roughness tester (Tesa-rugosurf 90- G). The most important parameters affecting the surface roughness are highlighted. Statistical tests used for this study including the analysis of variance (ANOVA) and the response surface methodology (RSM) are presented. The results show the significance of cutting depth and feed rate in the reduction of surface roughness. The machining conditions producing the optimum roughness of the surface, in the experiment range, were investigated using desirability-function-approach for the optimization of multiple-factors of response. The results showed that the quality of surface derived with the wiper carbide insert has significant improvement in comparison to the conventional carbide insert. The maximum improvement of 3.5 times between the wiper insert and conventional insert was achieved at a surfacing speed of 75 meter/min and is limited to 3, 2.5 and 2 times for a surfacing speed of 100, 125 and 150 meter/min, respectively.

Keywords: wiper inserts, surface roughness measurement, turning operation, process parameters

1. Introduction

The machining process of hard turning has drawn increased attention in research and development. Hardened steels exhibit many attractive properties, such as wear resistance, high strength, and high thermal stability, and have a variety of uses, including dies, automobiles, bearings, and gears. Due to the industrial importance of hardened steels, there is a need for advanced technologies in the machining of these materials (Srithar, Palanikumar, & Durgaprasad, 2014), namely, in the development of processes with a higher material removal rate (MRR). Hard turning works on materials by making use of various solid cutting tools. The development of wiper inserts started in recent years with the fabrication of cemented carbide inserts. Wiper inserts can eliminate many of the operations in grinding and achieve excellent surface finish. Wipers also improve the roundness and overall quality of the component in comparison to grinding. The main factor affecting turning operation performance is the feed. It is estimated that the cutting time can be reduced to half when the feed is doubled; consequently, it is possible to achieve twice the number of components within the same time duration (Sandvik, 2014).

There is increased competition in the manufacturing industries to achieve high-quality products and high productivity. Finished products of desired sizes and shapes are conventionally manufactured by turning performed blanks with the aid of cutting tools. During this production, a key requirement is the quality of the surface, as surface roughness can affect fatigue strength, wear rates, assembly tolerance, coefficients of friction, clean-ability, corrosion resistance, and aesthetics (Sahoo, Baral, Rout, & Routra, 2012). In addition, the manufacturing industry requires higher rates of material removal for feasible mass production while retaining the product quality. Higher removal rates can be obtained by adjusting the process parameters such as cutting speed, cutting depth, and feed rate.

Higher cutting speeds would require greater power input that could eventually exceed the power capacity of the machine tool. Likewise, increasing other process parameters could result in higher cutting temperatures, which would be harmful to the product, as well as to the tool. It could lead to dimensional inaccuracies via thermal deformation, impact the machined surfaces through built-up edge formations and thermal defects, and impair the movement and sharpness of the tool. Given their pivotal role in the efficiency, effectiveness, and overall economy of manufacturing, it is important to select appropriate process parameters (Sahoo, Baral, Rout, & Routra, 2012).There are various benefits of hard machining in comparison to the process of grinding. The implementation of the technology requires producing surfaces that meet the requirements of integrity and surface finish. At the same time, the technology should be economically viable. The development of harder work materials has promoted the technology of tool material at a faster rate. The technology has the potential of achieving machining of the materials with reliable performance and higher material removal rate (Suresh, Basavarajappa, Gaitonde, Samuel, & Davim, 2013). This study aims to compare the surface quality achieved using wiper and conventional carbide inserts in the turning of high-strength steel. This study investigates whether the surface quality derived with the wiper carbide insert exhibits improvement in surface roughness in comparison to the conventional carbide insert.

2. Literature Review

Researchers have conducted various investigations into the performance of ceramic tools and cubic boron nitride (CBN) in the machining of hard materials. However, there has been no systematic study describing the statistical effects of turning parameters and other associated conditions. Çaydas (2010) in this study, it was concluded that increasing the tempering temperature resulted in a corresponding decrease in the hardness of the samples. These findings showed that the best surface roughness was achieved when CBN tools were followed by ceramic and then grade carbide tools. Conventionally, the finish machining of hardened steel is accomplished through the process of grinding. In this process, the grinding wheel's shape is a major limitation in its application to geometries of regular workpieces; this is one of the many reasons why finish hard turning is preferred over grinding. Other benefits include reduced setup time and increased metal removal rate. CBN tools are appropriate for hard machining of steels because they are chemically inert to steel (Dureja, Gupta, Sharma, & Dogra, 2010).

2.1 Hard Turning and Grinding

Steel parts are used for carrying critical loads in jet engines and automotive drive trains. These parts are conventionally manufactured by a series of processes, which include costly and time-consuming polishing and grinding. A less costly and efficient method to produce these parts involves precisely forging hot metal into the parts that are net-shaped, followed by hardening and machining via hard turning. The process of hard turning differs from the process of conventional turning mainly in the hardness of the workpiece, the chip formation mechanism involved, and the cutting tool required. The unique feature of hard turning is the direct machining of parts once they are hardened. As a result, the process has lower equipment costs, fewer process steps, shorter setup time, and greater flexibility for part geometry. It is estimated that hard turning can reduce the costs of manufacturing by up to 30% (Dureja, Gupta, Sharma, & Dogra, 2010).

Gaitonde et al. (2009a) Hard turning with a ceramic cutting tool has several advantages over the process of grinding. These include no need for a coolant, improved material properties, reduced processing costs, reduced power consumption, and increased productivity. However, the benefits of hard turning do not allow it to fully replace grinding because there is still insufficient data concerning tool wear and surface quality. This paper is an attempt to identify the impacts of machining time and cutting depth on the dimensions of machinability. The experiments were carried out according to full factorial design (FFD). The parametric analysis showed that the CC650 conventional insert was beneficial in reducing the machining force, specific cutting force, and power, while the performance of CC650WG wiper insert was superior with respect to tool wear and surface roughness.

Gaitonde et al. (2011) studied the influence of cutting speed, feed rate, and machining time on machinability aspects. Specifically, the cutting force, surface roughness, and tool wear in AISI D2 cold work tool steel hard turning have been investigated for three different ceramic inserts, namely, CC650, CC650WG, and GC6050WH. A multilayer feed-forward artificial neural network (ANN), trained using error back-propagation training algorithm has been employed for predicting the machinability. The input–output patterns required for the ANN training and testing are obtained from the turning experiments planned through full factorial design. The simulation results demonstrate the effectiveness of ANN models to analyze the effects of cutting conditions as well as to study the performance of conventional and wiper ceramic inserts on machinability.

Gaitonde et al. (2009b) established the relationships between the cutting conditions (cutting speed, feed rate, and machining time) and machinability aspects (machining force, power, specific cutting force, surface roughness, and tool wear). The surface response methodology-based mathematical models are proposed for modeling and analyzing the effects of process parameters on machinability during turning of high chromium AISI D2 cold work tool steel using CC650WG wiper ceramic inserts. The experiments have been planned as per full factorial design. From the parametric analysis, it is revealed that the power increases with the rising of the feed rate, while the specific cutting force decreases, whereas the requirement of machining force is low at small values of feed rate and machining time. The surface response analysis also indicates that the surface roughness can be reduced at lower values of feed rates and machining time with higher values of cutting speeds, while the maximum tool wear occurs at a cutting speed of 150 m/min for all values of feed rates.

Quiza, Figueira, and Davim (2008) carried out an experimental investigation using ceramic cutting tools, composed approximately of Al2O3 (70%) and TiC (30%), on cold work tool steel D2 (AISI) heat treated to a hardness of 60 HRC. Two models were adjusted to predict tool wear for different values of cutting speed, feed and time. One of them is based on statistical regression, and the other is based on a multilayer perceptron neural network. Parameters of the design and the training process, for the neural network, have been optimized using the Taguchi method. Outcomes from the two models were analyzed and compared. The neural network model has shown better capability to make accurate predictions of tool wear under the conditions studied.

Davim et al. (Davim & Figueira, 2007; Davim, 2011) considered the influence of the wiper inserts when compared with conventional inserts on the machinability parameters (cutting forces, surface roughness, and tool wear) obtained in hard turning of AISI D2 hardened steel. A plan of experiments was established considering hard turning with predetermined cutting parameters. An analysis of variance (ANOVA) was performed to investigate the machinability parameters using conventional and wiper inserts. With wiper ceramics inserts, machined surfaces with $Ra < 0.8$ μm were achievable. Consequently, surface qualities (dimensional accuracy) in a workpiece of mechanical precision, IT < 7, are possible.

2.2 Impact of the Radius of the Cutting Edge

Fulemova and Janda (2014) discussed the impact of the radius of the cutting edge and the preparation of the edge on the tool life, machined surface's roughness, and cutting forces. The preparation of a cutting edge was accomplished by using a wiper geometry for the inserts. The inserts were used to machine a steam turbine casing of steam turbine during the process of dividing planes. The cutting inserts were fabricated by grinding, drag-finishing and laser ablation. The radii of the edges of the drag-finished tools were 5, 10 and 15 μm; laser-finished tools had an edge radius of 5 μm. The workpiece material was a dual phase ferrite-martensite steel, consisting of 1% chromium and 9% molybdenum. Submicron sintered carbide was used as the cutting insert material, and in the milling cutter, only one cutting insert was present.

2.3 Machining Technology

Fulemova and Janda (2014) looked into the surface roughness of the workpiece, which the authors considered to be a crucial factor in the technology of machining. Wiper inserts have been gaining popularity due to their significance in cutting tools and have been utilizing for the past several years. This research examines the impact of the wiper inserts surface roughness in comparison to conventional inserts. Experiments were conducted for AISI 1045carbon steel, due to the steel's importance in manufacturing industry. Different amplitude parameters were used to quantify the roughness of the surface (Ra, Rz, Rq, Rt). Machined surfaces with Ra less than 0.8 μm (micron) were achieved at a high feed rate and with the use of wiper inserts. Thus, it is possible to obtain a precise and high-quality workpiece surface without cylindrical grinding.

2.4 Predictive Model

Ozel, Karpat, Figueira, and Davim (2007) the design of the tool nose impacts the productivity of and surface finishing from hard turning. The setup utilizes neural network modeling as a prediction tool. Predictions were made for tool flank wear and surface roughness. In addition, the arrangement also employed multiple regression models to provide the predictions. The results from experiment showed that surface roughness values *R*a as low as 0.18–0.20 μm were attainable using wiper inserts. Analysis of the tool flank wear attained a tool life criterion value of VBC = 0.15 mm due to the elevated temperatures at high cutting speeds at up to 15 minutes of cutting time. The comparison was made with experimental data that were non-training. The findings indicated that the neural network models were appropriate for predicting the surface roughness and tool wear patterns for a range of cutting conditions.

2.5 Grinding Alternative

Guddat, M'Saoubi, Alm, and Meyer (2011) Due to its high productivity and flexibility and no need for coolants, hard turning is currently considered (in the finishing of hardened steel workpieces) to be an alternative to grinding. The use of wiper PCBN inserts results in the reduction of surface roughness compared to the use of conventional inserts for machining experiments at enhanced feed rates. Although these are the well-known benefits, the impact of wiper PCBN inserts on the surface integrity during hard turning has not been discussed in detail in the literature. This study aims to investigate the impact of wiper PCBN inserts on the finishing cutting forces and surface integrity of hardened AISI 52100 (100Cr6) (58-62 HRC). A statistical model was developed to predict the surface roughness values and cutting forces for a range of cutting parameters. The assessment of subsurface and surface integrity revealed that, in comparison to conventional geometrical inserts, the implementation of PBCN wiper inserts led to higher compressive residual stresses and marked improvement in the surface roughness.

2.6 Desirability Function Approach

Research from Elbah, Yallese, Aouici, Mabrouki, and Rigal (2013) compared the surface roughness (Ra, Rt, Rz) produced using wiper and conventional inserts. The comparison was made for the hard turning of AISI 4140 hardened steel (60 HRC), and the experiments were planned using a Taguchi (L27) orthogonal array. The analysis of variance (ANOVA) and response surface methodology (RSM) were employed to check the validity of the quadratic regression model. The tests were also used to identify the significant parameter impacting the roughness of the surface. The findings showed that the cutting depth and rate of feed significant impacted the roughness of the surface. The best machining conditions for producing the optimum surface roughness in the ranges considered in these experiments were determined by using desirability function. The findings demonstrated that the quality of the surface finished with the ceramic wiper inserts was radically improved compared to the quality from the conventional ceramic inserts by a factor of 2.5. The measurements of roughness indicated a dependence on CC6050WH tool wear. However, although the wear reached up to the permitted flank wear value of 0.3 mm the roughness Ra was always less than 0.9 μm.

2.7 Green Manufacturing

According to Grzesik and Wanat (2006), much progress has already made in the domain of green manufacturing, including hard and dry (often high-speed) machining technologies. For example, the need for increased productivity has given rise to the wider application of PCBN and ceramic tools with special multi-radii/wiper geometries. This study mentions some key features in the surface roughness developed in the turning of hardened low-chromium alloy steel. The study used mixed alumina–titanium carbon (TiC) ceramic cutting tools, which were equipped with both wiper and conventional inserts. The key geometrical characteristics of the surfaces obtained from both tools were extracted using leading two-dimensional (2D) parameters of the surface roughness and 3D visualization, which provided comprehensive surface topography and prediction of the properties of service. The findings indicated that for equivalent rates of feed, i.e., 0.2 mm/rev for wiper tools and 0.1 mm/rev for conventional tools, the resultant surfaces had equivalent 3D roughness height parameters and comparable figures of skew/kurtosis. For pre-determined parameters of cutting, surfaces finished with wipers consisted of blunt peaks with clearly smaller slopes, compared with surfaces finished with conventional inserts, which promoted in superior bearing properties. Only minor changes in Ra were noted among the 15-minute machining trials.

2.8 Updated Technique of Optimization

Paiva et al. (2012) describes an experimental study using AISI 52100 hardened steel employing Multivariate Robust Parameter Design (MRPD). The hardened steel was turned with wiper inserts made from mixed ceramic (Al2O3+TiC) and coated with TiN. The key features of MRPD is that it considers both noise (zi) and controllable (xi) variables in the hard turning process to determine the parameter values that would minimize the response distance (yi) from the respective targets (Ti) while maintaining the minimum variance value for each variable. By using a crossed array, a response surface design was submitted to the impact of four cases developed with a 22 full factorial design of two noise factors - tool flank wear (Z2) and workpiece hardness decreasing (Z1). The response survey design was consisted of the feed rate (f), cutting speed (Vc), and cutting depth (d). This arrangement of the experiment enabled the calculation of the variances, means and mean square errors (MSE) for the five surface roughness parameters (Ra, Ry, Rz, Rq, and Rt). As there was a positive correlation in these responses, to apply and extrapolate the information, the study used Principal Component Analysis (PCA). The researcher used the multivariate mean square error (MMSE) as the criterion for optimization to find a robust solution. The experimental and theoretical results were confirmed and found to be convergent. for f=0.191 mm/rev, Vc=199.9 meter/min, and d=0.190 mm, the five surface roughness parameters and the variances were reduced. The results were better than the results obtained with individual optimization.

2.9 Milling Cutter Systems

de Souza Jr, Sales, Santos, and Machado (2005) used two face milling cutter systems encountered in the shop floor under cutting condition in the high-speed cutting of gray cast iron. The first system 'A' had 24 Si3N4 inserts with square wiper edges. The second system 'B' was a mixed tool material system with 24 wiper inserts, 20 of which were Si3N4 interspersed with 4 inserts of PCBN. The constants included speed of cutting (Vc), feed rate per tooth (fz), and depth of cut (doc). The parameters of surface roughness (Rt and Ra) and waviness (Wt), burr formation (length of the burr, h), and tool life (based on flank wear, VBB max) were used to compare the two systems. The performance of system B was better than the performance of system A based on all parameters, although only the end-of-life criterion on the basis of Rt parameter has been marked.

2.10 Energy-dispersive X-ray (EDX) Micro-analysis

Grzesik (2009) analyzed the mechanisms of wear from using mixed ceramic inserts on 60-HRC steel (equivalent to DIN 41Cr4 or AISI 5140) specimens in hard and dry finish turning operations. Various tests were executed under different rater of feed, a small cutting depth of 0.2 mm, and a constant speed of cutting of 100 m/min. Light optical microscopy (LOM), back-scattered electron (BSE) techniques, and scanning electron microscopy (SEM) were carried out to analyze the morphological characteristics of worn surfaces developed under different mechanisms of wear. Moreover, the elemental compositions of the products of wear were determined by energy-dispersive X-ray (EDX) micro-analysis (also known as energy-dispersive X-ray spectroscopy (EDS)). Thus, the microstructure and microscopic dimensions of tool wear were considered. In summary, the wear mechanisms analyzed in these tests included fracture, abrasion, BUE, plastic flow and transferred layers, and tribochemical impacts. Signatures of different wear mechanisms appeared depending on the thermal and mechanical conditions produced in the contacted areas. Particularly, two forms of material transfer with varying morphologies occurring at the rake-chip interface were especially pronounced.

2.11 Wide use of wiper inserts

Today, wiper inserts are the top choice for a wide variety of application to ensure maximum productivity and versatility. With productive wiper inserts there are alternative also available when cutting conditions change. Wiper inserts enable improved machining economy, excellent surface finishes, the elimination of expensive grinding operations, and undisturbed production and require less supervision and reduced inventory. Figure 1 shows the edge preparation for different wiper inserts and conventional finishing inserts (Sandvik, 2014).

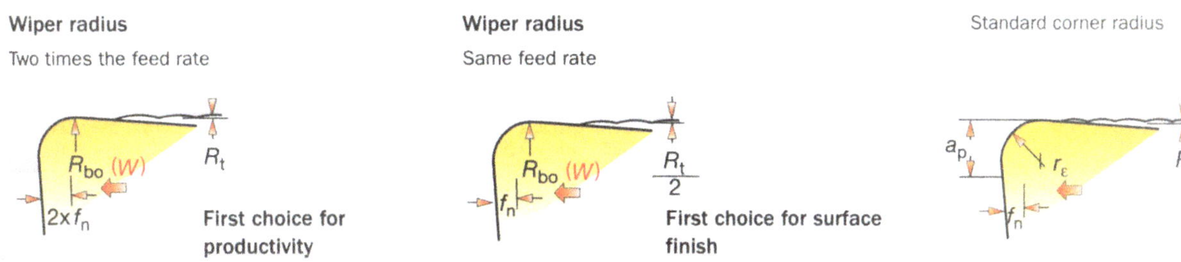

Figure 1. Edge preparation for different wiper inserts and regular finishing insert

3. Materials and Methods

The chemical compositions of the high-strength steel material considered here is shown in Table 1. The material was austenitized at 900°C for 5 hr., air-cooled, heated at 880°C for 5 hr., quenched in oil, tempered at 590 to 600°C for 8 hr., and finally air-cooled once more. The hardness was HV 400 to 420.

Table 1. Chemical composition for high strength steel material

Range	C	Si	Mn	Ni	Cr	Mo	V	S	P
From	*0.32*	*<0.25*	*0.55*	*2.5*	*0.9*	*0.45*	*0.09*	*≤0.004*	*≤0.008*
To	*0.36*		*0.65*	*3.2*	*1.1*	*0.55*	*0.12*		

The test specimens were made with an initial diameter of 50 mm and length of 100 mm, with 40 mm of the length used for chuck clamping, 10 mm for clearance grooving, and the remaining 50 mm for testing. At the other end of test specimen, a standard conical center was created for supporting the center of the tail stock.

An EMCO Concept Turn 45 CNC lathe equipped with Sinumeric 840D was used to conduct the experiments. Wiper and conventional carbide inserts were clamped into the tool holders. The specifications for the wiper and conventional inserts were DCMX11T304-WF, DCMT11T304-PF, respectively, while the tool holders were the same, SDJCL 2020K11. The clearance angle, cutting edge angle and nose radius were maintained at 7°, 55° and 0.4 mm, respectively. All the experiments were conducted in wet conditions. The surface roughness was measured and reported for a length 50 mm and evaluated using a surface roughness tester Tessa. All cutting parameters were controlled using a CNC part program. Figure 2 shows the test rig for machining the test specimens and measuring the surface roughness.

Figure 2. Test Rig for machining the workpieces and measuring the surface roughness

The test plan was carried out using (128) test specimens. The specimens were divided into two equal divisions (A and B), each division consisting of (64) test specimens and tested using wiper inserts and conventional inserts, respectively. Each division was further divided into 16 groups, where each of the four groups was machined at a fixed speed. Groups 1-4, 5-8, 9-12 and 13-16 were processed at speeds of 75 meter/min, 100 meter/min, 125 meter/min and 150 meter/min, respectively. Each sample in each group was machined to a different cutting depth (0.1, 0.15, 0.20 and 0.25 mm), and each depth was processed at a different feed rate (0.05, 0.1, 0.15 and 0.20 mm/rev, respectively).The cutting parameters and surface roughness (R_a, R_t and R_z) for Groups 1-4, 5-8, 9-12 and 13-16 are listed in Appendix A, where R_a= arithmetic mean roughness (μm), R_t = total roughness (μm), R_z= mean depth of profile (μm).

4. Results and Discussion

4.1 Analysis of Variance

The processes parameters considered here are; speed, feed and depth of cut while the response variable is the surface roughness (R_a, R_t and R_z). The results of the study were entered into SPSS for analysis of variance. Division (A) samples was finished using wiper inserts and division (B) samples were finished with conventional inserts. There was a total 16 groups in each division: groups 1 to 4 were machined at 75 meter/min, groups 5 to 8 at 100 meter/min, groups 9 to 12 at 125 meter/min, and groups 13 to 16 at 150 meter/min.

The statistical analysis was conducted to determine if there were significant differences in the mean values in division A and division B for the different machining speeds. The results are as shown in Table 2a and b.

For the mean Ra in division A, the ANOVA analysis calculated a significance level of 0.039, which is below 0.05. Hence, there is a statistically significant difference in the mean values of Ra between the different machining speeds with the wiper insert. The mean value of Ra increased for machining speeds up to 125 meter/min, after which it began to decrease, as shown in Table 3a and b.

Table 2a. Descriptive Ra for wiper insert

	N	Mean	Std. Deviation	Std. Error	95% Confidence Interval for Mean		Minimum	Maximum
					Lower Bound	Upper Bound		
75	16	.41394	.192303	.048076	.31147	.51641	.129	.687
100	16	.44569	.156611	.039153	.36224	.52914	.235	.660
125	16	.56438	.151701	.037925	.48354	.64521	.326	.786
150	16	.54388	.177148	.044287	.44948	.63827	.311	.788
Total	64	.49197	.178050	.022256	.44749	.53644	.129	.788

Table 2b. ANOVA Ra for wiper insert

	Sum of Squares	df	Mean Square	F	Sig.
Between Groups	.259	3	.086	2.976	.039
Within Groups	1.739	60	.029		
Total	1.997	63			

Table 3a. Descriptive Ra for conventional insert

	N	Mean	Std. Deviation	Std. Error	95% Confidence Interval for Mean		Minimum	Maximum
					Lower Bound	Upper Bound		
75	16	1.45150	.676214	.169054	1.09117	1.81183	.454	2.402
100	16	1.36769	.474170	.118543	1.11502	1.62035	.746	1.990
125	16	1.41869	.378276	.094569	1.21712	1.62026	.817	1.960
150	16	1.10838	.360905	.090226	.91606	1.30069	.629	1.645
Total	64	1.33656	.496010	.062001	1.21266	1.46046	.454	2.402

Table 3b. ANOVA Ra for conventional insert

	Sum of Squares	df	Mean Square	F	Sig.
Between Groups	1.168	3	.389	1.630	.192
Within Groups	14.332	60	.239		
Total	15.500	63			

For the mean Ra in division B, the ANOVA analysis calculated a significance level of 0.192, which is not below 0.05. Hence, there was no statistically significant difference in the mean values of Ra among the different machining speeds with the conventional insert, as shown in Table 4a and b.

Table 4a. Descriptive Rz for wiper insert

	N	Mean	Std. Deviation	Std. Error	95% Confidence Interval for Mean		Minimum	Maximum
					Lower Bound	Upper Bound		
75	16	1.80800	.865630	.216407	1.34674	2.26926	.639	3.068
100	16	2.35163	.801152	.200288	1.92472	2.77853	1.053	3.531
125	16	3.22644	.609059	.152265	2.90189	3.55098	2.173	4.036
150	16	3.09925	.475806	.118952	2.84571	3.35279	2.386	3.909
Total	64	2.62133	.900516	.112564	2.39639	2.84627	.639	4.036

Table 4b. ANOVA Rz for wiper insert

	Sum of Squares	df	Mean Square	F	Sig.
Between Groups	21.261	3	7.087	14.256	.000
Within Groups	29.828	60	.497		
Total	51.089	63			

For the mean Rz in division A, the output of the ANOVA analysis yielded a significance level less than 0.0001. Hence, there was a statistically significant difference in the mean values of Rz between the different machining speeds with the wiper insert. The mean values of Rz increased with machining speeds up to 125 meter/min, after which it began to decrease, as shown in Table 5a and b.

Table 5a. Descriptive Rz for conventional insert

	N	Mean	Std. Deviation	Std. Error	95% Confidence Interval for Mean		Minimum	Maximum
					Lower Bound	Upper Bound		
75	16	6.66250	3.869956	.967489	4.60035	8.72465	1.442	13.058
100	16	7.12263	3.280621	.820155	5.37451	8.87074	2.904	12.045
125	16	6.49319	3.180011	.795003	4.79868	8.18770	2.641	11.218
150	16	6.10806	3.272275	.818069	4.36439	7.85173	1.878	10.846
Total	64	6.59659	3.349665	.418708	5.75987	7.43332	1.442	13.058

Table 5b. ANOVA Rz conventional insert

	Sum of Squares	df	Mean Square	F	Sig.
Between Groups	8.487	3	2.829	.243	.866
Within Groups	698.389	60	11.640		
Total	706.876	63			

For the mean Rz in division B, the output of the ANOVA analysis yielded a significance level of 0.866. Therefore, there was no statistically significant difference in the mean values of Rz between the different machining speeds with the conventional inserts, as shown in Table 6a and b.

Table 6a. Descriptive Rt for wiper insert

	N	Mean	Std. Deviation	Std. Error	95% Confidence Interval for Mean		Minimum	Maximum
					Lower Bound	Upper Bound		
75	16	1.90013	.873654	.218414	1.43459	2.36566	.664	3.193
100	16	2.51056	.830200	.207550	2.06818	2.95294	1.114	3.655
125	16	3.58206	.536010	.134002	3.29644	3.86768	2.696	4.269
150	16	3.48344	.464557	.116139	3.23589	3.73098	2.658	4.163
Total	64	2.86905	.980649	.122581	2.62409	3.11401	.664	4.269

Table 6b. ANOVA Rt for wiper insert

	Sum of Squares	df	Mean Square	F	Sig.
Between Groups	31.251	3	10.417	21.307	.000
Within Groups	29.334	60	.489		
Total	60.585	63			

For the mean Rt in division A, the ANOVA analysis calculated a significance level less than 0.0001. Therefore, there was a statistically significant difference in the mean values of Rt between the different machining speeds with the wiper insert. The mean values of Rt increased with the machining speed up to 125 meter/min, after which it began to decrease, as shown in Table 7a and b.

Table 7a. Descriptive Rt for conventional insert

	N	Mean	Std. Deviation	Std. Error	95% Confidence Interval for Mean		Minimum	Maximum
					Lower Bound	Upper Bound		
75	16	6.87019	3.912362	.978090	4.78544	8.95494	1.596	13.153
100	16	7.32213	3.315463	.828866	5.55544	9.08881	3.074	12.145
125	16	6.72962	3.213627	.803407	5.01720	8.44205	2.760	11.479
150	16	6.36525	3.345241	.836310	4.58270	8.14780	2.074	11.191
Total	64	6.82180	3.391790	.423974	5.97455	7.66904	1.596	13.153

Table 7b. ANOVA Rt for conventional insert

	Sum of Squares	df	Mean Square	F	Sig.
Between Groups	7.514	3	2.505	.210	.889
Within Groups	717.254	60	11.954		
Total	724.767	63			

For the mean Rt in division B, the output of the ANOVA analysis yielded a significance level of 0.889. Hence, there was no statistically significant difference in the mean values of Rt between the different machining speeds with the conventional insert.

Overall, the results of the study showed significant differences in the mean values in the wiper insert cases; however, there were no significant differences in the conventional insert cases. At 75 meter/min, the improvement in the wiper insert groups was 3.5 compared to conventional inserts. At 100 meter/min, the difference between the two insert cases was 3.0. At 125 meter/min, the improvement was a value/factor of 2.5 between the two cases. At 150 meter/min, the improvement factor/value was 2.0. From the results, it is evident that the level of improvement decreased with the increase in machining speed.

4.2 Response Surface Methodology

The response surface methodology was used to determine the relationship between the required response and the independent parameters of the process. There were six steps involved. The quadratic model for the roughness of the surface is given as follows:

$$Y = b_0 + \sum_{i=1}^{k} b_i X_i + \sum_{i,j}^{k} b_{ij} X_i X_j + \sum_{i=1}^{k} b_{ii} X_i^2$$

In the above equation, the desired response is denoted by Y, Xi is the coded variable related to the cutting parameters, b_o is a constant, and bj, bii, and bij are coefficients. Figure 3a and 3b show the impact of feed rate and cutting depth on the roughness of the surface at 75 meter/min using the wiper insert (DCMX11T304-WF) and the conventional insert (DCMT11T304-PF), respectively. The measurements indicated that the tool wear was heavily dependent on DCMX11T304-WF, as depicted by the direction of the graph.

Figure 4a and b show the impact of feed rate and cutting depth on the mean depth of profile at 75 meter/min using the wiper insert (DCMX11T304-WF) and the conventional insert (DCMT11T304-PF), respectively. The measurements indicate that the tool wear was heavily dependent on DCMX11T304-WF as depicted by the direction of the graph.

Figure 5 shows the values of Ra, Rt, and Rz for the wiper insert cases at the speed of 75 meter/min.

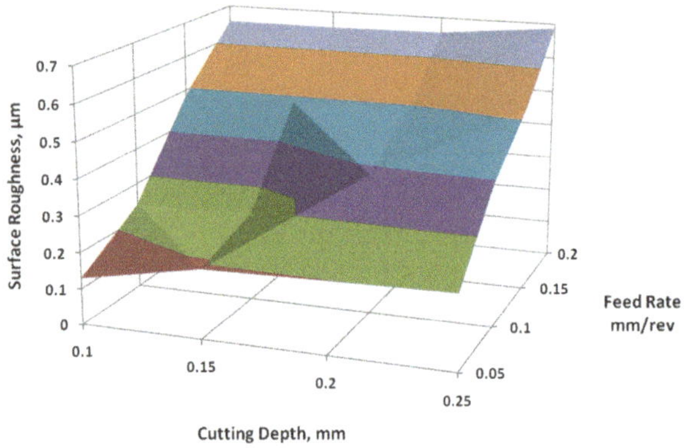

Figure 3a. Impact of feed rate and cutting depth on the roughness of the surface for machining speed of 75 meter/min (Wiper Insert)

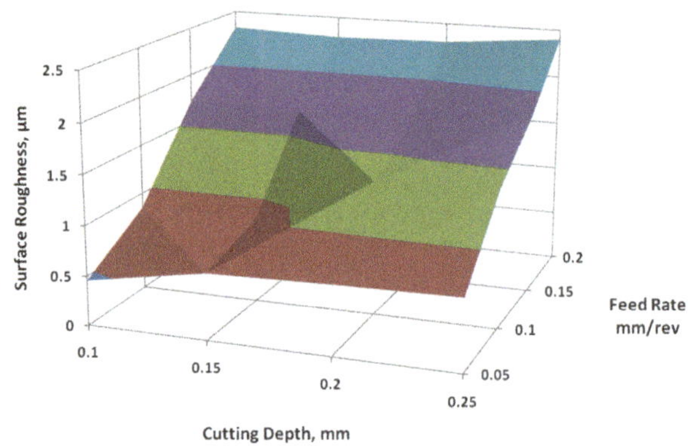

Figure 3b. Impact of feed rate and cutting depth on the roughness of the surface for machining speed of 75 meter/min (Conventional Insert)

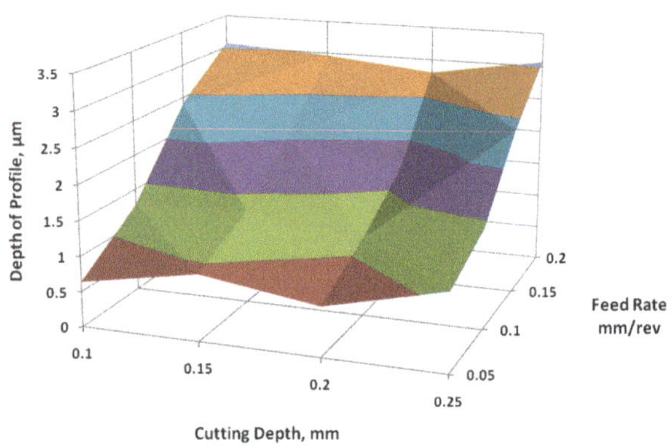

Figure 4a. Impact of feed rate and cutting depth on the mean depth of profile for machining speed of 75 meter/min (Wiper Insert)

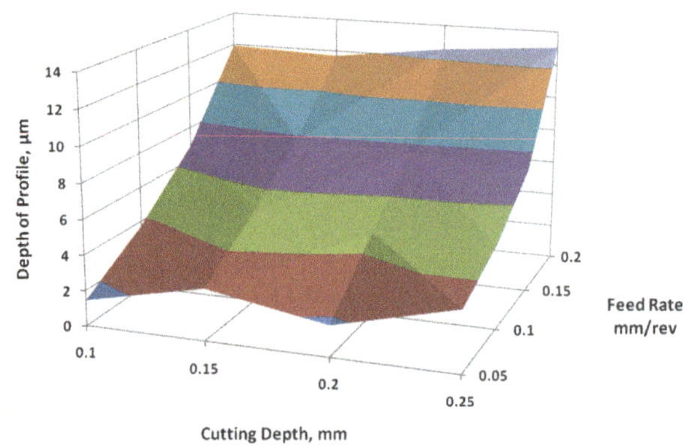

Figure 4b. Impact of feed rate and cutting depth on the mean depth of profile for machining speed of 75 meter/min (Conventional Insert)

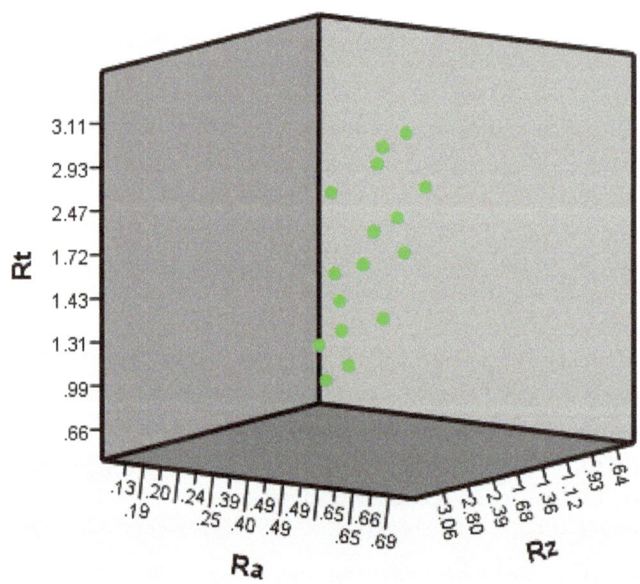

Figure 5. Values of Ra, Rt, and Rz for wiper insert for machining speed of 75 meter/min

4.3 Comparison with Literature Review

The review of the literature has highlighted the benefits of hard turning over the process of grinding. Wiper inserts were presented to reduce many of the operations of grinding and to achieve excellent surface finishes. Wiper inserts were determined to be superior with respect to tool wear and surface roughness. The findings of this study provided further support for the results of earlier studies (Davim & Figueira, 2007; Davim, 2011). The statistical analysis showed that there were significant differences in the values of Ra, Rz, and Rt when using wiper inserts at different machining speeds. The mean values of Ra, Rz, and Rt increased with the increase in machining speed up to a certain level, beyond which the mean value decreased.

5. Conclusions

This study involved a comparative analysis of the effects of wiper inserts and conventional inserts on surface roughness in hard turning. The experiments were performed on 128 test specimens. The arithmetic mean roughness, the total roughness, and the mean depth of profile were calculated, and analyses of variance and the surface response were performed. The optimal machining conditions to yield the best surface characteristics for the given experimental range were investigated using a desirability function approach. The results indicated that the quality of the surface obtained from a wiper carbide insert was significantly improved in comparison to that from a conventional carbide insert. The maximum improvement of 3.5 times between the wiper insert and conventional insert was achieved at a surfacing speed of 75 meter/min and is limited to 3, 2.5 and 2 times for a surfacing speed of 100, 125 and 150 meter/min, respectively. These results can be considered as useful guide to the manufacturing engineers as they have great impact on the quality of the final products.

References

Çaydas, U. (2010). Machinability evaluation in hard turning of AISI 4340 steel with different cutting tools using statistical techniques. *Proceedings of the Institution of Mechanical Engineers, 224*, 1043-1055.

Correia, A. E., & Davim, J. P. (2011). Surface roughness measurement in turning carbon steel AISI 1045 using wiper inserts. *Measurement*, *44*(5), 1000-1005.

Davim, J. P. (Ed.). (2011). *Machining of hard materials*. Springer Science & Business Media.

Davim, J. P., & Figueira, L. (2007). Comparative evaluation of conventional and wiper ceramic tools on cutting forces, surface roughness, and tool wear in hard turning AISI D2 steel. *Proceedings of the Institution of Mechanical Engineers, Part B: Journal of Engineering Manufacture*, *221*(4), 625-633.

de Souza Jr, A. M., Sales, W. F., Santos, S. C., & Machado, A. R. (2005). Performance of single Si 3 N 4 and mixed Si 3 N 4+ PCBN wiper cutting tools applied to high speed face milling of cast iron. *International Journal of Machine Tools and Manufacture*, *45*(3), 335-344.

Dureja, J. S., Gupta, V. K., Sharma, V. S., & Dogra, M. (2010). Wear mechanisms of TiN-coated CBN tool during finish hard turning of hot tool die steel. *Proceedings of the Institution of Mechanical Engineers, Part B: Journal of Engineering Manufacture, 224*(4), 553-566. http://dx.doi.org/10.1243/09544054JEM1664

Elbah, M., Yallese, M. A., Aouici, H., Mabrouki, T., & Rigal, J. F. (2013). Comparative assessment of wiper and conventional ceramic tools on surface roughness in hard turning AISI 4140 steel. *Measurement*, *46*(9), 3041-3056.

Fulemova, J., & Janda, Z. (2014). Influence of the cutting edge radius and the cutting edge preparation on tool life and cutting forces at inserts with wiper geometry. *Procedia Engineering*, *69*, 565-573.

Gaitonde, V. N., Karnik, S. R., Figueira, L., & Davim, J. P. (2009a). Machinability investigations in hard turning of AISI D2 cold work tool steel with conventional and wiper ceramic inserts. *International Journal of Refractory Metals and Hard Materials*, *27*(4), 754-763.

Gaitonde, V. N., Karnik, S. R., Figueira, L., & Davim, J. P. (2009b). Analysis of machinability during hard turning of cold work tool steel (type: AISI D2). *Materials and Manufacturing Processes*, *24*(12), 1373-1382.

Gaitonde, V. N., Karnik, S. R., Figueira, L., & Davim, J. P. (2011). Performance comparison of conventional and wiper ceramic inserts in hard turning through artificial neural network modeling. *The International Journal of Advanced Manufacturing Technology*, *52*(1-4), 101-114.

Grzesik, W. (2009). Wear development on wiper Al 2 O 3–TiC mixed ceramic tools in hard machining of high strength steel. *Wear*, *266*(9), 1021-1028.

Grzesik, W., & Wanat, T. (2006). Surface finish generated in hard turning of quenched alloy steel parts using conventional and wiper ceramic inserts. *International Journal of Machine Tools and Manufacture, 46*(15), 1988-1995.

Guddat, J., M'Saoubi, R., Alm, P., & Meyer, D. (2011). Hard turning of AISI 52100 using PCBN wiper geometry inserts and the resulting surface integrity. *Procedia Engineering, 19*, 118-124.

Huang, Y., Chou, Y. K., & Liang, S. Y. (2007). CBN tool wear in hard turning: a survey on research progresses. *The International Journal of Advanced Manufacturing Technology, 35*(5-6), 443-453. http://dx.doi.org/10.1007/s00170-006-0737-6

Özel, T., Karpat, Y., Figueira, L., & Davim, J. P. (2007). Modelling of surface finish and tool flank wear in turning of AISI D2 steel with ceramic wiper inserts. *Journal of materials processing technology, 189*(1), 192-198.

Paiva, A. P., Campos, P. H., Ferreira, J. R., Lopes, L. G. D., Paiva, E. J., & Balestrassi, P. P. (2012). A multivariate robust parameter design approach for optimization of AISI 52100 hardened steel turning with wiper mixed ceramic tool. *International Journal of Refractory Metals and Hard Materials, 30*(1), 152-163.

Quiza, R., Figueira, L., & Davim, J. P. (2008). Comparing statistical models and artificial neural networks on predicting the tool wear in hard machining D2 AISI steel. *The International Journal of Advanced Manufacturing Technology, 37*(7-8), 641-648.

Sahoo, A. K., Baral, A. N., Rout, A. K., & Routra, B. C. (2012). Multi-objective optimization and predictive modeling of surface roughness and material removal rate in turning using grey relational and regression analysis. *Procedia Engineering, 38*, 1606-1627. http://dx.doi.org/10.1016/j.proeng.2012.06.197

Sandvik (2014). *Turning Tools Manual.* Cutting tools from Sandvik coromant (2014).

Srithar, A., Palanikumar, K., & Durgaprasad, B. (2014). Experimental Investigation and Surface roughness Analysis on Hard turning of AISI D2 Steel using Coated Carbide Insert. *Procedia Engineering, 97*, 72-77. http://dx.doi.org/10.1016/j.proeng.2014.12.226

Suresh, R., Basavarajappa, S., Gaitonde, V. N., Samuel, G. L., & Davim, J. P. (2013). State-of-the-art research in machinability of hardened steels. *Proceedings of the Institution of Mechanical Engineers, Part B: Journal of Engineering Manufacture, 227*(2), 191-209. http://dx.doi.org/10.1177/0954405412464589

Appendix– A Results of Test Experiments

Test No.#	Group #	Speed M/min	Depth of cut mm	Feed mm/rev	Surface finish µm					
					Wiper insert(A)			Conventional insert (B)		
					R_a	R_t	R_z	R_a	R_t	R_z
1	A1-B1	75	0.1	0.05	0.129	0.664	0.639	0.454	1.596	1.442
2				0.10	0.240	1.428	1.272	0.851	3.803	3.707
3				0.15	0.485	2.496	2.408	1.713	7.030	6.853
4				0.20	0.653	3.112	3.062	2.320	12.370	11.897
5	A2-B2	75	0.15	0.05	0.191	0.987	0.934	0.663	3.045	2.900
6				0.10	0.245	1.425	1.355	0.858	5.335	5.243
7				0.15	0.488	2.105	2.031	1.704	7.890	7.809
8				0.20	0.652	3.095	2.961	2.280	11.795	11.685
9	A3-B3	75	0.20	0.05	0.199	0.808	0.695	0.696	1.867	1.739
10				0.10	0.387	1.308	1.118	1.349	4.839	4.731
11				0.15	0.489	1.719	1.675	1.714	7.775	7.628
12				0.20	0.656	2.932	2.800	2.303	12.637	12.534

13	A4-B4	75	0.25	0.05	0.205	1.157	1.093	0.715	3.536	3.433
14				0.10	0.398	1.504	1.424	1.388	4.912	4.778
15				0.15	0.519	2.469	2.393	1.814	8.340	7.163
16				0.20	0.687	3.193	3.068	2.402	13.153	13.058
17	A5-B5	100	0.1	0.05	0.235	1.114	1.053	0.746	3.621	3.477
18				0.10	0.309	2.191	1.926	0.938	5.314	5.251
19				0.15	0.490	2.804	2.588	1.521	8.655	8.465
20				0.20	0.612	3.068	2.972	1.862	12.145	12.045
21	A6-B6	100	0.15	0.05	0.248	1.206	1.151	0.780	3.629	3.521
22				0.10	0.351	2.060	1.969	1.094	5.519	5.335
23				0.15	0.504	2.828	2.688	1.577	8.844	8.540
24				0.20	0.637	3.403	3.211	1.989	12.117	12.000
25	A7-B7	100	0.20	0.05	0.248	1.384	1.263	0.752	3.484	3.328
26				0.10	0.381	2.126	1.972	1.165	5.370	5.242
27				0.15	0.557	3.070	2.795	1.695	9.053	8.491
28				0.20	0.645	3.294	3.063	1.964	12.004	11.716
29	A8-B8	100	0.25	0.05	0.268	1.906	1.777	0.827	3.074	2.904
30				0.10	0.423	2.516	2.258	1.279	5.170	4.952
31				0.15	0.563	3.655	3.531	1.704	8.234	8.057
32				0.20	0.660	3.544	3.409	1.990	10.921	10.638
33	A9-B9	125	0.1	0.05	0.326	2.732	2.173	0.817	2.922	2.684
34				0.10	0.524	3.722	3.588	1.331	4.958	4.696
35				0.15	0.608	3.616	3.508	1.558	8.552	8.257
36				0.20	0.652	3.907	3.705	1.654	11.345	11.063
37	A10-B10	125	0.15	0.05	0.338	2.957	2.484	0.852	3.046	2.828
38				0.10	0.553	4.257	3.767	1.377	4.894	4.751
39				0.15	0.615	3.647	3.503	1.544	8.033	7.924
40				0.20	0.737	4.033	3.876	1.827	11.479	11.218
41	A11-B11	125	0.20	0.05	0.345	2.696	2.239	0.867	2.760	2.644
42				0.10	0.567	3.574	3.132	1.451	4.642	4.504
43				0.15	0.646	3.941	3.617	1.629	7.744	7.468
44				0.20	0.752	4.269	3.585	1.877	11.218	10.764
45	A12-B12	125	0.25	0.05	0.349	2.804	2.467	0.873	2.913	2.641

46				0.10	0.580	3.563	2.819	1.451	5.091	4.778
47				0.15	0.652	3.396	3.124	1.631	7.977	7.741
48				0.20	0.786	4.199	4.036	1.960	10.100	9.930
49	A13-B13	150	0.1	0.05	0.311	2.658	2.442	0.629	2.499	2.306
50				0.10	0.356	3.038	2.548	0.716	5.706	5.294
51				0.15	0.672	3.941	3.649	1.348	7.184	6.926
52				0.20	0.688	3.561	3.308	1.384	10.824	10.565
53	A14-B14	150	0.15	0.05	0.333	2.840	2.386	0.678	2.432	2.280
54				0.10	0.391	3.425	2.945	0.785	5.418	5.243
55				0.15	0.782	4.045	3.847	1.581	7.387	7.088
56				0.20	0.788	3.266	3.157	1.645	11.191	10.846
57	A15-B15	150	0.20	0.05	0.380	3.038	2.644	0.783	2.142	2.066
58				0.10	0.388	3.664	2.925	0.802	4.573	4.309
59				0.15	0.619	4.161	3.570	1.253	7.057	6.722
60				0.20	0.709	3.466	3.156	1.474	10.960	10.628
61	A16-B16	150	0.25	0.05	0.429	3.666	2.938	0.864	2.074	1.878
62				0.10	0.447	3.687	3.275	0.956	3.909	3.794
63				0.15	0.676	3.116	2.889	1.361	7.299	7.018
64				0.20	0.733	4.163	3.909	1.475	11.189	10.766

22

A Study of Technical Measures for Increasing the Roof-Contacted Ratio in Stope and Cavity Filling

Kouame Joseph Arthur Kouame[1], Yu Feng[1], Fuxing Jiang[1] & Sitao Zhu[1]

[1] School of Civil and Environmental Engineering, University of Science and Technology Beijing, Beijing, China

Correspondence: Kouame Joseph Arthur Kouame, School of Civil and Environmental Engineering, University of Science and Technology Beijing, Beijing, China, 30 Xueyuan Road, Haidian District, Beijing 100083, China. E-mail: josepharthurkk.ustb@yahoo.com

Abstract

Due to the increasing depth of mines, fill mining is increasingly widely used in metal mines. As a result, supporting pit roofs has become an important issue to which an increasing number of people are devoting attention. This paper analyses the factors affecting the rate of supporting pit roofs under stope filling conditions, including filling slurry properties and the filling process. Building on this, the paper examines a number of potential measures for improving the rate of supporting pit roofs, including creating good conditions for filling, optimizing filling slurry properties, eliminating the problems caused by water and improving filling technology. Mining companies must select from these measures according to the particular conditions in which their mines operate, but given the right conditions all of the measures examined have the potential to increase the rate at which pit roofs are supported.

Keywords: filling mining, rate of supporting pit roofs, stope filling conditions, filling process, filling slurry properties

1. Introduction

The increasing depth of mining in recent years and instability in roadway and fields of rock has precipitated a rise in the potential for geological disasters such as rockburst or seismic rockburst (Fall, Célestin, Pokharel, & Touré, 2010). This in turn, in combination with broader public awareness of issues relating to the environment has increased the application of stoping in metal mines and more widely. Cemented filling, layered filling and fill mining are the most commonly used methods (Zhou, Chen, Zhang, & He, 2012). Fill mining is not only applicable when the ore body creates complex mining conditions, but also as a method of minimizing mining dilution, improving ore recovery, controlling press activities, increasing efficiency of production, ensuring mine safety and reducing the environmental and ecological damage. (Zhang, Yan, & Liu, 2012; Gürtunca, Leach, York, & Treloar, n.d.; Zhou, 2010). But we found that there are still issues relating to poor filling stoping, low mining intensity, labor-intensiveness and other practical matters in the mining process, which filled most obvious pick top rate (Trist, Higgin, Murray, & Pollock, 2013).

The 'filling top rate' is a general indicator of how much filling has occurred, and should be strictly defined in relation to the filling paste and the area of the roof ceiling. Under realistic conditions it is difficult to calculate the total area of roof space more accurately than the top rate value. In order to facilitate the calculation of this indicator, Zhang Yongliang proposed the following three methods to calculate the rate of the top pick:

(1) The ratio of the average height of representation

Before filling begins, measure the height of an empty field in different locations and record the average height (H). After a complete end is filled out, measure the average H of the filling body, or alternatively install a viewing window in the wall looking into the space which is to be filled before filling it. Select a stope roof similar in height to the average H of the measuring point for the mark after filling has been completed, and then measure the H of the upper surface markers and filling body. We can then use the following formula to calculate the results to approximate the rate of filling body top pick, ε:

$$\varepsilon = (h / H) \times 100\%, \text{ or } \varepsilon = [(Hh') / H] \times 100\%.$$

H: vertical height of the area before filling, h: vertical ascent of the area after filling

(2) Cross-sectional area ratio method;

The total area with top pick rate ε is expressed as backfill of the area s of top pick, before filling the empty field with the roof of the S ratio (ε = s / S).

(3) Volume ratio of representation. The aim of filling slurry is to form stope backfill volume, and for the volume of the ore mined ratio to represent the rate of filling on the top pick.

This theory does not take into account the strength of the filling materials. In order to achieve the effect of a higher filling body to support the roof rock, and then filling the top rate as high as possible (Ford, Price, Cooper, & Waters, 2014)—especially in mining, workplaces such as roof pillars and under other unusual conditions—effectively supporting the control stope roof is pressed to make the filling body. In addition to filling body strength and the required standards of technical performance, it also meets the requirements of backfill compaction top. When the top is defunct then filling the empty roof top area of the supporting roof meets or exceeds the effective maximum allowable storm drain area. Once this has been filled with ore body or caving, security staff will pose a greater threat (Wang & Park, 2001). After mining, part of the survey showed that the use of filling method and filling the top rate is not as high as is commonly a problem in mines, and the higher rates of some mines are empty on top. Although according an alternative mining method dictates the reduction of the width of the road into the concrete after filling remedies, this only increases the cost of filling mining and reduces botg production efficiency and the strength of the filling body.

2. The Top Pick Effect—Stope Effect Factor

A number of factors influence the effect of top pick filling; including specific, random, unpredictable, multifaceted and other kinds of factors. Making reference to the relevant literature, we summarize the following aspects of first the lead mining field, then empty area filling, in the top rate factors:

2.1 Stope Conditions

(1) The roof is not flat. Due to uneven roof space area, it is difficult during the filling process to ensure the slurry and depression top pick in the filling material do not block spout. In addition, there are likely to be some blind spots which remain unfilled, resulting in a partially empty top.

(2) It is difficult to set the discharge port plate in the highest possible position. If stope is constrained by existing conditions and cutting points are arranged at the highest point of the roof space area, the higher stope roof is difficult to pick at the top.

(3) In the filling mining approach; the top rate is affected by the sectional shape of the stope (Nasir & fall, 2009). Route type stopping system, flat arch arc tops and influenced spandrel arches can still lead to slurry settling in part of the top pick. In gravity-delivered filling systems it is easier to pick the top. Where there is a flat-shaped roof with an empty top, the empty top area is significant and it becomes difficult to guarantee the safety of mining operations.

Pulp slurry leakage does not generally lead to the closure of the drilling project (Vaszita, 2014). Whether or not this is not closed after the construction of the project, there with still be unavoidable empty drilling rock fissures in which pulp slurry filling will have to be run. Plasma leakage phenomenon leads to the top of the filling area appearing empty.

2.2 Filling Slurry Characteristics

(1) A low concentration of filling slurry indicates or precipitates a high water content. When filling slurry concentration is too low, therefore, the filling material will segregate and stratify. If the material is mortar, if its coarse particle deposition is too concentrated, the natural slope angle increases and the water moves to occupy a specific area space. It then stops filling slurry filling this area.

(2) Natural sedimentation of filling slurry (Liu, Dang, & He, 2012; Zhang, Yan, & Liu, 2012). Once the area has been filled by slurry, dehydration and solidification sclerosis cause the volume of he filling material to decrease, leading to an empty top. The presence of water (capillary water in the rock or gravity-pooled water) is a major factor in causing slurry to appear as natural subsidence.

(3) The filling slurry gravity gradient (Zhang, Yan, & Liu, 2012): Gravity flow in filling is used when slurry flows due to the settlement of filling aggregate in the filling pipe discharge mouth-edge, and this position appears to stope the gravity gradient. Poor filling slurry fluidity means a greater gravity gradient, resulting in the filling tube becoming attached to the top of the discharge port; the slope under the influence of gravity cannot pick the top edge of the stope.

(4) Poor physical properties of filling material (Wenbing, 2004): Filling slurry particle size, grain size distribution, clay content and other factors such as fluidity, permeability, sedimentation rate and natural shrinkage greatly affect the natural slope angle. The natural slope angle has significant impact on the top of the flat geometry of the mined areas; large gob makes it hard to pick the top. Settlement and natural shrinkage inevitably occur between the backfill and the roof space.

2.3 Filling Process

(1) Since the filtration of water from filling slurry is a slow process, the excess water which slurry generally contains results in slurry taking a long time to fill. If water *is* completely filtered out, a large gap is left. During the filling process, the larger the body to be filled, the more difficult the process (Ford, Price, Cooper, & Waters, 2014; Wang & Park, 2001), and this difficulty has a serious impact on the top area in the resulting fill.

(2) Strawed direction of water to the filling area and then into the headspace area: in order to prevent clogging of the pipeline during filling, before and after each filling there is often an injection of water for 5-10 minutes into the filling tube. The resulting entry of water into the top pick area to be filled causes fine particles of cement and suspended fine mud to transfer, resulting in segregation and lower body mass filling. The excess water quickly becomes difficult to filter out, seriously affecting the filling top pick.

(3) Single-point cutting: this refers to forming a large natural slope angle (Ford, Price, Cooper, & Waters, 2014). In filling processes using a single point cutting, the cutting-point filling slurry forms a large natural slope angle. In large empty field areas, this phenomenon is more pronounced and results in an empty field which cannot support a pick at the top edge.

(4) Unreasonable filling cutting position: if you do not fill the pipeline located in the highest point of the roof or near the top pick at the spout empty area, and yet still higher than the top, it is difficult to take effective measures to improve the effect of top pick filling.

2.4 Human Factors

Inevitably, the level of experience of the mine operators, their understanding of filling time and their flexibility to use and adapt top technology will impact on top pick filling.

3. Analysis of Measures to Improve Top Rate Access

Gob factors in the pipeline filling of top pick present many challenges, and numerous measures have been developed to overcome these. But each of these measures only applies under certain circumstances and tackles particular factors: gob is a world filled with top pick technical problems (Liu, Dang, & He, 2012). Thus in the implementation of top pick filling operators must adopt a variety of measures to achieve the best results.

3.1 Creating Good Conditions for Stope Filling Top Pick

(1) The application of smooth blasting on neat roof space area (Smooth blasting on neat roof space area, Han, Kamber, & Pei, 2011): in order to meet roof space specifications, the method of horizontally stratified cemented fill mining is used in last tier recovery. Once the top pick has been inserted into the hole slightly inclined blast holes are arranged in a row, according to the requirements of the particular blast. Coupled charge pressure-mining should *not* be used, as it will empty the flat roof area.

(2) Improved outstanding shape headspace region (Santamouris, 2014): this method has long been applied where the ore body is particularly thick. A tilt approach is recovered from a hierarchical approach roadway to the bottom end, approaching at 3 to 8 degrees. The angle stope filling roadway ends in slim contact with the filling material. Easy-filling slurry is used in order to better fill up to the top of the field.

3.2 Preparation of Filling Slurry

(1) Select the appropriate slurry concentration and packing materials. After filling the slurry into the stope there will be natural subsidence due to slurry dewatering which will lead to some shrinkage. As such, the operator should select an appropriate grain size distribution with a ratio that provides good permeability and sufficient liquidity in the filling slurry. Higher concentrations of filling material are needed which provide reasonable mobility, though as it stands the concentration of the slurry is inversely proportional to mobility. The ideal concentration will provide high slurry fluidity.

(2) The use of intumescent material additives (expansion additives added to the filling material) (Bonchis, Duff, Roberts, & Bosse, 2014): aluminum calcium sulfate, gypsum, iron, lime, alum stone are used as expansion agents. When the expanding agent is incorporated in the slurry, hydration reaction occurs causing the volume of slurry to increase. The gap filling material is filled, so that the filling body is denser and effectively reduces shrinkage.

Studies have shown that with the addition of 10% lime, suction effect increases in the slurry and the expansion can effectively maintain the slurry pick of the top spot rate. This method has only been applied for the guarantee of expansive roof restorations, in part because of the lack of volume contraction from the roof, and in part because the effect of missing filling body on the gap already filled is not obvious.

(3) Using high-water rapid hardening materials: high-water quick-setting materials, also known as high-water cured materials (such as high alumina cement, betonies and other raw materials) are subdivided into two categories of material, A and B. In use, A and B are combined with water or mortar and transported to the site near the filling mixture. After mixing, the wet slurry quickly becomes a highly crystalline solid. These materials have high strength, but on the downside, additional engineering costs and increasing transportation costs make them increasingly less economical (Bonchis, Duff, Roberts, & Bosse, 2014).

3.3 Eliminating the Influence of Water

(1) Human treated patio set (Fu, Li, Qiao, & Li, 2011). Within each patio set, where a person carries out filling line treatment, there are setbacks from the use of round wood (generally external box with nailed back logs, wrapped in gauze and with slats nailed).

(2) With a three-way stopcock and water wash treatment lead pipe (Wang & Park, 2001): a three-way stopcock is installing in the filling wall. This leads the way to achieving the water filling of OTC emissions. Before each filling, the valve opens downward so that water from the filling valve exhaust can lead the way into the outer space area. Next, a period of time is left for the wet sand to slowly close the valve filling the empty area with slurry (which is injected into another outlet of the three-way valve). After filling the pipe is washed: the concentration of slurry in the pipeline is allowed to thicken, and the valve is slowly opened while flushing water discharge through the pipe to the outside wall.

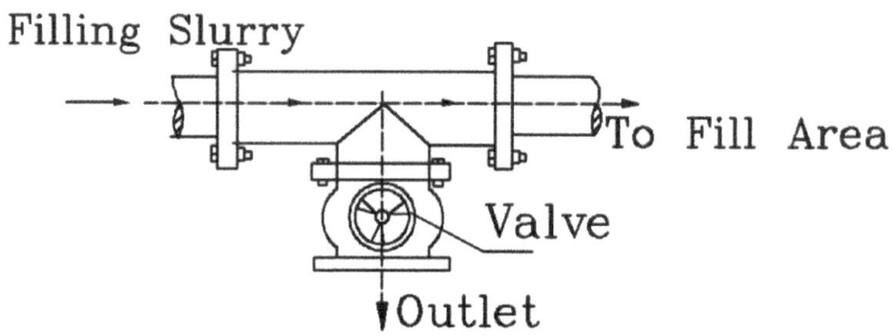

Figure 1. Three-way plug valve schema for stope backfilling

(3) Using filling wells to pump water upward: once filled with water, the pump will empty the area in the upper part and excess water in the slurry will promptly withdraw, continuing to the top of the filling.

(4) Using filter pipes downstream of the discharge: drainage pipes embedded in an empty area in the filling process allow for water in the filling slurry to prolapse through the filter outflow pipes.

3.4 Improving the Filling Process

(1) Partition filling (Liu, Dang, & He, 2012): when filling a stope with a large area, in order to minimize the empty area that fills with the slurry produced by the impact of natural slope angle docking at the top rate, it is necessary to fill the empty area to partition the top pick. The key to success in this process is the size of the top pick partition rather than quality. If the partition is too big the filling process becomes complex and labor-intensive, requiring a large amount of preparatory work. The filling becomes inefficient. In order to reduce the size of the filling hose so that it can access the work area, the partition volume should meet a filling volume. Filling to ensure access is done on the top premise, to minimize the number of partitions. The partitioning method uses natural terrain isolation, separating wood columns and stacked sand bag isolation.

(2) Fractional filling: fractional filling is used to carefully control filling time. The filling takes one day to cure. These time restrictions limit filling into the empty area of slurry volume. It will not fill a single plasma surface because of too much rise, which is greater than the wall can support. Each increase in the height of the filling material is small. At the final stage, the filling material will not shrink due to large natural subsidence and leaves an empty top (Chen, Ji, Zhang, & Wu, 2014).

(3) Multi-point cutting filling (Chen, Ji, Zhang, & Wu, 2014): this method is used for filling slurry with poor liquidity, in order to avoid filling blind. In the case of conditions that increase the number of cutting points, where the gravity gradient of filling slurry is small, the top filling rate increases accordingly. Multi-point cutting can be divided into three seedforms:

First, the stope roof is filled with hanging tubes by drilling holes in the filling tube

Second, multi-root canal filling is used

Third; multiple holes are drilled, filling holes in the middle of the roadway to the stope and filling the stope by filling holes

(4) Forced top pick:

[1] The mortar pumps method. This is applicable where at the end of the filling and contraction settlement the gob missed the top, but for whatever reason the top pick system cannot continue to be used. Air and mortar are pumped into the top area that has been missed, due e.g. to a crimp in the top of the slurry pump tube. The system puts out a short range, high concentration, filling wash. The water precipitates better control or facilitates the cancellation of top pick filling.

[2] The blasting method. This is used to further improve the quality of the filling. It is done via a controlled blasting of the upper rock caving and rock hulking (using a broken expansion coefficient of about 0.3-0.7). This ensures timely roof caving and compact top pick.

(5) Erecting a temporary retaining wall to settle compensation (Zhong et al., 2015): since, at the end of the filling and solidification process, the slurry settlement will shrink, the erection of a temporary filling retaining wall (and filling part of the slurry under pressure) is a way to compensate for this decline in filling body volume.

(6) Filling the main area with the primary filling system and simultaneously filling the top pick with a small filling system (Zhou et al., 2012, Gürtunca, Leach, York, & Treloar, n.d.): this method is applicable where rock is less stable and the filling of the empty area requires a higher top. After filling several times with a small underground mortar pump, consisting of a small, homemade mixing bucket system and with access to high concentrations of top filling, this method can work where there is too low a working efficiency of top pick in the local supplement applications.

(7) Noting the empty top pressure, or high concentrations of concrete filling material (Ford, Price, Cooper, & Waters, 2014): this involves using drilling and concrete pouring machines, and injection under pressure of concrete or high concentrations of filling material into the empty top. This is applicable in situations where the filling material has low water content. The solidification process after pressure injection is almost constant, effectively enhancing the quality of top access.

(8) Post artificial top creation (Santamouris, 2014): in the case of gently and slightly inclined deposits, artificial top picks are more common. When filling a body only 0.5-0.8 meters from the roof, masonry mortar can be used and subsequently wiped with a stone top. Of course, waste rock and mine timber can also manually be thrown into this top space; but this method is labor-intensive, inefficient, and subjects workers to poor working conditions.

3.4 Management Methods

(1) Improving operating procedures and filling management: specially trained mine staff who understand the physical and mechanical properties of filling slurry can deploy a variety of cutting-edge technologies and means of access. The development of relevant operating procedures, the use of set bonuses, and other techniques are essential to motivate workers and mobilize their enthusiasm. It is critical that they are encouraged to take the initiative and to innovate, striving to improve the process of top pick filling.

(2) Reducing worker error and strictly quality-controlling projects:

[1] Drilling projects must be carrier out in a timely manner after construction blocks, to prevent running pulp (Vaszita, 2014).

[2] Solid wall construction and the process of wall filling must occur according to design requirements, to avoid filling instability and collapse. Once this work has been carried out in a timely manner and approved, the next step can be carried out.

[3] Improving roof monitoring (Santamouris, 2014): holes can be drilled to serve as observation wells. After filling, the top pick can be filled with water or slurry spill in the roof of an empty field in order to determine whether the filling is fully completed. It is also possible to install a viewing window to monitor filling progress.

4. Conclusions

As a result of the top of stope filling, a problem has arisen which affects the rate of top pick filling in many multi-level surfaces. Relevant factors can be broadly divided into stope filling conditions, filling slurry characteristics (including the filling process) and human factors. In order to improve top filling rate it is important to take examples from different angles corresponding to treatment measures.

This paper has analysed proposals in five areas: creating good stope conditions; optimizing the preparation of the slurry filling process; eliminating the effects of water; improving the filling process; and improving the management of mine workers.

By presenting research on improving the rate of top pick filling relevant to all of the various conditions under which different mines produce empty tops, and suggesting appropriate technical measures in response, this paper contributes to improving top pick filling rate in mines internationally.

Acknowledgments

This research paper is made possible through the help and support from everyone, including: parents, teachers, family, friends, and in essence, all sentient beings.

Especially, please allow me to dedicate my acknowledgment of gratitude toward the following significant advisors and contributors:

First and foremost, I would like to thank Prof. Jiang Fuxing for his most support and encouragement. He kindly offered invaluable detailed advices and financial support.

Second, I would like to thank Prof. Wang Hongjiang to read my paper and to provide valuable advices and to reproof the paper, as well as all the other professors who have taught me about backfill over the past two years.

Finally, I sincerely thank to my parents, family, and friends, who provide the advices and prayers and encouragements. The product of this research paper would not be possible without all of them.

References

Bonchis, A., Duff, E., Roberts, J., & Bosse, M. (2014). Robotic explosive charging in mining and construction applications. *Automation Science and Engineering, IEEE Transactions on, 11*(1), 245-250. http://dx.doi.org/10.1109/TASE.2013.2241425

Chen, Z. J., Ji, H. G., Zhang, Z. J., & Wu, Z. H. (2014). Research on the variation rules of stress in deep exploitation of hard rock metal mines based on dynamic mine pressure monitoring. *Transit Development in Rock Mechanics: Recognition, Thinking and Innovation*, 59.

Fall, M., Célestin, J. C., Pokharel, M., & Touré, M. (2010). A contribution to understanding the effects of curing temperature on the mechanical properties of mine cemented tailings backfill. *Engineering Geology, 114*(3), 397-413. http://dx.doi.org/10.1016/j.enggeo.2010.05.016

Ford, J. R., Price, S. J., Cooper, A. H., & Waters, C. N. (2014). An assessment of lithostratigraphy for anthropogenic deposits. *Geological Society, London, Special Publications, 395*(1), 55-89. http://dx.doi.org/10.1144/SP395.12

Fu, L., Li, Q., Qiao, L., & Li, Y. (2011). Numerical study and application on technology of slotting and blasting down the roof for handling mined-out areas. *Rock Mechanics: Achievements and Ambitions, 277.*

Gürtunca, R. G., Leach, A. R., York, G., & Treloar, M. L. (n.d.). *In situ* performance of cemented backfill in a deep-level South African gold mine. *Minefill,* 93, 121-8.

Han, J., Kamber, M., & Pei, J. (2011). *Data mining: concepts and techniques*. Elsevier.

Liu, Z. X., Dang, W. G., & He, X. Q. (2012). Undersea safety mining of the large gold deposit in Xinli District of Sanshandao Gold Mine. *International Journal of Minerals, Metallurgy, and Materials, 19*(7), 574-583. http://dx.doi.org/10.1007/s12613-012-0598-y

Nasir, O., & Fall, M. (2009). Modeling heat development in hydrating CPB structures. *Computers and Geotechnics, 36*(7), 1207-1218. http://dx.doi.org/10.1016/j.compgeo.2009.05.008

Santamouris, M. (2014). Cooling the cities–a review of reflective and green roof mitigation technologies to fight heat island and improve comfort in urban environments. *Solar Energy, 103*, 682-703. http://dx.doi.org/10.1016/j.solener.2012.07.003

Trist, E. L., Higgin, G. W., Murray, H., & Pollock, A. B. (2013). *Organizational choice*. Routledge.

Vaszita, E. (2014). Environmental risk of mining. *Engineering Tools for Environmental Risk Management: 1. Environmental Deterioration and Contamination-Problems and their Management, 1*, 113.

Wang, J. A., & Park, H. D. (2001). Comprehensive prediction of rockburst based on analysis of strain energy in rocks. *Tunnelling and Underground Space Technology, 16*(1), 49-57. http://dx.doi.org/10.1016/S0886-7798(01)00030-X

Zhang, R., Yan, C., & Liu, J. (2012). Effect of mangroves on the horizontal and vertical distributions of rare earth elements in sediments of the Zhangjiang Estuary in Fujian Province, Southeastern China. *Journal of Coastal Research, 29*(6), 1341-1350.

Zhong, R., Li, W., Chen, Y., Ji, J., Yang, Y., & Hu, C. (2015). Significant Zn–Pb–Cu remobilization of a syngenetic stratabound deposit during regional metamorphism: a case study in the giant Dongshengmiao deposit, northern China. *Ore Geology Reviews, 64*, 89-102. http://dx.doi.org/10.1016/j.oregeorev.2014.06.015

Zhou, L. X. (2010). The Review and Prospects of 50 years in EMM Industry of China. *China's Manganese Industry, 1*, 002

Zhou, Y. J., Chen, Y., Zhang, J. X., & He, Q. (2012). Control principle and technology of final compression ratio of backfilling materials. *J Min Saf Eng, 3*, 351-356.

23

Switching Direction of Liquid Flow by Amplification of Wetting Properties of Partilly Roughened Prism Array

Hyemin Lee[1] & Hyunsik Yoon[1]

[1] Department of Chemical and Biomolecular Engineering, Seoul National University of Science & Technology, Seoul, 139-743, Korea

Correspondence: Hyunsik Yoon, Department of Chemical and Biomolecular Engineering, Seoul National University of Science & Technology, Seoul, 139-743, Korea. E-mail: hsyoon@seoultech.ac.kr

Abstract

Superhydrophilic and superhydrophobic surfaces inspired from nature have been developed for anti-fogging, anti-reflecting, water-repelling, anti-icing and self-cleaning films. The amplification of the hydrophilicity and hydrophobicity originates from surface roughness, particularly from their multiscale geometry. In this paper, we present a strategy to exploit the enhancement of wetting properties on roughened surfaces for switching the direction of unidirectional liquid flow on an opened microchannel. We exploit a unidirectional liquid spreading on a hierarchical two-face prism array, which have the symmetry in microscale while having the asymmetry in nanoscale. Based on the switching the direction of the lower critical angle, the transition of the surface chemistry results in the change of flow directions in an open channel.

Keywords: prism, superhydrophobicity, superhydrophilicity, directional flow, polymer, contact angle

1. Introduction

Superhydrophilic and superhydrophobic surfaces can be observed in daily life and nature such as a paper, sponge, plant leaf, leg of water strider. These surfaces have been developed for functional application including anti-fogging, anti-reflection or anti-sticking, self-cleaning and anti-icing films (Barthlott et al., 1997; Blossey et al., 2003; Choi et al., 2011; Feng et al., 2005; Quéré et al., 2008; Verplanck et al., 2007; Xia et al., 2008). It is well known by Wenzel and Cassie-Baxter models (Quéré et al., 2008) that the wetting properties such as hydrophilicity and hydrophobicity can be amplified by increasing surface roughness. Especially, hierarchical structures inspired from lotus leaves have been developed for their superhydrophobicity and water repellency (Barthlott et al., 1997; Blossey et al., 2003). Although the surface with superhydrophilicity or superhydrophobicity can be used as a special coating by themselves, there still remains needs for new and emerging applications such as microfluidic devices, cell arrays, and separation of oil and water (Ueda et al., 2013). Here, we propose a strategy exploiting a hierarchical structure with the symmetry in microscale while having the asymmetry in nanoscale to manupulate liquid flow direction on an open channel. By using the mechanism of the critical contact angle on an edge (Kim et al., 2013; Oliver et al., 1977; Yoon et al., 2015), we could switch the flow direction due to the change in the direction of lower contact angle. Since the contact angle difference originates from the difference of roughness between adjoining surfaces, the strategy has an advantage of the wide range of material choices for the manipulation of liquid flow.

2. Methods

2.1 Materials

We used a homogeneous UV-curable liquid prepolymer by mixing multi-functional acrylated prepolymer (Ethoxylated Trimethylolpropane triacrylate, M3130, Miwon Specialty Chemical Co., Ltd.) with acrylated-functionalized polysiloxane (Ac-PSi) (Silicon polyether acrylate, Rad2200N, Degussa). As a photoinitiator, we added Darocur 1173 (2-hydro-2-methyl-1-phenyl-1-propane, Ciba Specialty Chemicals, Switzerland). The detail blend ratio could be found elsewhere (Choi et al., 2011).

2.2 Polymeric Prism Array

We prepared a nickel prism master by conventional mechanical machining. After preparing a master, liquid prepolymer mixture of acrylated prepolymer and acrylated-functionalized polysiloxane (Ac-PSi) was dropped onto a master and a transparent PET film was placed on the prepolymer. The UV-curable prepolymer mixture was crosslinked by ultra-violet (UV) light exposure. Then, the crosslinked polymeric prism array was detached from the master.

2.3 Two Face Prism Array

The polymeric prism array (20 or 50 μm in period, 45° in the prism angle) was deposited with 20 nm thick gold film only on one face of the prism array by oblique metal deposition. Gold films deposited by a conventional thermal evaporator and we use inclined loader to get an oblique angle of around 50°. During the evaporation, metal atoms were guided vertically, with the inclined loader defining the oblique angle. The detail experimental procedure can be found elsewhere (Kim et al., 2013; Yoon et al., 2009, 2015).

2.4 Two Face Prism Array

The prepared two-face prism array was roughened by oxygen plasma. The gold-deposited faces are not etched by oxygen plasma because the metal films behaves as an etch mask. On the other hand, polymeric face has nanoroughness because of selective etching of polymer while the silicon moiety acted as another etch barrier. The detail experimental data could be found elsewhere (Yoon et al., 2015).

2.5 Surface Treatments

To make surface hydrophobic, the multiscale two-face prism arrays were treated with a liquid phase self-assembled monolayer (SAM) species of fluoroalkylsilane under an ambient atmosphere. The patterned substrate was dipped for several minutes into tridecafluoro-1,1,2,3-tetrahydrooctyl-1-triethoxysilane that was mixed with perfluorocarbon fluid, FC-40 (3M Co., Ltd, USA) as a diluent. Then, the substrate was placed in an oven at 120 °C for 10 min to remove excess diluent completely as well as to ensure bonding of –OH functional groups on oxidized surfaces.

2.6 Measurements

To measure wetting properties, advancing and receding contact angles (CAs) of water droplets were measured at ambient atmosphere using a contact angle analyzer (DSA100, Kruss GmbH., Germany) by the sessile drop method with increasing or decreasing droplet volume. The CA values presented in this experiment were averaged over five different positions.

3. Results and Discussion

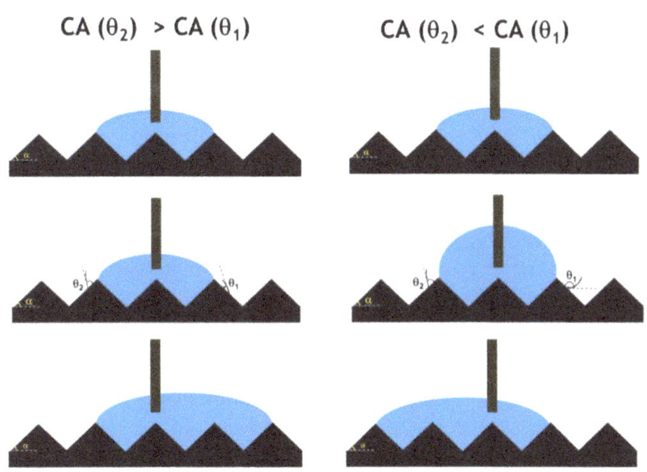

Figure 1. Schematic illustration for switching the flow direction in a prism array

Figure 1 is a schematic illustration of the strategy of switching the flow direction by the concept of critical contact angles with the three-phase contact line across the ridges of channels created by the prism array. As we discussed in our previous works (Kim et al., 2013, Yoon et al., 2015), the liquid-front is pinned on prism edges and the pinning is continued during increasing liquid volume until the contact angle of the liquid exceeds its critical contact angle on the edge. The critical contact angle θ_c can be derived as follows (Kim et al., 2013, Oliver, et al., 1977; Yoon et al., 2015):

$$\theta_c = \theta_* + \alpha \quad (1)$$

where θ^* is the contact angle on a planar surface and α is the prism angle. In the hierarchical two-face prism system, the contact angles of the two-faces are different because of their different morphologies. As a result, when the contact angle on the right face (θ_1) is lower than the angle on the left face (θ_2), the liquid flows to the right direction because the critical angle is lower in the right face. When we change the contact angle (θ_1) of the right face to be greater than θ_2 by proper chemical treatments, the direction of lower critical angle is switched to the left face. The novelty of the hierarchical two-face prism arrays is that the capability of switching liquid flow direction by exploiting the transition of the surfaces from hydrophobic to hydrophilic to switch the direction of lower critical contact angle.

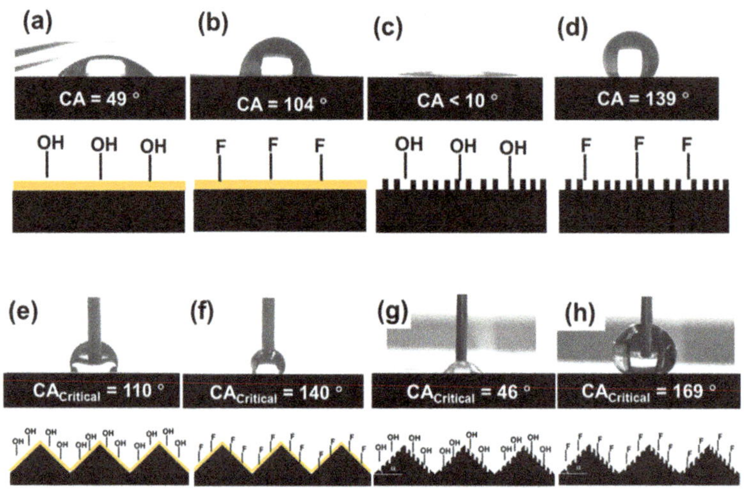

Figure 2. Pictures representing contact angles on planar surfaces (a-d) and critical angles (e-h)

As shown in Figure 2(a) and 2(b), the contact angle on a planar gold surface after oxygen plasma is ~50° and changed to ~100° after a fluorinated self-assembled monolayer (SAM) treatment. The prism surface then becomes hydrophilic again by treating it with oxygen plasma at the mild condition (25 W for 10 sec). We also noted that the hydrophilic transformation by oxygen plasma treatment involves the ozone cleavage of C-F bonds and the substitution of –OH groups or functionalization to –OH groups. As shown in Figure 2(c) and 2(d), the contact angles on roughened surfaces are significantly changed from 5° to 139°by switching the surface chemistry from hydrophilic to hydrophobic. The amplification of hydrophobicity and hydrophilicity could be explained by Wenzel's model (Quéré et al., 2008). Figure 2(e-h) shows pictures representing the critical contact angles on symmetric prism arrays coated by Au film and roughened by the oxygen plasma. The average of the differences between critical contact angles (Figure 2(e-h)) and contact angles (Figure 2(a-d)) is 43.25°, which is in good agreement with the prism angle (45°) derived from Eq. (1).

When the two faces assembled together as a form of the multiscale two face prism array, we can exploit the changes of the direction of lower critical contact angle to switch the flow directions. Figure 3 shows the graph of contact angles on planar surfaces switched by chemical treatments. It is notable that repeating the cycles of SAM/oxygen plasma treatments for more than 5 times shows excellent reversibility in contact angle. As shown in Figure 3, the lower contact angle after hydrophobic treatment is in the smooth side of the Au-coated surface because the contact angle on the roughened surface is greater than that on the smooth surface. On the other hand, the lower contact angle after hydrophilic treatment is in the side of the roughened surface because it is more hydrophilic

From the change of the direction of lower contact angle by transition of the surface from hydrophilic to hydrophobic or vice versa, we expect that the liquid flow direction can be switched. Previously, we reported a scheme of two-face prism array based on stimuli-responsive polymers to switch the flow direction (Kim et al., 2013). Since the range of the contact angles of the polymer should be restricted (45°~72°), however, it is difficult to choose a compatible metal film with a contact angle within the range. In the case of the multiscale two-face prism array, it has adjoining smooth and rough surfaces; the contact angle on the rough surface is lower when the entire surface becomes hydrophilic. On the other hand, the contact angle on the smooth surface becomes lower when it is changed to a hydrophobic surface. By exploiting the aforementioned wetting characteristics, the change in flow direction could be achieved with the wide ranges of material choices.

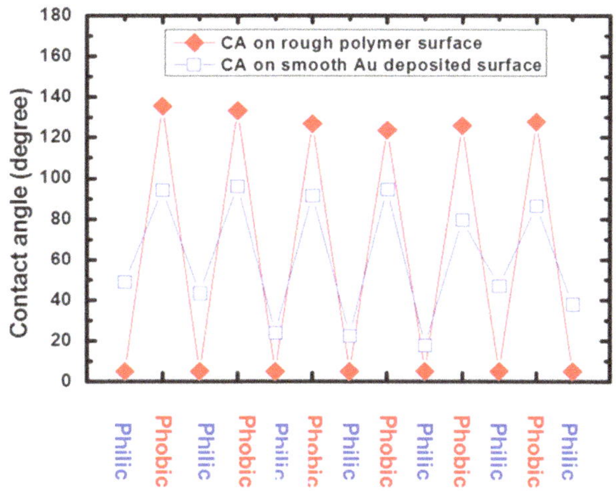

Figure 3. A graph showing reversible switching of contact angles on planar surfaces

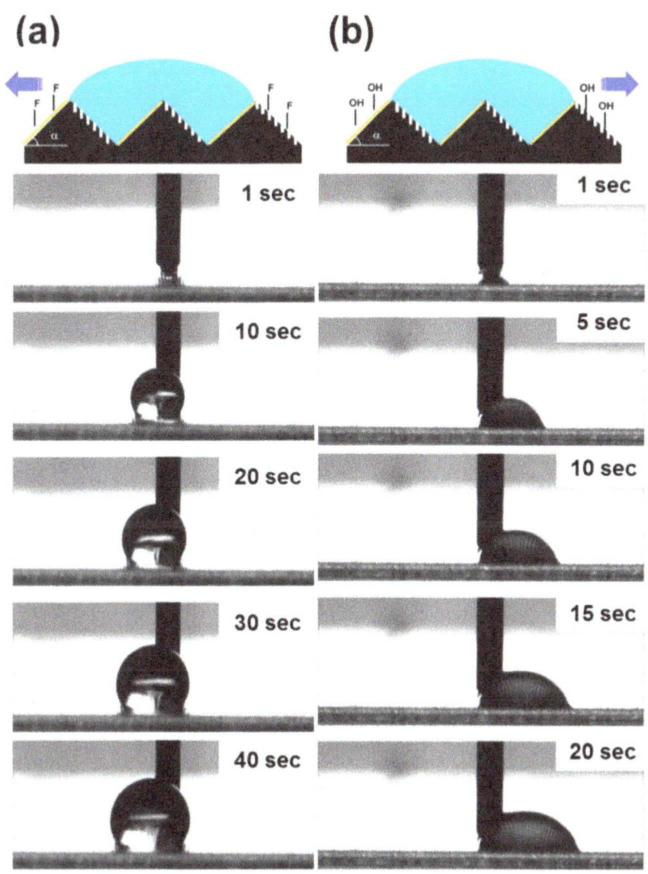

Figure 4. Movie cuts of liquid flow (a) after fluorinated SAM treatment and (b) oxygen plasma treatment

Figure 4 shows the movie cuts for the change in the direction of liquid flow on hydrophobic and hydrophilic two-face prism arrays, respectively, with the same micro- and nano-structures maintained. As shown in Figure 4(a), the liquid flows towards the smooth metal face with a relatively slow speed at a feeding rate of 5 μl/min because of its high critical contact angle in the moving direction. Remarkably, the moving direction was reversed towards the roughened polymer face after the oxygen plasma, as shown in Figure 4(b). Due to the hydrophilic nature of the surfaces, the liquid flow speed was increased at the same feeding rate. Such a reversal of flow direction was repeated more than five times, in good agreement with the similar wetting behavior.

4. Conclusions

In summary, we propose a scheme to switch a flow direction by hierarchical two-face prism arrays which are symmetric in microscale but asymmetric in nanoscale. The prism arrays were used to switch the direction of water flow in reversible manner without deterioration of the hierarchical structures. It was observed that the liquid flows to the direction of the lower critical contact angle. By exploiting the transition of the roughened surfaces from hydrophobicity to hydrophilicity, we further demonstrated the switching of flow direction on pre-defined prism arrays, which would enable the control of the direction of liquid flow on demand.

Acknowledgments

This study was supported by Seoul National University of Science and Technology.

References

Barthlott, W., & Neinhuis, C. (1997). Purity of the sacred lotus, or escape from contamination in biological surfaces. *Planta, 202*(1), 1-8. http://dx.doi.org/10.1007/s004250050096

Blossey, R. (2003). Self-cleaning surfaces—virtual realities. *Nature materials, 2*(5), 301-306. http://dx.doi.org/10.1038/nmat856

Choi, S. J., Choi, M. K., Tahk, D., & Yoon, H. (2011). Fabrication of a hierarchical structure by oxygen plasma etching of a photocured microstructure containing a silicon moiety. *Journal of Materials Chemistry, 21*(38), 14936-14940. http://dx.doi.org/10.1039/c1jm12305j

Feng, X., Zhai, J., & Jiang, L. (2005). The fabrication and switchable superhydrophobicity of TiO2 nanorod films. *Angewandte Chemie International Edition, 44*(32), 5115-5118. http://dx.doi.org/10.1002/anie.2005 01337

Kim, S. M., Koh, J. H., Suh, H. S., Yoon, H., Suh, K. Y., & Char, K. (2013). Thermoresponsive switching of liquid flow direction on a two-face prism array. *Soft Matter, 9*(16), 4145-4149. http://dx.doi.org/10.1039/c3sm27901d

Oliver, J. F., Huh, C., & Mason, S. G. (1977). Resistance to spreading of liquids by sharp edges. *Journal of Colloid and Interface Science, 59*(3), 568-581. http://dx.doi.org/10.1016/0021-9797(77)90052-2

Quéré, D. (2008). Wetting and roughness. *Annu. Rev. Mater. Res., 38*, 71-99. http://dx.doi.org/10.1146/annurev.matsci.38.060407.132434

Ueda, E., & Levkin, P. A. (2013). Emerging Applications of Superhydrophilic - Superhydrophobic Micropatterns. *Advanced Materials, 25*(9), 1234-1247. http://dx.doi.org/10.1002/adma.201204120

Verplanck, N., Coffinier, Y., Thomy, V., & Boukherroub, R. (2007). Wettability switching techniques on superhydrophobic surfaces. *Nanoscale Research Letters, 2*(12), 577-596. http://dx.doi.org/10.1007/s11671-007-9102-4

Xia, F., & Jiang, L. (2008). Bio-inspired, smart, multiscale interfacial materials. *Adv. Mater, 20*(15), 2842-2858. http://dx.doi.org/10.1002/adma.200800836

Yoon, H., Jeong, H. E., Kim, T. I., Kang, T. J., Tahk, D., Char, K., & Suh, K. Y. (2009). Adhesion hysteresis of Janus nanopillars fabricated by nanomolding and oblique metal deposition. *Nano Today, 4*(5), 385-392. http://dx.doi.org/10.1016/j.nantod.2009.08.007

Yoon, H., Sung, S. H., Koh, J. H., Kim, S. M., Choi, S. J., Suh, K. Y., & Char, K. (2015). Directional step flow across ridges on multiscale two-face prism array. *Macromolecular Research, 23*(2), 145-148. http://dx.doi.org/10.1007/s13233-015-3026-0

Mechanical Properties Investigation of Multi-Lap Stainless Steel Sheets

Qasim M. Doos Al-Attaby[1] & Shaymaa Abdul Khader Al-Jumaili[1]

[1] Baghdad University, College of Engineering, Mechanical Engineering Dept., Baghdad, Iraq

Correspondence: Shaymaa Abdul Khader Al-Jumaili, Baghdad University, College of Engineering, Mechanical Engineering Dept., Baghdad, Iraq. E-mail: sh.iraq79@yahoo.com

Abstract

The mechanical testing and microstructers for austenititc stainless steel have been carried out to indicate the effect of the localized heating on the spot welds. The results indicated that the increasing of the heating will increase the tensile shear strength rapidly up to the expulsion point. Therefore, the hardness increases with the increasing of the weld nugget area. Hence, the fracturing tendencies have been observed inside the nugget. Nevertheless, the outer surface cracking is developed by adding a thin layer of pure metals between the two stainless steel 304L sheets. It was shown that the addition of pure metals, 3-lap sheets decreases the joint strength as compared with the 2-lap joints for the same parameters. The effect of micro cracking and the intensive brittleness decreases the joint strength. The excessive cracking was observed in 3-lap as compared with the 2-lap joint.

Keywords: austenititc stainless steel, joint strength, lap joint, resistance spot welding, stainless steel, solidification cracking

1. Introduction

Resistance welding (RW) is a fusion-welding process that uses a combination of heat and pressure to produce a localized heating what called nugget weld area, see Figure 1. The heat being generated by electrical resistance to current flow at the bulk workpieces and at the interfaces due to electrical resistance (Wei & Wu, 2012) (Groover, 2007). Resistance spot welding is effective way to join metal sheets in various manufacturing, automobile, aerospace and packaging industries (M. Vural, 2004; Aslanlar, 2006; Cullen, Athi, Al-Jader, Shaw, & Al-Shamma'A, 2007; Mukhopadhyay, Bhattacharya, & Ray, 2009; Ambroziak & Korzeniowski, 2010; Wei & Wu, 2012). A typical car body contains thousands of spot welds for joining the mixture types of metal material and thickness of sheet (Welding, Industry, & Practice, 2002; Vural, Akkuş, & Eryürek, 2006). Hence, the process is used extensively for joining low carbon steel, stainless steels and galvanized steel components due to their wide applications as combined with the required properties(Agashe & Zhang, 2003; Pouranvari, Abedi, Marashi, & Goodarzi, 2008; Marashi, Pouranvari, Amirabdollahian, Abedi, & Goodarzi, 2008; A. M. Al-Mukhtar & Doos, 2013). In addition, aluminum alloys are more frequently used also in automotive industry especially as an alternative material for car-bodies with high impact strength and the weight ratio (Ambroziak & Korzeniowski, 2010; Han, Thornton, Boomer, & Shergold, 2010). Nevertheless, dissimilar materials are also welded. The amount of heat in the joint is determined by the material thickness and type, amount of the current and cross-sectional area of the welding tips and contact surfaces.

Resistance spot welding, like any process, affected by several factor such as; electrode force, electrode geometry, work piece and/or electrode misalignment, electrode surface roughness, electrode wear, temperature of electrodes and work piece, micro-movement at any of the surfaces (Welding et al., 2002), current flow, resistance of the circuit, and length of time the current is applied (Groover, 2007). Nevertheless, the main factors having an effect represent by; current, time, and electrode force (Groover, 2007; Aslanlar, Ogur, Ozsarac, Ilhan, & Demir, 2007). The welding current, electrode pressure and hold time affected the thermo mechanical interactions of the welding process and changed the final nugget geometry volume, see Figure 1 (Tsai, Jammal, Papritan, & Dickinson, 1992). The cracking and insufficient depth is the reason of the most toughness reduction (A. Al-Mukhtar & Doos, 2013). The properties of most importance which influence weldability have been investigated. Therefore, the selection of suitable welding conditions, the fusion zone characteristics, and heat affected zone, HAZ play a significant role in any spot weldability investigations. The current study provoked the heat generation and resistance effect by inserting a high thermal sheet metal. Hence, the strength and fracturing were observed.

2. Experimental Work

2.1 Material

Welding was performed on a specimen prepared per the AWS standard, while changing the welding current and welding force in cycles of welding time (A. M. Al-Mukhtar, 2002; Kim, Lee, Lee, & Rhee, 2004). In this study, we choose AISI 304 L type of austenitic stainless steel with thickness 0.8 mm was welded. For several years ago, austenitic stainless steel 304L have been employed in industrial applications such as pipes and pressure vessels for structural purposes. The influences austenitic stainless steel 304L on the industry are mainly due to its mechanical strength as well as excellent corrosion (Park et al., 2004; N R Anand1, 2013). The chemical composition and the mechanical properties are given in Table 1and Table 2, respectively (MatWeb, n.d.).

Table 1. Chemical composition of stainless steel

	C	Cr	Ni	Mn	V	Mo	Si	Ti	Nb	Cu	S
Austenitic Stainless Steel 304L	0.016	17.8	8.10	1.04	0.101	0.02	0.466	0.008	0.008	0.062	0.012

Table 2: Mechanical properties of stainless steel

	σy (MPa)	Tensile strength σu (MPa)	Elongation EL%	HV. Vickers
Austenitic Stainless Steel 304L	241	586	44	161

2.2 Geometry and Specimen Preparation

The geometry of specimen is rectangular with dimension based on American National Standards Institute (ANSI) and American Welding Society (AWS) (Connor, 1987; Zhou, Hu, & Zhang, 1999; A. M. Al-Mukhtar, 2003). The dimensions for each plate to be spot welded are 78 mm length, 19 mm width, and overlap is 19 mm, see Figure 1. After cutting the specimens, it is necessary to their surfaces were chemically cleaned by acetone before resistance spot welding to eliminate surface contamination (Kahraman, 2007). The lap tensil shear specimen widly used as systematic testing for investigation joint strength, failure mode, microstructres and welding defects, see Refs (Vural et al., 2006; Shamsul, Hisyam, & Muhibbah, 2007; Triyono, Purwaningrum, & Chamid, 2013). The inserting a sheet in the middle of lap specimens was used in some case, but for the same metals, see Sin, Yang, Yu, Kim, and Kang (2007). Therefore, this work aims to show the effect of thermal and physical properties of different metals on the mechanical properties of the spot join. Thin layers of pure copper and pure aluminium with 0.09 mm and 0.1 mm thickness, respectively, were used as a transition material between the two sheets of austenitic stainless steel 304L, to investigate the joint strength, see Figure 1 (right). Three sets of samples were welded by fixing electrode force, welding time and using three values of the welding current has been changed sequentially for 4000, 6000 and 8000 Amp, by increasing 2000 Amp, respectively.

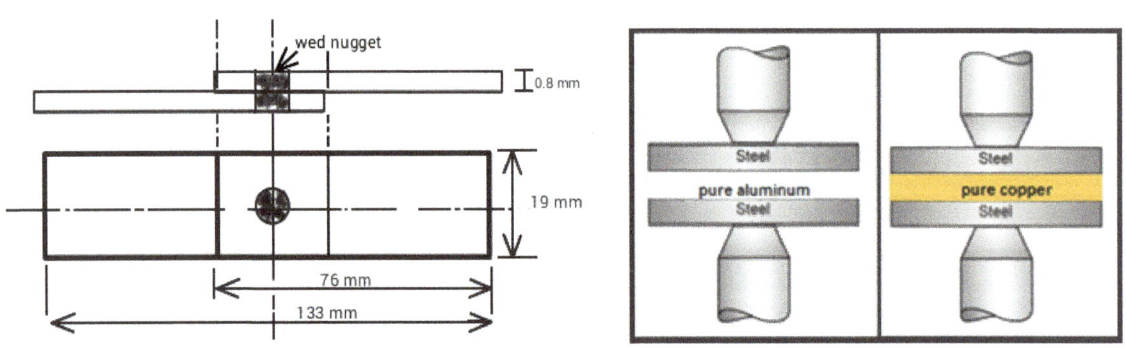

Figure 1. Dimension of tensile shear test (mm), ANSI/AWS standard

2.3 Materials Test

Static strength was measured using the conventional tensile shear lap specimens that shown in Figure 1 (left) according to AWS practice (Zhou et al., 1999). All welds were tested in tensile shear using LARYEE computer control electronic universal testing machine. It was equips with high precision servo motor control system. Control system is full closed loop measuring and control system. Control software can automatically figure out routine data, such as load-extension and stress-strain curves. The available function of the machine is to maintain both the peak load and the weld strength. The lap shear specimen of two flat plates of 0.8 mm thickness with an overlapping distance of 19 mm was welded, see Figure1 (left). The geometry was approximated as closely as possible to pure shear made on the joint (A. M. Q. D. Al-Mukhtar & Doos, 2013). The microhardness of cross section weld nugget was tested. The hardness measurements were carried out using a Digital Hardness Vickers Tester TH714 machine. Three hardness measurements were made and then take the average of them.

3. Results and Discussion

3.1 Effect of Current on Joint Strength and Weld Nugget Area

Welding current is the most parameter among all the parameters affecting the amount of heat obtained. According to Joule law, $Q = I^2Rt$; where Q is generated heat, I is current, R is electrical resistance of the circuit, and t is time that the current is allowed to flow through the circuit (Zhang & Senkara, 2006; Nodeh, Serajzadeh, & Kokabi, 2008; Mirsalehi & Kokabi, 2010). So when the weld current increases the joint strength (maximum load) increases due to the weld nugget area increasing until the maximum current value that achieved by the welding machine. The nugget size increasing until expulsion occurs. The results agree with different studies, see Ref (Aslanlar et al., 2007; Pereira, Ferreira, Loureiro, Costa, & Bártolo, 2010; Moshayedi & Sattari-Far, 2012). The increasing current, i.e. increasing the heat input will increase the melting area, see Figure 2. This usually leads to increase the strength of the joint. However, the increasing beyond the maximum ranges causing the splashing and expulsion of metal due to the complicated mechanism between the electrode force and nugget growing pressure between the sheets, see A. Al-Mukhtar and Doos (2013). Moreover, the continuing increased welding currents will cause the nugget diameter to become shorter as the result of the splashing in the interlayer (Shamsul et al., 2007).

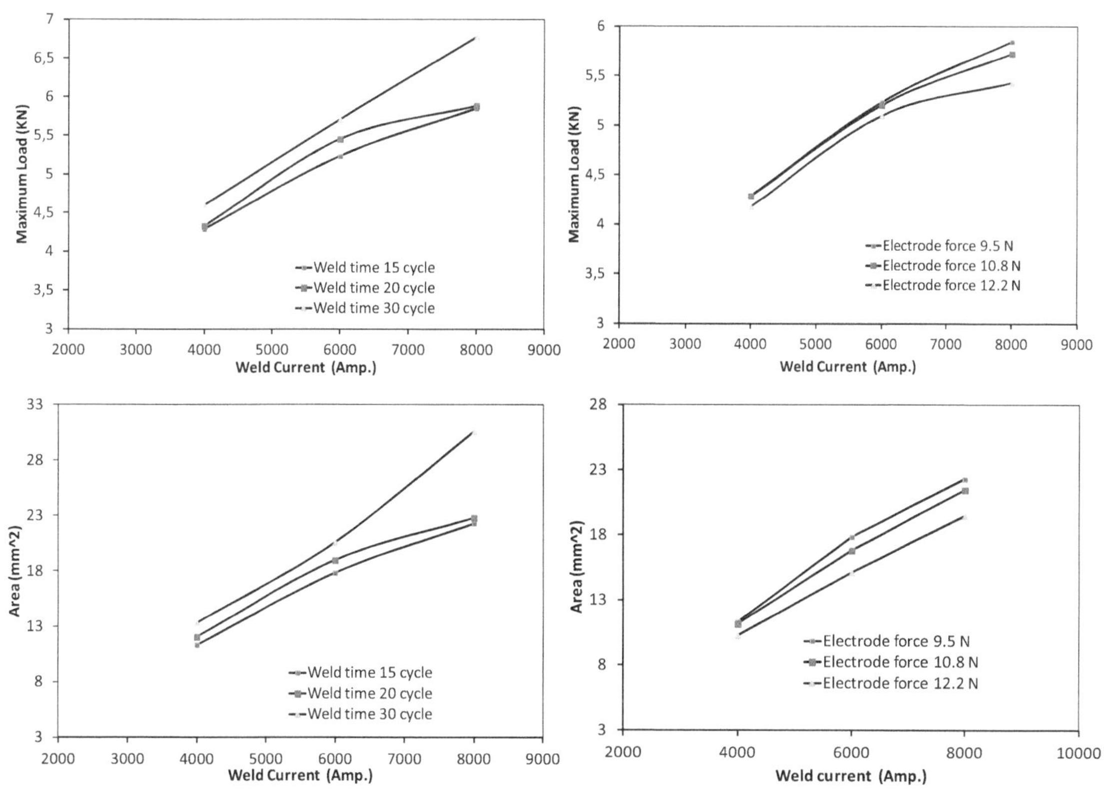

Figure 2. Effect of welding current on the tensile shear strength and weld nugget area

That is agrees with almost works (Shamsul et al., 2007; A. M. Al-Mukhtar & Doos, 2013). The weld area increase rapidly with weld current, from low current levels until expulsion occurs at the higher levels. A similar behaviour occurs with the weld strength, due to the direct relation between the weld area and its strength. Effect of welding current on the tensile shear strength and weld nugget area was shown in Figure 2.

The maximum load was obtained before the failure. The expulsion occurs at that maximum current. Hence, the cracking is initiated around the weld nugget. Subsequently, a pull out fracture will be developed.

3.2 Effect of Weld Time on Joint Strength and Weld Area

In the light of results obtained from the experimental work indicated that increasing weld time causes to enhance high heat input to weld zone, hence, increase weld nugget area (Nodeh et al., 2008). So the tensile-shear strength of joints will be increased. The effect of weld time less than the effect of weld current. The relations of weld time with joint strength and weld nugget area show in the Figure 3.

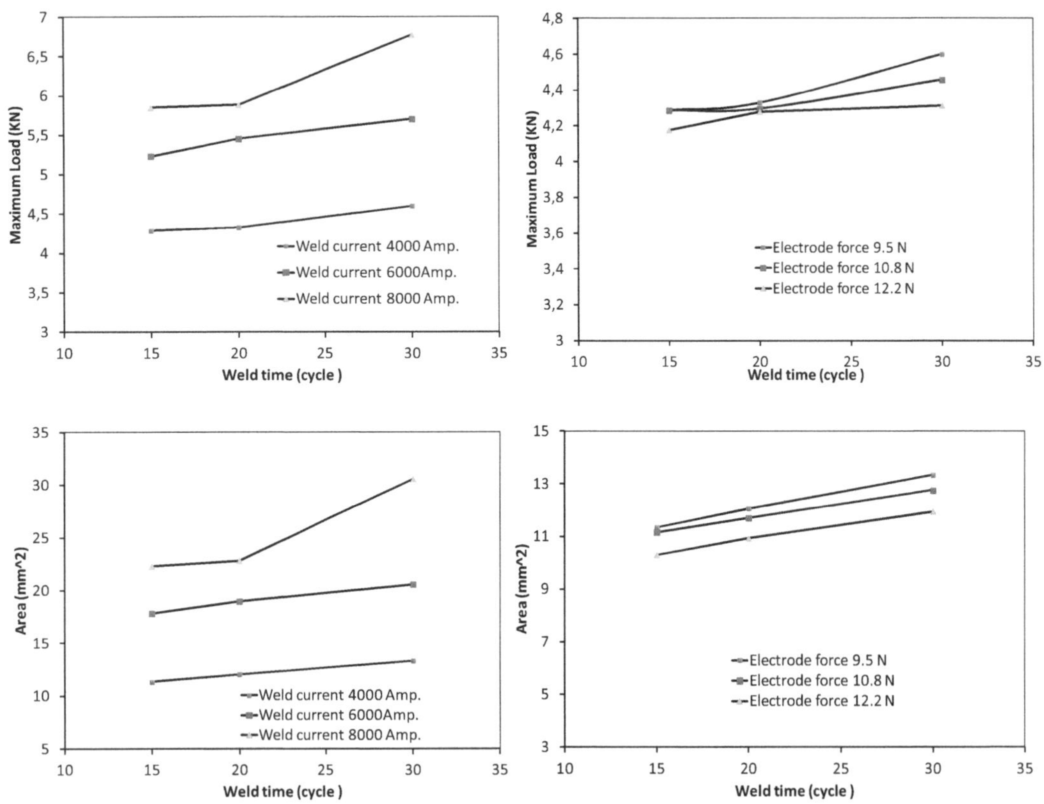

Figure 3. Effect of welding time on the tensile shear strength and weld nugget area

3.3 Effect of Electrode Force on Joint Strength and Weld Area

The effect of electrode force on joint strength and weld nugget area is investigated. Results showed that the increasing electrode force decreases fusion zone size and therefore leads to reduce weld nugget area and joint strength. The reason for the adverse effect of electrode force due to decreases the electric resistance and thus reduces the generated heat at the sheet-sheet interface. Effect of electrode force on the joint strength and weld nugget area were shown in Figure 4.

3.4 Hardness Profile

Microhardness for 2-lap and 3-lapspecimens has been measured. Generally the values of hardness distribution include the maximum values in HAZ. The less values were observed in fusion zone (weld nugget). HAZ and the weld zone hardness are mainly higher that the base metal hardness (Mukhopadhyay et al., 2009). The observed hardness profiles are similar to the ones reported by (Park et al., 2004; Aslanlar, 2006). The microhardness of as-welded specimen has values a little higher than the microhardness of transition pure copper specimen. The microhardness of specimen using pure aluminum as transition material has the highest values due to formation of intermetallic components under the same welding conditions. The reason of increase the microhardness of the specimen AL-ASS returns to formation solid compounds such as oxide aluminum. The microhardness distribution is shown in Figure 5.

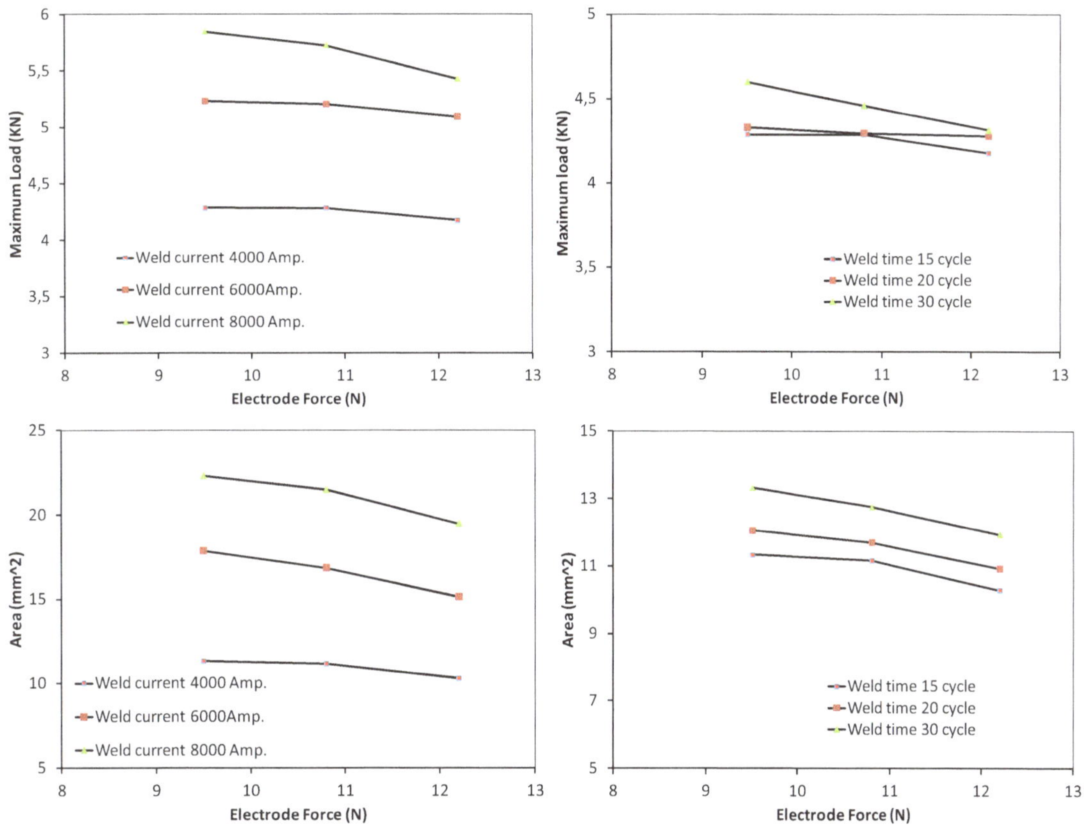

Figure 4. Effect of electrode force on the tensile shear strength and weld nugget area

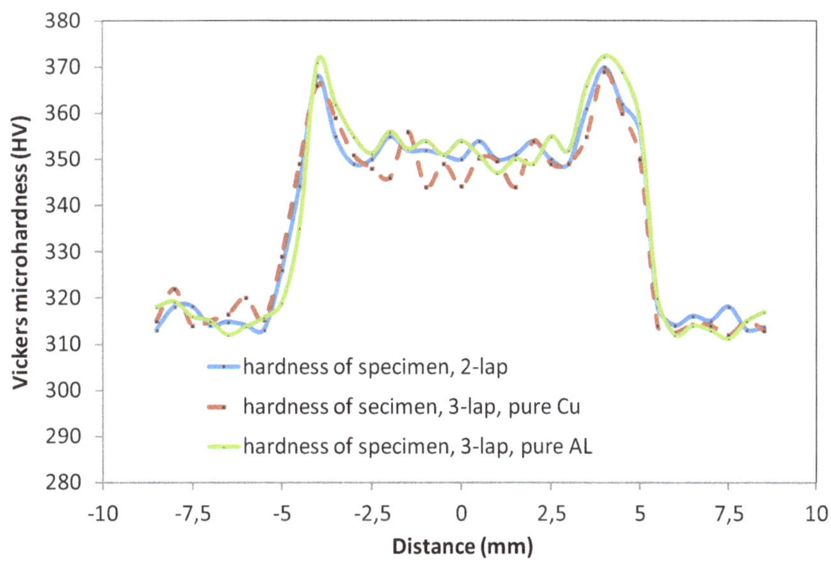

Figure 5. Vickers microhardness profile for spot weld, I= 6000 Amp, time=30 cycles, and electrode force=9.5 N, longitudinal direction

3.5 Mechanical and Metallurgical Properties Due to Transition Material

The use of pure copper as a transition metal between two sheets of austenitic stainless steel 304L reduce the joint strength. The maximum load of joint strength under the conditions (welding current of 8000 Amp, welding time of 30 cycles and electrode force of 9.5N) was 6.775 KN. The maximum load of joint strength of using transition pure copper was 5.149 KN. However, welding under the same condition with the pure aluminum the expulsion occurs in the weld nugget due to the high thermal conductivity. The maximum load of joint strength under the conditions (welding current of 6000Amp, welding time of 30 cycles and electrode force of 9.5N) was 5.743KN, and the maximum load of joint strength of using transition pure copper was 4.5 KN; while the maximum load of joint strength of using transition pure AL was 4.13 KN. In other words, the percentage of maximum load decreasing was approximately 24%, and 26%, see Figure 6. This reduction in maximum load value is due to many reasons; one of the most important of these reasons is the difference in the metals solidification rate and melting temperature change, which leads to shrinkage and cracking. These cracks are in different directions, locations and lengths. However, the intical crack length was determined using fracture mechanic tools, see (A. M. Al-Mukhtar, Biermann, Hübner, & Henkel, 2010). If the cracks perpendicular to the direction of applied load in the tensile shear test, the maximum load will be low. In the austenitic stainless steel, the austenitic matrix shows relatively low alloying element diffusion rates, a high thermal expansion coefficient and high electric resistance (Hasanbaşoğlu & Kaçar, 2007).

The fracture stress of spot welding of 2 and 3-lap sheet was investigated by Sin et al. (2007). It was shown that the fracture stress was decreased by inserting a middle sheet due to the decrease in weld ability according to the increase in welding thickness. However, the effect of the multiple interfaced nugget, and the microstructural cracking effect play a significant role in the current work.

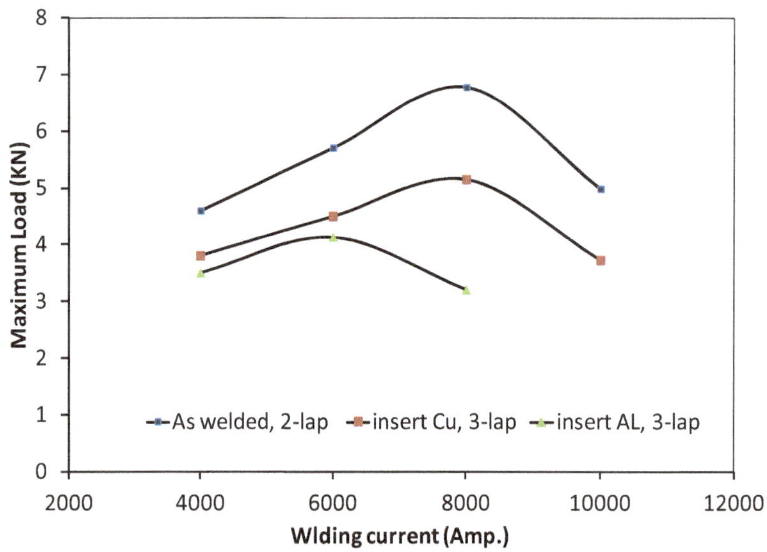

Figure 6. Weld joint strength of 2-lap and 3-lap specimens. Electrode pressure=0.483 MPa, Time=30 Cycle

High conductivity materials require more current than mild steels to give an equivalent heat input for two reasons, first, due to the high electrical conductivity and secondly, due to the high conduction of heat away from the weld area. Therefore, the investigation of intermediate metals such copper and aluminum has been introduced. In practice, high conductivity materials such as aluminum are welded for short times with exceptionally high currents (Brandes & Brook, 1992). The aim of adding the intermediate sheet or the presence of the intermediate layers of different alloys with different electrical and thermal conductivities, is to investigate the substantially heat inputs. Therefore, and however the using of lower current values, a plug fracture was obtained for the 3-lap, see Figure 7. The pressure has to be considered during the welding and precautions to prevent the cracking and cavitations have to be taken. Hence, a lower strength was obtained from the tensile test.

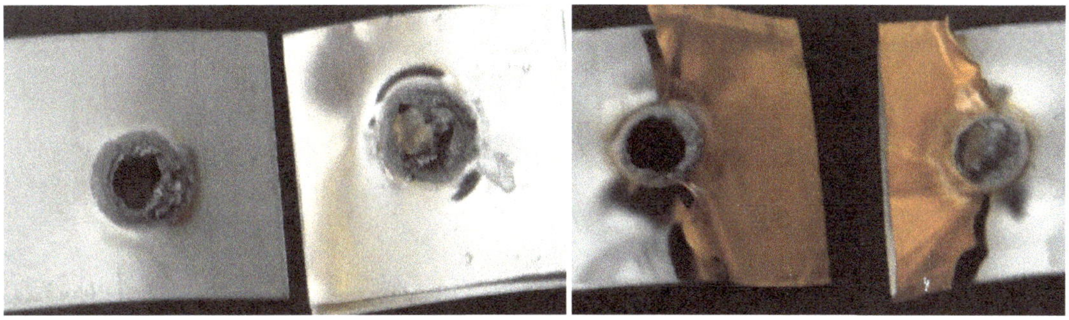

a) b)

Figure 7. Fracture plug of 3-lap sheets spot welded metal, a) almuimin- stainless steel fracture, b) cupper stainless steel fracture

4. Conclusions

In this study the microstructural and mechanical behavior of resistance spot welds done on stainless steel 2-lap and 3-lap sheets, welded at different welding parameters, is carried out. The properties are the most importance which influence weldability is examined. Therefore the suitable welding conditions have been investigated. In this study, two plates of stainless steel were placed as a lap joint (2-lap), and spot welded using varied welding current. In addition, 3-lap specimens were welded. The results show that the increasing welding current increased the nugget size for the 2-lap specimens up to expulsion limit. The nugget has a maximum area before the expulsion. Thus, the strength increases when the nugget diameter is increased. The electrode force decreasing the interface resistance and decreasing the heating and the nugget area. Hence, the maximum carrying load was decreased. In case of 3-lap, the microstructural investigation revealed that the excessive cracking and the intensive melting occurs before the expulsion limit in case of 2-lap. Hence a lower strength was observed. The high thermal conductivity of inserting metals increases the burning and the melting. Thus, the electrode indentation increases. Therefore, a plug fracture was obtained. Using transition pure copper and aluminum reduces the maximum load and microhardness values under the same conditions compare with as- welded specimen.

Microhardness of weld zones increase when insert thin pure aluminum sheet between two austenitic stainless steel sheets due to formation solid compounds such as oxide aluminium.

Acknowledgments

The authors would like to thankfully appreciate the support received from Dr.-Ing. Ahmed Al-Mukhtar, Guest Scientist at Technische Universität Bergakademie Freiberg, Faculty of Geosciences and Geoengineering and staff member in Al-Khawarizmi College of Engineering, Baghdad University for his fruitful discussions and guidance.

References

Agashe, S., & Zhang, H. (2003). Selection of schedules based on heat balance in resistance spot welding. *Welding Journal, 1*(2), 179-183. Retrieved from http://www.aws.org/wj/supplement/07-2003-AGASHE-s.pdf

Al-Mukhtar, A., & Doos, Q. (2013). Cracking Phenomenon in Spot Welded Joints of Austenitic Stainless Steel. *Materials Sciences and Applications, 2013*(October), 656-662. Retrieved from http://www.scirp.org/journal/PaperInformation.aspx?paperID=38635&

Al-Mukhtar, A. M. (2002). *Spot Welding Efficiency and its Effect on Structural Strength of Gas Generator and its Performance*. Baghdad University.

Al-Mukhtar, A. M. (2003). Investigation of Some Welding Parameters in Resistance Spot Welding of Austenitic Stainless. In *5th Sc. Conference of the College of Engineering, Baghdad University*.

Al-Mukhtar, A. M., Biermann, H., Hübner, P., & Henkel, S. (2010). Determination of Some Parameters for Fatigue Life in Welded Joints Using Fracture Mechanics Method. *Journal of Materials Engineering and Performance, 19*(9), 1225-1234. http://dx.doi.org/10.1007/s11665-010-9621-5

Al-Mukhtar, A. M., & Doos, Q. (2013). The Spot Weldability of Carbon Steel Sheet. *Advances in Materials Science and Engineering, 2013*, 1-6. http://dx.doi.org/10.1155/2013/146896

Ambroziak, A., & Korzeniowski, M. (2010). Using Resistance Spot Welding for Joining Aluminium Elements in Automotive Industry. *Archives of Civil and Mechanical Engineering, 10*(1), 5-13. http://dx.doi.org/10.1016/S1644-9665(12)60126-5

Aslanlar, S. (2006). The effect of nucleus size on mechanical properties in electrical resistance spot welding of sheets used in automotive industry. *Materials and Design, 27*(2), 125-131. http://dx.doi.org/10.1016/j.matdes.2004.09.025

Aslanlar, S., Ogur, A., Ozsarac, U., Ilhan, E., & Demir, Z. (2007). Effect of welding current on mechanical properties of galvanized chromided steel sheets in electrical resistance spot welding. *Materials & Design, 28*(1), 2-7. http://dx.doi.org/10.1016/j.matdes.2005.06.022

Brandes, E. A., & Brook, G. B. B. T.-S. M. R. B. (1992). 33 - Welding. Oxford: Butterworth-Heinemann. http://dx.doi.org/10.1016/B978-0-08-051730-8.50038-8

Connor, L. P. (1987). Welding handbook. *Society, 1.*

Cullen, J. D., Athi, N., Al-Jader, M. A., Shaw, A., & Al-Shamma'a, A. I. (2007). Energy reduction for the spot welding process in the automotive industry. *Journal of Physics: Conference Series.* http://dx.doi.org/10.1088/1742-6596/76/1/012022

Groover, M. P. (2007). *Fundamentals of modern manufacturing: materials processes, and systems.* John Wiley & Sons.

Han, L., Thornton, M., Boomer, D., & Shergold, M. (2010). Effect of aluminium sheet surface conditions on feasibility and quality of resistance spot welding. *Journal of Materials Processing Technology, 210*(8), 1076-1082. http://dx.doi.org/10.1016/j.jmatprotec.2010.02.019

Hasanbaşoğlu, A., & Kaçar, R. (2007). Resistance spot weldability of dissimilar materials (AISI 316L-DIN EN 10130-99 steels). *Materials & Design, 28*(6), 1794-1800. http://dx.doi.org/10.1016/j.matdes.2006.05.013

Kahraman, N. (2007). The influence of welding parameters on the joint strength of resistance spot-welded titanium sheets. *Materials & Design, 28*(2), 420-427. http://dx.doi.org/10.1016/j.matdes.2005.09.010

Kim, T., Lee, Y. S., Lee, J., & Rhee, S. H. (2004). A Study of Nondestructive Weld Quality Inspection and Estimation during Resistance Spot Welding. *Key Engineering Materials, 270-273*, 2338-2344. Retrieved from http://www.scientific.net/KEM.270-273.2338

M. Vural, A. A. (2004). the resistance spot weldability of galvanized interstitial free steel sheets with austenitic stainless steel sheets. *J Mater Process Technol, 1-6.*

Marashi, P., Pouranvari, M., Amirabdollahian, S., Abedi, A., & Goodarzi, M. (2008). Microstructure and failure behavior of dissimilar resistance spot welds between low carbon galvanized and austenitic stainless steels. *Materials Science and Engineering: A, 480*(1-2), 175-180. http://dx.doi.org/10.1016/j.msea.2007.07.007

MatWeb. (n.d.). *Your Source for Materials Information.* Retrieved from http://www.matweb.com/Page1/1.(n.d.),1

Mirsalehi, S. E., & Kokabi, a. H. (2010). Fatigue life estimation of spot welds using a crack propagation-based method with consideration of residual stresses effect. *Materials Science and Engineering: A, 527*(23), 6359-6363. http://dx.doi.org/10.1016/j.msea.2010.06.070

Moshayedi, H., & Sattari-Far, I. (2012). Numerical and experimental study of nugget size growth in resistance spot welding of austenitic stainless steels. *Journal of Materials Processing Technology, 212*(2), 347-354. http://dx.doi.org/10.1016/j.jmatprotec.2011.09.004

Mukhopadhyay, G., Bhattacharya, S., & Ray, K. K. (2009). Strength assessment of spot-welded sheets of interstitial free steels. *Journal of Materials Processing Technology, 209*(4), 1995-2007. http://dx.doi.org/10.1016/j.jmatprotec.2008.04.065

N R Anand1, V. M. C. * and N. K. S. (2013). The effect of shielding gases on mechanical properties and microstructure of austenitic stainless. *International Journal Of Mechanical Engineering And Robotics Research, 2.*

Nodeh, I. R., Serajzadeh, S., & Kokabi, a. H. (2008). Simulation of welding residual stresses in resistance spot welding, FE modeling and X-ray verification. *Journal of Materials Processing Technology, 205*(1-3), 60-69. http://dx.doi.org/10.1016/j.jmatprotec.2007.11.104

Park, S. H. C., Sato, Y. S., Kokawa, H., Okamoto, K., Hirano, S., & Inagaki, M. (2004). Corrosion resistance of friction stir welded 304 stainless steel. *Scripta Materialia, 51*(2), 101-105. http://dx.doi.org/10.1016/j.scriptamat.2004.04.001

Pereira, a. M., Ferreira, J. M., Loureiro, a., Costa, J. D. M., & Bártolo, P. J. (2010). Effect of process parameters on the strength of resistance spot welds in 6082-T6 aluminium alloy. *Materials & Design, 31*(5), 2454-2463. http://dx.doi.org/10.1016/j.matdes.2009.11.052

Pouranvari, M., Abedi, A., Marashi, P., & Goodarzi, M. (2008). Effect of expulsion on peak load and energy absorption of low carbon steel resistance spot welds. *Science and Technology of Welding and Joining*. http://dx.doi.org/10.1179/174329307X249342

Shamsul, J. B., Hisyam, M. M., & Muhibbah, T. (2007). Study Of Spot Welding Of Austenitic Stainless Steel Type 304. *Journal of Applied Sciences, 3*(11), 1494-1499.

Sin, S. R., Yang, S. M., Yu, H. S., Kim, C. W., & Kang, H. Y. (2007). Fatigue Analysis of Multi-Lap Spot Welding of High Strength Steel by Quasi Static Tensile- Shear Test. *Key Engineering Materials, 345-346*, 251-254. http://dx.doi.org/10.4028/www.scientific.net/KEM.345-346.251

Triyono, T., Purwaningrum, Y., & Chamid, I. (2013). Critical Nugget Diameter of Resistance Spot Welded Stiffened Thin Plate Structure. *Modern Applied Science, 7*(7), 17-22. http://dx.doi.org/10.5539/mas.v7n7p17

Tsai, C. L., Jammal, O. a, Papritan, J. C., & Dickinson, D. W. (1992). Modeling of Resistance Spot Weld Nugget Gro*wth, 71*(2), 47s-54s.

Vural, M., Akkuş, a., & Eryürek, B. (2006). Effect of welding nugget diameter on the fatigue strength of the resistance spot welded joints of different steel sheets. *Journal of Materials Processing Technology, 176*(1-3), 127-132. http://dx.doi.org/10.1016/j.jmatprotec.2006.02.026

Wei, P. S., & Wu, T. H. (2012). Electrical contact resistance effect on resistance spot welding. *International Journal of Heat and Mass Transfer, 55*(11-12), 3316-3324. http://dx.doi.org/10.1016/j.ijheatmasstransfer.2012.01.040

Welding, R. S., Industry, A., & Practice, G. (2002). Joining - Resistance welding 2 Resistance welding, 1-19.

Zhang, H., & Senkara, J. (2006). *Resistance Welding Fundamentals and Applications*.

Zhou, M., Hu, S., & Zhang, H. (1999). Critical specimen sizes for tensile-shear testing of steel sheets. *Welding Journal-New York-*. Retrieved from http://aws.org/wj/supplement/sept99/ZHOU.pdf

25

Lignin-Laccase-Mediator-Systems (LLMS) for the Production of Binderless Medium Density Fiberboards (MDF)

Markus Euring[1], Alexander Kirsch[1], Patricia Schneider[1] & Alireza Kharazipour[1]

[1] Faculty of Forest Science and Forest Ecology, Büsgen-Institute, Department of Molecular Wood Biotechnology and Technical Mycology, Georg-August-University of Goettingen, Germany

Correspondence: Markus Euring, Faculty of Forest Science and Forest Ecology, Büsgen-Institute, Department of Molecular Wood Biotechnology and Technical Mycology, Georg-August-University of Goettingen, Germany. E-mail: meuring@gwdg.de

Abstract

Medium density fiberboards (MDF) are produced mainly by urea-formaldehyde resins (UF) as binding agent, which are synthesized from finite fossil resources. Those boards may emit critical amounts of formaldehyde, which can influence the health of humans and animals. In recent times the wood panel board industry is looking for alternative glues which contain less or no formaldehyde. In order to avoid potential formaldehyde emissions altogether it would be preferable not to use binders which are formaldehyde based at all. One possibility is to use natural binders or to activate the wood fibers' own binding forces by applying Laccase-Mediator-Systems (LMS). As a support of the LMS interactions with wood fibers technical lignin can be added. In this study it was found out that the addition of technical lignin intensified the fiber to fiber bindings. Two Lignin-Laccase-Mediator-Systems (LLMS) were analysed by Gel-Permeation-Chromatography and Cyclic voltammetry. Later the LLMSs were tested in the pilot scale production of MDF. The determination of the physical technological properties revealed that the LLMS treated MDF have higher dimension stabilities than only LMS treated MDF and approximately the same thickness swelling after 24 h. The results indicate that the application of LLMSs have a high potential for natural bonded MDF.

Keywords: technical lignin, laccase, mediator, Laccase-Mediator-System, Lignin-Laccase-Mediator-System, Medium density fiberboard, Gel-Permeation-Chromatography, Cyclic voltammetry

1. Introduction

Approximately ten million square meters of medium density fiberboards (MDF) are produced yearly in Europe (EPF, 2013). About 20% of the production costs can be attributed to the use of mainly petrochemical binders (Kharazipour, 2004). Here we refer to mainly the use of urea-formaldehyde resins (UF) and melamine-urea-formaldehyde resins (MUF) which are synthesized from finite fossil resources. Due to fluctuating raw oil prices, as well as the dependence of the wood-based materials industry on these petrochemical binding agents, there has been a noticeable price increase in wood-based products in recent years (Kharazipour, 2004). In addition the formaldehyde emissions from wood-based building materials are regarded critically and can be attributed to the use of formaldehyde based binders. At certain concentrations formaldehyde can induce irritation of the eyes, mucus membranes and skin (Yang et. al., 2001; IARC, 2004). In 2004 formaldehyde was classified as a carcinogen by the International Agency for Research on Cancer (IARC, 2004). Accordingly, increased research attention has focused on appropriate measures of reducing formaldehyde emissions in products in which it is contained (Oberdorfer, 2008).

At the beginning of 2009 in California, the first stage in reducing the formaldehyde emissions in the field of wood based products came into force. The "CARB" standards (CARB = California Air Resources Board) as an example, required for medium density fiberboards (MDF) a reduction of the formaldehyde emissions from the local norm of 0.3 ppm to 0.21 ppm (CARB, 2008). In 2011 the second stage in the reduction of emissions to an upper limit of 0.11 ppm came into force. For economic reasons, the European manufacturers of wood-based products oriented themselves on CARB and EN13986 with the emissions class E1 and E2 and also followed the recommendations of IKEA IOS MAT 003 as well as EPF-S (European Panel Federation-standard) in which the actual permitted formaldehyde emissions are lower than the top limit recommended by the CARB standards (SUBSPORT, 2013). It remains to be seen what the future will bring in terms of regulations and measures to be taken with regard to formaldehyde emissions, considering that in 2014 formaldehyde was upgraded in terms of carcinogenicity from category 2 (suspected of carcinogenic properties) to category 1B, and in terms of mutagenicity from category 3 (possibly causing genetic mutations) to category 2 according to EU regulation 605/2014 (2014). Category 1B means that formaldehyde now is considered to be a probable cause of cancer in humans and that the cancer causing effect has been proven in animals (EG 1272/2008, 2008). Category 2 in terms of mutagenicity, classifies formaldehyde as a material potentially harmful for humans in that it could possibly cause genetic mutations in germ cells (EG 1272/2008, 2008). Many manufacturers of petrochemical products containing formaldehyde are exploring new ways for handling formaldehyde, while taking into consideration all current regulations (Formacare, 2015). On the one hand discussions are taking place regarding a reduction of formaldehyde in these products and on the other a reduction in the use of formaldehyde in the actual production process (Wiechmann & Kohla, 2014). In local industry, no harmful emission levels for employees in the manufacturing process of wood-based materials with binders containing formaldehyde have been measured. However various possibilities to reduce emissions are being evaluated (Wiechmann & Kohla, 2014). In order to avoid potential formaldehyde emissions altogether it would be preferable not to use binders which are formaldehyde based at all.

In this respect several wood processing factories, in particular those for wood-based products, are currently occupied with testing low-emission or emission free binders. A possible solution can be found in the development of natural binding agents (Müller et al., 2007, Euring et al., 2011, 2013, and 2015). The Lignin-Laccase-Mediator-System is a naturally based binding agent and can offer the wood-based product industry a chance to establish environmentally acceptable production methods.

Lignin occurs in quantities of approximately 50 million tons worldwide as a byproduct of cellulose manufacture (15 million tons of lignin-sulfonates and 35 million tons of kraft-lignin) (Puls, 2009). In fact lignin is a mass product for which there are hardly any uses. According to Glasser (2009) only 2% of the technical lignins worldwide are used in industrial processes, although they belong to the most inexpensive of raw materials (approx. 0.15 €/kg). The remaining lignins are burnt for energy generation, or not used at all. For this and other reasons, technical lignins which are available widely and in large quantities, are the most promising polyphenol for the development of binding agents.

It makes sense to use these technical lignins where the natural lignins are already present on the wood fibers due to the high content of phenol groups. This is amongst others, the case with fibers which have been through a thermo-mechanical pulp process (TMP). The middle-lamella lignins in the wood fibers are incrusted on the fiber's surface after the thermo-mechanical process, an effect which is recognized as a significant chemical transformation (Sixta, 2006). The laccase can activate these surface lignins through oxidation resulting in free lignin fractions, which when hot pressed to fiber board cause the fibers to adhere to one another creating a fiber to fiber bond (Hüttermann et al., 2001, Kharazipour et al., 1997, 1998). The utilization of technical lignins and mediators as catalyzers should accelerate and intensify this process, which should lead to a more potent adhesion of the wood fibers (Euring, 2008).

This study will investigate formaldehyde free wood fiber products, specifically medium density fiber-boards (MDF). These wood fiber products should be produced without the addition of conventional, petrochemical binding agents, using a Lignin-Laccase-Mediator-System (LLMS). The use of Laccase-Mediator-Systems has already been comprehensively tested. The results showed that when using an LMS of laccase and 4-hydroxicbenzoic acid, not all mechanical and physical properties were satisfactory (Euring et al., 2011, 2013). The use of technical lignins in LLMS activates additional bonding strength. The wood fiber surface lignins which have been activated with the laccase and mediator should be supported by the use of technical lignin components. For the first time this study will test the applicability of the mediators caffeic acid (CA) and vanillyl alcohol (VA) in the production of MDF.

LLMS should make a stronger bond between the individual wood fibers possible. The implementation of LLMS should be integrated seamlessly into the conventional fiberboard production process. In preliminary stage to the production and testing the physical-technological properties of the MDF, analyses of lignin molecular weight through gel permeation chromatography (GPC) and also of the redox potential of Laccase-Mediator-Systems using a potentiostat, will take place. Those analyses are helpful to understand LMS or LLMS actions. The results and experiences e. g. the optimal LMS or LLMS concentrations will directly be transferred to the MDF production

2. Materials and Methods

2.1 Materials

2.1.1 Wood fibers

The TMP (thermo-mechanical pulping) -fibers originate from the company Gutex (Waldshut-Tiengen) and consist of 80% spruce and 20% fir fibers.

2.1.2 Lignin

Borrement Ca 120 (Calcium-Ligninsulfonate, Borregaard LignoTech, Germany)

2.1.3 Laccase

Novozym 51003 (from the company Novozymes, Bagsvaerd, Denmark) fermented from *Trametes vilosa* in recombination with *Aspergillus oryzae,* at hand in liquid form and brown in color due to stabilizer. Its specific activity was measured by determining the oxidation of ABTS (Matsumura et al. 1986). The activity of the laccase stock was about 1,000 U/ml.

2.1.3 Mediators

Caffeic acid (CA): from the company Sigma-Aldrich (Seelze, Germany) with a chemical purity of 95%.

Vanillyl alcohol (VA): 4-Hydroxy-3-methoxybenzyl alcohol, from the company Sigma-Aldrich (Seelze, Germany) with a chemical purity of 98%.

2.1.4 McIlvain Buffer

Composed of 0.2 M di-potassium hydrogen phosphate (K_2HPO_4) and 0.1 M citric acid ($C_6H_8O_7$), buffered to pH 6.0. Both components were from AppliChem, Darmstadt, Germany.

2.1.5 Sodium Acetate Buffer

To carry out the potentiostatic trial, a sodium acetate buffer was used in order hold a pH level of 6 constant throughout the tests. The buffer was made by producing 0.1 M sodium acetate (NaAc) solution. Subsequently the pH value was brought to 5 using acetic acid.

2.2 Methods

2.2.1 Determination of the molecular weight of the lignin using Gel-Permeation-Chromatography (GPC).

Gel-permeation chromatography was used to determine the molecular weight distribution of the lignin prior to, and following, incubation with laccase or respectively laccase and mediator. The measurements were made using a high pressure liquid chromatography (HPLC) unit from *Waters.* The HPLC method used here is referred to as Gel-permeations chromatography as reference to the use of gel columns in the HPLC instrumentation. In this case it is a special method of column chromatography which is suitable in particular for separating molecules and therefore for establishing molecular weights.

The HPLC instrumentation is comprised of the following:

Waters e2695 Separation Module, *Waters* 2489 UV-Detector,

GPC- columns: SHODEX KF-803, SHODEX KF-804, SHODEX KF-G

Computer for recording and analyzing the results, Software: *Waters* Empower 3 Personal

Eluent (mobile phase): Wasser-Acetonitrile 80 : 20 (v : v), pH 7.8

Standards: Polystyrene Standards with molecular weights von 2000–150000 Daltons

Sample preparation was as follows:

0.5 g Ca-Ligninsulfonate were added to 27 ml McIlvain buffer (pH 6.0). Depending on the type of sample either 10 U/ml laccase or 10mM mediator were added to the buffer solution. The samples were incubated for 15, 30 and 60 minutes in order to compare the molecular weights. Following the incubation time the samples containing laccase were stopped through being briefly brought to the boil.

Subsequently 20 µl from each sample was taken, put into vials, and placed into the GPC, where measurement of the samples took place. The previously recorded values for the polystyrene standards were used as reference values for the evaluation software in calculation of the lignin molecular weights.

Measurements of both controls and samples took place in a UV detector at a wavelength of 254 nm. The eluent used was a water/acetonitrile mixture (80 : 20, v : v, pH 7.8). All results were recorded and calculated using software from Waters, and can be portrayed as spectra using the weight average molecular weight (Mw), the number average Molecular weight (Mn) and the poly-dispersity (Mw/Mn).

2.2.2 Electrochemical measurements of LMS by means of Potentiostat

Cyclic voltammetry is an electrochemical method of investigation used to analyze redox reactions. Not only simple parameters, such as redox potential, but also reaction kinetic parameters such as electron transfer stages and chemical reactions which are potentially taking place, can be recorded and analyzed by means of cyclic voltammetry.

For the purpose of this study the electrochemical measurements were used solely to analyze the redox potential of the mediators caffeic acid and vanillyl alcohol with laccase, in order to assess their reaction-kinetic suitability.

The potentiostat used was a PCSTAT101 from Metrohm Autolab B.V. from Utrecht in the Netherlands. It exhibits a three-electrode system with a carbon working electrode and a reference electrode which is immersed in a 3-molaric potassium chloride solution. The software used was also from Metrohm Autolab B.V. in the version 1.10.4.

For the analysis of the mediators caffeic acid and vanillyl alcohol with a potentiostat, the mediators were first dissolved in a sodium acetate buffer and diluted to a concentration of 0.2 mM. The laccase used in the trials diluted 100 times. For the samples with lignin 2 mM Ca-Ligninsulfonate were added to evaluate the effect of the lignin within the LLMS expecting changing redox-potentials. The measurements were made in a range from 100 mV to 1000 mV with a feed velocity of 5 mV/s. The output potential was held for 10s, before the voltage change occurred. 5 mL of mediator solution was used in each measurement. At first three measurements as a control were taken using pure mediator solution. In order to record the interaction between mediator and laccase, the laccase solution was dripped onto the working electrode. After the solution had completely dried out, a measurement was taken using a 5 ml mediator solution and the same parameters as with the pure mediator solution. Here it is important to note that after each trial the electrode was cleaned completely. Each measurement was repeated three times. The results were reported in a voltage-current strength diagram.

2.2.3 Production of LLMS bonded MDF

LLMS bonded MDF were produced under the following procedural conditions:

- Wood fiber amount per treatment: 20 kg
- Laccase concentration: 100 U/ml per 0.001 kg absolutely dry fibers
- Mediator concentration: 10 mM per 0.001 kg absolutely dry fibers
- Ca-Ligninsulfonate: 6% referred to 1 kg absolutely dry fibers
- Raw density: 750 kg/m³
- Board dimensions: 600 mm x 800 mm
- Board thickness: 10 mm
- Produced at an outset moisture content of the incubated fibers of 10-15%
- Comparison of the resulting physical/mechanical properties with those of the untreated fibers and those of MDF produced with UF resin.
- Press temperature: 190 °C
- Press Time: 12 s/mm

The LLMS treated wood fibers were directly transferred from the Blender to the tube dryer to guarantee an uninterrupted production of the fibers to a fiber mat formation and then pressing to MDF (equivalent to the industrial process of MDF).

The MDF were produced using the following steps.

1) Blending of fibers with LLMS

The combination of fibers and each LLMS in buffer solution (McIlvain) was achieved using the Blender process in the MDF pilot plant of the Biotechnikum (Büsgen-Institute, Goettingen).

2) Drying of the treated fibers

As known from previous studies a fiber moisture content of 10-15% was aimed for in the drying process, in order to support the natural binding forces, but also to avoid a loss of MDF quality associated with higher moisture levels (>15%) in the following hot-pressing process (Euring, 2008, Felby et al., 2011, Euring et al. 2011). The drying temperature was adjusted so that the incubation process with LLMS would not be stopped or interrupted.

3) Fiber Mat formation

Monolayer fiber mats were generated.

4) Pre-pressing and hot-pressing

Hot-pressing of the fiber mat followed in the hot press at 190 °C for 12 s/mm board thickness after pre-pressing. Both temperature and time were determined as optimal parameters for pressing of LLMS bonded MDF in pilot-scale. Each type of MDF was replicated six times.

5) Determining the physical-technological properties of the pilot scale MDF

Determination of the following properties was carried out after a conditioning phase of each MDF in a standard climate:

- Determination of bending strength (EN 310, 1993), sample size: 5 x 20 x 0,1 cm^3, 6 samples/MDF
- Determination of internal bond strength (EN 319, 1993), sample size: 5 x 5 x 0,1 cm^3, 8 samples/MDF
- Determination of the hygroscopic behavior of the material (thickness swelling and absorption) (EN 317, 1993), sample size: 5 x 5 x 0,1 cm^3, 8 samples/MDF

Significant difference was calculated by using Tukey's HSD (Honestly Significant Difference) test with a significance level of $P < 0.05$.

6) Production of the reference MDF boards.

The boards used as references were produced in dry-process using the same process parameters and target sizes, as used in the production with LLMS treated wood fibers. The reference boards were produced with inactivated LLMS (cooking for 10 minutes), only with laccase, with technical Lignin and laccase, with simple LMS and with 10% UF resin like in a typical MDF in Europe. Lastly the physical-technological properties were analyzed and compared.

3. Results and Discussion

3.1 GPC Measurements

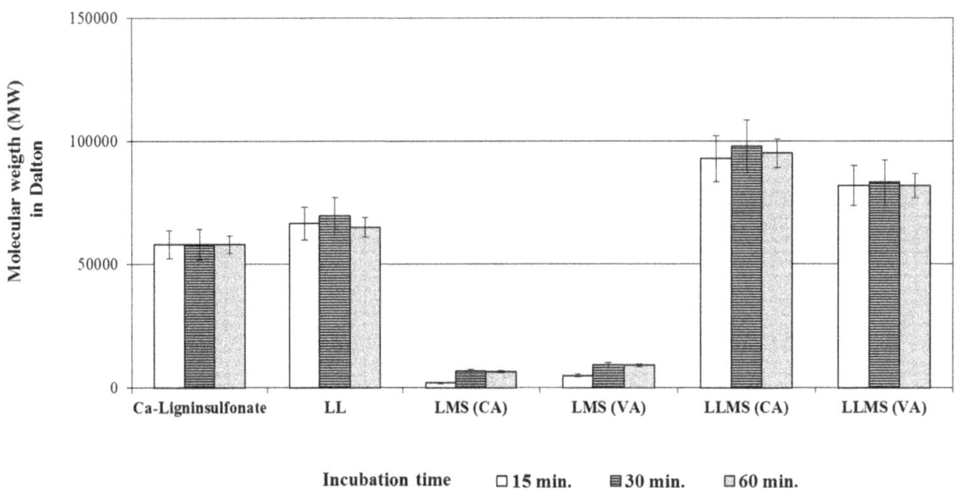

Figure 3. Average molecular weight (Mw) of Ca-Ligninsulfonate with the mediators caffeic acid and vanillyl alcohol (Laccase concentration 10 U/ml, mediator concentration 10 mM in McIlvain buffer, pH 6,0) after incubations of 15, 30 and 60 minutes. Average values from 6 trial series. Sample identifiers: Ca-Ligninsulfonate = Ca-Ligninsulfonate in buffer solution, LL = Ca-Lignosulfonate + Novozym 51003 in buffer solution; LMS (CA): Novozym 51003 + caffeic acid in buffer solution; LMS (VA): Novozym 51003 + vanillyl alcohol in buffer solution; LLMS (CA): Ca-Ligninsulfonate + Novozym 51003 + caffeic acid in buffer solution; LLMS (VA): Ca-Ligninsulfonate + Novozym 51003 + vanillyl alcohol in buffer solution

For the first time in the course of this study lignin molecular weights were determined using a water-based eluent in a GPC instrumentation unit. Up till now GPC runs of lignin were only carried out using dangerous solvents such as tetrahydrofuran (THF) (Euring, 2008, Majcherzyk & Hüttermann, 1996). During the trials of Ca-Ligninsulfonate treated with laccase and laccase mediator with incubation times of 15, 30 and 60 minutes, which were successful right from the start, it was observed that both treatment methods resulted in an increase in lignin molecular weight (see Figure 3). Untreated Ca-Ligninsulfonate kept a molecular weight of 55,000 Dalton throughout the entire analysis time, while in the case of incubation with laccase the weights increased to over 70,000 Dalton with the highest value being reached after 30 minutes. It can be assumed that the increase in molecular weight is due to oxidation of the phenol group of the lignin to phenoxy radicals (Ortner et al., 2015), which would then be available to facilitate the subsequent bonding of the fibers to MDF. After treatment with LLMS (CA) and LLMS (VA) even higher lignin molecular weights were achieved, which can be associated to a higher effectiveness of LLMS than laccase alone. With results almost as high as 100,000 Daltons after 30 minutes incubation with LLMS containing caffeic acid, this is the most effective LLMS. The GPC tests with LMS (CA) and LMS (VA) resulted in only low molecular weights because here there is no lignin present. However, the detected molecular weights are a result of the partly phenolic structures of the mediator. The fact that all laccase, LMS or LLMS samples showed molecular weights are highest at 30 minutes incubation and sink after 60 minutes incubation can be explained by the fact that phenoxy radicals possibly react with each other and are deactivated. Thus it can be seen that the optimal incubation time is that of 30 minutes, especially when considering the process of MDF board production which usually requires 30-40 minutes from the gluing to fiber mat formation and pressing (Euring et al., 2011).

3.2 Cyclic Voltammograms of LLMS

The tests with caffeic acid show a significant change in the curve after addition of laccase (Figure 4). The electric current is elevated during oxidation and a shift into the negative range occurs during reduction. The cyclic voltammogram of the tests with caffeic acid show a significant interaction between laccase and mediator (CA and LMS (CA). It can be seen that there is a reduction of current during the oxidation process and on the other hand a shift of the reduction peak into the negative. Together with Ca-Ligninsulfonate the currents become more intensive caused by the presence of lignin (Lignin + CA and LLMS (CA)). The oxidative effect of laccase is indicated by these results. Furthermore, the oxidation and therefore the radical formation are partly reversible. There is virtually a reversible electron transfer present, and the so formed radicals are partially stable.

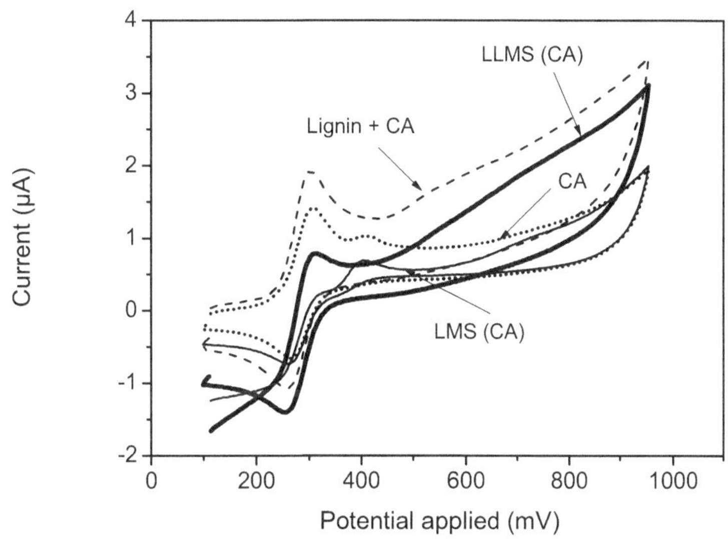

Figure 4. Cyclic voltammograms with CA = caffeic acid in buffer solution, LMS (CA) = Novozym 51003 + caffeic acid in buffer solution, Lignin +CA = Ca-Ligninsulfonate + caffeic acid in buffer solution, LLMS (CA) = Ca-Ligninsulfonate + Novozym 51003 + caffeic acid in buffer solution

The potentiostatic test of vanillyl alcohol as mediator shows a decrease of measured current during oxidation and stronger negative current during the reduction process independently if in presence of lignin (Figure 5). The cyclic voltammographic results of the mediator vanillic acid, display similar results to those with caffeic acid. The oxidative effect of the laccase on the mediator can be recognized by the decreased current during the oxidation process. The change in current during the reduction is however slightly weaker than with caffeic acid, which indicates less stable radicals (Diaz Gonzalés et al., 2009).

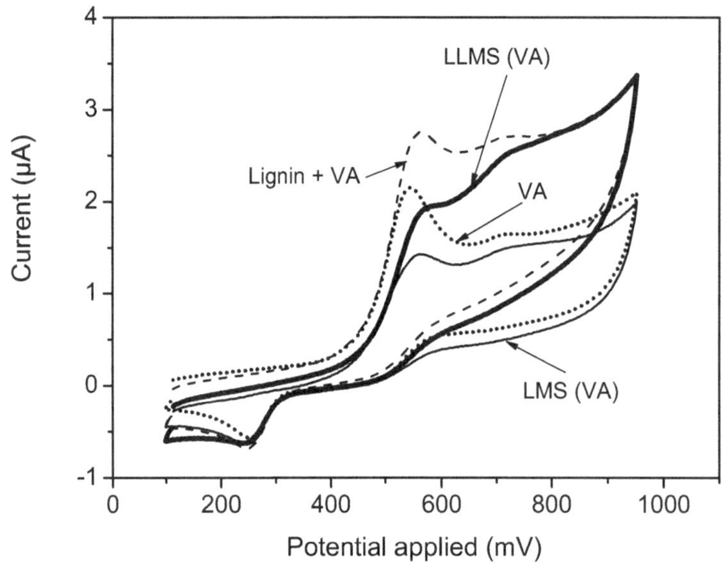

Figure 5. Cyclic voltammograms with VA = vanillyl alcohol in buffer solution, LMS (VA) = Novozym 51003 + vanillyl alcohol in buffer solution, Lignin +VA = Ca-Ligninsulfonate + vanillyl alcohol in buffer solution, LLMS (VA) = Ca-Ligninsulfonate + Novozym 51003 vanillyl alcohol in buffer solution

Based on the higher stability of the radicals, caffeic acid would appear to possess a higher potential to function in LLMS (Siebert, 2015).

3.3 Physical-Technological Properties of MDF Made with LLMS in Pilot Scale

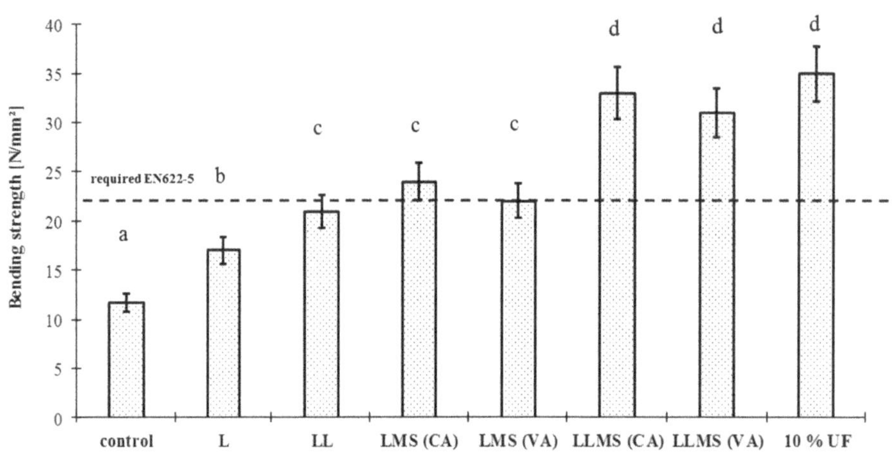

Figure 6: Bending strengths of MDF (raw density 750/kg/m³, board thickness 10mm), control (not activated LLMS), L = Novozym 51003, LL = Ca-Ligninsulfonate + Novozym 51003, LMS (CA) = Novozym 51003-caffeic-acid-system, LMS (VA) = Novozym 51003-vanillyl alcohol-system, LLMS (CA) = Ca-Ligninsulfonate-Novozym 51003-caffeic acid-system, LLMS (VA) = Ca-Ligninsulfonate-Novozym 51003-vanillyl alcohol-system, 10 % UF = reference with 10 % (in absolutely dry fibers) urea formaldehyde (UF)-resin. Means marked with different letters (a, b, c, d) are significantly different at $P < 0.05$

The resulting bending strengths of the MDF differed depending on the treatment of the fibers (Figure 6). In both LLMS, (CA) and (VA), the required bending strength norm of at least 23 N/mm² was exceeded by far, and almost equaled that of the control MDF with 10% UF. When applying only LMS, the bending strengths fluctuate around the required norm. The control boards, the laccase and also the laccase-Ligninsulfonate bonded MDF did not reach the required norm value.

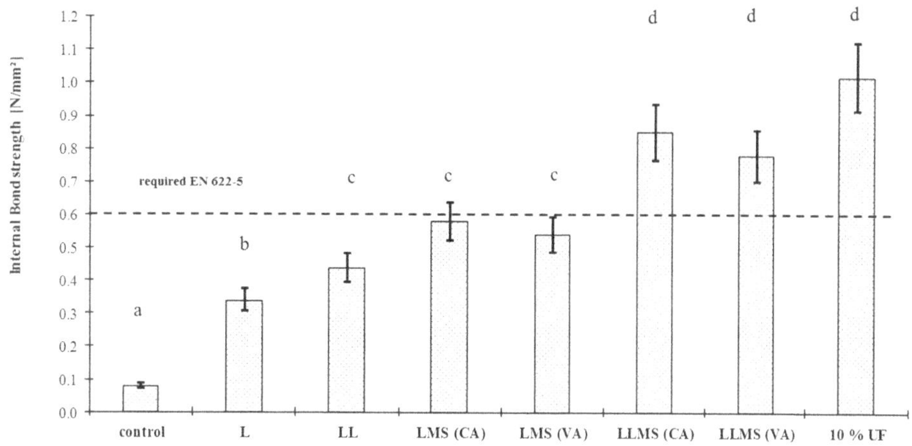

Figure 7. Internal bond strengths of MDF (raw density 750/kg/m³, board thickness 10mm), control (not activated LLMS), L = Novozym 51003, LL = Ca-Ligninsulfonate + Novozym 51003, LMS (CA) = Novozym 51003-caffeic-acid-system, LMS (VA) = Novozym 51003-vanillyl alcohol-system, LLMS (CA) = Ca-Ligninsulfonate-Novozym 51003-caffeic acid-system, LLMS (VA) = Ca-Ligninsulfonate-Novozym 51003-vanillyl alcohol-system, 10 % UF = reference with 10 % (in absolutely dry fibers) urea formaldehyde (UF)-resin. Means marked with different letters (a, b, c, d) are significantly different at $P < 0.05$

The values for internal bond strength of MDF bonded with 10% UF exceed those of the required norm of 0.6 N/mm² (Figure 7). The required norm is also reached distinctly with LLMS with the mediator caffeic acid or vanillyl alcohol. In the case of the MDF with the two different LMS, it didn't quite reach the norm values for IB and therefore do not comply with the required norm. The control samples and the laccase bonded MDF fall short of the norm.

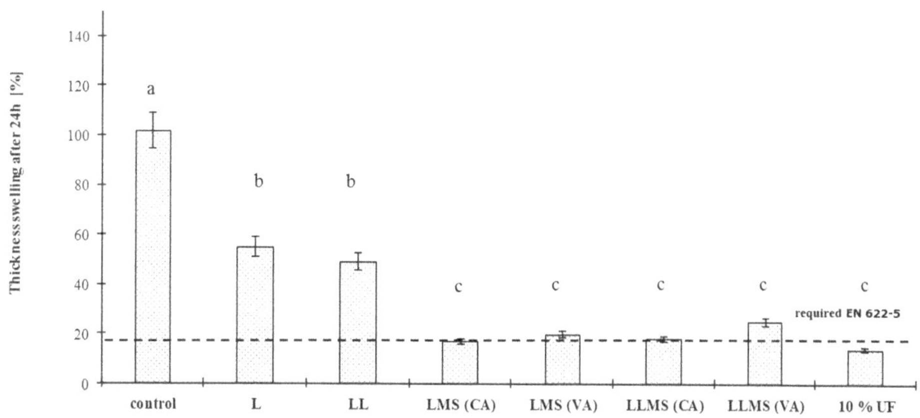

Figure 8. Thickness swellings of MDF after 24 hours in water. (raw density 750/kg/m³, board thickness 10mm), control (not activated LLMS), L = Novozym 51003, LL = Ca-Ligninsulfonate + Novozym 51003, LMS (CA) = Novozym 51003-caffeic-acid-system, LMS (VA) = Novozym 51003-vanillyl alcohol-system, LLMS (CA) = Ca-Ligninsulfonate-Novozym 51003-caffeic acid-system, LLMS (VA) = Ca-Ligninsulfonate-Novozym 51003-vanillyl alcohol-system, 10 % UF = reference with 10 % (in absolutely dry fibers) urea formaldehyde (UF)-resin. Means marked with different letters (a, b, c) are significantly different at $P < 0.05$

Thickness swelling values after 24 h water immersion are still too high in the LMS or LLMS bonded MDF (Figure 8). However the values are almost as good as those with 10% UF bonded MDF. All other MDF in the study showed thickness swelling values which are too high. As has been discovered in earlier studies, the hydrophilic properties of technical lignins have a negative effect on water absorption (Müller et al. 2007). The use of hydrophobic agents would be of assistance in this case. Another possibility could be to choose higher pressing temperatures than 190 °C to improve the physical-technological properties of the boards, but in industrial process temperatures of 200 °C and above are considered uneconomic, as well as too long pressing times (Dunky & Niemz, 2002).

4. Conclusion

The study demonstrates that the Lignin-Laccase-Mediator-System as a natural bonding agent is very suitable for the production of MDF. In both the analytical tests and in the production of LLMS bonded MDF, significant advantages over the previous use of LMS can be seen (Euring et al., 2011, 2013).

The GPC analysis results of LLMS in comparison with the previously investigated LMS already show a significant increase in the lignin molecular weight after 15 minutes incubation, with the highest results after 30 minutes. 30 minutes incubation coincides with the duration of MDF production from blending to the pressing process of the completed product. In production terms therefore, the use of LLMS in the production of MDF is appropriate.

As shown in the positive technological mechanical properties of the so produced MDF, laccase and mediator activated wood fiber surface lignin are supported by the additional phenoxy radicals contributed by the technical lignins, and therefore facilitate stronger bonding of the individual wood fibers. Caffeic acid proved to be more effective than vanillyl alcohol in LLMS. Further tests need to follow in order to optimize the LLMS particularly with regard to thickness swelling properties.

Acknowledgements

This work was supported by the Deutsche Forschungsgemeinschaft (DFG), grant EU 124/2-2 to Markus Euring.

References

CARB. (2008). California Air resources Board. Airborne Toxic Control Measure to Reduce Formaldehyde Emissions from Composite Wood Products, April 2008, Final Regulation Order. California Environmental Protection Agency. Retrieved from http://www.arb.ca.gov/regact/2007/compwood07/fro-final.pdf

Díaz González, M., Vidal, T., & Tzanov, T. (2009). Electrochemical Study of Phenolic Compounds as Enhancers in Laccase-Catalyzed Oxidative Reactions. *Electroanalysis, 21*, 2249–2257. http://dx.doi.org/10.1002/elan.200904678

Dunky, M., & Niemz, P. (2002). Holzwerkstoffe und Leime - Technologie und Einflussfaktoren. Springer-Verlag Berlin Heidelberg. http://dx.doi.org/10.1007/978-3-642-55938-9

EG 1272/2008. (2008). Verordnung (EG) Nr. 1272/2008 über die Einstufung, Kennzeichnung und Verpackung von Stoffen und Gemischen. Retrieved from http://www.gaa.baden-wuerttemberg.de/servlet/is/16495/1_1_08.pdf. Accessed 19 Sep. 2015

EN 310. (1993). Holzwerkstoffe; Bestimmung des Biege-Elastizitätsmoduls und der Biegefestigkeit. Deutsche Fassung, Beuth-Verlag.

EN 317. (1993). Spanplatten und Faserplatten; Bestimmung der Dickenquellung nach Wasserlagerung. Deutsche Fassung Beuth-Verlag.

EN 319. (1993). Spanplatten und Faserplatten; Bestimmung der Zugfestigkeit senkrecht zur Plattenebene. Deutsche Fassung, Beuth-Verlag.

EU 605/2014. (2014). Verordnung (EU) Nr. 605/2014 der Kommission vom 5. Juni 2014 zur Änderung der Verordnung (EG) Nr. 1272/2008/EG des Europäischen Parlaments und des Rates über die Einstufung, Kennzeichnung und Verpackung von Stoffen und Gemischen zwecks Einfügung von Gefahren- und Sicherheitshinweisen in kroatischer Sprache und zwecks Anpassung an den technischen und wissenschaftlichen Fortschritt. Retrieved October 15, 2015 from http://www.umwelt-online.de/regelwerk/ eu/14a/14a_0605.htm

Euring, M. (2008). *Einsatz von Mediatoren bei der enzymatischen Aktivierung der fasereigenen Bindekräfte zur Herstellung von enzymgebundenen, bindemittelfreien Holzwerkstoffe*. Dissertation, Faculty of Forest Sciences and Forest Ecology, Georg-August University Goettingen, 211 pages.

Euring, M., Rühl, M., Ritter, N., Kües, U., & Kharazipour, A. (2011). Laccase-mediator-systems for eco-friendly production of medium-density fiberboard (MDF) in pilot scale: Physico-chemical analysis of the reaction mechanism. *Biotechnol. J., 6*(10), 1253-1261. http://dx.doi.org/10.1002/biot.201100119

Euring, M., Trojanowski, J., & Kharazipour, A. (2013). Laccase-Mediator catalyzed modification of wood fibers - Studies on the mechanism and medium-density fiberboard (MDF) making. *For. Prod. J., 63*(1/2), 54-60. http://dx.doi.org/10.13073/FPJ-D-12-00075

Euring, M., Kirsch, A., & Kharazipour, A. (2015). Hot-Air/Hot-Steam Process for the Production of Laccase-Mediator-System Bound Wood Fiber Insulation Boards. *BioResources, 10*(2), 3541-3552. http://dx.doi.org/10.15376/biores.10.2.3541-3552

Felby, C., Hassigboe, J., & Lund, M. (2002). Pilot-scale production of fiberboard's made by laccase oxidized wood fibres: board properties and evidence for cross-linking of lignin. *Enz. Microbial Technol., 31*, 736-741, http://dx.doi.org/10.1016/S0141-0229(02)00111-4

Formacare. (2015). REACH Regulation and Formaldehyde. Retrieved October 16, 2015, from http://www.formacare.org/regulatory-information/reach

Glasser, W. G. (2009). Lignin-Retrospect and prospect. In: Stoffliche Nutzung von Lignin. Gülzower Fachgespräche, Nand 31, Fachagentur für Nachwachsende Rohstoffe e.V. (FNR), 42-62

Hüttermann, A., Mai, C., & Kharazipour, A. (2001). Modification of lignin for the production of new compounded materials. *Appl. Microbiol. Biotechnol., 55*, 387-394. http://dx.doi.org/10.1007/s002530000590

IARC (2004): International Agency for Research of Cancer): *IARC classifies formaldehyde as carcinogenic to humans.* Press release nr. 153, 15. April 2004.

Kharazipour, A, Hüttermann, A., &Lüdemann, H. D. (1997). Enzymatic activation of wood fibres as a mean for the production of wood composites, *J. Adhes, Sci. Technol., 11*, 419–427. http://dx.doi.org/10.1163/156856197X00796

Kharazipour, A. (2004). Holz als Werkstoff. Im Jahrbuch 2004/2005 Nachwachsende Rohstoffe, Wirtschaftsfaktor Biomasse, C.A.R.M.E.N, 335-337

Kharazipour, A., Schindel, K., & Hüttermann, A. (1998). Enzymatic activation of wood fibers for wood composite production. *ACS Symposium Series, 687*, 99-115. http://dx.doi.org/10.1021/bk-1998-0687.ch009

Majcherzyk, A., & Hüttermann, A. (1996). Size-exclusion chromatography of lignin as ion-pair complex. *J. Chromatogr, A 764*, 183-191. http://dx.doi.org/10.1016/S0021-9673(96)00909-0

Matsumura, E., Yamamoto, E., Numata, A., Kawano, T., Shin, T., & Murao, S. (1986). Structures of the laccase-catalysed oxidation products of hydroxybenzoic acids in the presence of ABTS. *Agric. Biol. Chem., 50*(5), 1355–1357.

Müller, C., Kües, U., Schöpper, C., & Kharazipour, A. (2007). Natural Binders. In U. Kües (Ed.), *Wood Production, Wood Technology, and Biotechnological Impacts* (Chapter 16, pp. 347-382). Universitätsverlag Göttingen.

Oberdorfer, G. (2008). Formaldehydemissionen aus Holzwerkstoffen. *Holzforschung, 2008, 4*, 8-9.

Ortner, A., Huber, D., Haske-Cornelius, O., Weber, H. K., Hofer, K., Bauer, W., ... Guebitz, G. M. (2015). Laccase mediated oxidation of industrial lignins: Is oxygen limiting? *Process Biochem, 50*(8), 1277-1283, http://dx.doi.org/10.1016/j.procbio.2015.05.003

Puls, J. (2009). Lignin-Verfügbarkeit, markt und Verwendung: Perspektiven für schwefelfreie Lignine. In: Stoffliche Nutzung von Lignin. Gülzower Fachgespräche, Band 31, Fachagentur für Nachwachsende Rohstoffe e.V. (FNR), 18-41

Siebert, J. (2015). Untersuchungen zu neuen Mediatoren im Laccase-Mediator-System. Master Thesis. Faculty of Forest Sciences and Forest Ecology, Georg-August University Goettingen, 60 pages

Sixta, H. (2006). *Handbook of pulp. 2.* Aufl., Wiley-VHC.

SUBSPORT. (2013) Specific *substances alternatives assessment formaldehyde.* Retrieved November 6, 2015, from http://www.subsport.eu/wp-content/uploads/data/formaldehyde.pdf

Wiechmann, B., & Kohla. M. (2014). Verbände-Positionspapier zur Neu-Einstufung von Formaldehyd in die TA Luft. Retrieved October 29, 2015 from http://vhi.de/wp-content/uploads/sites/2/2014/07/Verbaende-Positionspapier-Formaldehyd.pdf

Yang, A., Zhang, Y. P., Chen, D., Chen, W. G., & Wang, R. (2001). Eye irritation caused by formaldehyde as an indoor pollution – a controlled human exposure experiment. *Biomed. Environ. Sci., 14*(3), 229-36.

26

Nanotechnology: A Journey towards Finding Solutions

Satya Pal Singh[1]

[1] Department of Applied Sciences, Madan Mohan Malaviya University of Technology, Gorakhpur, UP, India

Correspondence: Satya Pal Singh, Department of Applied Sciences, Madan Mohan Malaviya University of Technology, Gorakhpur, UP, India. E-mail: singh.satyapal@hotmail.com, spsinghaps@gmail.com

Abstract

Nanotechnology is the understanding and control of matter at the diemnsions ranging between 1-100 nm. One nanometer is one billionth of a meter. Nanotechnology involves manipulation of atoms, imaging, measuring and modelling at nano scale. Its potentials were first highlighted by Richard Feynman in the American Physical Scociety meeting in 1959. Though, he did not coin the world nanotechnology himself but he explored the possiblities of functional materials at the bottom of the scale. In last two decades this technology has been commercialized to great extent and gaining importance day by day influencing the economies of different countries and henceforth enforcing the policy makers to address the issues like environment, health and safety. Governments are regularisaing and monitoring its research, uses, applications and technology transfer which includes intelluctaul property rights. This paper addresses the dimensions and trends of nanotechnology covering economic aspects. The paper is focussed on the changes in the functional properties of nanomaterials as physical, chemical, optical, electronic, electrical, magnetic etc. in comparision to those of the bulk of material. It has been discussed how the basic and advance research in nanoscience could be explotiedfor making technologies for its commercial and industrial applications for the benefit and safety of the soceity. Thin film magnetism is demonstrated using Monte Carlo simulation method. Experimental synthesis of some of the nanorods and qunatum dots are also discussed.

1. Introduction

According to USA National Nanotechnology Initiative mission variuos aspects of nanotechnology can be encapsulated in the definition-" The essence of nanotechnology is the ability to work at the molecular level, atom by atom, to create large structures with fundamentally new molecular organiziation. Behavior of structural features with a typical dimension in the range of 1-100 nm (i.e . 1000 times smaller than the diameter of the human hair) exhibit important changes in comparison to those of the bulk of the materials. Nanotechnology is concerned with materials and systems whose components and systems exhibit novel and significantly improved physical, chemical and biological properties, phenomena and processes due to their nanoscale size. The goal is to exploit these properties by gaining control of structures and devices at atomic, molecular and supramolecular levels and to learn efficiently manufacture and use these devices."(National Nanotechnology Initiative, 2000) In word nanoscience the prefix nano has the Greek origin which means "dwarf". A nanomaterial is defined as a material which as atleast one of its three dimensions in the range of 1-100 nm. The electrical, optical and elctronic properties of nanomaterials are changed drastically because of the fact that these materails are composed only of few hundreds of molecules. In a loose sense, one can say that these materials may allow electron conduction with more ease because of the absence of the bulk of the materials. The assembly of only few hundreds of molecules have a different band gap than those of atoms in a bulk of material altering the semiconducting properties. These bands also changes their optical response. Manipulation of atoms and molecules at nanoscale can help us to bring desired change inproerpties of our interest. CNT, MWCNT and C_{60} (i.e. fullerene) doped with atoms of Khave been widely used as nanomaterials for their electrical as well as mechanical properties(Poole & Owenes, 2009). The number of atoms on surface in comparison tothose of the whole volume is also very high in the case of nanomaterials which allows for better chemical and catalytic reactions. This happens because of having more number of active sites. As for example for a FCC nanoparticle of n layers, the total number of atoms in nanoparticle is given by- $N = \frac{1}{3}[10n^3 - 15n^2 + 11n - 3]$ where as those on the surface is given by-$N_{surf} = [10n^2 - 20n + 12]$(Poole & Owenes, 2009). There exist three generic types of nanostructures- 1. Zero-dimensional (0D) systems (Quantum dots) 2. One-dimensional (1D) systems (Qunatum wires) and 3. Two-dimensional (2D) systems (Quantum well).

The dimensions of the nanomaterial restrics the degrees of freedom available for electrons and holes. If the physical dimension(s) of the nanomaterials become smaller than the charecteristic length of electrons and holes in material (i.e. thermal de-Broglie wavelength for electrons and the coulombic range of interaction for electron-hole pair called exiciton), the carriers reflect strong confinement effect. The electrons or holes experience a completely different potential field inside a material offew hundred of atoms or molecules. Schrodinger equation solution for these carriers drastically changes and shows quantum confinement effect on density of states and the mobilities of electrons and holes (Poole & Owenes, 2009).

US is the first country to take National Nanotechnology Initiatives in the year 2000. The investment in nanotechnology has been blooming since then. In 14 years the total investment has increased by appproximately four folds reaching to an amount of 1.7 billion dollors in US. Following US example most of nations including India have taken nanotechnology initiatives some way or other and have started harnessing it to boost variuos sectors as energy, defense, aerospace, and information technology, health, transportation, internal security, agriculture, food and drugs etc. Nanotechnology is capable of providing solutions as green and clean energy production, efficient transfer and storage of energy. Nanotechnology based products presently in the market include faster computer processors and mobiles, higher density memory devices, lighter-weight auto parts, more energy efficient computer and television displays, stain-resistant clothings, antibiotic bandages, cosmetics and drugs etc. Nanotechnology is evolutionary in nature offering incremental improvements in performance cost, size, and weight and energy efficiency. Not only this, it is also said to be an exponential technology which means that products incorporating nanotechnology can be used for making manufacturing technologies better, more efficient and cheaper which in turn reduces the cost of manufacturing. Thus the growth in nanotechnology sector is expected to grow exponentially. In 1965, Gordon Moore, co-founder of Intel observed and stated that the number of transistors per square inch on Integrated Circuits (IC) had doubled every year since the IC was invented.This is known as Moore's law Figure 1.The global investment through these products incorporating nanotechnology has produced revenues of 200 billion dollars in 2008 of which US alone has captured 80 billion dollars. The estimated global revenue is 3.1 trillion dollars in 2015 of which half would be generated from semiconductors.

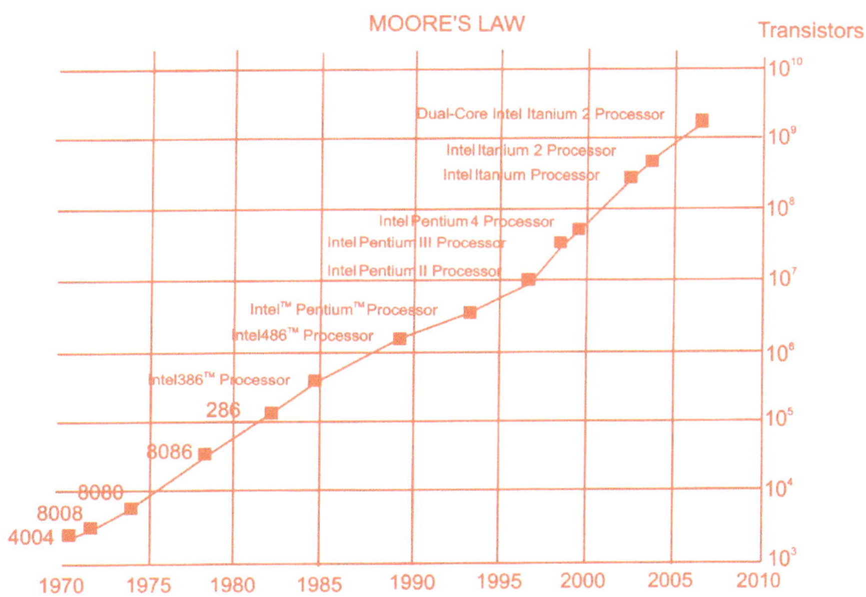

Figure 1. Moore's law shwoing exponetial growth

Technolgies are always driven by the advent of materials. The different ages for materials can be broadly classified into different categories as stony age, bronze age, iron age, steel age, semiconductor age, advance materials (ceramic, polymer and metal matrix composites) age and the present day nanomaterials and nanocomposites age. Nanomaterials have the properties as high tensile strength, light weight, greater water, dust and chemical resistant, high surface to volume ratio, more energy efficiency and power saving. These properties make them unique choice for possible applications (Figure 2) ranging as croceries, bathtubs, toothpaste, sun creams, mobile phones, TV screens, LED (light emitting diodes), articles paintings, cosmetics etc., aerospace engineering, biotechnology and drugdelivery, satellite, communication and remote sensing, pollution control and envrionment protection, energy and data storage, fast computing processors and weapons and agroproducts (National Nanotechnology Initiative, 2000). Nano-motors, bio-senors, bio-photonic devices, spintronic, photonic and mechanical (i.e. nano electromechanical systems NEMS) devices have emerged as recent thrust areas of naotechnology.

Nanomaterial based devices and instruments are widely being used for the prification of water. The waste water treatment becomes mandatory for the reasons well known. The industrial waste water can cause heart, lungs and skin diseases. It has been observed to cause cancer among routine users. The industries like cement, leather and tanning produce waste water containing metals and other harmful chemicals. Many of these are toxic and harmful to living being.

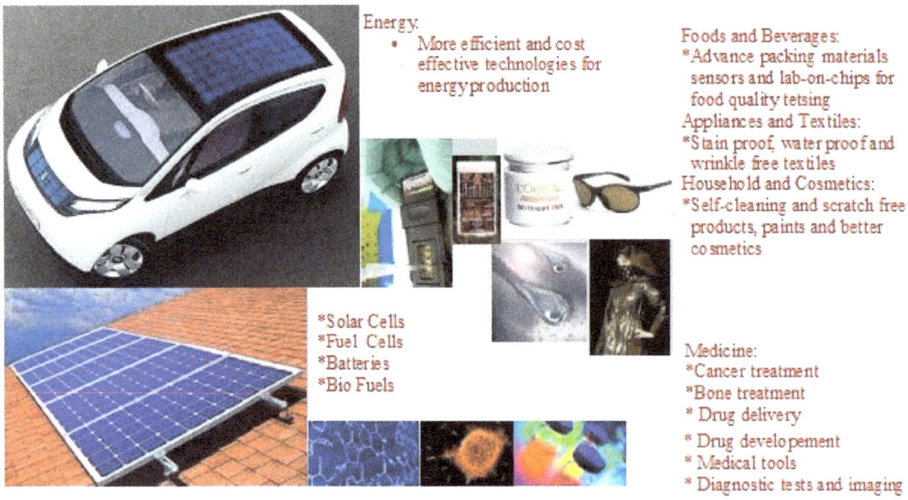

Figure 2. Applications of Nanotechnology

Nanomaterials as Zeolite,Carbon Nano Tubes (CNTs), very thin membranes and other mesoporous materials have been used for separation,filtration and removal of meatls as Cr, As etc from contaminated water (Wang & Lo, 2009; Rana, Mohan, & Rajagopal, 2004; Gupta & Ali, 2004; Ali & Gupta, 2006; Gupta, Gupta, & Sharma, 2001; Singh, Gupta, & Gupta, 2007; Shevade & Ford, 2004; Rengaraj, Choi, & Yi, 2004; Singh, 2014). Cr(VI) is adsorbed by CNTs and is separated from containimated water but reusing CNTs have been a challenge. Recently, - A novel idea has been discussed and presented for recycling such CNTs in Singh (2014). The issue of interest is that whether the in and out vitro cleansing of adsrobed Cr(VI) from such CNTs can be achieved by flowing these CNTs in a narrow streamline jet of some fluid of interest inside a channel (Singh, 2014). Not ony that the molecular level tweaking force or exertions at CNTs surface can be produced using ultrasonic transducers by producing alternate denser and rarer regions inside medium (i.e. accoustic grating) Figure 3. This may be technologically very challenging because of the fact that the precision and control required is of the molecular scale (Singh, 2014).

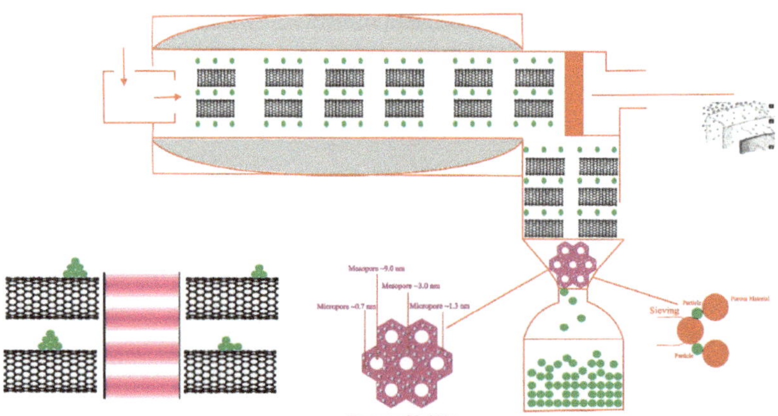

Figure 3. A novel mechanism to produce tweaking molecular forces fro cleansing of CNTs and re-cycling

2. Dimensions of Nanotechnology

a) The Particle in Box: A Pandora Box

The resulting wave function and energies are respectively given by –

$$\Psi(n_x, n_y, n_z) = \sqrt{\frac{8}{L_x L_y L_z}} . \sin\frac{n_x \pi x}{L_x} \sin\frac{n_y \pi y}{L_y} \sin\frac{n_z \pi z}{L_z} \tag{1}$$

$$E(n_x, n_y, n_z) = \left[\left(\frac{\pi n_x}{L_x}\right)^2 + \left(\frac{\pi n_y}{L_y}\right)^2 + \left(\frac{\pi n_z}{L_z}\right)^2\right]\frac{\hbar^2}{2m} \tag{2}$$

$$= \left(k_x^2 + k_y^2 + k_z^2\right)\frac{h^2}{8\pi^2 m} \tag{3}$$

$$= \frac{\hbar^2 k^2}{2m} \tag{-4}$$

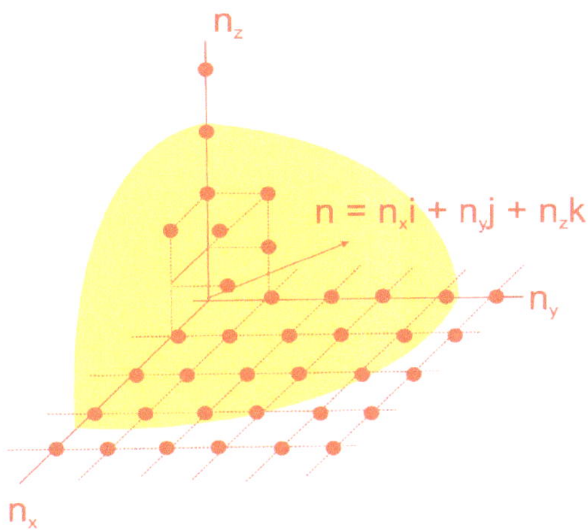

Figure 4.Shows the the lattice space composed of qunatum number n_x,n_y and n_z

Energy density in a three dimensional box is given by-

$$g_{3D}(E)\,dE \simeq \frac{1}{8}\left[4\pi n^2 dn\right] \tag{5}$$

$$n^2 = n_x^2 + n_y^2 + n_z^2 \tag{6}$$

$$g_{3D}(E) \simeq \frac{\pi n^2}{2}\left(\frac{dn}{dE}\right) \tag{7}$$

$$E = \frac{\hbar^2}{2m_{ef}}\left[\frac{\pi^2}{L_0^2}\left[n_x^2 + n_y^2 + n_z^2\right]\right] \tag{8}$$

$$\Rightarrow n = \sqrt{\frac{8m_{ef} L_0^2}{h^2}}\sqrt{E} \tag{9}$$

$$\frac{dn}{dE} = \frac{1}{2}\sqrt{\frac{8m_{ef}.L_0^2}{h^2}}.\frac{1}{\sqrt{E}} \tag{10}$$

$$g_{3D}(E) \simeq \frac{\pi}{4}.\left(\frac{8m_{ef}.L_0^2}{h^2}\right)^{3/2}\sqrt{E} \tag{11}$$

The density of states is defined by dividing $g_{3D}(E)$ by the physical volume of the box-

$$\rho_{energy}(3D) \simeq \frac{\pi}{4}.\left(\frac{8m_{ef}.L_0^2}{h^2}\right)^{3/2}\sqrt{E} \tag{12}$$

Figure 5. Quantum Dots with gradually stepping emission from violet to deep red. Just by taking dimensions of few nanometers one can see that energy states of the electrons under such confinement becomes as high as to form Plasmon which scatter light in this fashion

Source: https://en.wikipedia.org/wiki/Quantum_dot#

b) Particle in a Two Dimensional Box-

Repeating similar exercise but in two dimension, we can write-

$$\rho_{energy}(3D) \simeq \frac{\pi}{4}.\left(\frac{8m_{ef}.L_0^2}{h^2}\right)^{3/2}\sqrt{E} \tag{13}$$

$$g_{2d}(E)dE \simeq \frac{1}{4}(2\pi n)dn \tag{14}$$

$$E = \frac{n^2h^2}{8m_{ef}.L_0^2} \Rightarrow n = \sqrt{\frac{8m_{ef}L_0^2}{h^2}}\sqrt{E} \tag{15}$$

$$\frac{dn}{dE} = \frac{1}{2}\sqrt{\frac{8m_{ef}L_0^2}{h^2}}.\frac{1}{\sqrt{E}} \tag{16}$$

$$g_{2D}(E) \simeq \frac{2\pi m_{ef}L_0^2}{h^2} \tag{17}$$

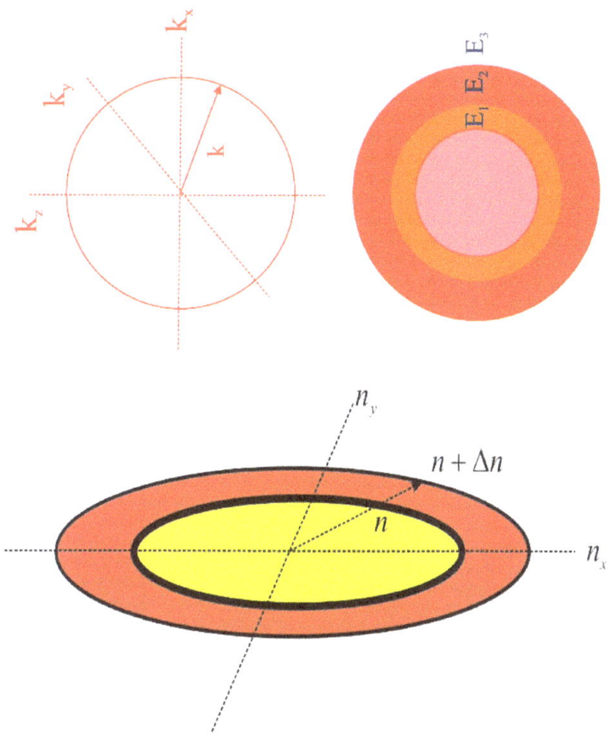

Figure6. Shows the the lattice space composed of qunatum number n_x,n_y in 2 dimensions

Density of states is obtained by dividing the physical volume of the system-

$$\rho_{energy}(2D) \simeq \frac{2\pi m_{ef}}{h^2} \tag{18}$$

Density of states as obtained for a particle in a two dimensional box is a constant.

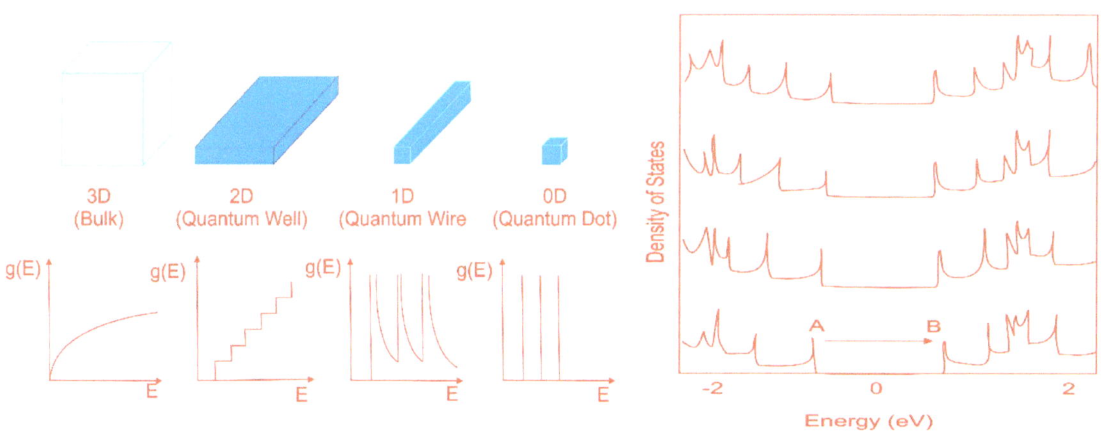

(a) (b)

Figure 7. a) Shows the density of states vs E for different type of materials b) Shows the one dimensional density of states for (8,8), (9,9), (10,10) and (11,11) armchair carbon nanotubes

c) Particle in a One Dimensional Box-

The energy of the particle is given by –

$$E = \frac{n^2 h^2}{8 m_{ef} L_0^2} = \frac{\hbar^2 k^2}{2 m_{ef}} \tag{19}$$

$$k = n\pi / L_0 \tag{20}$$

$$n = \sqrt{\frac{8 m_{ef} L_0^2}{h^2}} \sqrt{E} \tag{21}$$

$$g_{1D}(E) dE \simeq dn \tag{22}$$

$$g_{1D}(E) \simeq \frac{dn}{dE} \simeq \frac{\Delta n}{\Delta E} \tag{23}$$

$$g_{1D}(E) \simeq \frac{1}{2} \sqrt{\frac{8 m_{ef} L_0^2}{h^2}} \cdot \frac{1}{\sqrt{E}} \tag{24}$$

$$\rho_{energy} 1D = \sqrt{\frac{2 m_{ef}}{h^2}} \frac{1}{\sqrt{E}} \tag{25}$$

3. Trends of Nanotechnology

Though, the importance of assembling atoms at nanoscales was first highlighted by famous physicist Prof. Richard Feynman in a meeting of American Physical Society on 29th December 1959 at Caltech, it is still estimated to be an emerging area in general from the point of view of the developments in the area of nanotechnology. We can group different countries in three categories on the basis of their achievements and advancement in this area-National Activity Nations, Current R & D Empowerment Nations and Demonstration of Interest Nations. USA, China, Japan and Russia are the major players in National Activity Nations. Countries like India, Brazil, Malaysia, Thailand, Singapore and South Africa are at Current R & D Empowerment Scale. Many other countries which have started their Nano initiative programs are at the lowest level of Demonstration of Interest Stage.

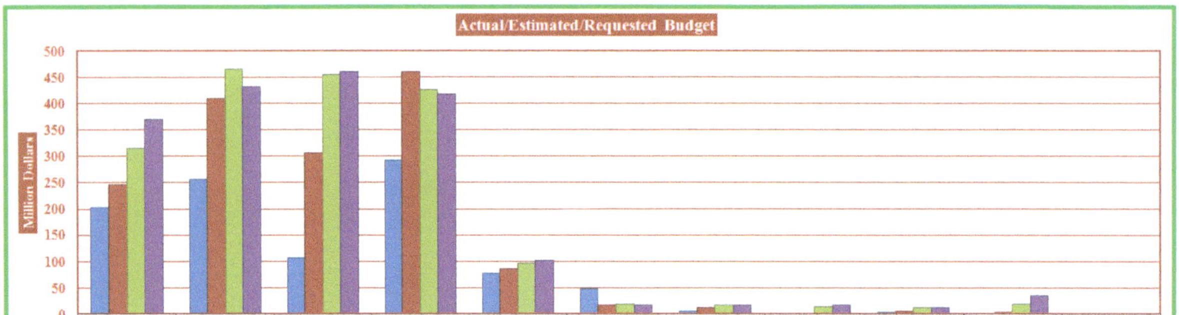

	Department of Energy	National Science Foundation	National Institute of Health	Department of Defense	National Institute of Standard & Technology	National Aeronautics and Space Agency	Environmental Protection Agency	Food and Drug Administration	National Institute of Food and Agriculture	Department of Homeland and Security	Agriculture Research Service
2004	202	256	106	291	77	47	5	0	2	1	0
2008	245	409	305	460	86	17	12	0	6	3	0
2012	313.8	466.3	456	426.1	95.4	18.6	17.5	13.6	11.3	18.7	2
2014	369.6	430.9	460.8	416.9	102.1	17.5	17.2	16.8	11.3	34.9	2

Figure 8. Shows the USA nanotechnology funding through different agencies

The USA National Nanotechnology Initiative (NNI) was launched in 2001 under the umbrella of National Science and Technology Council coordinating 15 federal funding agencies with specific nanotechnology budgets. USA has invested about 15.6 billion US dollars for nanotechnology during 2001 to 2012. The estimated budget of USA for the year 2013 was about 1.767 billion US dollars (Cientifica, 2011; Anton, Silberglitt, & Schneider, 2001; Shand & Wetter, 2007; Milanović & Bučalina, 2013). USA priorities on nanotechnology can be learnt from the increasing trends with enormous funding to various sectors through her different agencies in last 14 years from the Figure 8, Figure 9, Figure 10 and Figure 11. Figure 12 shows relative Indian participation through her National

Nano-initiative program. Figure 13 reflects that nano-enabled drug delivery have captured the market share to a reasonable extent and it will grow exponentially in coming future. That also reflects that nano-enabled drugs are more effective to cure from diseases because of its increasing demand in the market.

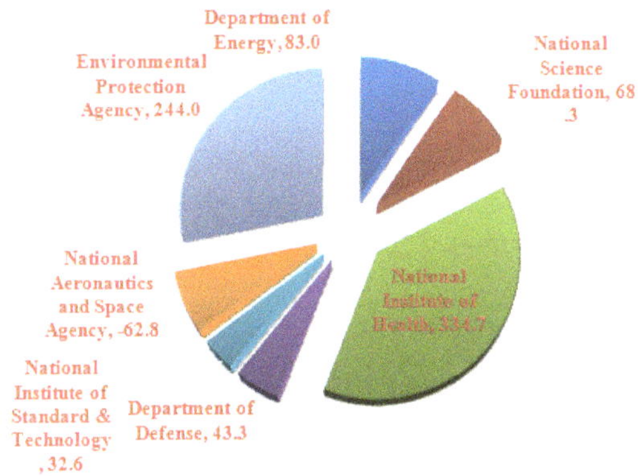

Figure 9. Shows the % change in USA budget of 2014 in comparison of the 2004 budget (Nanotechnology funding by different agencies)

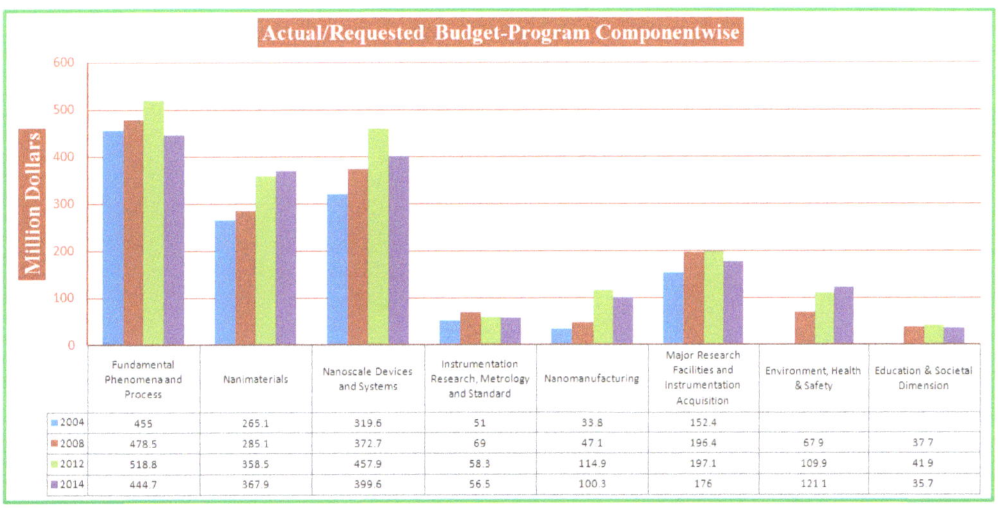

Figure10. The componentwise USA budget during 2004-2014

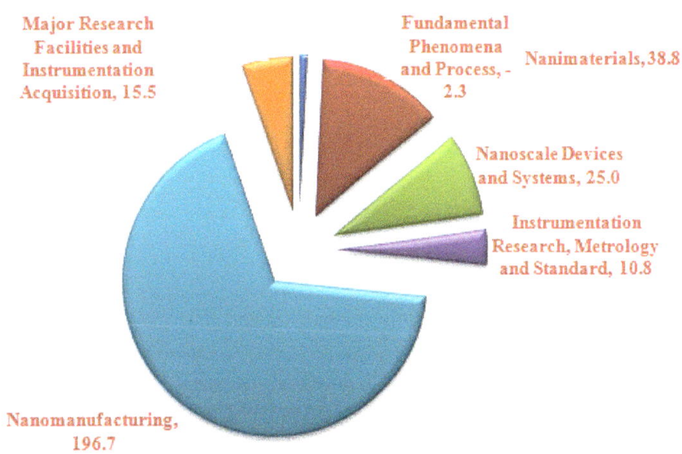

Figure 11. Reflects the % change in USA budget of 2014 in comparison to that of 2004 (Nanotechnology funding in variuos sectors)

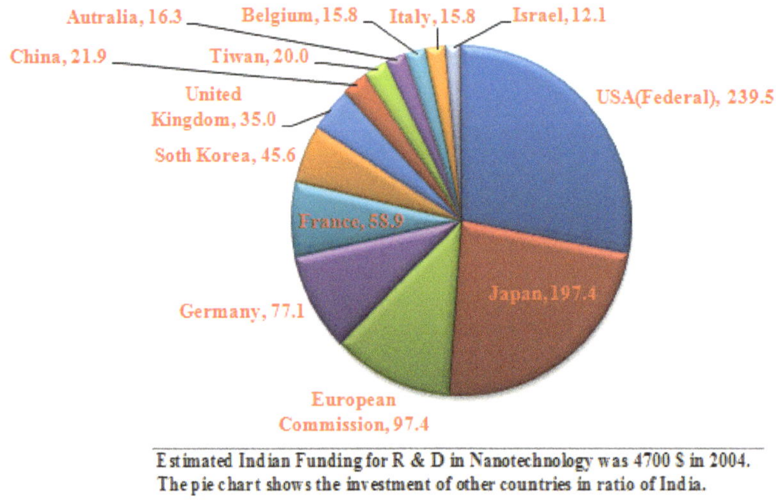

Figure12. Shows the ratio of the estimated nanotechnology funding in comaprison to that of Indian budget allocation by various National Activity Nations in 2004

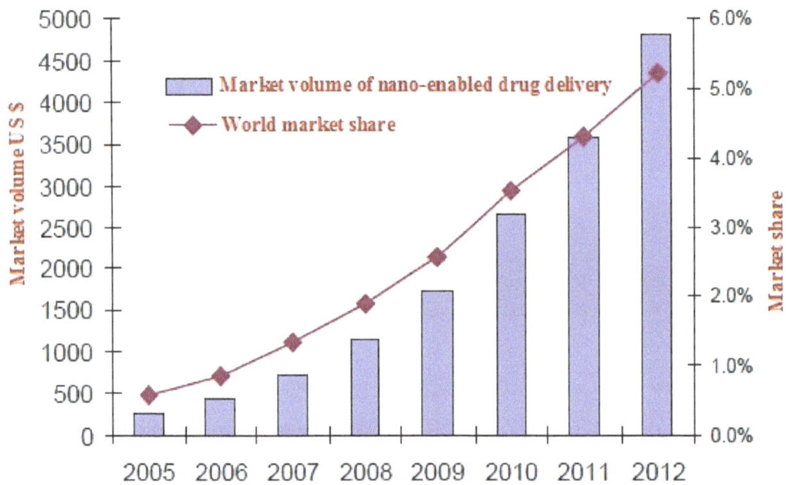

Figure 13. Global marketshare and market volume of nano-enabled drug delivery (Ezema, Ogbobe, & Omah, 2014)

India started her Nanoscience and Technology Initiative (NSTI) actively in her five year plan of 2007-2012 with an estimated budget of Rs. 1000.0 crores (The Energy and Resources Institute (TERI), 2010). This was aimed to establish center of excellence, targeting laboratories, infrastructure and human resource development in strong collaboration with foreign partners as USA, UK, Germany, Japan and Russia. A more specific detail of nano-initiative programs of many other countries as Japan, China, Russia (National Activity Group), Brazil, Malaysia, Thialand, Singapore and South Africa (Current R & D and Empowerment Scale) can be obtained in research paper (Ezema, Ogbobe, & Omah, 2014) and its refrences.

4. Methoddology and the investigations:

a) Synthesis, Characterization and Nanometrology

From earlier discussions, it is clear that nanotechnology gained econmic importance only few decades back. The reasons have purely scientific origins andcan be learnt in this section. Though, Feynman's visionary ideas on nanotechnology goes back to 1959 but there was little progress in that direction till 1981.That is the year which marked the invention of Scanning Tunneling Microscope (STM). STM was invented by a group of scientists led by Gerd Binnig and Heinrich Rohrer at IBM Zurich Research Laboratory. They published the first STM image of the Si(III)-7 x 7 surface in 1983 (Shong, Sow, & Wee, 2010). Gerd Binnig and Heinrich Rohrer shared the Nobel prize in physics in 1986 for their invention of STM. The Atomic Force Microscope (AFM) was invented in 1986. These two inventions have made greatest impact in the field of nanoscience and nanotechnology. Various top down and bottom up approaches (i.e. Chemical Vapor deposition (CVD), Molecular Beam Epitaxy (MBE), Colloidal method, Sol-gel method, Spin costingetc.) were developed in due course of time which definitely depended on the scientific progress in the area of nanometrology. The objectives of the nanometrology is to providecoordinates and identity of every constituent atoms in a nanostructured material, object or nanodevice. The important techniques and tools which play vital role in nanometrology can be listed as – X-rays diffraction (XRD), Atomic Force Microscope (AFM), Scanning Tunneling Microscope (STM), Transmission Electron Microscope (TEM) etc. The optical techniques for characterizing surface regularity are Ellipsometry, Scanning Angle Reflectometry (SAR), Surface Plasmon Resonance (SPR), Optical Waveguide Lightmode Spectroscopy (OWLS) etc (Shong, Sow, & Wee, 2010). The new developments and the modifications in these instuments and techniques after STM have helped scientists and engineers to manipulate, control and engineer materials and structures at nanoscale to great extent in last few decades.

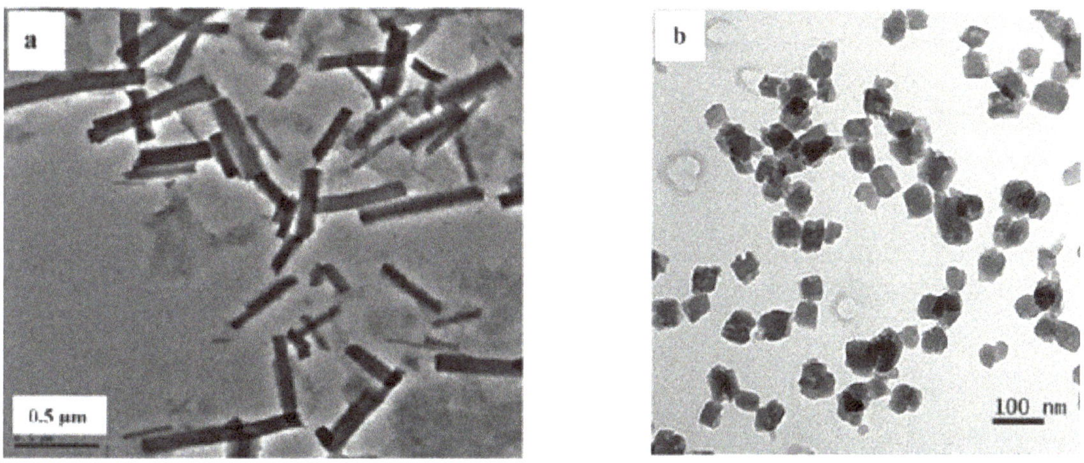

Figure 14. a) TEM micrographs fro nickel oxalate dihydrate synthesised usin a) CTAB/1-butanol/1-hexane and b) Tergito/1-octanol/cyclohexane Reprodcued by permission from (Ganguli, 2010)

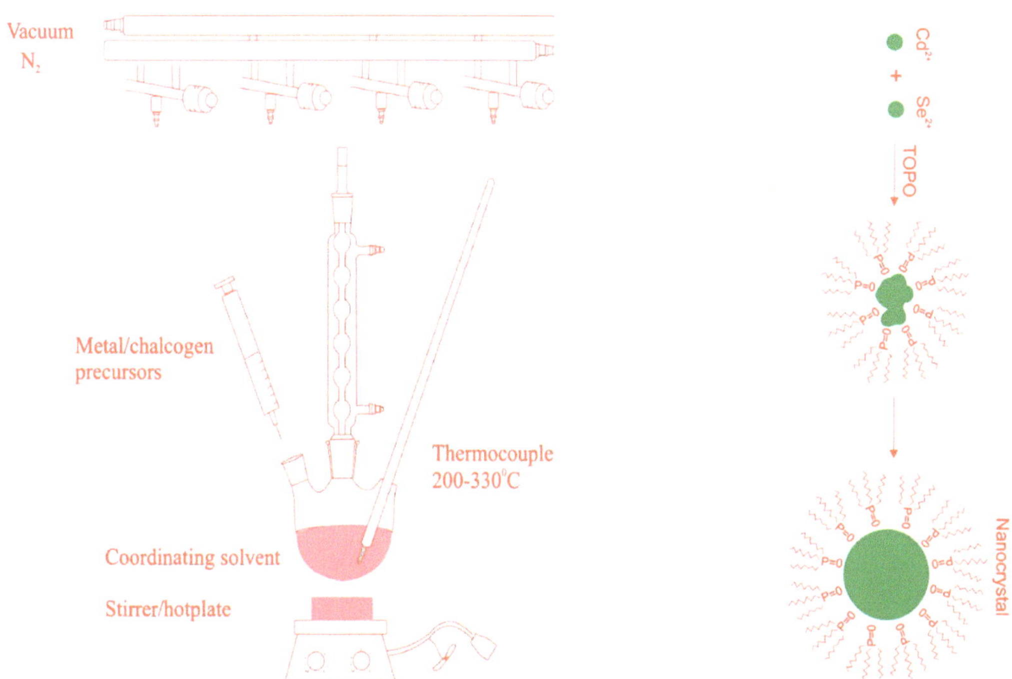

Figure 15 a) Shows the ray-sketch for synthesis Cd-Se quantum dots b) Shows the nucleation and crystal growth (CdMe2and Trioctylphosphine selenide are respectively usedas metal and chalcogen precursors)

The applications of nanoscale phenomena could start taking place only few decades back. I give you one example. The influence of spin on electron conductivity was first proposed by Nevill Mott in 1936 but it could be expoited for technological applications only after the discovery of Giant Magneto-Resistance (GMR) in 1988 (Ramsden, 2011). The discovery of GMR led to the barnch of spinotronics. It can be defined as the technology in which electron spin interplays with other properties of material as conductivity. To observe the phenomenon of GMR two ferromagnetic ultra thin films are separted by athin non-magnetic Fe or Cr film. There exists a ferromagnetic or antiferromagnetic interaction between the magnetic layers depending on the width of Cr or Fe layer which separates the two magnetic superlattices. The antiferromagnetic

state of the magnetic layers can be transformed back into the ferromagnetic state by an external magnetic field.The spin dependent scattering of the conduction electrons in non-magnetic layer cause a small resistance when spin alligns parallel in two films where as for anti-parallel spin allignment of spins in two separted films, the resistance is high. This technology is being used for read-write heads in computer hard drives. This type of highly polarized spin transport could be demonstrated as GMR and could be exploited for technological applications only after the development of methods for synthesis of high qulaity ultra-thin films of thickness <100 nm (Ramsden, 2011).

b) Computational Modelling

Computational modellinghas emergedas a new method of investigation at molecular scale bridging the gap between experiment and theory. It has been used by educators, scientists, engineers and researchers worldwide for understanding phenomena at nano and meso scales (Singh, Singh, & Roychoudhury, 2008; Saha, Singh, Singh, & Kwak, 2009; Singh, 2010, 2011, 2014; Singh, Singh, & Sharma, 2011, 2013). Molecular Dynamic (MD) simulation is the main atomistic simulation technqiue being used. In this simulation, the trajectories of the ith atom in an assembly of N interacting atoms is computed by numerically integrating Newton's second law of motion-

$$m_i \frac{d^2x_i}{dt^2} = \frac{\partial V}{\partial x_i}$$

Here m_i is the mass of the ith atom and V is the inter-atomic potential. MD simulation is a deterministic method. This method has a limitation because of using a model potential which may not be a very true representation of the natural forces in general yielding a non-realistic result. Another method is Monte Carlo simulation method which works on stochastic principle. MC simulation is of probabilistic nature. This works well for qualitative understanding but lacks exact quantitative results because of the fact that no input information as mass and velocity are fetched in the simulation program as input parameters. But it is competent to model and understand various physical phenomena at molecular scales. Figure 16 (a) Shows the precipitation of majority component drops after 10,000 MC Cycles in a non-conserved binary mixture on a lattice of 200 x 200 confined by two planar walls with moderate affinity E(wall)=-2.0 to the majority component using Lattice based Maxwell Boltaman (LMB) model (Singh, Singh, & Roychoudhury, 2008). The Figure 16 (b) shows the adsorption of the preferred component for non-conserved binary mixture for 2D case which condenses to the substrates relatively at lower temperature and weaker affinity of the surface. The stronger the affinity of the surface the lesser is the effect of spontaneous magnetization yielding lower adsorption on the surface. The stronger affinity allows for the formations of ridge like structures rather than spherical doamin. This effect is simliar to raising temperature which brings disorder in the system. Another simulation report shows the results of crystallization using 3D lattice confined by two chemically active planer walls Figure 16 (c). The coarse graining of the minority component B is shown in a slab of 40:40: 30 at $k_BT = 0.60$ and A: B = 3:2 with E (W - A) =+2.0 and E (W - B) = -2.0. The darker regions indicate the formation of nanodrops or the precipitates in the case of 3D lattice based NVT-MC simulation (Singh, 2012). These simulation reports were programmed and prepared using personal computer with Intel i3 processor.This result can be compared with crystallization process by bottom up approach and SEM image of an experimentalist. Self assembly of atoms or molecules can be well realised by the most common phenomena of nucleation and crystal growth. With the advent of fast computing computers, it has become possible to make trial moves so that a reasonable configuration (phase) space is visited and sampled in order to produce more realistic computer simulation results in past few decades.

Fig. 17 shows the spin alignment for magnetic materials.- Simulation results are presented here for the variation of magnetization in 2D thin film with temperature, coupling of neighbouring spins (i.e. J=0 paramagnetic case; J>0 Ferromagnetic case and J<0 Anti-ferromagnetic case). The same has been studied with change in applied external magnetic field on lattice size of 200 × 100 with peroidic boundary conditions and Metropolis algorithm. These results Figure 18 A & Figure 18 B are in well agreement with well known para, ferro and anti-ferromagnetic properties of materials (Singh, 2014; Tanner, 1995). The Figure 18 A & Figure 18 B right are plotted for antiferromagnetic film which also indicate that for low temperaure or higher copuling constant, there is more tendency to allign like spins (i.e. ferromagnetic behavior) for higher values of applied external magnetic fields. Even in the absence of any external magnetic field there exists a critical temperature at which the spontaneous magentization occurs. Each simulation was done for 5000 MC cycles and have taken around 10-12 minutes of time on Core 2 Duo intel processor. Another simulation reports the ferromagnteic property of 2D thin film (Singh, 2015). The film is annealed at high temprature at T=10.0 and abruptly cooled to T=0.50, 1.0 & 1.5 and is equilibrated for 20,000 Monte Carlo Cycles. The induced magnetisation curve is shown in Figure 19.

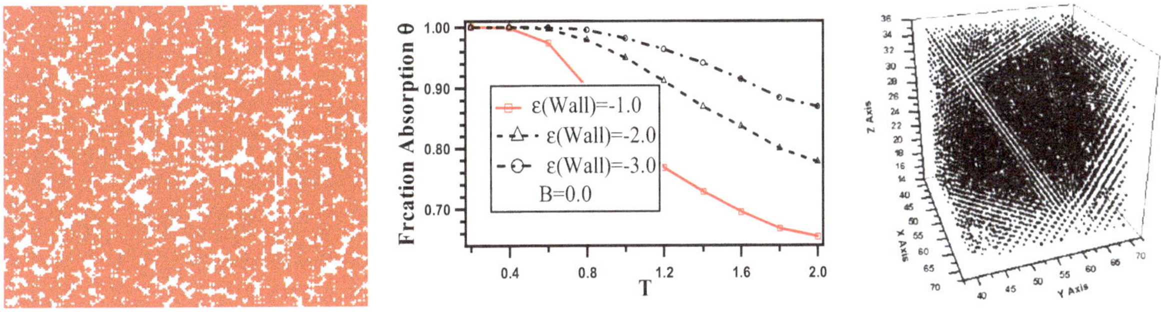

(a) (b) (c)

Figure 16. (a) Precipitation in a conserved binary mixture: Micrograph at kT=1.0 and e(wall)=-2.0 obtained by lattice based Monte Carlo Simulation for 2D case (b) fraction of surface adsorbed vs T at different surface spin affinity non-conserved spin cases & (c) Micrograph for 3D case Both micrographs are obtained for the case of confinement by two walls (Singh, 2011, 2012)

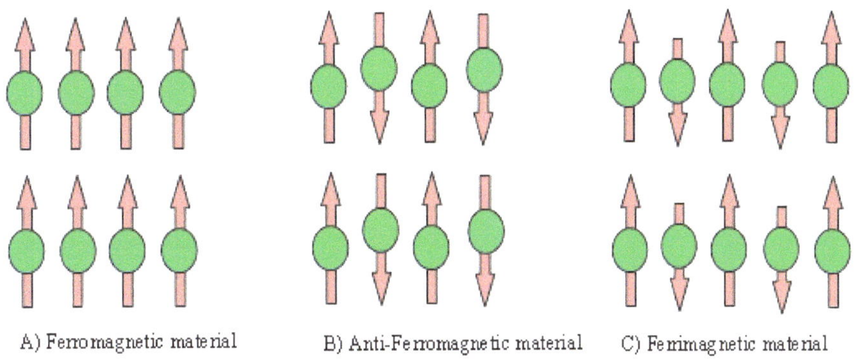

Figure 17. Spin allignment in ferromagnetic, anti-ferromagnetic and ferrimagnetic materials

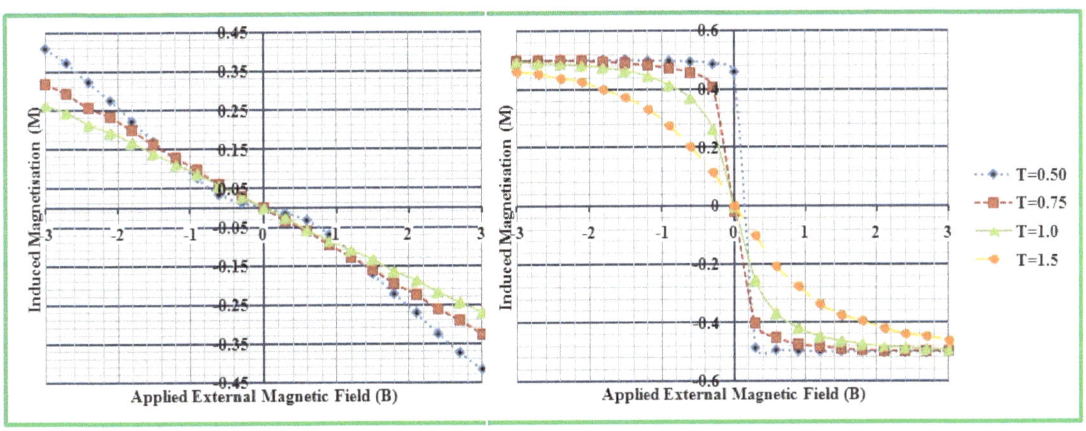

Figure 18 (A). Shows the simulation results of temperature dependence of induced magnetiation vs applied external magnetic field for anti-ferro (left) and ferromagnetic 2D thin film

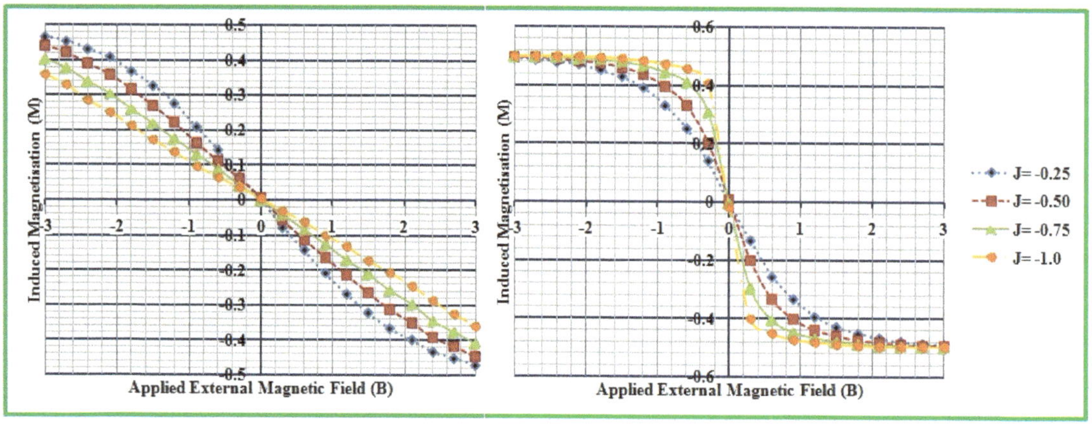

Figure 18 (B). Shows the simulation results of spin coupling dependence of induced magnetiation vs applied external magnetic field for anti-ferro (left) and ferromagnetic 2D thin film

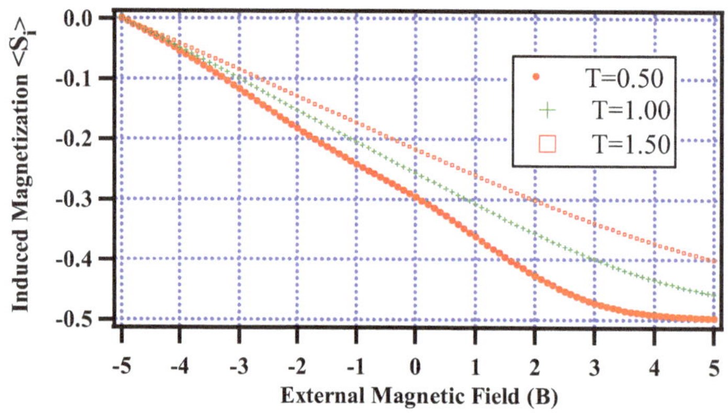

Figure 19. Shows the induced magnetization in a 2D thin film (100 × 100) lattice size at temperatures A) T=0.50 B) T=1.0 & c) T=1.50 with nearest neighbor spin coupling constant J=1.0. All data have been obtained after run cycle of 20,000 Monte Carlo Cycles (Singh, 2015)

In 1965, Gordon Moore, co-founder of Intel observed and stated that the number of transistors per square inch on Integrated Circuits (IC)had doubled every year since the IC was invented. This has been obeyed till now and known as Moore's law. In 1971, Intel introduced the first microprocessor with about 2000 transistors in it. IBM personal computer could appear after a decade in 1981. In 2000, Pentium 4 Chip with 1.2×10^6 transistors fabricated with 180 nm process technology appeared. The Dual Core Intel Itanium chip has about 1.7×10^9 transistors occupying an area of about 50 mm × 20 mm with a gate length of 90 nm.Remarkable fact is to notice is that a 45 nm transistor can switch 3×10^{11} times per second (i.e. 100 GHz). Graphene based devices have achieved more than one THz frequency experimentally (Shong, Sow, & Wee, 2010; Ramsden, 2011).

Table1. A comparison of china, Russia and India in nanotechnology paper publications and patents at United States Patents and Trademark Office (Liu et al., 2009)

Country	Years	SCI Papers	Years	USPTO Patents
China	1984-2007	49193	1989-2007	162
Russia	1976-2007	12307	1980-2007	30
India	1980-2007	9126	1983-2007	115

From the simulations reports presented in this paper, one may realize that the advancement in the area of nanotechnology has made the dreams and ideas of Prof. Feynman a reality as he had spoken that it could be observed during the end of 20th century. A new branch of computational method bridging the pure theory and experiment could emerge because of the advancement in the area of nano-electronics.

5. Conclusion

According to a report published in Nature Nanotechnology in 2008, more than 60 countries were running national programs in nanotechnology and hundreds of nanotechnology products were commercially available (Ermolov et al., 2007). According to the study, the 7406 patents were granted by United States Patent and Trademark Office (USPTO) in the area of nanotechnology. European Patent office (EPO) granted 3596 nanotechnology patents whereas Japanese Patent Office (JPO) has registered 1150 during the period of 30 years (1976-2006). The annual number of nanotechnology patents granted by USPTO and EPO increased quasi-exponentially whereas the number of the patents granted by JPO increased slightly slower (Chen, Roco, Li, & Lin, 2008). This gives a clear indication of the significant advancement in the area of nanotechnology. Not only that the nanotechnology papers published by various countries have also increased to great extent in past few decades and more and more institutions, private companies and researchershave joined and focused research in the area of nanotechnology (Ermolov et al., 2007; Liu et al., 2009; Chen, Roco, Li, & Lin, 2008). The society is going through changes in global world because of the vast and un-interrupted use of information resources through internet. and the network of entertainment TV channels have wide spread across the globe. This fact have forced governments of various countries to take new steps to strengthen their societal fabric and harmony, to provide personal security to their citizen and maintain their ethical values intact in this new scenario. The hard believers of nanotechnology think of finding all the solutions of the problem of human being using nanotechnology. They think of synthesis of food by self-assembling of atom by atom. They think of molecular robots which may enter human body and cure all diseases. But many of us may not agree to that extent, nevertheless we hope that nanotechnology has all the potentials to galvanize the daily life of ours, speed up the communication, banking, industrial and government tasks. It can mobilize the transportation and save huge amount of energy which one can say "Energy saved is energy produced".

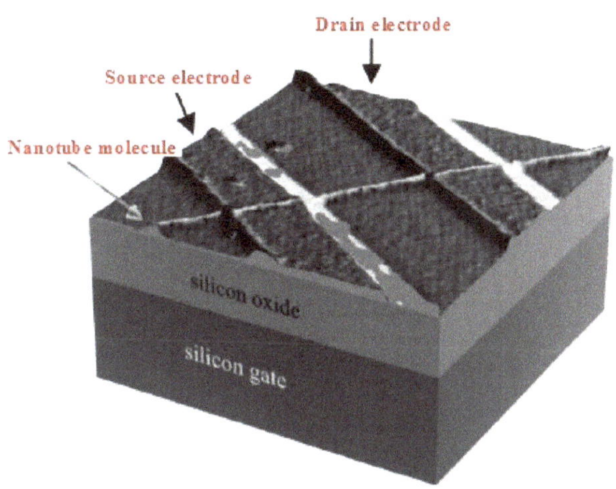

Figure 20. A single-molecule transistor that operates at room temperature, consisting of an individual semiconducting nanotube on two metal nanoelectrodes with the substrate as a gate electrode (right) (Mamalis, Vogtländer, & Markopoulos, 2004)

The nanotechnology has all the potentials to strengthen our internal and external security but poses new challenges as well because the same technology can be used by the terrorists and intruders. It is still under serious research that which amounts of different nanomaterials are harmful to living beings and to what extent? Some of nanomaterials are toxic where as some of these nano molecules are replicating which include genetic coding of animals and in future may be of individuals. This requires a constant monitoring and check by governments.The society is being driven by present day technology and the socio-economic ties have been strengthened in past few decades. Because of the reasons, most of governments have changed their policies for regulating, monitoring and controlling the various issues coming forth with the advancement of technologies for the goodness, safety and security of their people with their interest to gain economic profits and remain in world competition of the fast growing technologies. At the last but not the least, one must learn that the emerging area of nanotechnology may also reach to saturation in terms of new discoveries. The sizes of the single devices may get reduced to a few tens of angstroms Figure (20).The quantum mechanical effects may rule out the possibilities for any further miniaturization.

Acknowledgements

The author is thankful to Prof. Ashok K. Ganguli, Director, Institute of Nanoscience and Technology, Mohali, India for having discussion regarding Figure 14.

References

Ali, I., & Gupta, V. K. (2006). Advances in water treatment by adsorption technology. *Nature protocols*, *1*(6), 2661-2667.

Anton, P. S., Silberglitt, R., & Schneider, J. (2001). *The global technology revolution: bio/nano/materials trends and their synergies with information technology by 2015*. Rand Corporation.

Chen, H., Roco, M. C., Li, X., & Lin, Y. (2008). Trends in nanotechnology patents. *Nature nanotechnology*, *3*(3), 123-125.

Cientifica. (2011). *Global funding of nanotechnologies & its impact*. Retrieved July 2011, from www.cientifica.com

Ermolov, V., Heino, M., Kärkkäinen, A., Lehtiniemi, R., Nefedov, N., Pasanen, P., ... & Uusitalo, M. A. (2007, September). Significance of Nanotechnology for Future Wireless Devices and Communications. In PIMRC (pp. 1-5).

Ezema, I. C., Ogbobe, P. O., & Omah, A. D. (2014). Initiatives and strategies for development of nanotechnology in nations: a lesson for Africa and other least developed countries. *Nanoscale research letters*, *9*(1), 1-8.

Ganguli, A. K., Ganguly, A., & Vaidya, S. (2010). Microemulsion-based synthesis of nanocrystalline materials. *Chemical Society Reviews*, *39*(2), 474-485.

Gupta, V. K., & Ali, I. (2004). Removal of lead and chromium from wastewater using bagasse fly ash—a sugar industry waste. *Journal of colloid and interface science*, *271*(2), 321-328.

Gupta, V. K., Gupta, M., & Sharma, S. (2001). Process development for the removal of lead and chromium from aqueous solutions using red mud—an aluminium industry waste. *Water research*, *35*(5), 1125-1134.

Hullmann, A. (2006). The economic development of nanotechnology-An indicators based analysis. *EU report*.

Kuno, M. (2012). Introductory nanoscience: Physical and chemical concepts. *MRS Bulletin*, *37*(02), 169-170.

Liu, X., Zhang, P., Li, X., Chen, H., Dang, Y., Larson, C., ... & Wang, X. (2009). Trends for nanotechnology development in China, Russia, and India. *Journal of Nanoparticle Research*, *11*(8), 1845-1866.

Mamalis, A. G., Vogtländer, L. O. G., & Markopoulos, A. (2004). Nanotechnology and nanostructured materials: trends in carbon nanotubes. *Precision Engineering*, *28*(1), 16-30.

Milanović, V., & Bučalina, A. (2013). Position of the Countries in Nanotechnology and Global Competitiveness. *Management*, 68.

National Nanotechnology Initiative. (2000). *National Nanotechnology Initiative: The Initiative and its Implementation Plan* (pp. 19-20), Washington, D.C: National Science and Technology Council, Committee on Technology, Subcommittee on Nanoscale Science, Engineering and Technology.

Poole, C. P., & Owenes, F. J. (2009). *Introduction to Nanotechnology*. Wiley India Pvt. Ltd., India.

Quantum dot. (2015). In Wikipedia, the free encyclopedia. Retrieved from https://en.wikipedia.org/wiki/Quantum_dot#

Ramsden, J. (2011). *Nanotechnology: an introduction*. William Andrew.

Rana, P., Mohan, N., & Rajagopal, C. (2004). Electrochemical removal of chromium from wastewater by using carbon aerogel electrodes. *Water research*, *38*(12), 2811-2820.

Rengaraj, S., Choi, E. I., & Yi, S. J. (2004). Arsenic Removal Using Mesoporous Alumina Prepared via a Templating Method. *Environmental science & technology*, (3), 924-931.

Saha, A. K., Singh, S. P., Singh, J. K., & Kwak, S. K. (2009). Quasi-2D and prewetting transitions of square-well fluids on a square-well substrate. *Molecular Physics*, *107*(20), 2189-2200.

Shand, H., & Wetter, K. J. (2007). Trends in intellectual property and nanotechnology: implications for the global south. *Journal of Intellectual Property Rights*, *12*, 111-117.

Shen, Y. F., Tang, J., Nie, Z. H., Wang, Y. D., Ren, Y., & Zuo, L. (2009). Tailoring size and structural distortion of $Fe3O_4$ nanoparticles for the purification of contaminated water. *Bioresource technology*, *100*(18), 4139-4146.

Shevade, S., & Ford, R. G. (2004). Use of synthetic zeolites for arsenate removal from pollutant water. *Water Research*, *38*(14), 3197-3204.

Shong, C. W., Sow, C. H., & Wee, A. T. (2010). *Science at the nanoscale: an introductory textbook*. Pan Stanford Publishing.

Singh, A. K., Gupta, V. K., & Gupta, B. (2007). Chromium (III) selective membrane sensors based on Schiff bases as chelating ionophores. *Analytica chimica acta*, *585*(1), 171-178.

Singh, S. P. (2010, December). A review of molecular simulation: future scope and limitations in synthesis and characterization of nanostructured materials. In *Proceedings of the International Conference on Nanomaterials and Nanotechnology, at KSRCT, Erode Tamilnadu, India, Macmillan Advanced Research Series* (pp. 13-16).

Singh, S. P. (2011). Spinodal Theory: A Common Rupturing Mechanism in Spinodal Dewetting and Surface Directed Phase Separation (Some Technological Aspects: Spatial Correlations and the Significance of Dipole-Quadrupole Interaction in Spinodal Dewetting). *Advances in Condensed Matter Physics*, *2011*.http://dx.doi.org/10.1155/2011/526397

Singh, S. P. (2012). Spatial correlation in 2D and 3D thin films of conserved binary mixtures in the presence of wetting of substrates by the preferred majority component: interpretation in real scenario. *Applied Nanoscience*, *2*(3), 365-369. http://dx.doi.org/10.1007/s13204-012-0094-8

Singh, S. P. (2014). Regeneration process in removal of Cr (VI) and As(V) using MWCNT from aqueous solution, *Proceedings of National Conference on " Materials Science and Technology"* (pp. 93-97). 10-11 March 2014, Department of Applied Sciences, MMM University of Technology, Gorakhpur-273010. 012642 Singh S. P. et al, Regeneration process in removal of Cr (VI) and As(V) using MWCNT from aqueous solution, 2014 (Mar), 93-7, Proc Natn Conf Mater Sci Technol, ISA Volume:50 Issue:12 Date: 16/ June/2014.

Singh, S. P. (2014). Revisiting 2D Lattice Based Spin Flip-Flop Ising Model: Magnetic Properties of a Thin Film and Its Temperature Dependence. *European J of Physics Education*, *5*(3), 8-19.

Singh, S. P. (2015). Dimensions and Trends of Nanotechnology. *ISST Journal of Applied Physics, 1*, 73-79.

Singh, S. P., Singh, J. K., & Sharma, A. (2011). Investigating Bridge-Like Structures in a Square-Well Binary Mixture Using Nvt Monte Carlo Simulation. *International Journal of Nanoscience*, *10*(01n02), 329-333.

Singh, S. P., Singh, J. K., & Sharma, A. (2013). Adsorption of gas-like molecules to self-aligned square-well fluid channels under confinement of chemically patterned substrates. *Applied Nanoscience, 3*(3), 179-187.

Singh, S. P., Singh, K., & Roychoudhury, M. (2008). Monte Carlo simulation for diffusion limited surface directed phase separation. *Proceedings Of The National Academy of Sciences India Section A-Physical Sciences*, *78*, 79-83.

Tanner, B. K. (1995). *Introduction to the Physics of Electrons in Solids*. UK: Cambridge University Press.

The Energy and Resources Institute (TERI). (2010). *Nanotechnology development in India: The need for building capability and governing the technology*.

Wang, P., & Lo, I. M. (2009). Synthesis of mesoporous magnetic γ-Fe 2 O 3 and its application to Cr (VI) removal from contaminated water. *Water research*, *43*(15), 3727-3734.

Permissions

 Every chapter published in this book has been scrutinized by our experts. Their significance has been extensively debated. The topics covered herein carry significant findings which will fuel the growth of the discipline. They may even be implemented as practical applications or may be referred to as a beginning point for another development.

The contributors of this book come from diverse backgrounds, making this book a truly international effort. This book will bring forth new frontiers with its revolutionizing research information and detailed analysis of the nascent developments around the world.

We would like to thank all the contributing authors for lending their expertise to make the book truly unique. They have played a crucial role in the development of this book. Without their invaluable contributions this book wouldn't have been possible. They have made vital efforts to compile up to date information on the varied aspects of this subject to make this book a valuable addition to the collection of many professionals and students.

This book was conceptualized with the vision of imparting up-to-date information and advanced data in this field. To ensure the same, a matchless editorial board was set up. Every individual on the board went through rigorous rounds of assessment to prove their worth. After which they invested a large part of their time researching and compiling the most relevant data for our readers.

The editorial board has been involved in producing this book since its inception. They have spent rigorous hours researching and exploring the diverse topics which have resulted in the successful publishing of this book. They have passed on their knowledge of decades through this book. To expedite this challenging task, the publisher supported the team at every step. A small team of assistant editors was also appointed to further simplify the editing procedure and attain best results for the readers.

Apart from the editorial board, the designing team has also invested a significant amount of their time in understanding the subject and creating the most relevant covers. They scrutinized every image to scout for the most suitable representation of the subject and create an appropriate cover for the book.

The publishing team has been an ardent support to the editorial, designing and production team. Their endless efforts to recruit the best for this project, has resulted in the accomplishment of this book. They are a veteran in the field of academics and their pool of knowledge is as vast as their experience in printing. Their expertise and guidance has proved useful at every step. Their uncompromising quality standards have made this book an exceptional effort. Their encouragement from time to time has been an inspiration for everyone.

The publisher and the editorial board hope that this book will prove to be a valuable piece of knowledge for researchers, students, practitioners and scholars across the globe.

List of Contributors

Shaoqing Ge and Sunil Sinha
Department of Civil and Environmental Engineering, Virginia Tech, USA

W. H. El-Garaihy
Mechanical Engineering Department, Faculty of Engineering, Suez Canal University, Egypt

S. E. Oraby
Department of Manufacturing Engineering Technology, College of Technological Studies, PAAET, Kuwait

El-Sayed M. A. Rassoul
Production Engineering Department, Faculty of Engineering, Mansoura University, Egypt

H. G. Salem
Mechanical Engineering Department, Yousef Jamil Science and Technology Research Center (YJSTRC), American University in Cairo, Egypt

M. Klein, H. Seidlitz, C. Gerstenberger, P. Mayr and L. Kroll
Lightweight Structures and Polymer Technology, TU Chemnitz, Chemnitz, Germany

F. Podlesak and K. Höfer
Welding Engineering, TU Chemnitz, Chemnitz, Germany

Alexander Kirsch, Markus Euring and Kolja Ostendorf
Faculty of Forest Science and Forest Ecology, Buesgen-Institute, Department of Molecular Wood Biotechnology and Technical Mycology, Georg-August-University of Goettingen, Göttingen, Germany

Mohd Adib Yahya, C.W. Zanariah C. W. Ngah and M. A. Hashim
Faculty of Science and Technology, Universiti Sains Islam Malaysia (USIM), Malaysia

Z. Al-Qodah
Department of Chemical Engineering, Taibah University, Saudi Arabia

Raed Abendeh and Mousa Bani Baker
Al-Zaytoonah university of Jordan, Amman, Jordan

Khaled Ramadan
Applied Science University, Amman, Jordan

Rami Haddad
Jordan University of science and Technology, Irbid, Jordan

Jie Ren
School of Natural and Applied Science, Northwestern Polytechnical University, Xi'an 710072, PR China

Chemistry & Chemical Engineering College, Northwest Normal University, Key Lab of Bioelectrochemtry & Environmental Analysis of Gansu, Lanzhou 730070, PR China

Mengqi Yao, Jinfen Gu, Li Tao and Wu Yang
Chemistry & Chemical Engineering College, Northwest Normal University, Key Lab of Bioelectrochemtry & Environmental Analysis of Gansu, Lanzhou 730070, PR China

Guangcheng Zhang
School of Natural and Applied Science, Northwestern Polytechnical University, Xi'an 710072, PR China

L. M. Axelrod xelrod, I. G. Ma ryasev ryasev, A. A. Platonov and D. R D. R . Melnikova
Magnezit Group, Russia

Marcus Schulze and Holger Seidlitz
Chair of Lightweight Construction, Brandenburg University of Technology Cottbus - Senftenberg, Cottbus, Germany

Franziska König and Sabine Weiß
Department of Physical Metallurgy and Materials Technology, Brandenburg University of Technology Cottbus - Senftenberg, Cottbus, Germany

Antonio Sánchez-Fernández, Alejandra Leyva, Franz-Xaver Bulach, Alejandra Monsivais-Barrón and Jaime Bonilla-Ríos
Instituto Tecnológico de Estudios Superiores de Monterrey, Eugenio Garza Sada 2501, Col. Tecnológico, Monterrey, N. L., México

Laura Peña-Parás, Elisa M. Mendoza and Luis A. Bautista
Departamento de Ingeniería, Universidad de Monterrey, Ave. Morones Prieto 4500 Pte., Col. Jesús M. Garza, San Pedro Garza García, N. L., México

Luis E. Elizalde
Centro de Investigación en Química Aplicada, Blvd. Enrique Reyna Hermosillo 140 Saltillo, Coahuila CP25250, México

Mahmoud B. Alhasanat, Arabi N. S. Al Qadi and Loai A. Al Tarabulsi
Civil Engineering Department, College of Engineering, Al-Hussein Bin Talal University, Ma'an, Jordan

Omar A. Al khashman
Environmental Engineering Department, College of Engineering, Al Hussien Bin Talal University, Ma'an, Jordan

K. N. Vdovin, A. N. Zavalishchin, D. A. Gorlenko and N. A. Feoktistov
FSBEI HBE Nosov Magnitogorsk State Technical University, Magnitogorsk, Russia

Joseph R. Fernandez
University of South Florida, 4202 East Fowler Ave ENL 111, Tampa, Florida 33784, USA

S. Hembrom and B. N. Roy
Department of Metallurgical Engineering, BIT, Sindri, Dhanbad-828123, India

N. Khobragade and D. Roy
Department of Materials and Metallurgical Engineering, NIFFT, Ranchi-834003, India

J. O. Akinnifesi and O. O. Akinwunmi
Department of Physics and Engineering Physics, Obafemi Awolowo University, Ile-Ife, Nigeria

X. W. Zhou
Mechanics of Materials Department, Sandia National Laboratories, Livermore, California 94550, USA

M. E. Foster
Materials Chemistry Department, Sandia National Laboratories, Livermore, California 94550, USA

P. Yang
Electronic, Optical, and Nano Department, Sandia National Laboratories, Albuquerque, NM 87106, USA

H. Fan
Advanced Materials Laboratory, Sandia National Laboratories, Albuquerque, NM 87106, USA
Center for Micro-Engineered Materials, Department of Chemical and Biological Engineering, University of New Mexico, Albuquerque, NM 87106, USA

F. P. Doty
Radiation and Nuclear Detection Materials and Analysis Department, Sandia National Laboratories, Livermore, California 94550, USA

Che Dieudonne Tabong and Agwara Moise Ondoh,
Department of Inorganic Chemistry, Faculty of Science, University of Yaounde I, Yaounde, Cameroon

Divine Mbom Yufanyi
Department of Chemistry, Faculty of Science, The University of Bamenda, Bamenda, Cameroon

Josepha Foba
Department of Chemistry, Faculty of Science, University of Buea, Cameroon

Henrik L. Funke, Sandra Gelbrich and Lothar Kroll
Institute of Lightweight Structures, Technische Universität Chemnitz, Chemnitz, Germany

Adel Taha Mohamed Abbas
Department of Mechanical Engineering, College of Engineering, King Saud University, Saudi Arabia

Kouame Joseph Arthur Kouame, Yu Feng, Fuxing Jiang and Sitao Zhu
School of Civil and Environmental Engineering, University of Science and Technology Beijing, Beijing, China

Hyemin Lee and Hyunsik Yoon
Department of Chemical and Biomolecular Engineering, Seoul National University of Science & Technology, Seoul, 139-743, Korea

Qasim M. Doos Al-Attaby and Shaymaa Abdul Khader Al-Jumaili
Baghdad University, College of Engineering, Mechanical Engineering Dept., Baghdad, Iraq

Markus Euring, Alexander Kirsch, Patricia Schneider and Alireza Kharazipour
Faculty of Forest Science and Forest Ecology, Büsgen-Institute, Department of Molecular Wood Biotechnology and Technical Mycology, Georg-August-University of Goettingen, Germany

Satya Pal Singh
Department of Applied Sciences, Madan Mohan Malaviya University of Technology, Gorakhpur, UP, India